FUNGI

Their Structure and Development

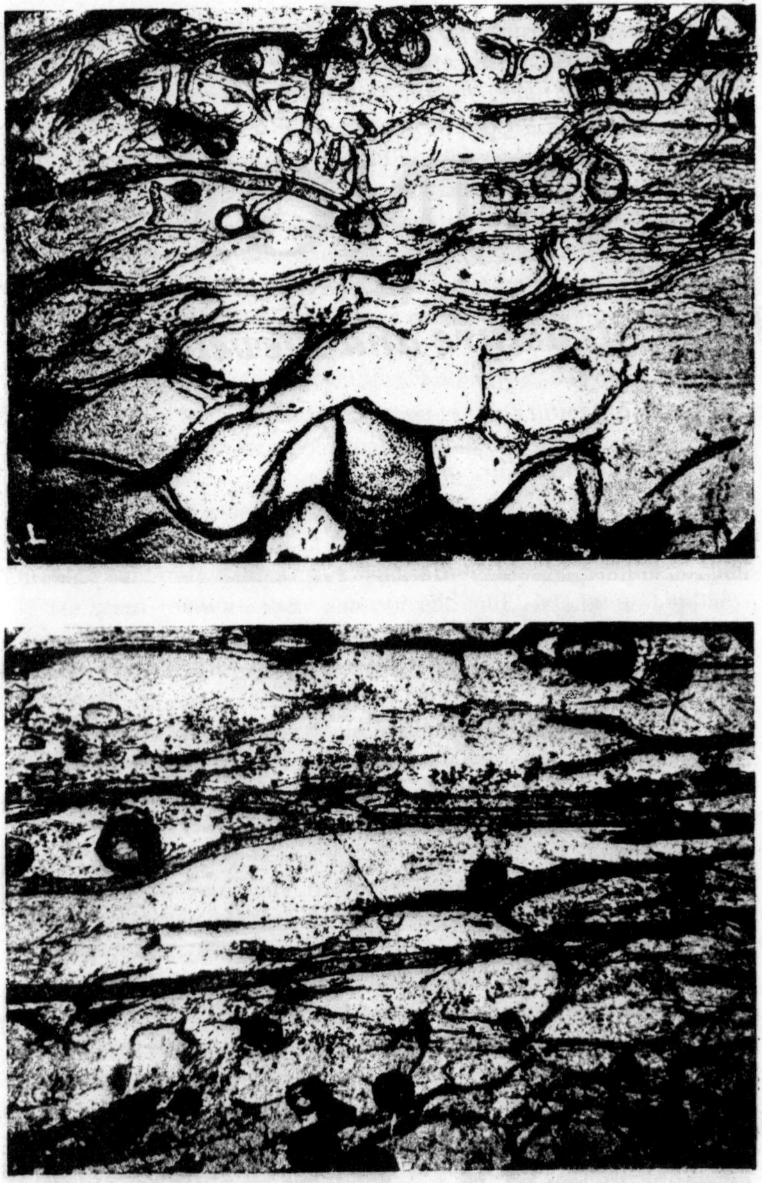

PALEOMYCES ASTEROXYLI

from the Old Red Sandstone, Muir of Rhynie, Aberdeenshire, × 100; after Kidston and Lang

FUNGI

Their Structure and Development

— 2nd Edition —

by

H.C.I. GWYNNE-VAUGHAN

G.B.E., L.L.D., D.Sc., F.L.S.

Professor of Botany in the University of London

&

B. BARNES

D.Sc., Ph.D., F.L.S.

Head of the Department of Biology,
Chelsea Polytechnic, London

BIOTECH

2014

Biotech Books

Delhi - 110 035

ISBN 81-7622-117-1

Published by : **BIOTECH BOOKS**
1123/74, Tri Nagar,
DELHI – 110 035
Phone: 27382765
e-mail: biotechbooks@yahoo.co.in

Showroom : 4762-63/23, Ansari Road, Darya Ganj,
NEW DELHI - 110 002
Phone: 23245578, 23244987

Printed at : **Chawla Offset Printers**
New Delhi – 110 052

PRINTED IN INDIA

To

THE STUDENTS OF
THE DEPARTMENT OF BOTANY
BIRKBECK COLLEGE

To

THE STUDENTS OF
THE DEPARTMENT OF BOTANY
BIRKBECK COLLEGE

PREFACE
TO FIRST EDITION

THIS book is in some sense a development of the volume on Fungi prepared by one of us and published in 1922 in the series of Cambridge Botanical Handbooks, for the text and figures have been freely drawn upon and the material revised and brought up to date. But it is also a new book; it includes the whole of the Fungi, not merely selected classes, and it is addressed to the student rather than to the investigator, with the consequence that the material has been somewhat abbreviated. It has been designed to meet the need for such a textbook in our own Department, and we hope it may be equally useful elsewhere.

Most of the illustrations are from published researches, and we desire to thank those authors whose figures we have copied. Illustrations the source of which is not stated are original. In the case of original figures the magnification and the authority for the specific name are given; this has also been done in other cases when the information was available.

We have appended a bibliography which makes no attempt to be exhaustive, it is rather a selection from mycological literature, and we have referred page by page to the relevant papers. There is a vast literature of pathology which we have left almost untouched, it is fully dealt with elsewhere; we are concerned primarily with the fungus, and only secondarily with its effect on other organisms.

It is a pleasure to express our indebtedness to several of our colleagues and to past and present students for specimens and information, and especially to those whose research or suggestions have helped us in the preparation of this book; wherever possible we have acknowledged their help in the text.

H. C. I. GWYNNE-VAUGHAN
B. BARNES

BIRKBECK COLLEGE
LONDON

PREFACE
TO SECOND EDITION

In the ten years that have passed since this volume first appeared, considerable advances have been made in the investigation of the fungi, and much new material has therefore been included in the present edition. In particular, the importance of flagellation as a guide to the interrelationships of the Phycomycetes has been emphasised, heterothallism has been discovered in the rusts, and many new life histories have been followed. To some extent the study of fungi is thereby simplified, but very much remains to be done.

We should like, once again, to express our thanks to those who have permitted us to use their figures; we hope that the book may be of use both to research students and undergraduates.

H. C. I. GWYNNE-VAUGHAN
B. BARNES

CONTENTS

CONTENTS

INTRODUCTION

The Fungi are saprophytic or parasitic members of the Thallophyta, entirely devoid of chlorophyll, reproduced by spores which may be sexual in origin, and possessing in most species a thallus made up of filaments, or **hyphae**,[1] which are usually colourless and together constitute the **mycelium**. The origin of the group is remote, the earliest undoubted fungi being found in the Old Red Sandstone; they consist (frontispiece[2]) of septate hyphae and of vesicles which were probably reproductive in character.

Fungal hyphae may be aseptate and coenocytic, or they may be divided by transverse septa into uninucleate, binucleate or multinucleate segments; longitudinal or oblique septation[3] is rare. The hyphae are for the most part richly branched. They elongate by apical growth[4] and, during vegetative development, spread loosely on and through the substratum. When fruit bodies are forming in the higher fungi, the hyphae often become woven into dense masses, resembling tissues; such masses are described as **pseudoparenchymatous**. When the mass contains more than one fructification it is termed a **stroma**. The hyphae may also aggregate to form a root-like strand, the **rhizomorph**, or a compact resting body, the **sclerotium**; in either of these, the outer hyphae are modified to form a rind, protecting the inner regions from desiccation.

In most of the Archimycetes, and in the yeasts, where a mycelium is not developed, the vegetative thallus is converted directly into a reproductive structure, without much change in its general form.

A mycelium begins its development as a germ tube put out from one of the numerous varieties of fungal spore (fig. 1). When the spore wall is very thin, as in germinating zoospores, the wall of the germ tube is continuous with it; more often the wall of the germ tube is continuous only with the inner layer of the spore wall. One or more germ tubes may push through the wall of the spore at points not previously recognisable, or they may find exit through special pits, the germ pores. As the germ tube elongates, the contents of the spore pass into it. Growth and branching follow, and

[1] Clarendon (thick type) is used to indicate terms specially or mainly applicable to fungi, and which it is therefore necessary to define.

[2] Kidston and Lang, 1921.

[3] Nichols, 1896; Kempton, 1919.

[4] de Bary, 1887; Smith, J. H., 1923.

reproductive organs may develop from some or all of the mycelial branches.

In many multicellular fungi, and occasionally in non-cellular forms, anastomoses take place between adjacent hyphae by means

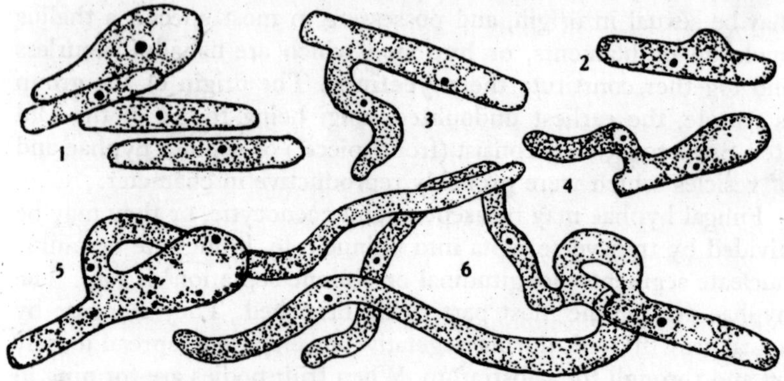

Fig. 1. *Ceratostomella fimbriata* (Ell. & Hals.) Elliott; 1, group of conidia; 2–6, stages in germination, × 1600.

of loops or branches, which, when short and straight, are known as **H-pieces**. These may connect not only branches of the same origin, but also unrelated germ tubes or older filaments (figs. 2, 3); sclerotia or fructifications grown under natural conditions, where several spores have germinated in close proximity, may accordingly be compound structures and the product of two or more spores. Such fusions are presumably nutritive in origin, and result in the pooling of the resources of several spores. Cells of the same hypha are sometimes brought into communication by means of anastomosing branches. The **clamp connections** (figs. 4, 228), common among Autobasidiomycetes, appear to fulfil a special function;[1] whatever nutritive significance they possess is limited by the fact that the passage is soon closed.[2]

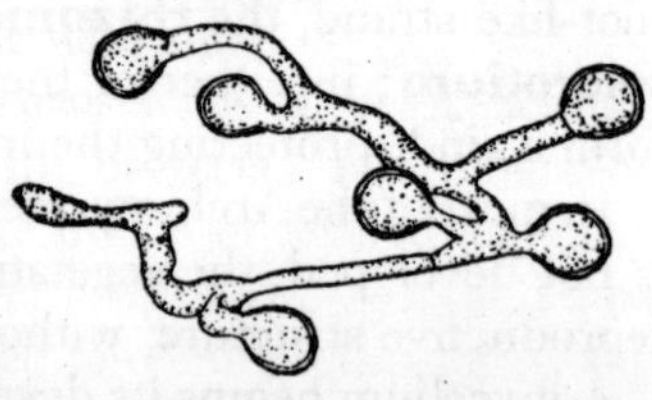

Fig. 2. *Lachnea* sp.; six conidia, the germ tubes of which have undergone fusion, × 720.

Nucleus. The fungal protoplast consists of granular or reticulate cytoplasm, which, in the older regions, leaves a vacuole in the

[1] Noble, 1937. [2] Hoffmann, W., 1856; Brefeld, 1877; and cf. p. 289.

middle of the cell or filament. The nucleus may show a delicate reticulum and one or more nucleoli, or its contents may be concentrated in a dense **chromatin body** surrounded by a clear area. It usually divides by mitosis, showing a well-marked spindle with centrosomes and asters; the origin of the spindle is commonly

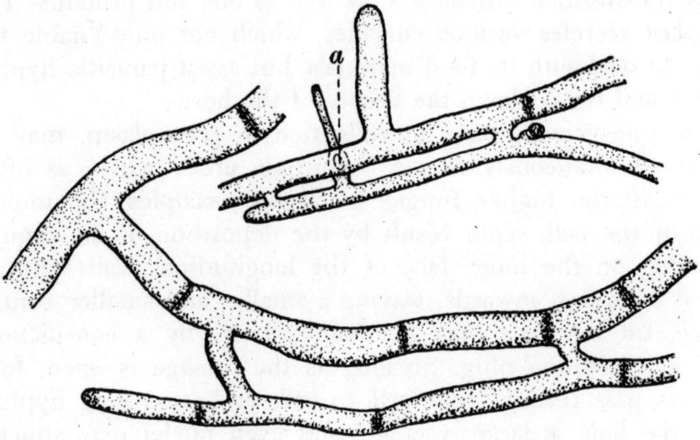

Fig. 3. *Lachnea abundans* Karst.; anastomoses between vegetative hyphae; *a*, hypha growing beneath another with a pore connecting the two, ×500.

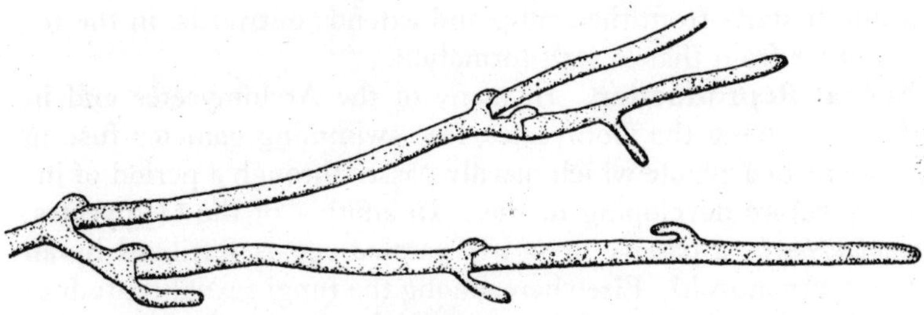

Fig. 4. *Coprinus* sp.; hyphae with clamp connections, ×675.

intranuclear, but it is extranuclear in the rusts. During the prophases of division, chromosomes are formed from the reticulum, or, if a chromatin body is present, it gives rise to chromosomes and to one or a few nucleoli. The nucleoli are thrown into the cytoplasm during karyokinesis and there disintegrate. The extrusion of additional chromatin masses has been observed in *Synchytrium* and other fungi.

Cell Wall. The cell wall varies in composition,[1] especially in fruit bodies. In many Archimycetes and Oomycetes, and in the yeasts, the wall consists largely of cellulose, but in the Zygomycetes and the higher fungi, cellulose is replaced by chitin and other substances of obscure character. Storage materials found in hyphae include a number of carbohydrates, particularly glycogen, which is readily transformed into sugars, as well as oils and proteins. The protoplast secretes various enzymes, which not only enable the fungus to deal with its food materials, but assist parasitic hyphae to enter and break down the tissues of the host.

New transverse septa,[2] in cells rich in protoplasm, may be formed simultaneously across the whole area; when, as often happens in the higher fungi,[3] a vacuole occupies the middle region of the cell, septa result by the deposition of an annular thickening on the inner face of the longitudinal walls and its gradual extension inwards, leaving a smaller and smaller central passage, till the hole may be finally closed by a conspicuous and deeply-staining plug. So long as the passage is open, food materials pass readily from cell to cell, and, in young hyphae, where the hole is large, vacuoles and even nuclei may squeeze through. The streaming of protoplast from cell to cell is readily observed.

If, as sometimes happens, an opening is made in a mature septum, it starts from the centre and extends outwards, in the reverse order from that of wall-formation.

Sexual Reproduction. In many of the Archimycetes and in *Allomyces* among the Oomycetes, free-swimming gametes fuse in pairs to form a zygote which usually passes through a period of inactivity before developing further. In another of the Oomycetes, *Monoblepharis*, a passive or nearly passive egg is fertilised by an active spermatozoid. Elsewhere among the fungi sexual reproduction occurs by the union of the contents of two uninucleate or multinucleate gametangia without the formation of separate gametes. To a multinucleate organ of this type the term **coenogametangium** is applied. The gametangia may be similar in behaviour and structure or may be differentiated as antheridia and oogonia. In some of

[1] Wettstein, F. von, 1921.
[2] Strasburger, 1880; Gwynne-Vaughan and Williamson, 1931.
[3] Wahrlich, 1893; Gwynne-Vaughan and Williamson, 1932; Buller, 1933.

the higher fungi the whole antheridium may be detached and carried by external agencies towards the female organ; it is then termed a **spermatium** and the receptacle in which spermatia are borne is known as a **spermogonium**.

Where fertilisation depends on the union of the contents of non-motile organs a considerable risk of failure is inevitable, while the existence of coenogametangia and of frequent anastomoses between vegetative cells makes possible its replacement by the fusion of gametangial or vegetative nuclei. Among the higher fungi such arrangements are not uncommon, and the complete disappearance of the sexual stage is also frequent.

Incompatibility. Fusion between nuclei of the same gametophytic thallus, whether in sexual or vegetative cells, provides none of the chances of variation inherent in exogamy. Dioecious fungi are rare, and, though spermatia of monoecious forms may be transported to different thalli, the power of movement of attached gametangia is very limited, so that, when endogamy is possible, it is likely to occur. It is prevented in a number of species by the condition of **self-incompatibility**. In such species the vegetative thalli are divided into two or more strains. Each strain, though it may bear gametangia, can seldom produce fructifications alone, but, when two appropriate strains meet and their nuclei become associated, the characteristic structures of the sporophyte are formed. Certain species have gone a step farther and are **self-sterile**, sexual organs being formed only when two appropriate thalli are present. Usually no morphological difference exists between such **compatible** or **complementary** strains, which are conveniently described as + and − or as *A* and *B*. An obvious interpretation of such a state of affairs is that *A* and *B* differ in sex; this is probably true in the Zygomycetes, where reproduction is by isogametangia, but, in the higher fungi, self-incompatible forms are known in which the *A* and *B* mycelia are alike monoecious, bearing both antheridia and oogonia. It is undesirable to speak of a difference in sex between strains bearing identical sexual organs. Rather we have here, as in certain angiospermous trees, a contrivance which prevents inbreeding in a monoecious form. Often, in these self-incompatible fungi, the sexual apparatus has partially or wholly disappeared, and the union between thalli of different complement depends on mycelial fusion.

The convenient terms **heterothallism** and **heterothallic** are used to distinguish the condition of those species which have more than one strain or form of thallus, whether the difference between the forms be sexual or otherwise. A species with one strain only is said to be **homothallic**, and is both self-compatible and self-fertile.

Meiosis. The sexual fusion or its equivalent is followed in all investigated cases by a nuclear reduction or meiotic phase, so that, as in other plants and animals, the number of chromosomes, which was doubled at the fertilisation stage, is subsequently halved in meiosis, and haploid and diploid phases follow one another. As in other organisms, meiosis is concerned with the distribution of hereditary characters, and these have now been shown, in diverse groups of fungi, to be inherited on mendelian lines.

Alternation of Generations. There is evidence that, in most of the lower fungi, meiosis takes place in the germination of the zygote, but in *Allomyces* the zygote gives rise to an independent sporophyte and an independent sporophyte is also found in the rusts, in other basidiomycetous fungi, and in the yeasts. In many of the higher fungi, however, the sporophyte is parasitic on the gametophytic mycelium and obtains nutrition through it. Often several sporophytes arise from a single gametophyte, and, in many fungi, the gametophyte sends out branches which grow around and protect the sexual organs and their products. When fertilisation or any equivalent process has wholly disappeared a morphological alternation of sporophyte and gametophyte may still be found though without the corresponding cytological changes, or, as in the *Fungi imperfecti*, a sporophyte may be developed no longer.

Spores and Spore Mother Cells. In the higher fungi, the characteristic spores of the sporophyte, with the development of which meiosis is associated, may be produced endogenously, as **ascospores**, in a mother cell termed an **ascus**, or exogenously, as **basidiospores**, on the exterior of a cell or row of cells known as a **basidium**. The asci or basidia are frequently arranged in parallel series, forming a fertile layer or **hymenium**, sometimes of considerable extent; they arise from a **sub-hymenial** region immediately below the hymenium, and among them are interpolated vegetative cells, or **paraphyses**, which are perhaps concerned in their nutrition, and which assist in the dispersal of the spores by keeping the mother cells apart. The ascus and basidium, and their products, have long been recognised as essential features in classification.

Accessory Spores. The accessory methods of multiplication have no relation to a sexual process, either normal or reduced, and have no significance in the alternation of generations. They are not homologous with the spores of the Bryophyta and Pteridophyta, but are devices for rapid increase and dissemination, comparable with the gemmae of *Marchantia* or the arrangements for vegetative propagation in the higher plants. The accessory spores may be borne on the sporophyte, as in the rusts and some Autobasidiomycetes, or on the gametophyte, as in most of the Phycomycetes and Ascomycetes.

In many lower fungi, zoospores (fig. 5) are developed in globular or tubular zoosporangia; this is especially common in aquatic forms. Related species living freely exposed to the air may shed the contents of their sporangia as walled non-motile spores, or may set free the whole sporangium, without division of its contents. The sporangium thus becomes a **conidium**, that is, a spore borne externally on a parent hypha (fig. 6), and not enclosed in any kind of sporangium. The conidium is the characteristic unit of accessory multiplication in fungi. Most often, the conidium germinates by the protrusion of one or more germ tubes, but, if the fungus has not completely lost the power of utilising fluid water as a means of dispersal, the conidium may give rise to zoospores when it falls in a wet place. Conidia are usually developed, either singly or in groups, upon stalks known as **conidiophores**; these show an almost endless variety of form and arrangement, reaching their best development in the *Fungi imperfecti*. Conidiophores may be freely exposed to the air, or they may be produced inside a special, flask-shaped receptacle, the **pycnidium**. The term **endoconidium** is applied

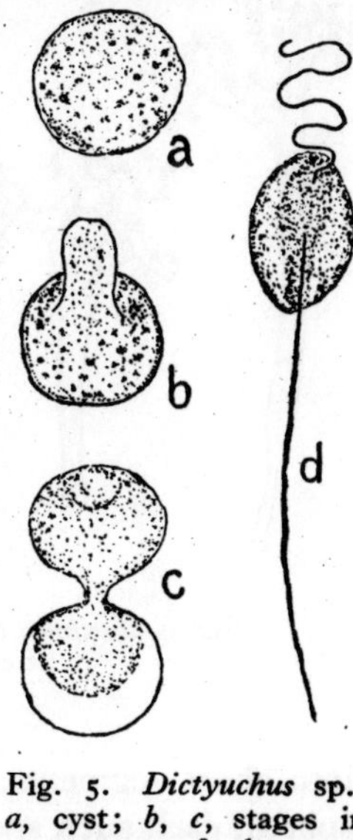

Fig. 5. *Dictyuchus* sp.; *a*, cyst; *b*, *c*, stages in emergence of the zoospore; *d*, zoospore, showing insertion of flagella; all × 1400; after Weston.

to conidia formed inside the conidiophore (fig. 7) and extruded from it.

Resistant **chlamydospores** are formed, either singly or in chains, in ordinary vegetative hyphae or in special branches. Portions of the contents of the hyphae contract, lose water, and, as the name implies, become surrounded by a thick wall.

In the Erysiphales and other fungi, vegetative hyphae may divide

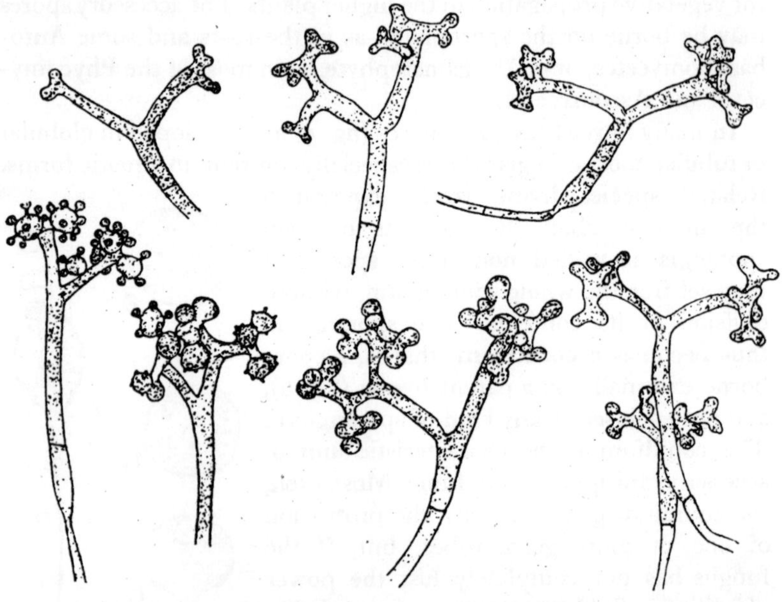

Fig. 6. *Lachnea cretea* (Cooke) Phil.; stages in the development of conidiophores and conidia, × 280.

into short segments, which break apart and function as asexual spores; each such spore is an **oidium**.

Many conidia, as well as other thin-walled spores, are able to **bud**, that is they give rise to lateral outgrowths which are nipped off as new cells. This method of propagation occurs in the ascospores of the Exoascaceae, and the basidiospores of the Ustilaginales. Vegetative cells may also increase by **budding**; the process is particularly characteristic of the yeasts, many of which are unable to form a mycelium; mycelia of *Mucor*, sunk in a fluid, may break up into yeast-like cells, but these resume mycelial characters on transfer to a solid substratum.

Morphology of the Spore. An asexual spore, whether it belongs to the principal or accessory fructification, is, when first formed, a single, transparent, colourless cell. It may divide to form a row of cells, as does the ascospore of *Geoglossum* (fig. 75), or it may give rise to a mass of cells, then being **muriform**, a condition found in the ascospores of *Pleospora* (fig. 76). As the spore ripens, pigments may appear in the wall or in the contents, where, too, oil globules may form. The wall of the spore is usually two-layered, consisting of a delicate endospore within an epispore which may be smooth or sculptured. In general, the sculpturing either takes the form of small isolated projections, the wall then being **verrucose**, or of anastomosing ridges separating more or less regular depressions; this is the **reticulate** condition. Spores often develop on short, special branches from the fertile hyphae, each such outgrowth being a **sterigma**.

Classification. The fungi are divided into three main classes according to the septation of the mycelium and the characters of the principal spores. The members of a fourth class, the *Fungi imperfecti*, multiply by conidia and lack a sexual process; their mycelia and conidial apparatus closely resemble those of the Ascomycetes, and it is probable that most of them are members of that class reduced or incompletely known.

The Myxomycetes are not fungi, and appear to have arisen separately from the Protista; the same is true of the Plasmodiophorales. A summary of the salient features of these groups will be found on pp. 43–48, as well as a brief description of the *Monadineae*

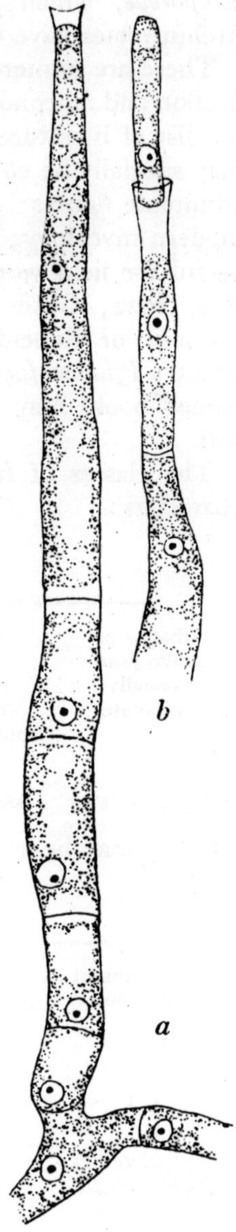

Fig. 7. *Ceratostomella fimbriata* (Ell. & Hals.) Elliott; *a*, endoconidiophore; *b*, escaping endoconidium, × 1600.

zoosporeae, which may represent the stock from which the Archimycetes have developed.

There are numerous works on the classification, or the classification and morphology of the fungi. Many of these are cited in the list of literature at the end of this book; among those which may specially be consulted are: Tulasne, 1861–5, of value for its admirable figures; de Bary, 1887, embodying the foundations of modern mycology; Swanton, 1909, and Ramsbottom, 1923, both useful for field work; W. G. Smith, 1908; Guilliermond, 1913; Rea, 1922; Coker, 1923; Gäumann, 1926; and the relevant portions of Rabenhorst's *Kryptogamen-Flora*, and Engler and Prantl's *Pflanzenfamilien*. The figures given in many of the above-named books may be used to supplement those in the present text.

The classes of fungi may be distinguished by the following characters:

FUNGI

- thallus or mycelium usually aseptate
  - PHYCOMYCETES
- mycelium septate
  - characteristic spores endogenous, ascospores
    - ASCOMYCETES
  - characteristic spores exogenous, basidiospores
    - BASIDIOMYCETES
  - ascospores and basidiospores lacking
    - *FUNGI IMPERFECTI*

They may be subdivided as follows:

PHYCOMYCETES

- mycelium absent or rudimentary
  - ARCHIMYCETES
    1. Chytridiales
    2. Ancylistales
    3. Protomycetales
- mycelium well developed
  - sexual reproduction by oospores; accessory spores often motile
    - OOMYCETES
      1. Blastocladiales
      2. Monoblepharidales
      3. Leptomitales
      4. Saprolegniales
      5. Peronosporales
  - sexual reproduction by zygospores; accessory spores non-motile
    - ZYGOMYCETES
      1. Mucorales
      2. Entomophthorales

ASCOMYCETES

- ascocarp, if present, either with no definite opening, or shield-shaped, or with asci irregularly arranged
  - PLECTOMYCETES
    1. Plectascales
    2. Erysiphales
    3. Exoascales
- ascocarp wide open when ripe, asci in parallel series
  - DISCOMYCETES
    1. Pezizales
    2. Helvellales
    3. Tuberales
    4. Phacidiales
    5. Hysteriales
- ascocarp flask-shaped, opening by an ostiole when ripe; asci in parallel series
  - PYRENOMYCETES
    1. Hypocreales
    2. Dothideales
    3. Sphaeriales
    4. Laboulbeniales

BASIDIOMYCETES

- number of basidiospores indefinite
  - HEMIBASIDIOMYCETES
    1. Ustilaginales
- number of basidiospores definite, usually four
  - basidium septate
    - PROTOBASIDIOMYCETES
      1. Uredinales
      2. Auriculariales
      3. Tremellales
  - basidium continuous
    - AUTOBASIDIOMYCETES
      1. Hymenomycetales
      2. Gasteromycetales

PHYSIOLOGY

SAPROPHYTISM, PARASITISM AND SYMBIOSIS

Since fungi never possess chlorophyll, they are dependent for their food on some sort of relation with other organisms. As **saprophytes** they may utilise organic stores or waste products, or they may break up dead tissues as a source of supply; as **parasites** they may prey upon living cells with consequences to the host that vary from trifling inconvenience to complete destruction; or as **symbionts** they may establish a relationship with another organism in which the advantages are not all on one side.

These physiological classes are connected by intermediate forms capable of changing their method of nutrition according to circumstances. A species which is strictly limited to one mode of existence is an **obligate** saprophyte, parasite or symbiont; a species which is usually saprophytic but capable of parasitic existence is described as a **hemi-saprophyte** or **facultative** parasite, and a form which is usually parasitic but sometimes saprophytic is a **hemi-parasite** or a **facultative** saprophyte.

SAPROPHYTISM

Most of the fungi are saprophytic. Many members of each systematic class, and especially a very large number of the Basidiomycetes, obtain food in this way.

Aquatic Fungi. Submerged organic materials, including the remains of freshwater and marine organisms, furnish substrata for the growth of aquatic fungi. To most of these, a plentiful supply of free oxygen is essential, so that they tend to develop in shallow water where little putrefaction is in progress. Growth does not seem to be affected in any significant manner by the presence of humus or of lime, nor does the hydrogen ion concentration[1] in the water seem to be of much importance. Suspended material, companions of iron, and the metabolic products of sulphur bacteria are, highly inimical to these fungi. Although shade does not cause

[1] Lund, 1934.

injury, aquatic fungi are most numerous in well-lit water, where the crowded population provides ample food.[1]

The development of the saprophytic forms is hindered when algae are abundant, and is usually checked when bacteria accumulate, probably because these use up the free oxygen rapidly. Some species, however, such as *Gonapodya prolifera* and *Macrochytrium botrydioides*, may occur in association with dense masses of bacteria;[2] they may be facultative anaerobes. Submerged twigs and fruits may bear colonies of several species of Oomycetes, though the fungi are seldom obvious on material freshly taken from the water, and require special treatment to ensure their strong development.[3] Substrata of animal origin are most often colonised by species of *Saprolegnia*.

Running water harbours a few characteristic forms. *Leptomitus lacteus* is sometimes abundant in water which is not acid to litmus and contains simple organic compounds of nitrogen. Water of similar reaction, but with little combined nitrogen and much carbohydrate in solution, may be inhabited by species of *Mucor*. In the presence of free nitric acid, heavy growths of *Penicillium fluitans* have been observed.[4]

Parasitic aquatic fungi, chiefly belonging to the Archimycetes, occur on or in members of almost every group of organisms inhabiting water;[5] though they are not usually abundant, they may at times cause epidemics of destructive character, particularly in hosts of gregarious habit, such as desmids and diatoms. The resting spores of the Conjugatae are occasionally destroyed in large numbers by fungal attacks.

On Wood. Ascomycetes and Basidiomycetes are important agents in the destruction of wood;[6] their hyphae absorb the con-

[1] Thaxter, 1895 i, ii, 1896 ii, iii; Petersen, 1903, 1910; Butler, E. J., 1907, 1911; Coker, 1923; Sparrow, 1932, 1933 i, ii, iii, 1934, 1935, 1936 i, ii; Barnes and Melville, 1932; Barnes, 1934 ii.

[2] von Minden, 1916.

[3] Barnes and Melville, 1932; Sparrow, 1933 i, 1936 i.

[4] Tiegs, 1919.

[5] Fischer, A., 1882; Church, 1893; Juel, 1901; von Deckenbach, 1903; Petersen, 1905; Cotton, 1908; Chatton and Roubaud, 1909; Apstein, 1911; Sommerstorff, 1911; Brierley, 1913; Moral, 1913; Sutherland, 1914–15; Dunkerly, 1914; Ferdinandsen and Winge, 1920; Mirande, 1920; Maire and Chemin, 1922; Arnaudow, 1923, 1925, Butler, J. B. and Buckley, 1927; Sparrow, 1929, 1931, 1934, 1936 i.

[6] Ward, 1898; von Schrenk, 1902; Waksman, 1931.

tents of the living cells of the xylem and medullary rays, and enter the non-living elements, either passing through the pits or penetrating the walls. These become delignified and give reactions for cellulose, and the middle lamella is dissolved, promoting the disintegration of the material; the wood loses weight and may be resolved into powdery touchwood. The enzymes responsible for these changes spread in a plane parallel to the surface of the walls, either from the pits, which become enlarged, or from the delicate passages left by the protoplasmic connections which traversed the walls of the xylem elements when these were young. In this way, considerable damage is done to timbers by the fungi of dry rot,[1] *Merulius lachrymans* and *Poria vaporaria*, and to paving blocks by *Lentinus lepideus*.[2] On the other hand, advantage ensues from the restoration to the soil of the material from fallen tree trunks, branches and twigs. The action of the higher fungi in this connection is especially important, for few other organisms appear to be able to break down woody material. Although many insects live on wood, some at least do so with the assistance of fungi which inhabit special regions in their alimentary canals and doubtless play a part in digestion, while others obtain material indirectly from the wood by devouring fungi which they have cultivated upon it.[3]

In Soil.[4] Fungi are largely confined to the top six inches of soil, where the forms most frequently met with in temperate climates are species of *Penicillium* and *Mucor*; *Aspergillus* and such Hyphomycetales as *Fusarium* and *Cladosporium* also abound. Oomycetes[5] are common, including, in addition to species of *Pythium* and *Phytophthora*, a number of specialised members of the Saprolegniales. Soil fungi are as diverse in their activities as they are in systematic position. Some, like subterranean species of *Pythium*, are facultative parasites which find sufficient organic material in the soil to permit of saprophytic growth; others, like the Mucorales, appear to be chiefly concerned with the breakdown of proteins, from which they obtain supplies of both carbon and nitrogen, since they seem to have little power of decomposing cellulose.

[1] Czapek, 1899; Wehmer, 1913–14; Barton-Wright and Boswell, 1931.
[2] Buller, 1905, 1906, 1909 i. [3] Buchner, 1930.
[4] Dale, 1912; Hall, 1920; Waksman, 1931.
[5] Butler, E. J., 1907; Harvey, 1925, 1927 i, ii; Coker, 1927; Coker and Braxton, 1926; Couch, 1927, 1930–1 i.

Most soil fungi however attack cellulose[1] vigorously, at the same time utilising amino-acids set free by bacteria as a source of nitrogen. If they are scantily supplied with carbohydrates, they may obtain both energy and nitrogen from proteins in the soil, but the process is extravagant, and surplus ammonia is set free.[2] Both ammonia and nitrates can be assimilated, but, apart from species of *Phoma*,[3] fungi do not seem able to fix gaseous nitrogen.

In addition to micro-fungi, some larger Discomycetes and many Autobasidiomycetes develop on soil; among these, the members of the Tuberales, Elaphomycetaceae, Terfeziaceae and Hymenogastraceae are completely subterranean, or **hypogeal**, in their development. They produce closed fructifications protected by a stout wall; as the spores approach maturity, the fruit emits a strong scent, which attracts animals, especially rodents. The fructification is eaten, the spores pass uninjured through the alimentary canal, and are thus distributed. The truffles, belonging to the genus *Tuber*, are the best known of such forms.

Species with subaerial sporophores are to be found both under trees and in open ground; their activities are well exemplified by the **fairy rings**[4] of dark green grass often seen in poor pastures. The ring is caused by the growth of a fungus, usually one of the Autobasidiomycetes, though rings due to the agency of the Ascomycetes, *Tuber* and *Morchella*, have been reported; both the structure of the ring and the distribution of the mycelium vary in detail according to the fungus present. The mycelium spreads from the centre outwards, dying off as the food materials in the soil are exhausted. At the advancing margin, ammonium compounds are set free, and converted by bacteria into nitrates, stimulating the formation of a ring of dense vegetation. Just behind this, where the growth of the fungus is especially energetic, the mycelium may form a dense felt in the soil; this impedes the upward passage of water and may cause local drought and the formation of a brown ring of dead plants. Still nearer the middle of the ring the soil contains degenerating mycelium, and the products of its decomposition favour the development of a circle of strong vegetation. The

[1] McBeth and Scales, 1913; Scales, 1915; Waksman, 1931.
[2] Waksman, 1918. [3] Duggar and Davis, 1916.
[4] Hutton, 1790; Gilbert, 1875; Lawes, Gilbert and Warrington, 1883; Molliard, 1910; Bayliss, 1911; Shantz and Piemeisel, 1917; Buller, 1922 ii.

soil in the central region has already passed through these changes, and bears a plant cover like that of the rest of the field. In wet seasons, and where the development of the mycelium is insufficient to cause local drought, the activity of the fungus is represented by a single ring of stimulated vegetation. The rings are often distinct enough to show well in photographs taken from the air; they may persist for hundreds of years. Fruit bodies of the fungus are not always present; they commonly form in autumn, from the vigorous young mycelium, and therefore in a circle.

Coprophilous Fungi. Fungi feeding on organic remains in the soil often benefit by the presence of natural manures, and help to break them up, rendering them available for the use of green plants.

From such fungi it is an easy transition to the extensive **coprophilous** flora, characterised by its habitat on the dung of animals, and especially of herbivorous species. The presence of cellulose in the dung is an important and probably a decisive factor in determining their occurrence, for many coprophilous fungi develop their fructifications on residues containing cellulose lying in the dung. In culture on dung agar, coprophilous fungi may fail to fruit unless scraps of filter paper or some other form of cellulose are included in the medium. The rich supplies of nitrogenous material provide a further important source of food, of special significance to the Zygomycetes which are the first fungi to colonise dung. Under natural conditions, the Zygomycetes are followed by Ascomycetes and then by Basidiomycetes, the last named no doubt playing an important part in the final disintegration of the vegetable substances in the substratum. Although the spores of many coprophilous fungi are dispersed by the wind, some of the commoner dung-inhabiting forms depend on a propulsive mechanism. The spores in the sporangium of *Pilobolus* are associated with gelatinous material, and, when the wall bursts, the upper part of the sporangium and the contained spores are shot off as a single mass and adhere to the body against which they strike. In many of the Ascobolaceae and Sordariaceae among Ascomycetes, the spores are surrounded by mucilage and together form a projectile, which, owing to its weight, can be shot to a much greater distance than a single spore. The sudden ejection of the mass seems to depend on the rapid formation of sugars of high osmotic value and the consequent absorption of water and bursting of the wall of the ascus. After ejection, the mass dries up and remains firmly attached to the grass or leaves on which it has fallen. Many coprophilous species, moreover, show positively phototropic reactions, the sporangiophores of *Pilobolus*, the necks of the perithecia of *Sordaria*, and the individual asci of some of the *Ascoboli* being regularly directed towards the light; discharge of the spores towards an open space is thus ensured.

Fungi on Fatty Substrata. Most fungi are able to utilise fats and oils; such substances are a common form of reserve food in spores, in mycelia, and especially in sclerotia, where the proportion of fat may reach as much as 35 per cent.; the fat-splitting enzyme, lipase, has been isolated from several fungi.

It is not surprising, therefore, that many fungi grow readily on fatty substrata.[1] *Eurotium* and *Penicillium* occur on the layer of oil placed over bottled fruits to check decomposition, and, together with members of other genera, are concerned in the "ripening" of cheese; the related *Monascus heterosporus* damages stored tallow in Australia and New Zealand. Fungal activity may reduce the oil content of oil cake from 10 per cent. to between 1 and 2 per cent. *Phycomyces* is of frequent occurrence on residues from oil mills, and the oily seeds of the brazil nut are often rotted by *Cunninghamella*.

Fungi producing Fermentation. Many fungi obtain nutriment from solutions of carbohydrates, and the yeasts also obtain energy by breaking up these substances without the intervention of free oxygen. When such anaerobic respiration is incomplete, ethyl alcohol, carbon dioxide and small quantities of other substances are formed. Many organisms are capable of this process as an emergency reaction, but in yeast it has become the normal routine, and may be regarded as the adaptation of a group of simple forms to a mode of life with a minimum of competition.

The reaction depends on the presence of the enzyme zymase, and is known as alcoholic fermentation. In its least complex form it may be represented by the equation:

$$C_6H_{12}O_6 = 2C_2H_6O + 2CO_2.$$

Zymase is secreted by the cells during their period of fermentative activity, but is not present in resting cells, being soon de-

[1] Biffen, 1899.

composed when the reaction comes to an end. It is made up of two co-enzymes; one is a soluble phosphate which enters into temporary combination with a part of the carbohydrate, but is ineffective in the absence of a second factor of unknown constitution. This phosphate is dialysable and is not destroyed by boiling; it may be separated from yeast juice by filtration under pressure, both filtrate and residue being inactive alone.

Only certain monosaccharides with the formula $C_6H_{12}O_6$, such as glucose and fructose, are capable of undergoing alcoholic fermentation; polysaccharides, such as cane sugar, lactose and maltose, must be hydrolysed to the appropriate monosaccharides before fermentation can occur. Hydrolysis is brought about by the excretion into the liquid of the enzyme invertase by the fungal cell.

A number of yeasts are utilised economically,[1] both in baking, where their value depends on the formation of carbon dioxide, causing the bread to "rise", and in brewing and other processes concerned with the production of alcohol. Wine yeasts ferment grape sugar and occur abundantly at vintage time on the grapes and their stalks; similarly, cider yeast occurs on apples. On the other hand, beer yeast, which acts on the sugar formed in germinating barley, is not known in the wild state.

Alcoholic fermentation may be caused by a yeast acting in association[2] with other fungi or with bacteria. The ginger-beer plant,[3] for example, which looks like lumps of soaked tapioca or sago, is made up of a yeast, *Saccharomyces piriformis*,[4] and a bacterium, *Bacterium vermiforme*. Added to commercial ginger, sugar and water, it causes the formation of ginger-beer. The bacteria utilise the products of metabolism of the yeast; the yeast benefits by the removal of these substances, the accumulation of which would inhibit its development. The two organisms thus live in symbiosis, each gaining by the association.

Zymase is secreted by several filamentous fungi. *Mucor racemosus* and other species,[5] when cultivated in a solution of sugar, form ovoid cells which multiply by budding and cause active fermentation, especially when the liquid is poorly aerated, and contains organic acids. *Mucor*-yeasts are widely used on the Continent

[1] Ramsbottom, 1936. [2] Kløcker and Schiönning, 1895, 1896.
[3] Ward, 1892. [4] Guilliermond, 1920 ii. [5] Nadson and Phillipov, 1925 i.

in preparing alcohol from maize; in the Far East they are responsible for the production of a number of characteristic alcoholic beverages.

PARASITISM

Facultative Parasites. Fungi which are able to pass through their whole development as saprophytes are sometimes found on living plants. They disintegrate the tissues in advance of their growing hyphae, so that they are not parasitic in the strict sense, but first kill the cells of their host and then live saprophytically on the dead remains.[1] If the spores of *Botrytis cinerea* are placed in a drop of nutrient fluid on a leaf of the broad bean, they show the first signs of a germ tube in two or three hours. The outer walls of the developing tube soon become converted into a mucilaginous sheath by means of which the fungus adheres to the leaf. If a drop of distilled water is used instead of some nutritive solution, substances diffuse into it from the host. These supply the food essential in the early stages of germination.

After growing for a while along the surface of the leaf, the germ tube turns down, and its tip, filled with dense protoplasm, is pressed against the cuticle of the host, where it may spread out to form a simple **appressorium**; as growth continues, the tube is held in place by its mucilaginous coat, and the cuticle is ruptured by the pressure of its tip. The fungus now either penetrates directly into an epidermal cell, or grows more or less horizontally in the subcuticular layers; these swell and appear to stretch the cuticle so that its penetration by other germ tubes is easier. As the hyphae travel through the epidermis, the palisade parenchyma is injured, the nuclei begin to disintegrate, the chloroplasts swell, and most of the starch disappears. A dark stain, which is one of the signs of death, spreads through the mesophyll in front of the hyphae; the latter secrete an enzyme which both disintegrates the cell walls of the host, and causes the death of the protoplasts, but it is unable to affect the cuticle, the penetration of which is always mechanical.

The manner of attack characteristic of *Botrytis* on the broad bean is by no means universal. Exosmosis of material into the drop of fluid containing the spores may, according to the character of the host and of the fungus, fail to stimulate development and may even

[1] Brown, W., 1915–23; Blackman and Welsford, 1916; Dey, 1919; Boyle, 1921.

inhibit it. Penetration of the cuticle does not always occur, for the hyphae of many fungi enter through stomata. Infection by the so-called **wound parasites** only takes place where the tissues have been exposed or damaged.

Once established inside the host, the fungus may send branches into the cells, or it may develop wholly in the intercellular spaces, killing those cells with which it comes into contact, and absorbing the food materials that diffuse out from the dead protoplasts; such forms are saprophytes in the same sense as is *Botrytis*.

Wood-destroying fungi flourish equally well on living and dead tissues; they harm the host chiefly by blocking the xylem and cutting off the water supply from regions above the infected area. This occurs in *Nectria cinnabarina*[1] on currant; since this species produces its ascocarps on dead stems, it is clear that the survival of the host is not essential to that of the fungus.

Obligate Parasites. In obligate parasitism,[2] on the other hand, the death of the host entails that of the fungus, and it is to the advantage of the latter that death should be postponed, at least until it has formed its spores.

The relations between parasite and host are exceedingly varied. Among **endophytic** forms, which invade the tissues of the host, unicellular parasites go through their whole development in a single cell, while filamentous species live mainly in the intercellular spaces or under the cuticle, and send short branches into the cells. These branches may become specialised as **haustoria** of limited growth and definite form.

Parasitic fungi which develop on the outside of the host are said to be **ectophytic**. The Erysiphaceae, for example, obtain their supplies by sending haustoria into the epidermal cells of the host, and the Laboulbeniales absorb food through the unbroken membranes of the infected insect, and seem to cause a minimum of inconvenience.

The susceptibility of the host, or its resistance to the attacks of the parasite, often varies with the environment. Umbellifers growing in damp, shady places are more liable to infection by *Protomyces*[3] than those in exposed habitats, and the same is true of the hosts of the Chytridiales,[4] though here the fact that the zoospores require water

[1] Line, 1922.
[2] Brooks, 1923, 1924, 1935.
[3] von Büren, 1915.
[4] Tobler, 1913.

in which to swim may be more important than the lessened resistance of the host. The wild sloe combats the fungus responsible for silver leaf, to which its presumed descendant, the Victoria plum, falls a ready victim; it is possible that the factor for resistance has been bred out in the course of cultivation.

The parasite may be strictly localised in relation to the host, as in many unicellular forms and in the Laboulbeniales, or the fungus may spread far beyond the point of infection, and may keep pace with the growth of new tissues. Perennial mycelia are not uncommon among rusts and other forms that infect trees; sometimes the mycelium lives for years in the tissues of a geophilous host, dying down to the organs of perennation when the host prepares for winter, and passing into the new shoots as they form in spring.

Invasion by a fungus may modify the tissues of the host, chiefly by causing abnormal growth or **hypertrophy**. This occurs in a simple form when single algal or fungal cells enlarge if they harbour chytridean parasites, and in a somewhat more complicated form when, as in species of *Synchytrium*, the neighbours of the infected cell enlarge and divide abnormally. Local deformations of this kind are common; thus peach leaf curl and many similar abnormalities are caused by *Exoascus* and its allies, and irregular rose-coloured blisters appear on the leaves of Ericaceae attacked by *Exobasidium*. More elaborate deformations are provoked by some of the Ustilaginales, and, in the form of witches' brooms, by the rusts and Exoascaceae.

A witch's broom is a bunch of modified twigs, due usually to attack by insects, but sometimes by fungi. These may enter a lateral bud, which, stimulated by the presence of the fungus, or by the food which the fungus deflects from its proper course, grows out to form a dense bush of twigs. The production of flowers is inhibited, the leaves appear earlier than those of healthy branches, and, even when the host is an evergreen conifer, fall at the end of each season. The affected shoot, in spite of the fungus within it, appears at first to flourish, though to the detriment of the rest of the tree. It is possible that something approaching local symbiosis has been established, but the fungus is clearly the gainer.

SYMBIOSIS

The physiological conditions under which the thallus of a lichen is built up are somewhat similar; the algal cells appear healthy; they are capable of vegetative multiplication, and sporulation occurs in the algal constituent of *Evernia prunastri*.[1] The fungus alone is concerned in the development of the fructification, a clear indication that it plays the dominant part.

The term **mycorrhiza** is applied to the structures formed by the association between the mycelium of a fungus and the roots or other organs of one of the higher plants.[2] The mycelium may be **endotrophic**, occurring within and between the cells of the host, and having little connection with a mycelium in the soil; or it may be **ectotrophic**, with a strong development on the surface of the root, and a limited distribution within the tissues. There is little evidence that nitrogen[3] is brought into combination by the fungus and utilised by the green plant; it may be that the activity of the fungus raises the concentration of the dissolved carbohydrate[4] in the cell sap, and that this change provokes reaction in the host, stimulating active growth which leads to the consumption of carbohydrate in respiration, or inducing the formation of storage organs where superfluous material is immobilised as starch. However this may be, it seems clear that the fungus takes the first step in the formation of the association, acts for a time as a parasite, and eventually succumbs to the higher plant. The latter may be little affected; it may, as in orchids and Ericaceae, depend on the co-operation of the fungus for the successful negotiation of a crisis in its life history; or, as in *Gastrodia* and other saprophytes, it may be unable to survive in the absence of the fungus.

Endotrophic Mycorrhiza. Fungal hyphae occur casually in the cortex of the roots of most perennial flowering plants and vascular cryptogams. The fungus enters from the soil and lives as a parasite in the outer layers of the root; in deeper lying and presumably better nourished cells its progress is arrested. Tangled, dichotomously branched tufts of hyphae are formed in this re-

[1] Paulson, 1921.
[2] Reissek, 1847; Frank, 1885, 1891; Weiss, 1904; Mason, 1928.
[3] Ward, 1888 ii; Ternetz, 1907; Burgeff, 1909; Goddard, 1913; Rayner, 1915.
[4] McLennan, 1920, 1926.

gion; they are digested by the host and leave dead residues of rounded form. Penetration may extend to the cells lying just outside the endodermis, but in healthy roots never goes deeper than this. The fungus does not form spores within the host, and fails to establish itself permanently as a parasite; it is possible that the vascular plant benefits to some extent from the digestion of the hyphae. However, there is no constant association; many roots contain no fungus, and evidence is lacking that infected plants do better than non-infected ones, or that the fungus gains in any way.[1]

A similar condition obtains in the gametophyte of the liverworts,[2] where an endophytic mycelium may be found, usually in the ventral cells of the midrib; the fungus does not enter cells rich in starch, and the sporophyte remains uninfected.

Rounded wefts of hyphae occur in the cortical cells of the young roots of orchids;[3] fructifications are not produced within the healthy host plant, but the fungus may be isolated and grown in culture; it belongs to the form genus *Rhizoctonia*,[4] some members of which are conidial stages of *Hypochnus*. The distribution of the fungus within the orchid varies with the characters of the latter: *Cattleya* is rootless for a portion of the year, and at that time does not contain a fungus; *Vanda* has perennial, permanently infected roots; infection is irregular in *Listera*, a genus rich in chlorophyll, and strong in *Corallorhiza*, in which chlorophyll is feebly developed; *Neottia*, a brown saprophyte, is always heavily infected. The rhizomes of *Gastrodia elata*[5] are attacked by *Armillaria mellea*, and the creeping stems of *Goodyera repens*[6] are occasionally entered by the fungus abundant in the roots.

The seeds of the Orchidaceae are small, and contain an undifferentiated embryo when they are set free from the capsule. Such seeds, extracted under aseptic conditions from the fruit, and sown on sterilised media, swell somewhat and turn green, but do not germinate; when, however, seeds are sown in company with the fungus from the roots of the plant, seedlings are obtained.[4] A hypha grows into the suspensor end of the seed and invades the cells; in each cell a ball of hyphae is formed before the next is

[1] Gallaud, 1905. [2] Cavers, 1903–4; Ridler, 1922–3; Magrou, 1925.
[3] Magnus, W., 1900; Bernard, 1909–11; Ramsbottom, 1922; Burgeff, 1931.
[4] Bernard, 1909; Ramsbottom, 1922.
[5] Kusano, 1911. [6] Costantin and Dufour, 1920.

entered. While this is going on, divisions of the cells at the opposite end of the seed lead to the differentiation of the stem and root; general swelling occurs, and the embryo is transformed into a green protocorm. The base of the protocorm is occupied by cells containing balls of hyphae. At the apex, the shoot is formed from a meristem which is not invaded by the fungus; the intermediate region consists of cells which have a marked capacity for digesting hyphae, and so prevent general infection. The young, endogenous root of the seedling burrows its way through the protocorm; it emerges uninfected, penetrates the soil, and is there attacked by the endophyte.[1]

The seeds of some orchids will grow although no fungus is present. Those of *Bletilla hyacinthina*,[2] sown on sterile media, give slender seedlings very different from the protocorms developed in the presence of *Rhizoctonia*; the seeds of *Cattleya*[3] and *Laelia*[3] germinate in moderately concentrated solutions of carbohydrates; growth is slower than when the fungus is available, but protocorms are formed, and good seedlings obtained. *Rhizoctonia* is stated[2] to raise the concentration of nutrient solutions in which it grows; maybe, within the seed, the concentration of the cell sap is increased, and a stimulus given to germination similar to that afforded when seeds of *Cattleya* are placed in a solution of a carbohydrate. But, although it is possible, by suitable technique, to bring about the germination of the seeds of some orchids in the absence of the fungus, such orchids are normally dependent on the fungus at this stage of their life history.

The Ericaceae resemble the Orchidaceae in the frequent occurrence of hyphae in their roots, and in the possession of small seeds. *Calluna vulgaris*[4] lives in association with a species of *Phoma*. The mycelium occurs not only in the roots and other parts free from chlorophyll, but also in the stems, leaves, flowers and fruits; the demonstration of the fungus in these parts demands careful manipulation. When the seed germinates, hyphae present in the seed coat infect the seedling and normal development takes place. Seeds deprived of their fungus by sterilisation of the testa, and sown on sterilised media, give rise to seedlings with no roots; such weak

[1] Ramsbottom, 1922. [2] Bernard, 1909.
[3] Bernard, 1909; Knudson, 1922.
[4] Ternetz, 1907; Rayner, 1915–22; Dufrenoy, 1917; Christoph, 1921.

seedlings grow vigorously when brought into relation with the fungus after it has been isolated and grown in culture. Strong seedlings are obtained by sowing seeds and inoculating them with the fungus. The relation here is one of controlled parasitism, for, if a weak seedling is inoculated from a strong culture of *Phoma*, it is killed by the latter.[1]

Ovarial infection occurs in other members of the Ericaceae, and the mycelium may possibly permeate the whole plant, as it does in *Calluna*. The closely related Pyrolaceae[2] include some green plants and some saprophytes; all are associated with fungi, the saprophytes markedly so. Correlated with the tendency in the family to adopt the mycorrhizal habit, there is a tendency to increase the number of seeds and to reduce the size and complexity of the individual seed. A similar state of affairs is found in the Burmanniaceae and Gentianaceae,[3] though, in the latter family, the germination of the seeds does not appear to depend on the assistance of a fungus. With that exception, the dependence of the higher plant on the fungus is obvious; the successful germination of the seed cannot take place unless the fungus is present; in the saprophytic forms, and in forms which contain some chlorophyll, the production of flowers is stimulated by the presence of the endophyte.

The dependence of the seed plant on the fungus during vegetative development is shown in *Gastrodia elata*,[4] a non-chlorophyllous orchid found in oak woods in Japan. The vegetative part of the plant consists of a tuberous rhizome; this produces long branches, at the ends of which new tubers are formed; in autumn the latter become separated by the death of their stalks and of the parent body. In the outer cells of the tuber, the mycelium of *Armillaria mellea* acts as a parasite, but, farther in, the tables are turned, and the invading filaments are destroyed. Association with the fungus is essential to the well-being of *Gastrodia*; when no union takes place, the orchid fails to flower and the newly formed tubers fall off in size; when isolation from the fungus is maintained for several seasons, the tubers are so much weakened that they are unable to give rise to a fresh crop, and the plant dies. *Gastrodia* is evidently parasitic on the fungus, and depends on it for supplies of nutrient

[1] Rayner, 1915.
[2] Oliver, 1890; Henderson, 1919.
[3] Stahl, 1900; Weiss, 1933.
[4] Kusano, 1911.

material; *Armillaria* appears to benefit only occasionally when it succeeds in destroying the tubers it has entered.

The condition in *Lolium*[1] is different. The roots contain an endophytic fungus; in the outer cells it is a weak parasite; in deeper cells, the invader is digested with the liberation of fatty material. Hyphae also occur in the grain, between the aleurone layer and the outer covering; at about the time of fertilisation, hyphae may be seen in the embryo sac; they grow at the expense of the nucellus and the carpel wall. If fertilisation is prevented, a sclerotium-like body is formed within the ovule; when fertilisation occurs, the endosperm digests some of the hyphae, to the advantage of the developing embryo. This is invaded, the hyphae taking up a position in the apex of the young shoot; they move up with the growth of the stem, accumulate at the base of the young carpels, and increase in quantity when the ovules are formed. The fungus has not been isolated from *Lolium*, and it has not been shown that the hyphae in the roots belong to the same species as those in the fruit. It seems probable, indeed, that two fungi are present, one a weak parasite in the roots, ultimately digested by the grass, the other a degenerate seed parasite, persisting in a weakened condition, and so enslaved that it does little more than assist in the nutrition of the embryo.

Endophytic hyphae are well developed in the prothallus of *Lycopodium*, in the leafy plant of species of *Lycopodium* and *Selaginella*, in the roots of *Cyathea* and members of the Marattiaceae, and in the prothallus and sporophyte of the Psilotaceae and Ophioglossaceae. The presence of a fungus appears to have little effect on the sporophyte, but there is evidence that the germination of spores in the Ophioglossaceae and Lycopodiaceae depends on the fungus; in the presence of a mycelium, the prothalli may be subterranean and devoid of chlorophyll; protocorms and tuberous structures are formed.[2]

Fungi are associated with the roots of nearly all wild perennial plants, but not of annuals; when swollen perennating organs are formed, they remain free from the fungus, which is confined to the absorbing roots. The connection between the presence of a fungus and the development of tubers or rhizomes, is indicated by a comparative study of certain closely related annual

[1] McLennan, 1920, 1926.
[2] Stahl, 1900; Lang, 1902; West, 1917.

and perennial species.[1] Thus, *Orobus coccineus* is an annual, with a richly branched subaerial stem; tubers are not formed, and though young roots may be attacked by a fungus, the invader is soon completely digested. *O. tuberosus* is a perennial; the flowering branches are relatively simple and spring from a branching underground system of stolons, on which tubers freely develop; the roots always contain a fungus, even during the winter. If, however, seeds of *O. tuberosus* are sown in sterilised soil, after sterilisation of the testa, they give rise to seedlings very like those of *O. coccineus*; the plants branch freely above the soil, and tubers are not formed. The genus *Mercurialis* offers a suggestive parallel: *M. annua* is a much-branched annual which contracts no lasting union with a fungus; *M. perennis* branches little above ground, perennates by means of a rhizome, and harbours a fungus in its roots. Similar relations are found in wild species of *Solanum*, but efforts to ascertain the cause of tuberisation in the potato suggest that a stimulus to the formation of tubers, comparable with that normally exerted by the fungus, is afforded by cultivation in richly manured soil. *S. Maglia*, a possible ancestor of the potato, forms tubers, and contains an endophyte in its roots. It may be surmised that the production of an organ of perennation, apparently proof against the fungus, protects the plant from attack during periods of inaction, when the cells are unable to digest an invader.

A condition recalling that found in endotrophic mycorrhiza is encountered in some insects belonging to the Coccideae.[2] The bodies of *Lecanium hemisphericum* and related species contain large numbers of yeasts; the eggs are infected during development, so that the fungus is passed on from generation to generation. The insects do not seem to suffer injury; they feed on plant juices, and it may be that the yeasts afford assistance in the elimination of the large amount of carbohydrates taken in by the insects. Comparable conditions exist in beetles that live on wood, yeasts being harboured in special outgrowths from the alimentary canal, and, presumably, helping to digest the resistant lignin.[3]

Ectotrophic Mycorrhiza. Many woody plants possess an ectotrophic mycorrhiza,[4] appearing as a dense felt of hyphae over

[1] Bernard, 1911 ii; Magrou, 1921. [2] Conte and Faucheron, 1907.
[3] Buchner, 1930.
[4] Frank, 1885; McDougall, 1914; Melin, 1921; Peyronel, 1921; Paulson, 1924.

the apical parts of the infected roots; the development of root hairs is impeded, and short coralloid branches are formed. The fungus makes its way between the outer cells of the root, and also penetrates some of the cells, where digestion of the hyphae apparently occurs. Infection of the young roots may take place at any period in the life of the green plant, but all young roots are not attacked; mycorrhiza is developed most strongly on roots which spread horizontally in layers of decaying leaves, or in masses of moss, lying beneath the canopy of the branches; it is rare or absent when the soil lacks humus; the roots then bear hairs, and function normally.

The relation between an ectotrophic fungus and its host, at least when the latter is green, appears to be more casual than in endotrophic forms. The association probably arises as an attempt at parasitism on the part of the fungus; indeed, when external conditions are unfavourable to the fungus, the latter may become so strongly parasitic as seriously to damage the root. Probably the fungus gets more out of the association than do the fungi of endotrophic mycorrhiza; perhaps there is a connection between the constant occurrence of certain agarics beneath certain trees and the mycorrhizal habit, the fungus being unable to fruit unless it is associated with an appropriate root. Thus, *Amanita muscaria* is commonly found under birch, *Lactarius deliciosus*, *Russula fragilis* and *Boletus badius* are characteristic of pine woods, and *B. versipellis* and *B. scaber* occur under deciduous trees. The brick red colour of the mycelium of *Cortinarius rubipes*[1] has made it possible to trace a connection between the fructifications of this fungus and young roots of oak. On the other hand, birches and hornbeams appear specially liable to the attacks of *Melanconis stilbostoma* and *Pseudovalsa lanciformis*[2] during times of drought, when the mycorrhiza has been killed off; this susceptibility may be due to the breaking up of the association, or merely a result of general weakening caused by dry conditions.

It seems likely that the fungi of ectotrophic mycorrhiza take up material from the substratum and pass some of it on to the associated plant;[3] there is no evidence that these fungi influence germination.

[1] Kauffman, 1906. [2] Paulson, 1924. [3] Rexhausen, 1920.

SPECIALISATION

Fungi vary in the extent to which they are adapted or restricted to a particular habitat; the range may be wide, as in *Eurotium herbariorum* and *Penicillium glaucum*, which may occur, under suitable conditions of temperature and moisture, on almost any plant remains, on vegetable products, such as bread or jam, or on substances of animal origin, such as cheese or leather. Similarly, *Synchytrium aureum* attacks more than a hundred species of dicotyledons, and *Phyllactinia Corylea* occurs on the leaves of many woody plants. Other fungi have a narrow range; many species of *Hydnum* are found only in fir woods, *Pyronema confluens* and several other fungi occur in nature chiefly on burnt ground, *Onygena equina* is restricted to the derelict horns and hooves of animals, and many species are found only on dung. Some fungi are limited, either as saprophytes or parasites, to the members of a particular family, a particular genus, or even a particular species.

Biologic Species.[1] Such specialisation may be associated with definite morphological characters, or it may be purely physiological, so that a number of forms, identical in their minute structure, differ in their ability to infect particular hosts; similarly, distinct strains may exist within a given host species, some susceptible to a particular fungus, and others resistant. The strains included in a species, whether parasite or host, which differ in their capacity to infect or be infected, are termed **biologic species** or **biologic forms**. Immunity in the host may be a recessive character in the mendelian sense; thus, when a variety of wheat susceptible to the rust, *Puccinia glumarum*, is crossed with a resistant form, all the plants in F_1 are rusted, and three-quarters of those in F_2 also suffer infection, while the remaining quarter stand green and uninjured among them.[2]

The resistance of a plant, which, when healthy, is immune to a particular species of parasite, may break down[3] if the plant is growing in an unsuitable environment, or if it is injured mechanically, or by the presence of another parasite. Attack by weak

[1] Ward, 1902–5; Diedicke, 1902; Marchal, 1902–3; Marryat, 1907; Freeman and Johnson, 1911; Stakman, Piemeisel and Levine, 1918; Wormald, 1919; Sampson, 1925; Brooks, 1924.

[2] Biffen, 1907, 1912; Pole Evans, 1911; Vavilov, 1914; Hayes, Stakman and Aamodt, 1925.

[3] Salmon, 1904–5.

parasites is favoured by debility in the host; obligate parasites, however, flourish although the host is vigorous. It is doubtful if the spread of a pathogenic fungus from its usual host to a second species is assisted by the parasite first obtaining a foothold in a weakened plant of the new host, subsequently becoming educated to invade healthy plants of the same kind.[1] When such migrations from host to host occur, they are probably due to a mutational change in the infective powers of the pathogen.

Heteroecism. Most parasitic fungi pass through their life history in or on the tissues of a single host; such forms are said to be **autoecious** in contradistinction to the **heteroecious** forms which require two alternating hosts for the completion of their development. With the exception of the ascomycetous species, *Sclerotinia Ledi*, all known heteroecious fungi belong to the rusts; they pass through their gametophytic stage relatively early in the year, and the spores which are the products of the sexual apparatus are incapable of germinating on the host on which they were borne, but develop if they alight on the appropriate host of the sporophyte. Such a condition may have originated as a means of prolonging the growth period when the original host of the gametophyte died down early in the year; but, in many existing rusts, the host of the gametophyte grows throughout the year, as in *Puccinia Graminis*, the well-known rust of wheat, the gametophyte of which occurs on the barberry.

REACTIONS TO STIMULI

The response of fungi to external stimuli can usually be related to the distribution of spores or to nutrition. In addition, special reactions occur in the approach and fusion of the sexual organs and their products, bringing antheridia and oogonia together in the Oomycetes, influencing the movement of the trichogyne towards the male organ in *Zodiomyces*, and perhaps in other Ascomycetes. Nothing is known of the active substances concerned in these reactions.

Chemotaxis. The zoospores of *Saprolegnia mixta*[2] swim towards aqueous extracts of dead insects and towards aqueous solutions of their decomposition products, such as proteins, phosphates, urea and organic acids, if suitably diluted; a substance which

[1] Brooks, 1935. [2] Müller, 1911.

attracts in weak solutions repels in strong ones. The zoospores are very sensitive, and will react to a solution containing 0·0001 per cent. of orthophosphoric acid, or 0·00025 per cent. of protein. The minimum concentration of organic acids which will cause response varies inversely with the degree to which dissociation takes place in solution. Similar responses occur in the zoospores of other species. They have an obvious biological significance, enabling the spores to reach a suitable substratum for further growth. Repulsion is not caused by poisonous salts; the zoospores swim readily into a solution containing a protein and a lethal dose of mercuric chloride.

Chemotropism. The strongest chemotropic response of hyphae and germ tubes is a negative one; they grow away from regions which have been staled by the products of their own metabolism.[1] A simple example of this is familiar in the circular growth of mycelia both in nature and in artificial culture. In so far as a clear field is available, hyphae tend to grow equally in all directions from the point of infection. The same factor may account for the alternate dense and sparse zones which characterise many fungal colonies, and are independent of changes in light and temperature.[2] Energetic growth results in the accumulation of waste substances in the medium in front of the elongating hyphae, and growth is accordingly reduced till a few hyphae pass beyond the inhibiting influence and give rise to a new ring of richly branched mycelium on the unstaled medium; the renewed growth stales the medium in front of it, and so on, again and again. In old cultures, the regular formation of zones may be disturbed by the growth of mycelia from spores falling and germinating beyond the limits of the original colony.

Chemotropism is investigated by means of perforated plates of mica coated on both sides with a suitable agar medium.[3] When spores of *Rhizopus nigricans* are sown in approximately equal numbers on both sides of such a plate, the germ tubes show no special reaction towards the holes. When, however, one side only is sown, most of the germ tubes grow towards and through the holes; as the fungus develops, the medium is staled, and the products of metabolism diffuse through the perforations into the unsown layer. Since the germ tubes grow away from the staling substances, they

[1] Clark, 1902; Fulton, 1906; Balls, 1908; Graves, 1916; Brown, W., 1922 ii, 1923, 1925; Pratt, 1924.

[2] Stevens and Hall, 1909.

[3] Graves, 1916.

follow the falling diffusion gradient through the holes, and find their way to the fresh medium. A heavily staled medium may be obtained by growing *Rhizopus* for some weeks in turnip juice, filtering to remove the mycelium, and adding agar. If such a medium, which has not been heated above 40° C., is contrasted with fresh turnip juice agar, and spores sown only on the latter, the germ tubes grow away from the holes, for a strong diffusion current of waste material passes from the stale to the fresh medium. If, however, the stale medium has been heated to 100° C., the germ tubes do not avoid the holes so obviously, indicating that the preliminary heating has dissipated some of the injurious substances.

The progress of staling may be estimated by periodical measurements[1] of the diameter of a fungal colony grown on a solid medium. *Botrytis cinerea* does not stale potato agar readily; the colonies work up to a maximum rate of growth and tend to maintain it. On the other hand, species of *Fusarium* reach a maximum rate, then fall off, and may cease to grow before they have covered all the available space, since the medium has been spoiled, either by the rapid outward diffusion of staling substances, or by the absorption of gaseous products of metabolism. The staling of the medium may become sufficient to bring to an end the growth of the fungus.

In general, renewed growth of a fungus cannot be obtained by stripping the growth from a staled medium and reinoculating with fresh spores; spores which fall on such a medium, either fail to germinate, or at best give weak growth. Ammonia and potassium bicarbonate are among the substances that cause staling;[2] the former is driven off by heat, the latter is decomposed. When a staled medium is heated, cooled, and sown with spores, further growth may result;[3] this shows that the previous cessation of growth was not due to the exhaustion of nutritive material, and indicates that volatile or unstable staling substances have been removed by heating. The staling of cultures may be delayed by providing an atmosphere of carbon dioxide, which unites with ammonia; by ventilation, which permits the ammonia to escape; by using a large bulk of medium, so that the development of a concentration of staling substances sufficient to stop activity is deferred; and by cutting out the centre of the colony, thereby removing the reservoir from which diffusion takes place into the unstaled medium.

[1] Brown, W., 1923. [2] Pratt, 1924. [3] Asthana and Hawker, 1936.

Rise of temperature accelerates the rate of chemical change; in agreement with this, staling is more rapid at high than at low temperatures.[1] In estimating the effect of temperature upon the behaviour of a fungus, it is however necessary to bear in mind that a high temperature may cause direct injury in addition to the influence exercised upon the production of staling substances, and so on the rate of growth.

There is some evidence that germ tubes and hyphae grow towards salts, sugars and other nutritive substances, and away from acids, alkalis and alcohols.[2] It is difficult to distinguish between reaction due to staling and positive chemotropic response; the latter is always much weaker than the former, and is probably best seen when the culture is young, and the concentration of waste products low.

Aerotaxis. The zoospores of *Rhizophidium pollinis* and *R. sphaerotheca* are extremely sensitive to free oxygen; when mounted under a coverslip they move at once to the margin of the drop, or to air bubbles in the preparation; they crowd into capillary tubes containing oxygenated water, and have a tendency to form groups around globules of turpentine, a good oxygen carrier.[3]

Aerotropism. Growth in response to varying concentrations of free oxygen is rarely shown by fungi. The sporangiophores of *Phycomyces nitens*[4] bend towards pieces of iron placed close to them. Rusting iron liberates small quantities of ozone, and apparently the bending takes place in response to this. Similar curvatures may be obtained towards sheets of platinum foil, or glass, bearing adsorbed ozone.

Hydrotropism. The position perpendicular to the substratum assumed by very young sporangiophores of the Mucorales is due to negative hydrotropism. In saturated air, the sporangiophores of *Phycomyces* are oriented irregularly, and the divergence of the tufted sporangiophores of *Rhizopus* may be due to negative hydrotropism in response to exuded water, drops of which often appear in or near the region of elongation. Exposure to moist air increases the rate of growth of the sporangiophores of *Phycomyces nitens*;[5] they sometimes show weak curvatures away from wet paper placed vertically in their neighbourhood.

[1] Balls, 1908; Brown, W., 1922 ii. [2] Miyoshi, 1894; Graves, 1916. [3] Müller, 1911. [4] Elfving, 1917. [5] Walter, 1921.

It is probable that vegetative hyphae grow towards moisture, but there is no conclusive evidence to show that this is so, as it is difficult to discriminate between the effect due to water, and that due to staling substances.[1]

Phototaxis. The zoospores of *Rhizophlyctis vorax*[2] and of *Polyphagus Euglenae*[3] move towards light. Both these species attack motile green organisms, the former infecting *Sphaerella lacustris*, the latter, *Euglena viridis*. The hosts obtain their carbon supplies by photosynthesis, react to light, and the phototactic response of the parasite brings it into regions where the hosts are to be found. The zoospores of *Rhizophidium pollinis*[4] show a weak positive reaction; this is probably bound up with their need for abundant supplies of free oxygen.

Phototropism. Many fruit bodies are sensitive to light, and by means of this reaction adjust themselves in a position favourable to the distribution of their spores, the direction of light indicating the direction of open space.

The sporangiophores of species of *Pilobolus*, *Mucor* and *Phycomyces* bend towards the light. In the genus *Pilobolus*, the hemispherical sporangium is borne on an aseptate sporangiophore which develops a swelling just below the sporangium and another at its base. A new crop of sporangia matures daily, and discharge takes place in the morning or the early afternoon; growth is apical, the tips of the young sporangiophores bending towards the source of light as they grow. An unnamed species[5] showed a recognisable curvature about thirty minutes after exposure to light in a given direction. Curvature is arrested during the early stages of the growth of the sporangium, and is resumed again when the subsporangial bulb is beginning to form; during the later stages of development, curvature takes place below the bulb, which may perceive light.[6] In this way the bulb and the terminal sporangium are pointed in the direction of light, and some accuracy of aim is secured. In a series of experiments[7] involving some 20,000 specimens, illuminated by means of apertures 1 cm. in diameter, nearly 90 per cent. of the sporangia entered the aperture or hit the walls within 1 cm. of it. When illuminated by two equal

[1] Fulton, 1906; Robinson, 1914.
[2] Strasburger, 1878.
[3] Wager, 1914 i.
[4] Müller, 1911.
[5] Jolivette, 1914.
[6] Buller, 1921.
[7] Parr, 1918.

sources of light, the sporangiophores pointed to one or the other; when the sources of light differed, the sporangia were shot off in larger numbers towards the light with a greater proportion of blue rays.

Young sporangiophores of *Phycomyces nitens* may show curvature in fifteen minutes. Bending takes place at a point just below the developing sporangium, and elongation proceeds in a straight line towards the light. Illumination increases the rate of growth,[1] so that exposure to unilateral light might be expected to produce a negative curvature in the sporangiophore. Apparently, however, its transparency enables the structure to act as a lens, so that light falling on one side is concentrated on the opposite side, where growth is accelerated and a positive curvature induced.

The necks of the perithecia of *Sordaria* and other pyrenomycetous fungi show a positive response sufficiently delicate to cause a zigzag development of the neck if the direction of light is repeatedly changed. The asci of *Ascobolus immersus* and *A. stercorarius* are also positively phototropic so that an appropriate direction is obtained for the ejection of their large spore masses.

The well-marked positive response shown by immature fruit bodies of *Coprinus niveus* and *C. curtus*, both coprophilous species, enables them to push up from the irregular substratum.[2] The stipe ceases to be phototropic when the pileus begins to expand, and develops instead a negatively geotropic reaction; by these means, the apex of the stipe is brought into the light before the growth of the pileus begins, and the latter has space for its horizontal expansion. A similar succession of reactions takes place in the development of the sporophores of *Lentinus lepideus*;[2] the young stipe is positively phototropic and in the absence of light grows straight onwards, without giving rise to a pileus; with adequate illumination a pileus appears, and, as its development proceeds, the positive response to light is replaced by a negative reaction to gravity. In *Amanita phalloides* and *A. crenulata* there is a positively phototropic response[3] even after the appearance of the pileus.

Psalliota campestris and *Coprinus comatus* usually occur on turf

[1] Blaauw, 1914. [2] Buller, 1909 ii.
[3] Streeter, 1909.

in open places; they are insensitive to light, and the adjustment of the parts is secured by negative geotropism.[1]

A negatively phototropic reaction is comparatively rare. Germ tubes from the basidiospores of *Puccinia Malvacearum* and from conidia of a species of *Botrytis* turn away from unilateral light; those from the aecidiospores of *Puccinia Poarum* and from the conidia of *Peronospora parasitica* and *Penicillium glaucum* are indifferent.[2]

Formative Influence of Light. Many fungi are unable to develop in a normal manner unless they are adequately illuminated.[1] Young fruit bodies of *Polyporus squamosus*, though quite irresponsive to the direction of light, fail to give pilei in continuous darkness. The same is true of many other Hymenomycetes, including those which, like *Lentinus lepideus*, are positively phototropic. A brief period of illumination during the early stages of growth may be sufficient to initiate the normal development of the sporophore; once started, the differentiation of the pileus can be completed in the dark. Direct sunlight stimulates the formation of the ascocarps of *Pyronema domesticum* and other Ascomycetes, especially of coprophilous species. Cultures which remain sterile in dull light often fruit when placed in a sunny window; possibly the light and warmth thus provided may bring about the chemical changes essential to fruiting. On the other hand, *Eurotium herbariorum* appears to form perithecia as readily in darkness as in light. Ultraviolet light causes fruiting in *Glomerella cingulata*.[3]

Geotropism. The sporangiophores of *Phycomyces* and of *Mucor* emerge at right angles to the substratum, grow to a length of about 2 mm. and then become negatively geotropic. The mycelia seem to be indifferent to gravity. The fructifications of several of the Helvellales and the stromata of *Poronia punctata*[4] and *Xylaria Hypoxylon*[5] among the Pyrenomycetes, exhibit a negatively geotropic reaction from early stages of development. The Hymenomycetales show marked response to gravity. In stipitate forms the stipe is negatively geotropic as soon as the pileus develops, though, in species of *Coprinus*, and in *Lentinus lepideus*, this may be preceded by a phase in which light is the directive influence.[6]

[1] Brefeld, 1877; Buller, 1909 ii.
[2] Robinson, 1914.
[3] Stevens, 1928, 1930.
[4] Dawson, 1900.
[5] Freeman, D. L., 1910.
[6] Buller, 1909 ii.

The geotropic response is of the nature of a gradual adjustment. The growing stalk swings beyond the vertical line, changes its direction and swings across it again, passing the vertical two or three times before it comes to rest. In *Amanita crenulata*[1] the stipe reaches its full development in about twenty-four hours; this may not allow time for complete adjustment, so that the sporophore may come to rest out of the vertical. In this species, the young stipe elongates throughout its length until more than half grown; the zone of most rapid elongation is just below the pileus, and as growth slackens it becomes narrower and narrower. The perceptive region is in the stipe, and is so sensitive that a definite response may be obtained by placing the stipe in a horizontal position for one minute.

In the stemless Hymenomycetes, as in the stipitate forms, the orientation of the pileus takes place in response to gravity.[2] This is well shown by the horizontally arranged fruit bodies of the bracket-like species of *Thelephora*, *Stereum*, *Polyporus* and *Polystictus*.

Alike in the Hydnaceae, Agaricaceae and Polyporaceae, the trama plates of the fertile region are positively geotropic. In stemless forms this reaction is responsible for the orientation of the hymenium; it may be particularly well seen in some of the Hydnaceae, where the spines grow downward whatever the orientation of the sporophore, and it could doubtless be demonstrated also in *Tremellodon* among the Tremellaceae. In stipitate species it appears as a supplementary reaction coming into play if further adjustment is needed when the stipe is fully grown. The limitations of the method in the gill-bearing fungi are obvious, for, if the pileus is oblique and the gills undergo much curvature, they become crowded and interfere with spore dispersal.

Formative Influence of Gravity. A young fruit body of *Polystictus cinnabarinus*[3] was rotated in a moist chamber on a horizontal clinostat for several weeks. A cushion-like structure was formed with the pores all over the surface, and with hardly any difference between the dorsal and ventral faces. This suggests that the localisation of the pores to the ventral surface under ordinary conditions is directly due to the influence of gravity.

[1] Streeter, 1909.
[2] Buller, 1909 ii.
[3] Hasselbring, 1907.

Interaction of Stimuli. The life history of a fungus is the sum of its reactions to stimuli, and may, as a rule, be divided into three phases; a period of vegetative growth merges into one characterised by the appearance of organs of accessory multiplication, and this is followed by the development of the sexual apparatus.[1] The change from a vegetative condition is influenced by nutrition, light, temperature, aeration, moisture, and the encounter of mechanical obstacles; any of these may be the limiting factor if the remaining conditions are favourable.

Conditions which favour vegetative activity are opposed to the development of sexual organs; to a less extent, they discourage the formation of accessory spores.[2] Many fungi give rise to a vigorous mycelium on substrata rich in carbohydrates, but remain sterile, or bear a scanty crop of conidia; a moderate supply of carbohydrates favours the formation of conidia. When the carbohydrate is further reduced, and some source of nitrogen is present, the sexual organs appear. In other species, a moderate addition of glucose encourages sexual reproduction, perhaps by accelerating the rate at which the fungus passes through its vegetative phase. *Eurotium herbariorum*[3] gives heavy crops of perithecia on media with 40 per cent. of cane sugar; the concentration of sugar in the medium probably checks vegetative activity. The Phycomycetes appear to be less sensitive to the presence of carbohydrates than the Ascomycetes, but, in *Phytophthora*, an increase in the proportion of carbohydrate may limit the development of sexual organs. There is evidence that, although a supply of nitrogenous food is a necessary preliminary to the formation of gametangia, they do not appear[4] until the substratum is depleted of this material. Rich nitrogenous manuring is stated to check the development of the fruit bodies of Hymenomycetales growing on soil.[5] Aquatic fungi[6] are markedly sensitive to external conditions, and tend to form fructifications when the substratum is becoming impoverished, or when the surrounding water is fouled by the products of their activity and that of other organisms.

Inactive spores may resist extremes of temperature which kill

[1] Klebs, 1896–1900; Fraser and Chambers, 1907.
[2] Klebs, 1896–1900.
[3] Fraser and Chambers, 1907.
[4] Robinson, 1926 ii.
[5] Gilbert, 1875.
[6] Klebs, 1896–1900; Kauffman, 1908; Obel, 1910; Lechmere, 1910–11; Collins, M. I., 1920; Kanouse, 1925.

the mycelium, though the latter can grow at temperatures too high or too low[1] to permit the formation of reproductive structures. Doubtless the temperature relations of most fungi are adjusted to the climatic conditions of their usual habitat, but, among British fungi, *Cladosporium herbarum*[2] forms spores at $-6°$ C., and *Eurotium herbariorum* fruits abundantly at 37° C.

Moisture may have a definite influence upon development, perhaps by regulating transpiration; in damp air, *Sporodinia grandis*[3] produces sporangia, but gives zygospores when the air is almost saturated, while under humid conditions, the stromata of *Xylaria Hypoxylon*[4] tend to break up into separate hyphae.

The formation of the organs of multiplication is often encouraged when the mycelium comes into contact with obstacles such as the walls of the culture vessel, which can have little or no nutritive significance. Growth is checked, cytoplasm accumulates in the tips of the hyphae, and a new type of development is initiated.[5] When a growing mycelium encounters that of another fungus, the latter may act as a mechanical obstacle, or may affect nutrition by the secretion of staling products, or by changes in the composition of the substratum. *Lachnea abundans* forms apothecia when in contact with *Penicillium* upon a medium on which it will not fruit in pure culture, and bacteria have been observed to influence the fruiting of other Ascomycetes.

A relation may exist between the rate of respiration and the development of reproductive organs.[4] Respiration is high in the early stages of growth, and falls as the mycelium ages; when this has happened, sporangia appear, or, if other conditions permit, sexual organs are produced. In closed Petri dishes, rings of conidiophores are formed at the margin of the culture,[6] but when the culture is made in an open dish, they develop over the whole surface; this may indicate a relation to the supply of free oxygen, and so be connected with respiration.

The appearance of the organs of multiplication is accompanied by changes in the structure and contents of the cytoplasm, and, in many fungi, by the formation of definite pigments, indicating that the metabolism of the plant is undergoing a change; there is

[1] Weimer and Harter, 1923.
[2] Brooks and Hansford, 1923.
[3] Robinson, 1926 i.
[4] Freeman, D. L., 1910.
[5] Robinson, 1926 i, ii.
[6] Barnes, 1924–5.

doubtless an intimate connection between this change in internal conditions and the altered behaviour of the fungus.[1] The external conditions which affect fertility do not appear to do so directly, but by modifying the chemical processes in the protoplasm.

Heterothallism. Heterothallism may be regarded as a special instance of reaction to stimuli. It occurs in the three great groups of fungi and may be of separate origin in each or may suggest some more or less remote connection between the Ascomycetes and Basidiomycetes. The term implies the existence, in a given species, of more than one gametophytic strain or thallus and that two of these must come together to bring about, or to expedite, the formation of the diplophase. Such a definition does not exclude strains which differ in sex, one bearing male and the other female organs, and the term has frequently been used among Phycomycetes to indicate dioecism of the gametophyte. It was in this sense that the word was first employed by Blakeslee when he described the existence of two strains in *Mucor* and its allies, neither of which produced sexual organs alone, but both of which, in contact, formed isogametangia. Though minor differences were sometimes seen in the activity of the strains, there was at that time no evidence that one was male and the other female; they were arbitrarily distinguished as + and −, or later as *A* and *B*.

In the Ascomycetes dioecism is rare, but many monoecious species possess two strains, and sexual organs only appear, or only give rise to the diplophase, when contact is made between the two. In the former condition the species is self-sterile, in the latter self-incompatible. Thus, in *Ascobolus magnificus*, the *A* and *B* strains are alike capable of bearing both antheridia and oogonia, so that no morphological difference exists between them, but neither, if growing alone, forms sexual organs. When mycelia of the two strains are intermingled, sexual branches grow up, antheridia and oogonia are differentiated, male organs of *A* fertilise female organs of *B*, and male nuclei of *B* enter the oogonia of *A*. The sporophyte thus contains both *A* and *B* nuclei and yields both *A* and *B* spores. In the self-incompatible *Humaria granulata* antheridia are lacking; female branches appear in single spore culture both on *A* and *B* strains, but they die without the formation of a sporophyte. In the absence of male organs, union of *A* and *B*

[1] Bezssonof, 1918–19; Boas, 1918–19; Sartory, 1918; Molliard, 1920; Brown, W., 1922 ii, 1923, 1925, 1926; Leonian, 1925; Brown, W., and Horne, 1926.

through the sexual apparatus is impossible, but, when complementary thalli are associated, use is made of the mycelial fusions common in fungi. Where these have occurred, the female organ proceeds to the formation of a sporophyte and the asci which it bears contain *A* and *B* spores, showing that nuclei from both thalli have passed through the oogonium.

The asci of *Neurospora tetrasperma*, another form with mycelial fusions, of *Podospora anserina* and of other species produce four binucleate ascospores which give rise to homothallic or self-compatible mycelia. One of the binucleate spores in the ascus is occasionally replaced by two small uninucleate spores, which prove to be respectively *A* and *B*, so that, in these species, the homothallic character depends on the presence of two compatible nuclei. The *A* and *B* factors have been shown to be inherited on mendelian lines. The gametophytes of many heterothallic species liberate conidia and these, if they germinate near a mycelium of opposite complement, may bring about the union of these strains. This method is probably important under natural conditions.

In *Ascobolus stercorarius*, a species which like *Humaria granulata* is without antheridia, a single strain may fruit when growing alone, though only after long delay. In *Glomerella cingulata* asci are developed in single spore culture, but on one strain they are scanty and on the other malformed; they contain only *A* or only *B* spores. When the two strains meet abundant asci result and bear spores of both complements. The production of abundant asci can also be induced by exposure to ultra-violet light. Again in *Penicillium luteum* the existence of complementary strains is associated with differences in vigour, some being able to form sterile fruits alone, others being limited to vegetative growth.

Among the Basidiomycetes, rust fungi show stages of development similar to those of the self-incompatible Ascomycetes. Female organs and detachable spermatia are found on both *A* and *B* thalli and association only takes place, in heterothallic species, between nuclei derived from complementary strains. Such nuclei may be brought together by the transfer of spermatia to hyphae of opposite complement, or by the mingling in the host tissue of *A* and *B* mycelia followed by mycelial fusion. Each strain, at any rate in some species, is able to form sporophytes in isolation, though the process is slower than when both are concerned.

The other groups of Basidiomycetes have no sexual apparatus and the union of compatible strains is achieved wholly through mycelial fusion, often facilitated by the dispersal of conidia. In these, as in other fungi, the characters of the complementary strains show mendelian inheritance. Here, as elsewhere, fructifications may form on either thallus growing alone, though often after considerable delay.

There is much scattered evidence that the formation of fruits in single spore culture is affected by nutritive conditions, and it is not impossible that complementary strains may prove to differ in the food materials they can accumulate,[1] so that, when their resources are pooled, by fertilisation or by mycelial fusion, the needs of the sporophyte are met.

Variation. When fungi are grown on media containing high concentrations of mineral salts, their more obvious characters may be so greatly changed as to give the impression that new forms have been obtained. When, however, spores from these cultures are sown on a medium of ordinary composition, they yield colonies of normal type. Such temporary changes, due to some special character of the medium, and lasting only so long as the fungus is growing under these special conditions, are modifications; they are not known to have any genetical significance.

If spores or young colonies are treated with strong chemical or physical agents, such as poisons, X-rays or high temperatures, new forms may develop which retain their peculiarities when grown in ordinary cultures,[2] no longer under the influence of the agent which provoked their appearance. Variants of this kind, representing a more or less permanent alteration in the stock, have been obtained in *Chaetomium*[3] and in the Mucorales[4] by the use of X-rays, in *Aspergillus*[5] and *Penicillium*[6] by the addition of poisons to the medium, and in *Eurotium herbariorum*,[7] *Botrytis cinerea*[8] and *hamnidium elegans*[2] by exposure of spores to heat. It is possible at spores which have lain ungerminated for some months or ears may sometimes produce abnormal colonies. The morphogical and cultural characters of variants derived from specially eated material suggests that, in general, they are damaged forms of

[1] Gwynne-Vaughan, 1926, 1928, 1935.
[2] Barnes, 1935 i, ii.
[3] Dickson, 1932, 1933.
[4] Nadson and Phillipov, 1925 ii, 1928.
[5] Schiemann, 1912.
[6] Haenicke, 1916.
[7] Barnes, 1928.
[8] Barnes, 1930.

the original, weakened in growth, or unable to develop the normal pigmentation of the species. It is not yet clear to what extent fungi growing under natural conditions are liable to alteration by the chance action of some violent external agent; it seems probable that new forms arise from time to time, but there is no reliable information about the frequency of the phenomenon or the power of the new forms to survive in competition.

In addition to changes brought about under experimental conditions capable of some degree of control, many fungi, when grown in culture, give rise without special treatment to new forms, either as sectors of normal colonies, or more rarely, on the germination of spores. Such changes are described as **saltations**:[1] though they are of common occurrence, the mechanism of the process is still obscure.

FORMS RESEMBLING FUNGI

The members of several groups of simply organised creatures possess some characters similar to those of the lower fungi; the more important of these groups are the *Monadineae zoosporeae*, the Myxomycetes, and the Plasmodiophorales. The *Monadineae zoosporeae* differ from the lowest Archimycetes chiefly in their habitual ingestion of solid material; they may well be the modern representatives of the ancestral stock of the fungi. The Myxomycetes and the Plasmodiophorales are not fungi, and have doubtless arisen from lower forms along independent lines of development.

MONADINEAE ZOOSPOREAE

The *Monadineae zoosporeae* is a group of aquatic organisms, mostly parasites of algae or fungi. They devour the contents of the cells of their hosts, producing solid granules of undigested residues in the process, and, in these respects, differing from any known members of the Archimycetes. Multiplication is by means of zoospores; and, in some species, resting spores develop after the fusion of two motile gametes. Few have been investigated in detail.

Pseudospora Rovignensis[2] is a parasite of *Vaucheria piloboloides*. The uniflagellate zoospore settles on the host, pierces the wall, and enters the filament. It ingests the contents of the filament, and,

[1] Stevens, 1922; Chaudhuri, 1924; Brown, W., 1926; Horne and Das Gupta, 1929; Dickinson, 1932, 1933.
[2] Schussnig, 1929.

after a period of feeding, becomes amoeboid, still however retaining its flagellum. The single nucleus then divides, longitudinal division of the body follows, giving two zoospores. These feed, divide in duê course, and their progeny continue along the same lines, providing for the rapid multiplication of the organism. Walled resting spores may also be formed, maybe after a fusion of two zoospores, which are thus facultative gametes. The resting spores probably give rise to zoospores.

Pseudospora parasitica[1] occurs in dying filaments of *Spirogyra*. The globose zoospore bears one trailing flagellum. It enters the cell of the host through a tiny perforation, and, once inside, rapidly enlarges, forming a digestive vacuole as it grows. The ingestion of solid material is accompanied by the accumulation of brownish digestive residues in the vacuole; these provide the readiest means of recognising the organism. When the parasite is mature, a wall develops round the protoplast, which then divides into a number of zoospores. They escape into the lumen of the host, and either continue to develop there, or pass into the surrounding water and invade other cells. The zoospores may function as facultative gametes, fusing in pairs to give a rounded, walled zygote. The contents contract inside the wall, the digestive residues are ejected from the living material, and the protoplast, freed from rubbish, rounds up, yielding a small, walled resting spore within the original cyst. Zoospores probably escape when the zygote germinates.

The species of *Aphelidium*[2] and of *Aphelidiopsis*[2] have life histories similar to that of *Pseudospora parasitica*, with some differences in detail suggesting a closer approach to the Archimycetes. The uniflagellate zoospore of *Aphelidium* is covered by a thin wall which is left behind when a host cell is invaded; this apparently insignificant character occurs in many species of lower fungi, and may well be a reliable indication of affinity. *Aphelidium* usually attacks diatoms and other small algae. It does not form zoospores until the cell of the host is exhausted. When the zoospores escape from the dead host, they aggregate for a short time in a group, thus exhibiting in simple fashion a behaviour characteristic of several genera of Phycomycetes. *Aphelidiopsis* has biflagellate zoospores; before these are formed, the protoplast fragments into several portions, each of which becomes a walled zoosporangium; sori of zoosporangia are

[1] Schussnig, 1929. [2] Scherffel, 1925 i.

formed in somewhat similar fashion by *Synchytrium* and other members of the Chytridiales.

There are clear resemblances between the members of the *Monadineae zoosporeae* and such Archimycetes as *Sphaerita* and *Olpidium*; the only marked difference is that the latter are walled from an early stage in development, and therefore can utilise only substances in solution.

MYXOMYCETES

The organisms which are placed in the Myxomycetes[1] are isolated and show signs of specialisation which cannot be reconciled with a position at the base of the fungi.[2] In early stages of development they appear as small, naked uninucleate amoebae, the **myxamoebae**, with a contractile vacuole, and other vacuoles into which solid material is taken; this mode of nutrition is adopted by the organism throughout its active life. Mitosis is followed by the division of the whole body, and repetitions of this process give rise to a large number of individuals. In the lower Myxomycetes, myxamoebae alone occur, but, in more advanced types, the myxamoebae are converted into biflagellate pyriform zoospores; these retain some capacity for amoeboid movement, especially at the blunt posterior end, where pseudopodia are put out, and solid material is captured. The anterior end is narrowed to form a flagellum, with which a blepharoplast is associated.

After a short period of multiplication, the zoospores may either fuse in pairs,[3] or may resume the myxamoeboid form before fusing.[4] There is evidence that the spores may be differentiated into two physiological types, between which alone fusion may occur,[5] but this is not invariable.[6] The amoeboid zygotes come together in large numbers; in lower forms, a **pseudoplasmodium** of closely associated but distinct individuals results; in higher forms, the fusion of the cytoplasm of the zygotes gives origin to a **plasmodium**. Small plasmodia flow together to form larger ones, active growth accompanied by nuclear division occurs, and the plasmodium moves over or through the substratum, taking in and devouring solid material,[5] including spores, zoospores and myxamoebae which have

[1] Lister and Lister, 1925; Lister, 1933.
[2] Pascher, 1918.
[3] Wilson and Cadman, 1928; Cayley, 1929.
[4] Jahn, E., 1911.
[5] Skupienski, 1918, 1920.
[6] Cayley, 1929.

not fused. The plasmodium is a thin sheet of naked protoplasm, containing a network of rather dense strands in which streaming is well seen. As a rule, the area covered by an adult plasmodium does not exceed a few square inches; the plasmodium of *Fuligo varians* may cover more than a square yard.

During the time that growth is active, plasmodia tend to move towards moisture, and to avoid strong light. Exposure to drought causes the formation of a dense mass of walled bodies; each of these contains about a dozen nuclei, and the whole mass constitutes a sclerotium. This is tolerant of adverse conditions, and may remain quiescent for some years. When water is supplied, the cellulose walls are absorbed, and a plasmodium reappears; the time required for the resumption of activity is roughly proportional to the duration of the period of desiccation.[1]

Sooner or later the plasmodium changes its behaviour and moves towards a relatively dry position in good light. Protuberances form upon it in one or more places, and from these sporangia appear, singly or in groups. In the higher Myxomycetes, a mass of uninucleate spores develops within the sporangium, embedded in a system of threads or tubes, the **elaters**; these form the **capillitium**, in which hygroscopic movements bring the spores to the surface, whence the wind carries them away. As a rule, a single myxamoeba is produced when the spore germinates.

The nuclei of the spores, and of the myxamoebae, are haploid; the diploid condition is initiated when the zygotes are formed, and persists until spore formation, when meiosis occurs.[2]

PLASMODIOPHORALES

The Plasmodiophorales, with the single family Plasmodiophoraceae,[3] are endoparasites in higher plants or in insects. In the vegetative condition, the organism is a naked mass of protoplasm, the myxamoeba; increase in size is accompanied by nuclear division, and ultimately the thallus breaks up into a number of walled spores.

The earliest known stage of *Sorosphaera Veronicae*[4] is a uni-

[1] Jahn, E., 1919. [2] Jahn, E., 1911.

[3] Woronin, 1878; Maire and Tison, 1909, 1911; Winge, 1913; Schwartz, 1914; Cook, 1933.

[4] Blomfield and Schwartz, 1910; Maire and Tison, 1911; Winge, 1913; Webb, 1935.

nucleate myxamoeba lying in a meristematic cell of the stem of *Veronica*. Repeated nuclear division is followed by the cleavage of the body into a number of myxamoebae which grow and, after nuclear division, form new individuals. Meanwhile, the host cell is stimulated to divide, myxamoebae are thereby passively distributed to the daughter cells, further divisions follow, and groups of infected and hypertrophied cells come into existence.

All the nuclei in a myxamoeba divide at the same time, but division is not simultaneous in individuals associated in a host cell, even when they are so closely packed that they seem to be united into a plasmodium. The division[1] is mitotic. At metaphase, the chromosomes appear in a ring encircling a large nucleolus; later, the ring yields eight chromosomes, and four of these, with half the nucleolus, pass to each daughter nucleus.

As the myxamoebae become crowded in the host cell, oily material accumulates in their cytoplasm, the chromatin decreases in amount and stainable material is extruded from the nucleus as the nucleolus degenerates; the nuclear reticulum persists, and the nuclei now resemble those of other organisms. As these changes proceed in the construction of the nucleus, it is possible that nuclear fusions occur. The nuclei divide, together with the cytoplasm associated with them, giving, finally, uninucleate masses, which round off, assume walls, and become spores. Meiosis occurs during the formation of the spores. The whole living material of a myxamoeba is converted into a hollow sphere of walled spores enclosed in a common membrane; the spores are united laterally by the fusion of their walls. *Spongospora subterranea*,[2] a parasite of potato tubers, has a life history similar to that of *Sorosphaera Veronicae*, but the ripe spores lie in irregular spongy masses.

The germination of the spore has not been observed in *Sorosphaera*, but something of the process is known in other genera. The spore of *Spongospora subterranea* gives rise to a single uninucleate myxamoeba which moves by means of pseudopodia;[3] that of *Plasmodiophora Brassicae* liberates a naked, pyriform, uninucleate zoospore;[4] similar motile bodies have been seen in water containing mature and empty spores of *Tetramyxa parasitica*.

It is probable that part of the life cycle occurs outside the host.

[1] Webb, 1935.
[2] Horne, 1930.
[3] Kunkel, 1915.
[4] Chupp, 1917.

The myxamoebae of *Spongospora subterranea* encyst under dry conditions,[1] a possible adaptation to life in the soil; infection of the host occurs from soil which has been sterilised and subsequently inoculated with spores of *Plasmodiophora Brassicae*.[2]

During the vegetative phase, nutrient materials are taken in chiefly in solution, but the myxamoebae of *Ligniera radicalis* ingest algae,[3] and it is probable that the ingestion of solid material is usual when the myxamoebae are in the soil. The parasite may cause great changes in the cells of the host;[4] their nuclei are deformed, the cytoplasm degenerates, and finally the dead cell is occupied by masses of spores. Cells[5] containing chlorophyll are seldom attacked, and starch grains resist absorption for a long time. The host may be invaded through root hairs, through the surface of the root, or at growing points. The location of the parasite in meristematic cells provides good opportunities for the passive distribution of the parasite inside the host; moreover, it is probable that the myxamoebae pass[6] through the cell walls, entering hitherto healthy cells.

There is great variation in the effect on the host,[7] and in the degree to which penetration occurs. Virulent parasites, such as *Plasmodiophora Brassicae*,[8] induce much hypertrophy, and many abnormal divisions of the host cells; a root may be attacked by many myxamoebae, and turnips and cabbages invaded by this species consequently show extraordinary deformations, to which the popular name of club root is applied. In normal soil, the invasion of potatoes by *Spongospora subterranea*[9] is confined to the outer tissues by the formation of cork beneath the seat of infection, but in wet soil, the deposition of cork is interfered with and the parasite may enter deeply. The species of *Ligniera*[10] are weak parasites; they attack moribund root hairs and cortical cells; these are not stimulated to divide, and, as hypertrophy is not caused, the root shows no deformation.

[1] Kunkel, 1915.
[2] Woronin, 1878; Chupp, 1917.
[3] Maire and Tison, 1911.
[4] Blomfield and Schwartz, 1910; Lutman, 1913; Kunkel, 1915, 1918; Chupp, 1917.
[5] Woronin, 1878; Lutman, 1913.
[6] Lutman, 1913.
[7] Maire and Tison, 1909, 1911; Cunningham, G. C., 1912; Winge, 1913.
[8] Nawaschin, 1899; Favorsky, 1910.
[9] Osborn, 1911; Kunkel, 1915.
[10] Schwartz, 1914.

PHYCOMYCETES

The Phycomycetes include about 1000 species, almost all characterised by the aseptate vegetative thallus and by the production either of sporangia with motile or non-motile sporangiospores or of conidia more or less obviously derived from sporangia. Resting sporangia, oospores or zygospores are formed, sometimes after an act of fertilisation, sometimes without any preliminary fusion of gametes or gametangia.

The lowest species are mostly aquatic, living as parasites of plants and animals or as saprophytes on submerged material; some occur in the roots of higher plants, and a few species attack the leaves and shoots of angiosperms, especially those inhabiting wet places. These simple forms may complete their whole life history in a day or so, forming zoospores freely and spreading very rapidly in epidemic fashion. Rather more advanced Phycomycetes may be amphibious, occurring as parasites in many organisms, and abundantly as saprophytes in damp soil. When moisture is plentiful, the sporangia of these forms liberate zoospores, but under relatively dry conditions, the sporangia may function as conidia. The highest Phycomycetes include numerous parasites of terrestrial phanerogams, a few species parasitic on their own near relatives, and many common saprophytes; their sporangia may give rise to non-motile spores or may function as conidia. Zoospores are seldom produced; in general the spores of the highest Phycomycetes are disseminated by the wind.

The Thallus. It seems probable that two main factors have been concerned in the evolution of the thallus of the Phycomycetes—the extension of the thallus in relation to the need for searching wide areas for nutritive material, and the elaboration of specialised branches which raise the sporangia above the substratum and favour the escape and efficient distribution of the spores. Increase in the size of the thallus has been accompanied by a lengthening of the life of the individual, by the production of larger crops of sporangia, and by the assumption of the mycelial habit.

The lowest Phycomycetes have a rudimentary thallus, sometimes an ovoid sac, sometimes a simple or slightly branched filament, rarely a distinct branching mycelium. A few examples may be reviewed briefly. In *Olpidium* (figs. 9, 10, 11), the thallus is a rounded sac, without appendages; it is formed by the direct enlargement of the body of the zoospore, and is converted at maturity into a single reproductive structure. *Septolpidium*[1] is very like *Olpidium*, but the thallus becomes septate as it matures, and each chamber becomes a sporangium. *Pleotrachelus* (fig. 12 *c*, *d*) has a very general likeness to *Olpidium*, but in it, increase of surface in proportion to volume, and more thorough contact with the protoplast of the host cell, have been attained by the development of lobes; in the Ancylistales the same ends have been reached by elongation and slight branching of the thallus. It is possible that the mycelium of *Pythium* is a further advance in this line of development, marked by attenuation of the filaments and by the increased production of branches. In all these forms the vegetative organs are immersed in the substratum.

Another line of development may well begin with such genera as *Rhizophidium* (fig. 18), where the thallus consists of the enlarged body of the zoospore attached to the outside of the host, together with a weakly developed system of rhizoids buried in the substance of the host and gathering food from it. In related genera, the rhizoids become more important and may spread into more than one host cell. With increase in the development of the rhizoidal system, more of the thallus may inhabit the water outside the host cells, and extend to several such cells, as in *Polyphagus Euglenae*. Further, the swelling derived from the enlarged body of the zoospore may remain a purely vegetative structure, so that the sporangium develops as a branch from it, a condition found in *Polyphagus Euglenae* (fig. 23) and in *Macrochytrium botrydioides* (fig. 28 *a*). *Macrochytrium* may well be a link between some of the lower forms and genera such as *Rhipidium* (fig. 39) and *Sapromyces* (figs. 41, 42). Reminiscences of a similar organisation may be preserved in the thallus of the Monoblepharidales, Saprolegniales and some members of the Peronosporales.

The Peronosporales and the Zygomycetes are the highest groups of the Phycomycetes; they usually possess richly branched mycelia spreading on and in the substratum. These mycelia are relatively long-lived, and produce large crops of sporangia or of

[1] Sparrow, 1936 i.

conidia; in general, the asexual spores are dispersed by air-currents rather than by water.

Although the thallus of the Phycomycetes is usually aseptate in the vegetative condition, septum formation accompanies the development of reproductive organs in all but the simplest members of the group. In a few genera, such as *Allomyces* and *Basidiobolus*, the mycelium is regularly septate, and in *Dispira* and *Piptocephalis*, septa may occur freely, though not at regular intervals, but these conditions are exceptional. As is usual in fungi, the hyphae elongate by apical extension; the older parts are evacuated as the apex moves forwards, and an occasional septum may form between the active and the abandoned stretches of the hypha. In some species of *Saprolegnia* and *Mucor*, chlamydospores may develop by local condensations of the protoplast in older hyphae; when this occurs, septa delimit the chlamydospores from the rest of the hypha.

The Sporangium. In many of the Archimycetes the thallus is converted at maturity into one or more zoosporangia with little change in shape apart from the development of some simple means for the liberation of the zoospores. In a few species of *Pythium*, the sporangia form directly from unmodified hyphae, and in *Aphanomyces* there is little change apart from the deposition of a septum between the fertile and vegetative portions of the filament. The complicated branched sporangia of *Plectospira* show characters reminiscent both of *Pythium* and of *Aphanomyces*. In most of the Oomycetes and Zygomycetes, the sporangia are terminal, clavate to globose, and delimited from the stalks by basal septa; the sporangia may be solitary, at the end of a long sporangiophore, as in *Phycomyces*, or they may develop in complicated groups, often of great beauty, as in species of *Peronospora*, and in *Thamnidium elegans*.

In the Archimycetes, the activity of the organism ends in the production of one or more sporangia maturing simultaneously, but in Phycomycetes with a well-developed mycelium, vegetative growth and the formation of fresh sporangia may go on for some weeks. In these species, sporangia may appear in succession on the same hyphal system, a process characteristic of aquatic and amphibious forms, or they may ripen singly or in groups on specialised branches standing up from the mycelium. The formation of sporangia in succession on one branch system may be brought about by the proliferation of the fertile hypha, by the development of chains of sporangia, or by the building up of a sympodial system of fertile branches. **Proliferation** takes place by the growth of the sporangiophore through the base of an evacuated sporangium and the formation of a new sporangium inside or beyond its predecessor. The phenomenon is common in filamentous aquatic Phycomycetes, especially in *Saprolegnia* (fig. 8 *a*), but it is not confined to the

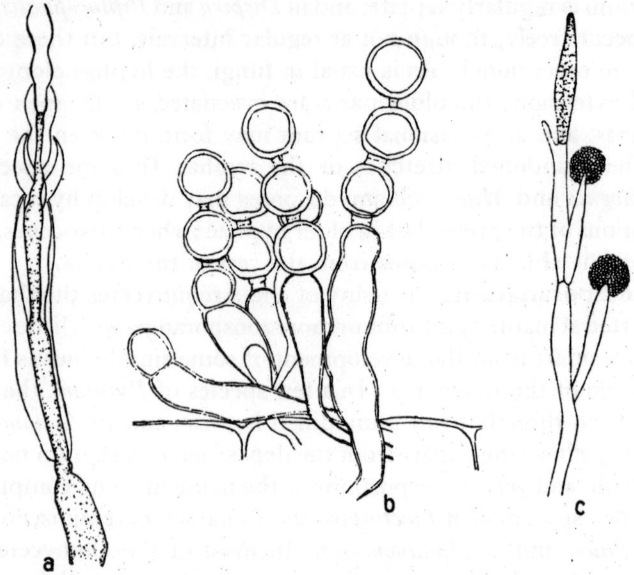

Fig. 8. *Saprolegnia monoica* Prings.; *a*, renewal of the sporangium by proliferation of the fertile hypha, ×225; after Pringsheim. *Cystopus candidus* (Pers.) Lév.; *b*, group of sporangiophores with chains of sporangia, ×450. *Achlya racemosa* Hildebr.; *c*, sympodium of sporangia, ×55; after Pringsheim.[1]

fungi, for it occurs in *Harpochytrium*,[2] a genus of Xanthophyceae formerly regarded as belonging to the Chytridiales, and in members of the Phaeophyceae.[3] In *Achlya* (fig. 8 *c*) and in some species of *Pythium*, sporangia are formed in sympodial branch systems, and in other species of *Pythium*, in *Monoblepharis*, and in the Albuginaceae (fig. 8 *b*), the sporangia develop in basipetal chains. In all these fungi, the position of a sporangium is influenced by that of its predecessors. Often, a differentiated sporangiophore is not present.

[1] Pringsheim, 1873.
[2] Scherffel, 1926 ii; Fritsch, 1935.
[3] Oltmanns, 1922.

Well-defined sporangiophores, distinguished from vegetative hyphae by their erect posture and often by their superior diameter, are usually formed by those species which do not form sporangia in succession. They are specially characteristic of the highest members of the group, particularly the Mucorales and Peronosporales, which depend on the wind for the dispersal of their spores.

In many aquatic and amphibious species, the sporangium liberates zoospores, or it may put out a germ tube before it falls from the sporangiophore; alternatively, the sporangium may drop off, and either function as a zoosporangium or germinate by means of a germ tube. This variation in behaviour in relation to moisture is transitional to the state where the sporangium normally acts as a conidium. The transformation of the sporangium into an air-borne conidium may be traced clearly in the Peronosporaceae, and, along rather different lines, in some families of the Mucorales.

The spores of the Phycomycetes are formed by cleavage[1] of the sporangial contents. In some of the simplest species, these contents may include many tiny vacuoles arranged in a series of intersecting planes. Cleavage occurs along these planes, dividing the sporangial contents into a number of uninucleate portions, each the rudiment of a zoospore. In most of the Phycomycetes, the small vacuoles flow together to give one large central vacuole from which, in many Archimycetes[2] and in the Oomycetes,[3] cleavage furrows proceed to the periphery of the young sporangium, cutting out the spore rudiments. These may then show a tendency to round off, but as they do so, the sporangium loses water and shrinks, so that the rudiments are pressed together and become indistinct as individuals. The rudiments lose more water, shrink further, round off again, undergo further differentiation, and become spores. In the Zygomycetes,[4] cleavage furrows arise at the periphery as well as in the middle of the mass, and the ripe spores may contain several nuclei.

Sporangia immersed in a substratum usually emit their spores through exit canals opening into the surrounding water. The sporangia of the lower Phycomycetes, when formed on the surface of the substratum, discharge their spores through one or more

[1] Harper, 1899 ii; Schwarze, 1922.
[2] Barrett, 1912 i.
[3] Rothert, 1903; Butler, E. J., 1907.
[4] Schwarze, 1922.

pores, produced by the local dissolution of the wall, or left after part of that wall has been lifted up as a lid; these methods occur also in many filamentous aquatic species. The sporangia of terrestrial species usually dehisce by means of irregular tears.

A number of aquatic and amphibious Phycomycetes discharge their zoospores by a process which seems not to occur in any other organisms. When the spores are nearly mature, the sporangium puts out a short hypha ending in a thin-walled vesicle. Meanwhile rudiments of zoospores appear in the sporangium, sometimes showing some power of independent movement as they develop. The rudiments migrate into the vesicle, complete their differentiation within it, and the spores are set free when the vesicle bursts. This method of spore discharge occurs in striking fashion in species of *Pythium*;[1] in *Phytophthora*[2] the vesicle is evanescent, and may burst before it has received all the contents of the sporangium.

A few genera of Phycomycetes employ special methods of spore discharge; in *Pythiogeton*, the whole contents of the sporangium may be ejected into the water before they are cleft into zoospores; in *Pilobolus*, the mass of spores is shot away from the sporangiophore; in *Basidiobolus* and related genera, the conidia are projected violently from their stalks.

The Spore. The zoospores of the Phycomycetes are of special interest, since they not only show likenesses to some simple flagellates which are probably the modern representatives of ancestral stock of the Phycomycetes, but they also furnish indications of the main lines of relationship within the group. Evidence is accumulating that the lower Phycomycetes fall into two main series, one possessing zoospores bearing a single posterior flagellum, the other possessing biflagellate zoospores with flagella often inserted in a lateral furrow. These differences in flagellation appear to be accompanied by a difference in the general physiology of the organisms, for the walls of the biflagellate forms usually give a bluish or violet coloration after treatment with chlor-zinc iodide, whereas those of the uniflagellate forms give the reaction very feebly, or not at all.

The zoospores of some aquatic Phycomycetes exhibit two periods of active movement separated by an encysted phase, a phenomenon

[1] Butler, E. J., 1907, 1913. [2] Rosenbaum, 1917.

known as **diplanetism**.[1] Diplanetism is shown very clearly by the zoospores of *Saprolegnia*. These spores emerge from the sporangium as pear-shaped bodies with two apical flagella. After swimming actively, perhaps for two or three minutes only, the spores lose their flagella, round off, and encyst. About twenty-four hours later, a bean-shaped zoospore bearing two lateral flagella emerges from the cyst, and, on a suitable substratum, puts out a germ tube and starts a new mycelium. In genera related to *Saprolegnia*, a gradual weakening of diplanetism may be traced. Diplanetism occurs also in the Chytridiales;[2] in these fungi, the encysted stage may be replaced by an amoeboid phase, during which solid material may be ingested and used as a source of energy for a second period of locomotion.

The zoospores of some aquatic fungi exhibit a form of behaviour which should be clearly distinguished from diplanetism. They escape from the sporangium, swim for a time, encyst, and subsequently emerge from the cysts without any change in the insertion of their flagella. The process may be repeated more than once, and is conveniently described as **repeated emergence**.[3] It may well be that the encystment, both in diplanetism and in repeated emergence, is due to the accumulation of insoluble carbohydrate waste resulting from the great metabolic activity of the spore during its formation and locomotion, and that the period of encystment enables the protoplast to free itself from this waste. In *Pythium adhaerens*,[4] repeated emergence may be accompanied by increase in the number of zoospores, each cyst emitting from two to five zoospores. In that species, therefore, repeated emergence provides a means of multiplication.

Sexual Reproduction. The Phycomycetes show as great diversity in their sexual processes as they do in other respects. Isogamy occurs in *Olpidium Viciae*, in *Synchytrium endobioticum*, and probably in many related forms. The gametes are zoospores which have been retarded in development by adverse conditions at the time of their differentiation. Motile gametes differing in size unite in *Allomyces javanicus*; in *Monoblepharis*, motile sperms swim to the oogonia and unite with a usually non-motile egg. In *Olpidiopsis*, union

[1] Leitgeb, 1869; de Bary, 1887; Butler, E. J., 1907; Atkinson, 1909 ii; Drechsler, 1930 i.
[2] Barrett, 1912 i; Curtis, 1921.
[3] Weston, 1919; Drechsler, 1930 i.
[4] Sparrow, 1931.

occurs between two thalli of unequal size, after each has undergone some development inside the host cell, and similar conditions appear to occur in *Monochytrium Stevensianum*. In *Zygorhizidium Willei* and *Polyphagus Euglenae*, uninucleate gametangia, attached to the outside of their hosts, unite by means of a conjugation tube put out by one of them. In the Ancylistales, antheridia and oogonia may form from segments of the same or of distinct thalli, the contents of the former passing into the latter through a pore or through a conjugation tube; the contents of the united gametangia round off to an oospore after union has taken place.

The Oomycetes develop well-differentiated antheridia and oogonia. In contrast to conditions in the Ancylistales, one or more oospheres are formed in the oogonium before fertilisation occurs. In the Oomycetes, the young gametangia are multinucleate; in most species, many nuclei degenerate during the preparations for fertilisation, though, in some species of *Albugo*, the union of many pairs of male and female nuclei appears to follow the mingling of the contents of the gametangia. In the Oomycetes, the contents of the antheridium are conveyed into the oospheres by a conjugation tube which passes through the oogonial wall.

The gametangia of the Zygomycetes are multinucleate, and many pairs of nuclei unite during the formation of the thick-walled zygospore. The gametangia which fuse may be alike in size and appearance, as in *Mucor*, or they may be noticeably different in these respects, as in *Zygorhynchus* and *Dicranophora*; they may develop on the same mycelium, as in *Sporodinia grandis*, or on separate mycelia, as in *Mucor* and *Rhizopus*.

The behaviour of members of the Saprolegniaceae in culture is greatly influenced by the conditions of the environment. It is possible to modify the manner in which the zoospores are discharged, to affect the development of the sexual organs, and even to bring about the conversion of an oogonial rudiment into a sporangium, or into an ordinary hypha; this variability makes difficult the identification of the species.

Phylogeny. The simplest Phycomycetes have many points of likeness with a number of non-pigmented flagellates[1] of generalised

[1] Ashworth, J. H., 1923; de Beauchamp, 1914; Caullery and Mesnil, 1905; Chatton, 1908; Chatton and Brodsky, 1909; Chatton and Roubaud, 1909; Dangeard, 1886; Dunkerly, 1914; Leger, 1908, 1914; Leger and Duboscq, 1909; Leger and Hesse, 1909; Němec, 1911 i, iii, 1913; Scherffel, 1925 i, 1926 i, ii; Schussnig, 1929.

character. If, as seems likely, the origin of the Phycomycetes is to be sought among these organisms, it becomes probable that the fungi have developed from lower forms along lines independent of those leading to green plants and to animals.

The Archimycetes are the simplest Phycomycetes; they are an assemblage of forms of simple character and probably of diverse origin. Members of the Archimycetes and Oomycetes agree in such characters[1] as the renewal of the zoosporangium by proliferation, the discharge of partly differentiated zoospores into a vesicle, the formation of a lid in the wall of the sporangium, and diplanetism in the zoospores.

The position is less clear in respect of the Zygomycetes, for their characteristic sexual processes cannot be paralleled satisfactorily in the other sections of Phycomycetes, and their sporangia are equally difficult to account for. They may have arisen from some simple form which conjugated by the union of independent thalli, but, though such forms are known, satisfactory transitional stages have not yet been found.

Classification. The Phycomycetes may be divided as follows:

Mycelium absent or rudimentary	ARCHIMYCETES
Mycelium well developed	
Sexual reproduction by oospores: accessory spores often motile	OOMYCETES
Sexual reproduction by zygospores: accessory spores non-motile	ZYGOMYCETES

This scheme has the advantage of simplicity, but it cannot be regarded as more than provisional. It may be possible in the future to construct a scheme dividing the Archimycetes and the Oomycetes into a uniflagellate and a biflagellate series, each going back to Protistan stocks. For the present, it is better to maintain a conservative attitude, and the Phycomycetes will be described within the three main groups set out in the table.

ARCHIMYCETES

The Archimycetes include about 500 species; many attack freshwater and marine algae, some cause disease in animals, and a few are well-known parasites of land plants; saprophytic forms have been less studied, but they appear to be fairly numerous.

The simplest species have a rounded, elongated or lobed

[1] de Bary, 1881 i.

thallus, formed by the enlargement of the zoospore; it bears no appendages, and, when growth ceases, is converted into one or more reproductive organs, formed simultaneously. In relatively advanced forms, rhizoids assist in nutrition, but take no part in the construction of the reproductive apparatus. The highest Archimycetes possess a mycelium normally consisting of tapering filaments comparable with the rhizoids of the lower forms rather than with the hyphae of the higher fungi.

The Archimycetes may be divided into three alliances, of which the first is the largest.

Vegetative body variable in form, used up in the production of one or more sporangia or gametangia; zoospores uniflagellate or biflagellate	CHYTRIDIALES
Vegetative body a stout filament, transformed at maturity into a row of reproductive organs; oogonia and antheridia present; zoospores biflagellate	ANCYLISTALES
Vegetative body a mycelium with terminal or intercalary resting sporangia; spores non-motile	PROTOMYCETALES

CHYTRIDIALES

In the Chytridiales the thallus[1] is at first unicellular, and either naked or walled; it may be a rounded, elongated or irregular sac, without appendages, and buried in the protoplast of the host; it may be a rounded, flask-shaped or elongated body, provided with rhizoids, and either lying inside the host, or more often attached to its surface; it may be filamentous, with local swellings associated with the formation of the reproductive organs, or clearly differentiated into a basal region concerned with nutrition and an apical reproductive portion.

The zoosporangia may be formed from the whole of the thallus; they may arise from the inflated portion of the thallus, with degeneration of the rhizoids, or they may form as swellings on the filaments. Sporangia lying inside the substratum liberate their zoospores through one or more papillae or exit canals, but when the sporangia lie on the surface of the substratum, the zoospores usually escape through one or more pores, or by means of an aperture formed by the casting off of a lid.

The body of the zoospore is usually globose; it contains one

[1] Karling, 1930, 1932; Sparrow, 1935.

nucleus and a single oil drop. When the spore is uniflagellate, the flagellum is posterior; the spore moves erratically, in a zigzag course, periods of rest frequently interrupting progress. The spores of the Woroninaceae move rather more smoothly; they are laterally biflagellate, and resemble the zoospores of the Oomycetes.

Freshly liberated zoospores are probably always naked; those of the lower forms remain so, entering the substratum in that condition, but in more advanced species the zoospore forms a wall before it germinates. If the species lives inside the substratum, the wall is abandoned as the protoplast of the zoospore makes its entry, but if development takes place on the surface of the substratum, the wall of the zoospore is retained and provides the foundation of the wall of the thallus and of the reproductive organs that form later.

Diplanetism is shown in the zoospores of *Olpidiopsis*,[1] *Pseudolpidium*[2] and *Synchytrium*[3] by the alternation of amoeboid phases and periods of motion by means of flagella.

Resting spores develop in similar situations to the zoosporangia, or, as in *Chytridium*, the zoosporangia may be superficial, the resting spores immersed in the host. Resting spores are usually globular, with a thick two-layered wall which may bear warts or spines. Despite the difficulty of observation of these minute fungi, fertilisation has been seen[4] in a number of species, and probably occurs in many more. Isogamy occurs in *Olpidium*[5] and *Synchytrium*,[3] union between young thalli in *Olpidiopsis*,[1] within the host cell, and in *Sporophlyctis*,[6] outside the host. In *Polyphagus*[7] and *Zygorhizidium*,[8] a conjugation tube unites two uninucleate individuals and allows fertilisation to proceed.

The Chytridiales include the following families:

Thallus at first naked and amoeboid, usually becoming walled in the course of growth, never constituting a mycelium

Thallus soon becoming walled, clearly distinct as a rule from the contents of the host cell. Zoospores uniflagellate

Thallus usually yielding one reproductive structure — OLPIDIACEAE

[1] Barrett, 1912 i. [2] Butler, E. J., 1907. [3] Curtis, 1921.
[4] Petersen, 1903, 1910; Scherffel, 1925 ii; Sparrow, 1933 iii.
[5] Kusano, 1912. [6] Serbinow, 1907.
[7] Nowakowski, 1876; Dangeard, 1900; Wager, 1913.
[8] Loewenthal, 1904–5.

Thallus yielding a sorus of zoosporangia or a resting sporangium — **Synchytriaceae**

Thallus not clearly distinct from the contents of the host cell. Zoospores biflagellate — **Woroninaceae**

Thallus walled from the beginning and differentiated into a fertile and a sterile portion, the latter usually feebly developed and ephemeral

Mycelium delicate

Mycelium represented by a system of rhizoids attached to the sporangium — **Rhizidiaceae**

Mycelium richly branched and spreading, with terminal or intercalary swellings — **Cladochytriaceae**

Mycelium coarse, divided into a main axis and a system of rhizoids; sporangia terminal on lateral branches — **Hyphochytriaceae**

Olpidiaceae

The Olpidiaceae are parasites in freshwater and marine algae, in higher plants and in animals; they are distinguished by the sac-like or lobed thallus, formed by the enlargement of the contents of the zoospore, and entirely transformed at maturity into a reproductive structure.

The uniflagellate zoospore of *Olpidium Viciae*[1] (fig. 9 *a*) swims for a time and then settles down on the surface of a young stem or leaf of *Vicia unijuga*. After a period of amoeboid movement, during which the flagellum is withdrawn, a wall is formed and an infection tube conveys the contents of the zoospore into an epidermal cell, where they accumulate as a naked, spherical mass. This enlarges, nuclear divisions occur, and growth proceeds until all the available space is occupied. As growth slackens, the thallus forms a wall, and, with suitable temperature and adequate moisture, the contents of the thallus are cleft into zoospores, no change in external form accompanying this conversion of the thallus into a zoosporangium. One or more exit canals pass from the sporangium to the exterior of the host (fig. 9 *b*, *c*), their tips become gelatinised, they burst, and the zoospores escape.

The formation and escape of the zoospores are hindered when the surface of the host is dry; zoospores checked in development by drought may still emerge if the arrest does not last for more than a few days. Retarded zoospores frequently fuse in pairs (fig. 9 *d*) to

[1] Kusano, 1912.

form a biflagellate zygote (fig. 9 *e*) which swims for a time before invading a host cell (fig. 9 *f*) in the same manner as a zoospore. The two nuclei of the zygote remain distinct as growth proceeds, fusing only after the zygote has become a thick-walled structure capable of prolonged rest. Under suitable conditions, the zygote is converted into a sporangium, emitting zoospores of the same kind as those liberated from ordinary zoosporangia.

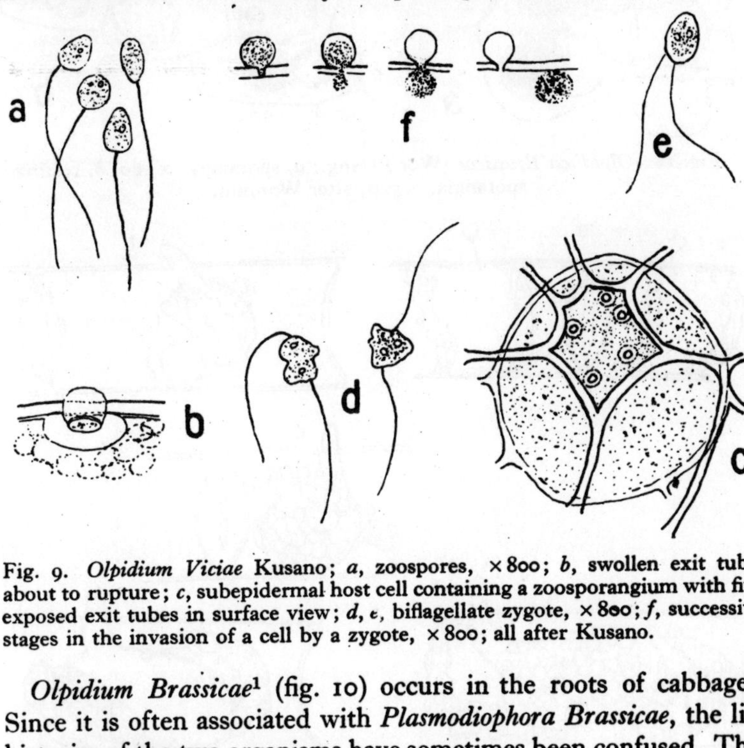

Fig. 9. *Olpidium Viciae* Kusano; *a*, zoospores, ×800; *b*, swollen exit tube, about to rupture; *c*, subepidermal host cell containing a zoosporangium with five exposed exit tubes in surface view; *d*, *e*, biflagellate zygote, ×800; *f*, successive stages in the invasion of a cell by a zygote, ×800; all after Kusano.

Olpidium Brassicae[1] (fig. 10) occurs in the roots of cabbages. Since it is often associated with *Plasmodiophora Brassicae*, the life histories of the two organisms have sometimes been confused. This species forms zoosporangia each bearing one exit canal; the zoospores may fuse,[2] as in *Olpidium Viciae*. Other species of *Olpidium* (fig. 11) attack algae, such as *Spirogyra* and *Oedogonium*.

Small uninucleate, amoeboid bodies, the young thalli of *Monochytrium Stevensianum*,[3] occur in the superficial cells of *Ambrosia artemisifolia*; they associate in pairs, cytoplasmic fusion follows,

[1] Woronin, 1878; Favorsky, 1910; Němec, 1912.
[2] Němec, 1922.
[3] Griggs, 1910.

the binucleate zygote enlarges rapidly, assumes a wall, and passes into a resting stage. Thalli which fail to fuse grow actively, and,

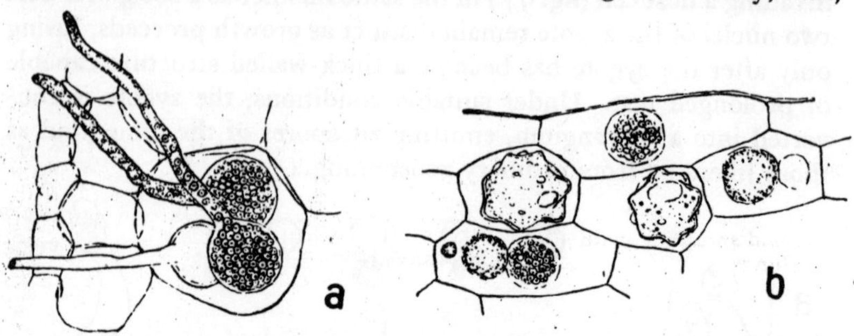

Fig. 10. *Olpidium Brassicae* (Wor.) Dang.; *a*, sporangia, × 160; *b*, resting sporangia, × 520; after Woronin.

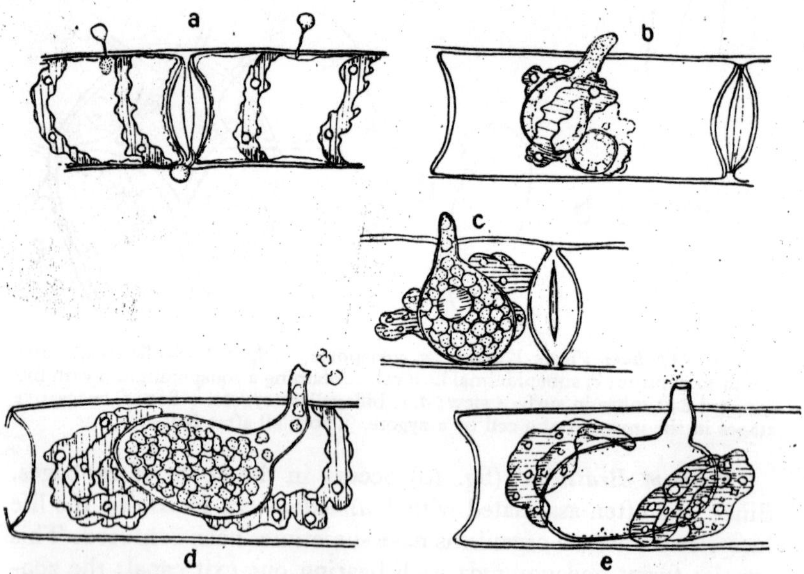

Fig. 11. *Olpidium* sp., in *Spirogyra*; *a*, zoospores invading the host; *b*, young sporangia with central vacuole; and zoospores in course of formation; *c*, sporangium about to discharge zoospores; *d*, escape of the zoospores; *e*, empty sporangium; all × 150.

after many nuclear divisions, their contents divide into bodies assumed to be zoospores; these have not been seen free from the sporangium. This union of thalli in a member of the Olpidiaceae provides an interesting parallel with similar phenomena in other families of the Chytridiales.

Several genera have been placed in the Olpidiaceae, chiefly because of their simple morphology, but it seems probable that some at least of these genera will be found to belong to other families; the principal genera of doubtful status are *Pleotrachelus*, *Ectrogella*, *Eurychasma* and *Pleolpidium*.

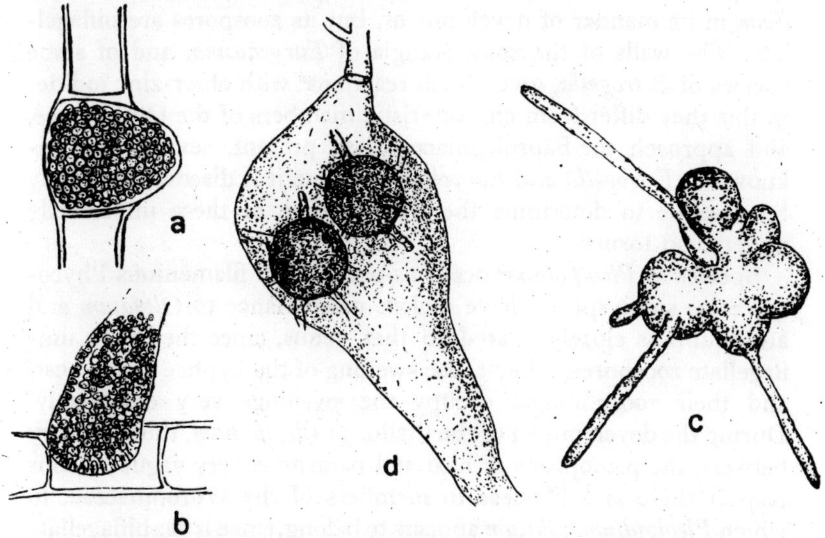

Fig. 12. *Eurychasma Dicksonii* (Wright) P. Magn.; *a*, developing sporangium in cell of *Ectocarpus*; *b*, liberation of zoospores. *Pleotrachelus Pollagaster* Petersen; *c*, lobed sporangium with four exit canals. All ×180; after Petersen. *Pleotrachelus fulgens* Zopf; *d*, two empty sporangia in aborted sporangiophore of *Pilobolus crystallinus*, ×250; after Zopf.

The simpler species of *Pleotrachelus*[1] differ from *Olpidium* mainly in the constant presence of several exit canals on the zoosporangium; other species of the genus have markedly lobate sporangia (fig. 12 *c*), with an exit canal to each lobe. Most of the species are parasites of marine algae, but *Pleotrachelus fulgens*[2] (fig. 12 *d*) occurs in *Pilobolus Kleinii*. There are indications that some species referred to *Pleotrachelus* have biflagellate zoospores;[3] if this is confirmed, these species will have to be transferred to another systematic position.

[1] Petersen, 1905.
[2] Zopf, 1884.
[3] Sparrow, 1934.

Ectrogella Bacillariacearum[1] closely resembles some species of *Pleotrachelus* in its morphology and life history. It lives in diatoms, and the elongated zoosporangium with two opposite linear series of exit canals is moulded to the form of the host. Some species referred to *Ectrogella* have biflagellate zoospores,[2] and show a form of diplanetism very like that of *Saprolegnia*; they may be simple members of the Saprolegniales. *Eurychasma Dicksonii*[3] (fig. 12 *a*, *b*) is a parasite of members of the Ectocarpales. It resembles *Olpidium* in its manner of development, but its zoospores are biflagellate. The walls of the zoosporangia of *Eurychasma*, and of some species of *Ectrogella*, give bluish reactions[4] with chlor-zinc iodide; in this they differ from characteristic members of the Olpidiaceae, and approach the Saprolegniaceae. At present, sexuality is unknown in *Ectrogella* and *Eurychasma*; when it is discovered, it may be possible to determine the true position of these imperfectly understood forms.

Species of *Pleolpidium*[5] occur as parasites in filamentous Phycomycetes; some species have a close resemblance to *Olpidium* and are doubtless closely related to that genus, since they have uniflagellate zoospores. They cause swelling of the hyphae of the host, and their zoosporangia occupy the swellings very completely. During the development of the thallus of *Pleolpidium*, the boundary between the protoplasts of host and parasite is very vague; in this respect there is a likeness to members of the Woroninaceae, to which *Pleolpidium inflatum* appears to belong, since it has biflagellate zoospores.

Many other incompletely investigated genera are placed in the Olpidiaceae on account of the simplicity of their life histories, so far as these are known. *Sphaerita endogena*[6] lives in Rhizopoda and other aquatic organisms; *Rhinosporidium Seeberi*[7] causes polypoid tumours in man; *Dermocystidium pusula*[8] is found in small swellings on the gills of trout and the skin of newts. Some at least of these genera may be intermediate between the more characteristic Olpidiaceae and such flagellate genera as *Pseudospora*[9] and *Aphelidium*.[10]

[1] Zopf, 1884; Petersen, 1905. [2] Scherffel, 1926 i; Sparrow, 1934.
[3] Magnus, P., 1905; Petersen, 1905.
[4] Scherffel, 1925 i. [5] Butler, E. J., 1907.
[6] Dangeard, 1886; Chatton and Brodsky, 1909; Jahn, T. L., 1933.
[7] Ashworth, J. H., 1923. [8] de Beauchamp, 1914; Dunkerly, 1914.
[9] Scherffel, 1925 i; Schussnig, 1929. [10] Scherffel, 1925 i.

SYNCHYTRIACEAE

The thallus of the Synchytriaceae is at first naked; it soon assumes a delicate wall, and is clearly distinct from the contents of the host cell as a whitish, yellowish or reddish globule. The mature thallus is converted into a sorus of sporangia or into a resting spore. The zoospores are uniflagellate.

The genus *Synchytrium*[1] contains a number of parasites in mosses, ferns and angiosperms; host plants in wet places are specially liable to attack. Epidermal cells are commonly entered, and, in general, cells containing chlorophyll are avoided; entrance may be made through a stoma, or directly through the cell wall. Warts and other deformations appear on infected parts.

Synchytrium endobioticum[2] is the most completely investigated species; it causes a serious wart disease of potatoes. The naked zoospore enters a young host cell, at or near ground level, and passes to the base of the cell, where it rounds off, enlarges, and forms a wall of two layers. In the meantime, the single nucleus enlarges enormously, and proceeds to give off stainable material into the cytoplasm.[3] A pore opens in the outer layer of the wall, and the contents emerge in a sac (fig. 13 *a*) furnished by the expanding inner layer; nuclear division begins during migration. When about thirty-two nuclei have been formed, small vacuoles appear in intersecting layers indicating the beginning of cleavage of the contents of the sac. Walls are laid down along the planes passing through the layers of vacuoles, cutting out four or five walled sporangia (fig. 13 *b*) inside the sac. Nuclear division continues until each sporangium contains two or three hundred nuclei, around which zoospores are organised. The mature sori absorb water, they swell, the envelope of the sorus and the wall of the host cell are ruptured, and the sporangia are forced out. They lie free on the surface of the host, and liberate zoospores through one or more rudimentary papillae, or through casual slits.

As in *Olpidium Viciae*, water supply is of great importance; the zoospores may be induced to function as gametes by keeping the

[1] de Bary and Woronin, 1863; Ludi, 1901; Tobler, 1913.

[2] Curtis, 1921; Collins, E. J., 1935.

[3] Stevens and Stevens, 1903; Stevens, 1907; Bally, 1912–19; Rytz, 1917; Curtis, 1921.

sporangia dry for a few days, so that the spores are old when they escape from the sporangium. Zoospores retarded in emergence swim for a time, pair off, and fuse to form naked biflagellate zygotes. The two nuclei fuse, and, after a host cell has been invaded, a walled resting sporangium is formed. In this, nuclei divide mitotically,[1] and finally, many zoospores are set free, and are

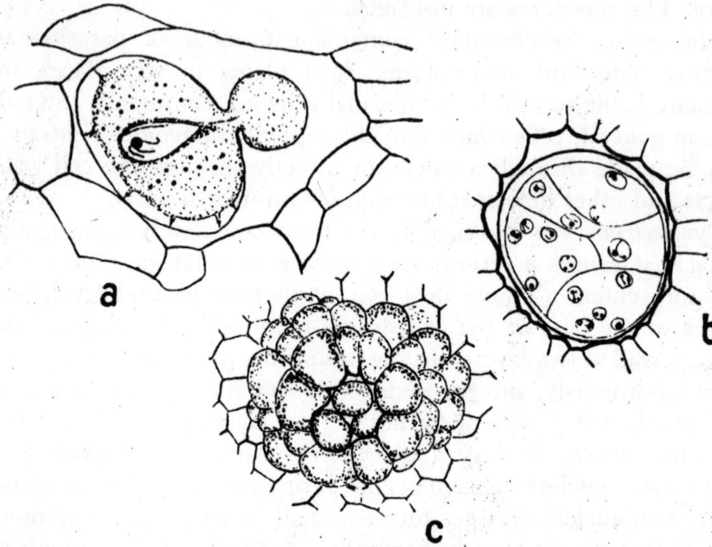

Fig. 13. *Synchytrium endobioticum* (Schilb.) Perc.; *a*, early stage in migration of the contents of the vegetative body, in preparation for the formation of the sorus; *b*, sorus of five sporangia shortly after the formation of the cleavage membranes; both ×400; *c*, surface view of mature sorus, ×170; all after Curtis.

apparently equivalent in all respects to those formed in the ordinary zoosporangia.

A cell invaded by a zoospore enlarges somewhat as the parasite develops; at the same time, neighbouring cells elongate and divide, so that the infected cell (fig. 13 *c*) comes to lie at the base of a funnel-shaped depression, surrounded by a raised rosette of hypertrophied tissue. Cells of the rosette may be entered by zoospores, with further enlargement, and this, together with abnormal divisions in diseased material, leads to the formation of large galls. Cells

[1] Welsford, 1921.

invaded by zygotes are stimulated to tangential division, and, as the zygote lies at the base of its host cell, the resting sporangia are buried in the warts; they are set free when the potato decays.

In *Synchytrium fulgens*,[1] a parasite of *Oenothera*, the sorus of sporangia develops inside the mature thallus, and not in a vesicle protruded from it. It seems that the zoospores of this species acquire the potentialities of gametes much more readily than the zoospores of *S. endobioticum*, so that fusions may be very abundant.

Synchytrium decipiens[2] is a parasite of members of the Leguminosae in America and in Eastern Asia. It does not form resting sporangia. The sorus of sporangia develops inside the mature thallus, and, when ripe, bursts to yield a powdery mass; infected plants often look as though they bore the aecidia of a rust. *S. aureum*[3] is widely distributed. It attacks more than one hundred species of phanerogams, of most diverse affinities, a character not found in other members of the genus. *S. aureum* forms resting sporangia only, maybe after a fusion of motile gametes. The contents of each resting structure are transformed intc a sorus of sporangia, enclosed in a vesicle.

The species of *Micromyces*[4] are parasites of members of the Conjugatae; their life histories are very like those of species of *Synchytrium*.

Woroninaceae

The Woroninaceae are mostly parasites of Oomycetes but one species[5] occurs in the rhizoids of *Riccia*. The thallus is naked at first, and difficult to distinguish from the contents of the host in which it is immersed. As growth proceeds, the thallus may divide[6] into a number of portions, each capable of independent vegetative development. At maturity, either a single sporangium, or else a resting structure, is formed from each vegetative unit. The sporangia often occur in groups, no doubt as a result of cleavage during growth, but the groups are not enclosed in a common envelope as

[1] Kusano, 1930. [2] Harper, 1899 ii; Stevens and Stevens, 1903.
[3] Rytz, 1907; Tobler, 1913.
[4] Dangeard, 1889; Denis, 1926; Couch, 1930–1 ii; Huber-Pestalozzi, 1931.
[5] du Plessis, 1933. [6] Serbinow, 1907.

are the sori of the Synchytriaceae. The zoospores bear two flagella of about equal length; these may be inserted apically, they may arise from a point a little behind the apex, or they may be distinctly lateral in insertion.

The species of *Olpidiopsis*[1] attack members of the Saprolegniales, often causing deformation of the hyphae of the host (fig. 14 *a*). The uninucleate zoospore is rounded, with two anterior flagella of about the same length attached close together. It swims for a time, comes to rest, the flagella shorten and again lengthen, and the zoospore resumes activity. Sooner or later a hypha is invaded and the naked contents of the zoospore become closely associated with the protoplast of the host. The parasite grows rapidly, nuclear divisions occur, and a wall forms as growth comes to an end, the whole thallus being converted into one zoosporangium, as in *Olpidium*. Indeed, this phase of the parasite cannot be distinguished from the corresponding stage of *Olpidium*. The biflagellate zoospores, however, provide a distinction; they escape into the water from one or more exit tubes.

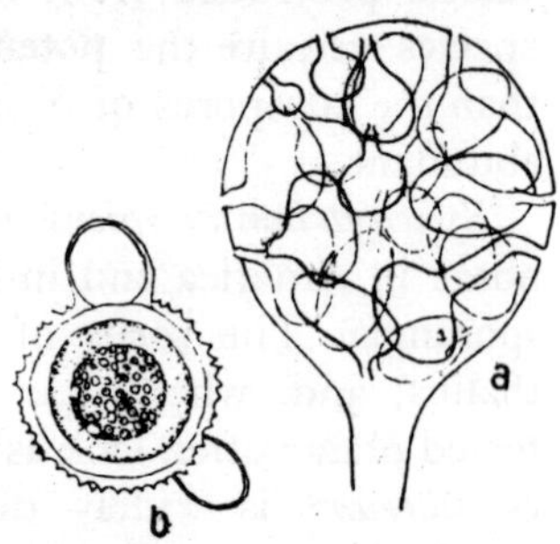

Fig. 14. *Olpidiopsis* sp.; *a*, empty sporangia in swollen filament of *Saprolegnia*; after Cornu. *Olpidiopsis luxurians* Barrett; *b*, mature resting sporangium with two empty antheridia attached; after Barrett.

A sexual process may occur in old cultures in foul water. Two thalli of different size develop side by side; the smaller has a smooth wall which gives reactions for cellulose. It becomes attached to the rough wall of its larger associate. Both thalli are multinucleate; they may be regarded as an antheridium and an oogonium. The contents of the antheridium pass through a pore into the oogonium, the nuclei form irregular groups, and fusions probably follow. A thick endospore appears inside the wall of the oogonium, enclosing an outer zone of finely granular material, and a central mass of coarser granules with one or more oil drops. The empty antheridium (fig. 14 *b*) remains attached to the rough wall of the oogonium; the oospore probably discharges biflagellate zoospores when it germinates.

[1] Zopf, 1884; Barrett, 1912 i.

Pseudolpidium[1] is very like *Olpidiopsis*, except that the resting structures, enclosed in spinous envelopes (fig. 15), do not bear empty antheridia.

The association between the parasite and the host is especially close in the species of *Rozella*,[2] *Woronina*,[2] and in *Pleolpidium inflatum*.[1] The latter is a parasite of the species of *Pythium* that inhabit the soil, and since it has biflagellate zoospores it appears to belong to the Woroninaceae. Its zoosporangium may be visible to the naked eye, and may contain some 7000 zoospores.

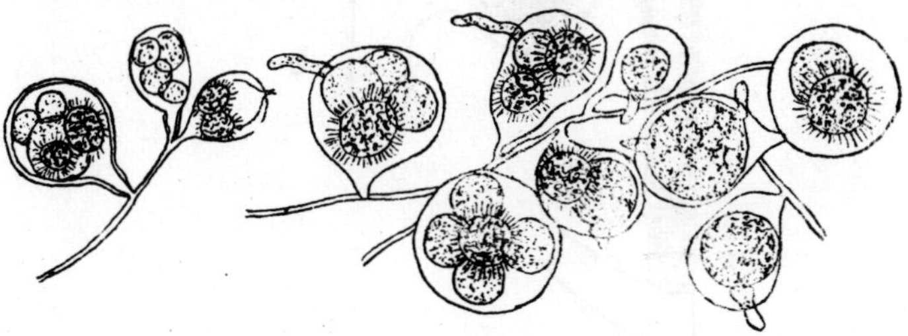

Fig. 15. *Pseudolpidium gracile*; sporangia and resting sporangia in swollen filaments of *Pythium*, ×300; after Butler, E. J.

Rozella and *Woronina* are parasites in members of the Saprolegniales; they form their sporangia in groups. In *Rozella septigena*[3] (fig. 16) the groups are linear, lying in slightly swollen hyphae of the host. The thallus divides to form a row of walled cylindrical sporangia, with their lateral walls firmly united with those of the surrounding hypha, which therefore appears to be septate. In *Woronina polycystis*[4] (fig. 17 *a*), another parasite of the Saprolegniales, the thallus fragments as it matures; each fragment rounds off to form a zoosporangium, so that a loose cluster of these organs is formed. Both genera have resting stages. Those of *Rozella* are globular, with thick, roughened walls not in contact with the walls of the host; those of *Woronina* form groups, which may occupy characteristic swellings of the hyphae of the host (fig. 17 *b*).

The simple vegetative structure of the Woroninaceae seems to ally them with the Olpidiaceae and the Synchytriaceae, but the biflagellate zoospores, and the indications of cellulose in the walls

[1] Butler, E. J., 1907.
[2] Cornu, 1872; Fischer, A., 1882.
[3] Fischer, A., 1882; Sparrow, 1932.
[4] Fischer, A., 1882; Cook and Nicholson, 1933.

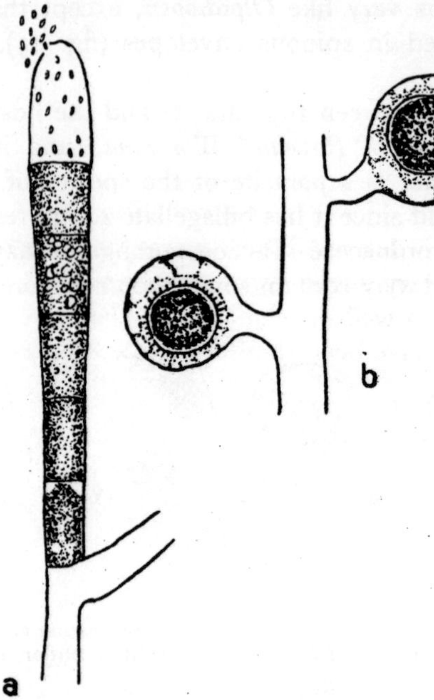

Fig. 16. *Rozella septigena* Cornu; *a*, zoosporangia in a linear series in a filament of *Achlya polyandra*, ×170; *b*, resting sporangia in outgrowths from a hypha of *Saprolegnia spiralis*, ×340; both after Cornu.

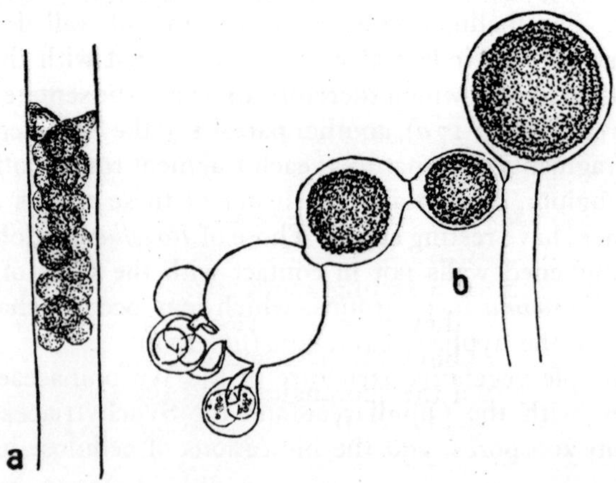

Fig. 17. *Woronina polycystis* Cornu; *a*, sorus of sporangia in the host filament, ×340; *b*, resting sporangia in swollen hyphae of *Achlya racemosa*, ×140; both after Cornu.

afford a sharp distinction. Probably, when more is known of these fungi, it will be necessary to place them in an alliance of the Archimycetes, separate from the Chytridiales.

RHIZIDIACEAE

Most of the Rhizidiaceae are parasitic on or in aquatic organisms; a few are saprophytes. The thallus is walled from the earliest stages of development; it is derived in the simpler forms from the enlarged zoospore, with a very slight development of rhizoids; more advanced forms have a well-developed rhizoidal system. Few members of the family have been thoroughly investigated.

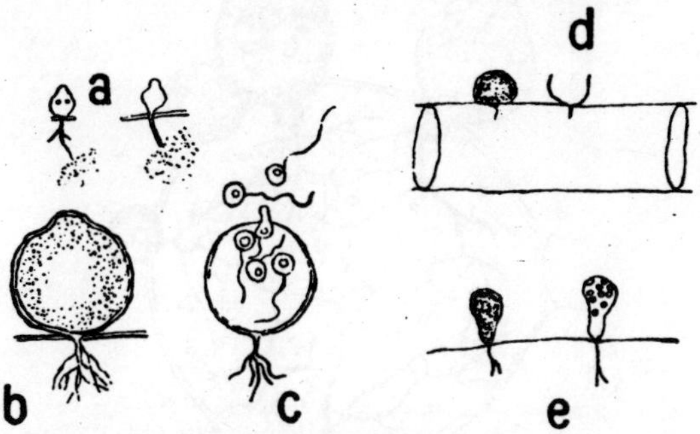

Fig. 18. *Rhizophidium brevipes*; *a*, zoospores invading the host; *b*, mature zoosporangium with rhizoids; *c*, escape of the zoospores; all after Atkinson. *Rhizophidium sphaerocarpus* (Zopf) A. Fischer; *d*, mature and empty sporangium; *Rhizophidium septocarpoides* Petersen; *e*, two sporangia; both after Petersen.

The species of *Rhizophidium*[1] (fig. 18) occur on unicellular and filamentous algae, on oospores of Oomycetes,[2] and on pollen grains and other vegetable material in water. A zoospore comes to rest on the substratum and penetrates it by means of a germ tube; the body of the zoospore, already clothed in a wall, expands to form either a zoosporangium or a resting sporangium. The germ tubes may remain simple, or they may give rise to a branched system of rhizoids arranged about the lengthened germ tube. A septum is not formed between the sporangium and the rhizoids. When the former is mature, the latter die. The zoospores escape through one or more pores in the wall of the sporangium.

[1] Zopf, 1884; Atkinson, 1909 i, ii. [2] Melhus, 1914.

Little is yet known of the conditions which determine the production of the resting sporangia; in *Rhizophidium goniosporum*[1] and some other species, two young plants of unequal size make lateral contact and the contents of the smaller pass through a pore into the larger. A resting spore then develops in the latter; it bears the empty wall of the male gametangium as an appendage. It is

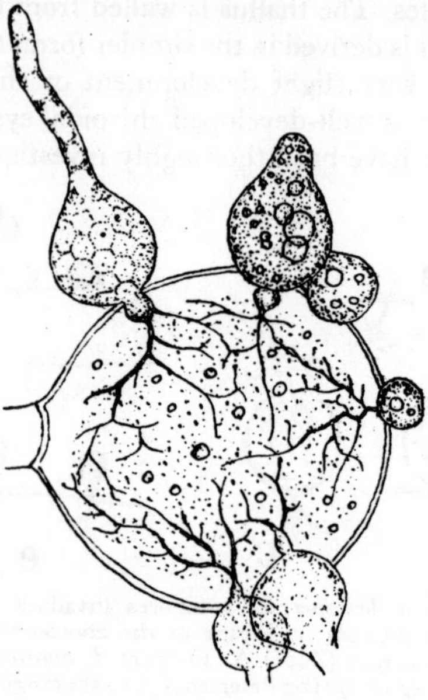

Fig. 19. ***Rhizidiomyces apophysatus*** Zopf; sporangia on an oogonium of *Achlya*, ×540; after Zopf.

probable that in other species of *Rhizophidium*, the formation of the resting stage is preceded by the union of motile gametes, as in *Olpidium*.

Sporophlyctis rostrata,[2] a parasite of *Draparnaldia*, agrees in its general life history with *Rhizophidium goniosporum*; when union occurs, however, the contents of the larger gametangium migrate into the smaller of the two associates.

Chytridium[3] is a genus of parasites of algae; the best known species

[1] Scherffel, 1925 ii; Sparrow, 1933 iii. [2] Serbinow, 1907
[3] Scherffel, 1926 i; Sparrow, 1933 ii, 1936 i.

is *Chytridium Olla.*[1] Zoosporangia are formed as in *Rhizophidium*, but the zoospores emerge after a lid has opened at the apex of the sporangium. The resting spores develop inside the substratum and it is possible that a sexual process is not concerned in their forma-

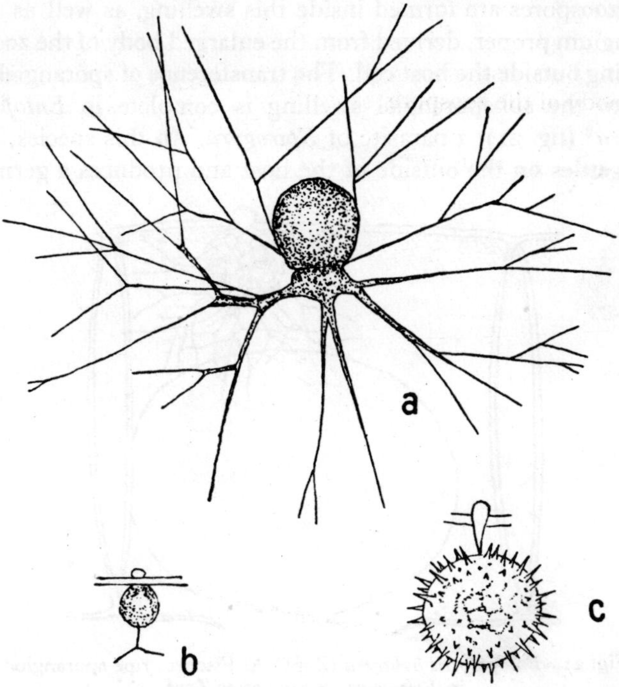

Fig. 20. *Rhizoclosmatium globosum* Petersen; *a*, mature plant. *Phlyctochytrium stellatum* Petersen; *b*, mature plant; *c*, resting sporangium. All after Petersen.

tion. They germinate by means of a short germ tube ending in a lidded sporangium.

Rhizidiomyces apophysatus[2] (fig. 19) attacks the oogonia of *Achlya*; the zoospore germinates on the surface of the host, and forms a flask-shaped sporangium bearing a subsporangial swelling and a branching system of rhizoids. After the zoospores are partly differentiated, the sporangium puts out a tube, the distal end of which swells into a vesicle. The contents of the sporangium pass

[1] de Bary, 1887; Sparrow, 1936 i. [2] Zopf, 1884; Coker, 1923.

into the vesicle, where the differentiation of the zoospores is completed; they escape by the rupture of the vesicle.

In *Phlyctochytrium stellatum*[1] (fig. 20 *b*) there is a well-developed subsporangial swelling lying inside the cell of the algal host; sometimes zoospores are formed inside this swelling, as well as in the sporangium proper, derived from the enlarged body of the zoospore and lying outside the host cell. The transference of sporangial functions to the subsporangial swelling is complete in *Entophlyctis bulligera*[2] (fig. 21), a parasite of *Spirogyra*. In this species, a zoospore settles on the outside of the host and produces a germ tube

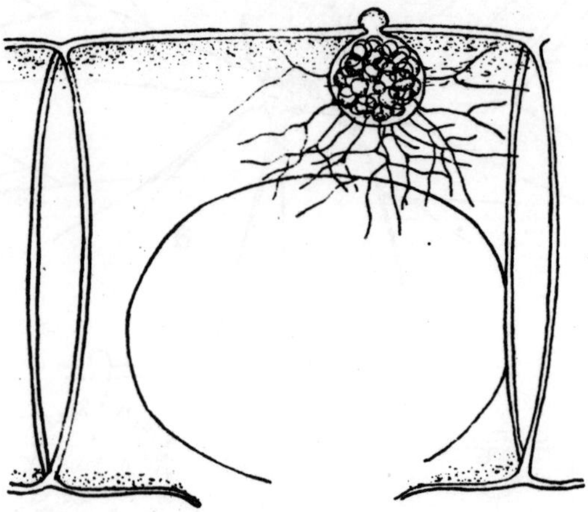

Fig. 21. *Entophlyctis bulligera* (Zopf) A. Fischer; ripe sporangium in *Spirogyra*, ×540; after Zopf.

in the usual way. The contents of the spore pass into the host cell, accumulating in the much swollen end of the germ tube, where the zoosporangium ultimately forms; it bears branched systems of rhizoids arising from several points on its surface.

Rhizoclosmatium globosum[1] (fig. 20 *a*) and *Asterophlyctis sarcoptoides*[1] are saprophytes on the empty chrysalis cases of aquatic insects. In each there is a swelling at the point where the rhizoidal system springs from the base of the zoosporangium; in *Asterophlyctis* the zoospores emerge from a pore at the base of the sporangium, close to the subsporangial swelling. In *Rhizoclosmatium globosum*, and the somewhat similar *Siphonaria variabilis*,[1] a peculiar

[1] Petersen, 1903, 1910. [2] Zopf, 1884.

process has been described; two thalli of unequal size become united by means of their rhizoids, and the contents of the smaller pass into the larger, where a resting spore develops.

The zoospores of *Zygorhizidium Willei*[1] (fig. 22 *a*) settle upon cells of *Cylindrocystis*; they may give rise to zoosporangia like those of *Chytridium*, opening by means of a lid (fig. 22 *d*). In these sporangia, nuclear divisions go on as growth proceeds, and many uninucleate zoospores are set free. On the other hand, the body of the zoospore

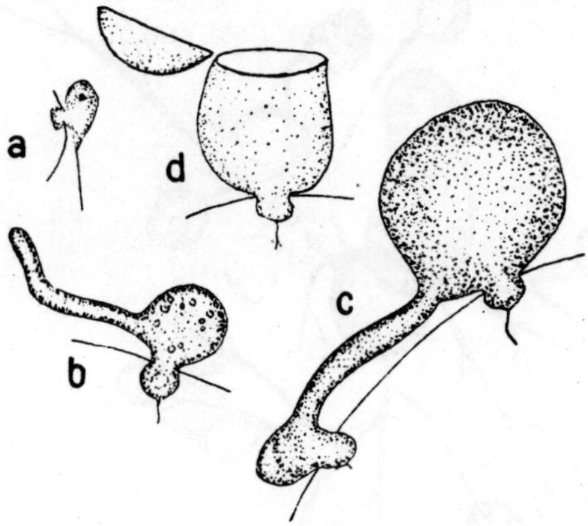

Fig. 22. *Zygorhizidium Willei* Loewenthal; *a*, zoospore invading the host; *b*, male plant with lateral branch; *c*, conjugation; *d*, empty sporangium, with lid; all × 1800; after Loewenthal.

may enlarge, but the whole remains uninucleate. Conjugation follows between two uninucleate individuals; one puts out a lateral filament (fig. 22 *b*), by means of which its contents are transferred to the other, which is generally larger (fig. 22 *c*). A thick-walled resting spore is formed; the subsequent fate of this is uncertain, but it is probable that zoospores are ultimately set free through a pore formed by the lifting of a lid. Like many other Phycomycetes, *Zygorhizidium* exhibits lability in its reproductive processes; zoosporangia bearing lateral hyphae have been observed, suggesting that, after preparations for conjugation have begun, a change may occur in the behaviour of the fungus, causing a potential gametangium to become a zoosporangium.

[1] Loewenthal, 1904–5; Scherffel, 1925 ii.

Polyphagus Euglenae[1] is sometimes abundant on the resting stages of *Euglena* on sewage beds and in foul ditches. The zoospore settles down in the water, develops a wall and puts out a richly branched system of rhizoids (fig. 23); contact is made with many host cells by the ends of the filaments of this system. At this stage the thallus is uninucleate and aseptate, the enlarged body of the zoospore being the most prominent feature. It puts out a lateral

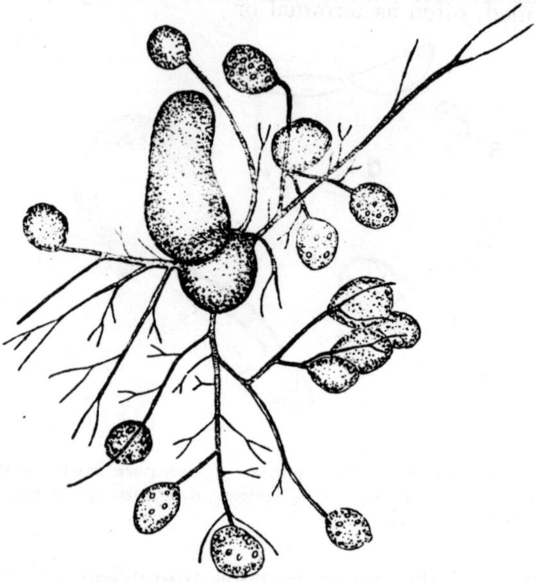

Fig. 23. *Polyphagus Euglenae* Nowak.; mature plant attached to fourteen Euglenae, and bearing a young sporangium, × 320; after Nowakowski.

branch, the contents of the thallus migrate into this, a basal septum cuts off the evacuated regions, and the branch becomes a sporangium. Many nuclear divisions precede the formation of the zoospores, which escape from an apical pore.

In old cultures, a sexual process occurs. As before, the thallus consists of the enlarged body of the zoospore containing a single nucleus, and bearing rhizoids. The preliminary stages of conjugation resemble those in *Zygorhizidium*, but, in *Polyphagus*, a swelling

[1] Nowakowski, 1876; Dangeard, 1900; Wager, 1913.

forms on the conjugation tube, close to the point of contact with the larger gametangium. The contents of the smaller gametangium migrate into the swelling, where they are joined by those of the larger (fig. 24), and the zygote forms a thick wall, sometimes smooth, sometimes spinous. After a resting period of some months, during which the two nuclei remain separate, the zygote puts out a branch, into which the contents pass. Then the two nuclei fuse, many nuclear divisions follow, and finally, zoospores are liberated from the sporangium formed as an outgrowth from the zygote. Thus, in either phase of the life history of *Polyphagus*, nuclear divisions occur only in the sporangia, not in the vegetative thalli or in the zygotes.

The direct contribution to the formation of the wall of the zygote made by the more active gametangium of *Polyphagus* is a point of special interest, since the initiation of the sexual process and the protection of the zygote are usually carried out by different participants. The functions which characterise male and female gametes however are not necessarily distinctive of gametangia, and fusion occurs between gametangia in *Polyphagus*.

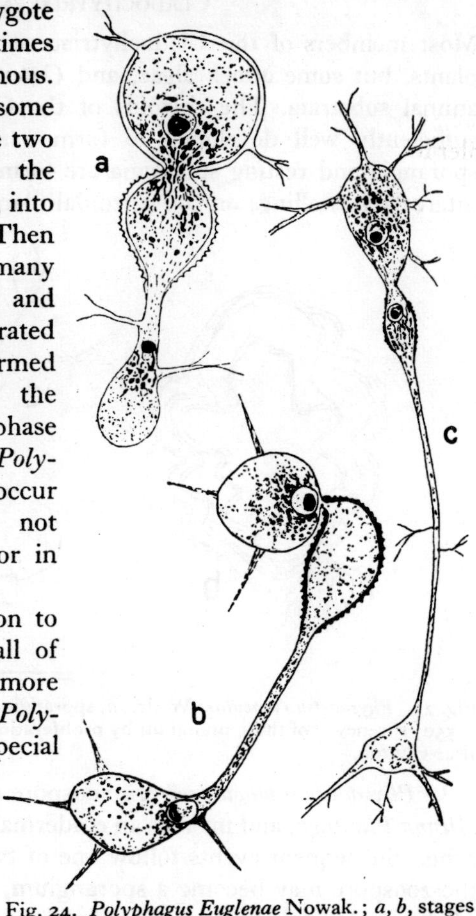

Fig. 24. *Polyphagus Euglenae* Nowak.; *a*, *b*, stages in formation of the conjugation tube and zygote; *c*, young zygote containing male nucleus; after Wager.

The genus *Harpochytrium*[1] has usually been included in the

[1] Atkinson, 1903; Scherffel, 1926 i; Fritsch, 1935.

Rhizidiaceae; some species however contain chlorophyll and the genus is now placed in the Xanthophyceae.

CLADOCHYTRIACEAE

Most members of the Cladochytriaceae[1] are parasites in higher plants, but some attack algae, and *Catenaria anguillulae* inhabits animal substrata. The rhizoids of the Cladochytriaceae may be sufficiently well developed to form a simple mycelium. Zoosporangia and resting sporangia are formed, often as terminal or intercalary swellings on the rhizoidal filaments.

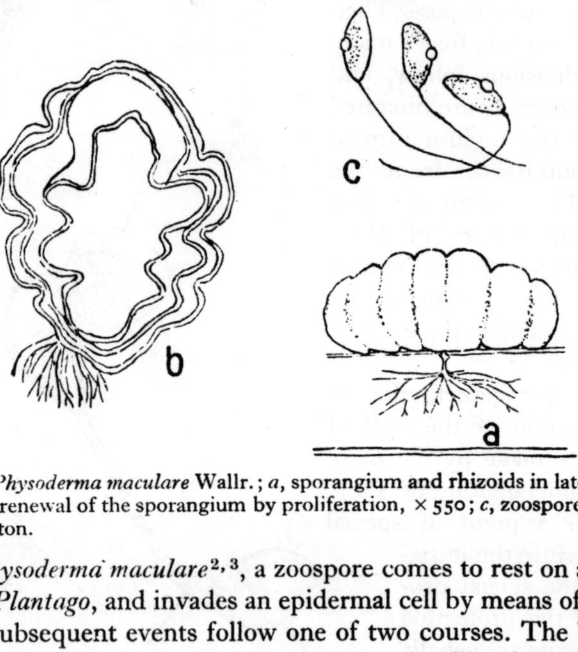

Fig. 25. *Physoderma maculare* Wallr.; *a*, sporangium and rhizoids in lateral view, × 550; *b*, renewal of the sporangium by proliferation, × 550; *c*, zoospores, × 850; after Clinton.

In *Physoderma maculare*[2,3], a zoospore comes to rest on a leaf of *Alisma Plantago*, and invades an epidermal cell by means of a germ tube. Subsequent events follow one of two courses. The body of the zoospore may become a sporangium, very like that of *Rhizophidium*, except that a septum lies between the sporangium (fig. 25 *a*) and the rhizoids inside the host cell. After the first sporangium has discharged its contents, a second may form inside it by proliferation of the apex of the rhizoidal system, and this may be repeated several times (fig. 25 *b*).

[1] Karling, 1931, 1935.
[2] *Physoderma maculare* Wallr. = *Cladochytrium Alismatis* Büsgen.
[3] Clinton, 1902.

Alternatively, all the contents of the zoospore pass into the enlarged end of the germ tube within the host cell. A transverse septum divides the swelling into a distal and a proximal segment. The distal segment is then divided into a number of cells, each of which puts out a filament; this elongates, makes its way into a fresh cell of the host, and repeats the behaviour of the germ tube and its terminal swelling. It follows that the fungus spreads freely in the substratum. In the meantime, the proximal segment produces rhizoids, absorbs nourishment from the host, and, when well fed, puts out a lateral outgrowth which becomes a resting sporangium. When this is mature, a lid forms from the upper part of the thick wall. The contents of the sporangium swell, the lid is lifted, the inner layer of the wall protrudes as a thin sac, and the zoospores escape through a small papilla.

Urophlyctis alfalfae[1] is a parasite of *Medicago sativa*, causing gall formation at or below soil level. The contents of the zoospore pass into a host cell, where they collect as a top-shaped body, surrounded by a wall. Oblique septa appear in this, cutting off a number of uninucleate peripheral segments, leaving a multinucleate central region. Each peripheral cell emits a filament which passes into a neighbouring host cell and swells at the end to a top-shaped enlargement (fig. 26 *a*). In this, segmentation occurs and the whole series of events is repeated. Meanwhile, the multinucleate central region gives rise to an apical outgrowth into which nuclei and cytoplasm pass. The outgrowth enlarges, develops a crown of short, branched rhizoids; it absorbs material from the host cell, and ultimately becomes a resting sporangium bearing a ring of scars indicating the former positions of the rhizoids.

A sexual process has not been observed in these genera. It is possible that the zoospores of *Physoderma maculare* are facultative gametes. If that be so, the simple zoosporangium may be formed when the host is attacked by a zoospore, while development inside the host is started by invasion by a motile zygote. Biflagellate zoospores have been observed after the germination of the resting spores of *Physoderma zea-maydis*,[2] but their significance is doubtful.

Nowakowskiella ramosa[3] attacks wheat plants floating in water. A swollen basal region bears eight to ten main hyphae, branching

[1] Jones and Drechsler, 1920.
[2] Ojerholm, 1934.
[3] Butler, E. J., 1907.

irregularly, and often anastomosing. Terminal pyriform swellings appear on the hyphae, a transverse wall forms, and the distal segment becomes the sporangium. A lid is formed from the outer layer of the wall, the inner layer of the wall protrudes as a papilla, and the zoospores emerge. Within the host, the ends of the hyphae swell (fig. 26 *b*, *c*) and divide to form large pseudoparenchymatous

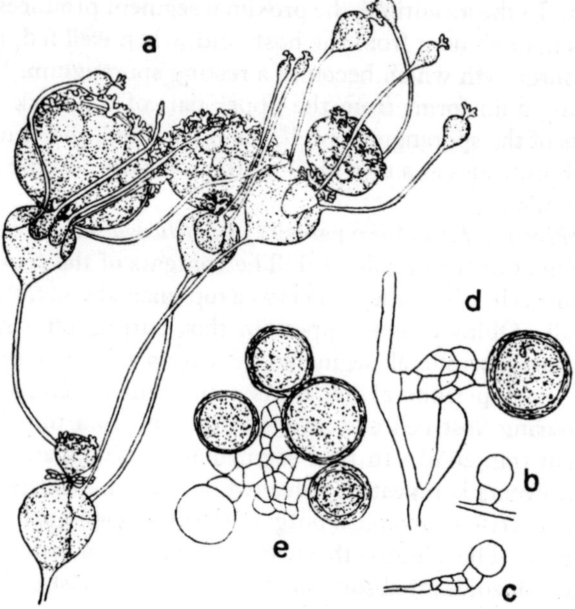

Fig. 26. *Urophlyctis alfalfae*; *a*, portion of thallus dissected from the host, showing top-shaped cells and resting sporangia, ×680; after Jones and Drechsler. *Nowakowskiella ramosa* Butler, E. J.; *b*, *c*, early stages in the formation of the resting sporangia; *d*, *e*, resting sporangia attached to a pseudoparenchyma; after Butler, E. J.

masses. In these, the marginal cells receive the contents of the remaining cells; they become globose, form thick walls (fig. 26 *d*), and are converted into a group of resting spores (fig. 26 *e*).

Megachytrium Westonii[1] is a parasite of *Elodea*. The zoospore settles on a leaf of the host, and puts out a rather thick hypha which branches freely on the outside of the leaf. Sooner or later, the leaf is entered. Sporangia and resting spores may develop both outside

[1] Sparrow, 1933 ii.

and inside the substratum; the resting spores are thick-walled expansions of the hyphae, and are not formed in the complicated manner characteristic of *Cladochytrium* and *Urophlyctis*.

Catenaria anguillulae[1] has a special affinity for animal substrata, including eelworms, and the eggs of the liver fluke and of rotifers. The thallus is filamentous, somewhat branched, with intercalary swellings. Simple or branched rhizoids spring from scattered points on the filaments. As the thallus ages, septa cut off the swellings from the unexpanded portions; the former are then organised into sporangia, each usually with one exit canal passing out of the host. *Catenaria* is not an obligate parasite, for it has been grown on boiled eggs of liver fluke, as well as on vegetable sub-

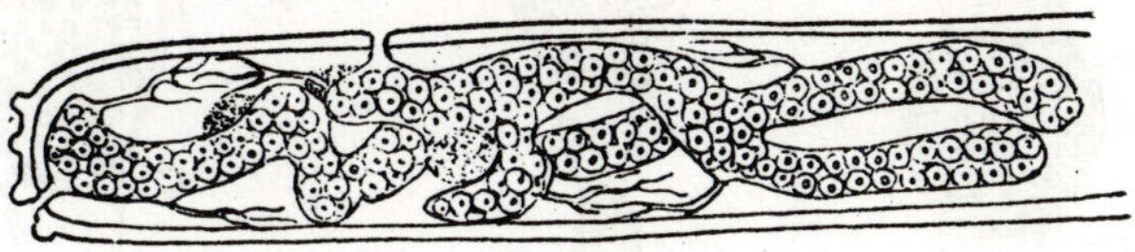

Fig. 27. *Mitochytridium ramosum* Dang.; mature sporangium; after Dangeard.

strata.[2] In cultures, the first sporangium is sometimes formed from the body of the zoospore, subsequent sporangia developing on filaments produced by it; evidently the mode of growth corresponds with a simplified version of that in *Physoderma* and *Urophlyctis*. When specimens of *Catenaria* are crowded in the host, the whole thallus may be converted into one sporangium; this gives a clue to the affinities of the rather puzzling *Mitochytridium ramosum*[3] (fig. 27), a parasite of the desmid *Docidium*. The thallus is a stout, slightly branched filament, bearing tufts of rhizoids here and there. At maturity, the whole thallus is converted without septation into a single sporangium from which uniflagellate zoospores are liberated. *Mitochytridium* shows a general likeness to *Catenaria*, and is more probably a member of the Cladochytriaceae than of the Ancylistales.

[1] Butler, J. B. and Buckley, 1927; Butler, J. B. and Humphries, 1932; Butler, E. J., 1928; Buckley and Clapham, 1929.

[2] Karling, 1934.

[3] Dangeard, 1911.

Hyphochytriaceae

The mycelium of the Hyphochytriaceae is coarse, and shows indications of differentiation into a basal and an apical region.

Macrochytrium botrydioides[1] grows on rotten fruit in foul water, sometimes associated with species of *Rhipidium*. The germinating zoospore forms a relatively massive main axis, attached to the sub-

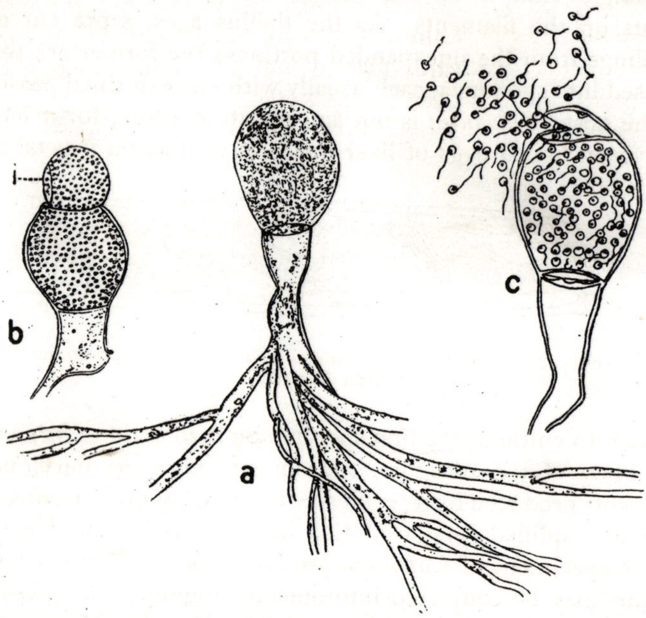

Fig. 28. ***Macrochytrium botrydioides*** **von Minden; *a*, mature plant, with sporangium; *b*, emergence of vesicle from the sporangium, i, lid; *c*, escape of the zoospores; after von Minden.**

stratum by a well-developed system of rhizoids; it bears a single lateral branch, from the end of which a globose sporangium is cut off (fig. 28 *a*). When the sporangium is mature, a lid forms in the outer layer of the wall (fig. 28 *b*), the contents emerge, surrounded by a delicate membrane, and the zoospores escape (fig. 28 *c*) when the latter bursts.

The habit of *Macrochytrium* recalls, on one side, that of the Rhizidiaceae, and on the other that of the Leptomitaceae. The

[1] von Minden, 1916.

zoospores have the characters usual in the Chytridiaceae, and the manner of dehiscence is reminiscent of *Cladochytrium*. The genus appears to be intermediate between the Chytridiales and the Leptomitales.

ANCYLISTALES

In the Ancylistales the thallus is a simple or branched filament, relatively wide in proportion to its length. At maturity, the thallus is divided into a chain of segments, each of which functions as a reproductive structure. The oogonium contains a single oospore; periplasm has not been recognised.

There are few species, all included in one family.

ANCYLISTACEAE

Most members of the Ancylistaceae attack Conjugatae.

The zoospore of *Lagenidium Rabenhorstii*[1] (fig. 29 *a*) is reniform, with two laterally inserted flagella; it comes to rest on a filament of

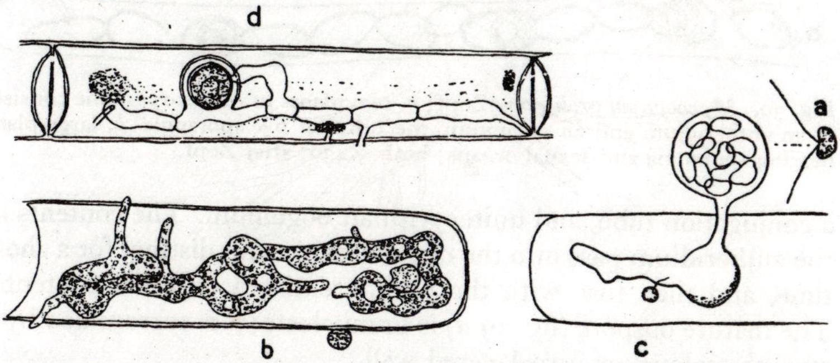

Fig. 29. *Lagenidium Rabenhorstii* Zopf; *a*, zoospore, ×720; *b*, thallus within the host, ×540; *c*, small plant with zoospores in the vesicle, ×720; *d*, monoecious plant, with an empty antheridium and an oospore, ×720; after Zopf.

Spirogyra, entering by means of a germ tube. Within the host, the tip of the germ tube swells and receives the contents of the zoospore. One or more filaments develop from the swelling, and give rise by apical growth to the short, thick, simple or branched tubes of the mycelium (fig. 29 *b*). The parasite does not spread to neighbouring host cells and the extent to which one thallus develops is strongly influenced by the presence of other parasites in the same

[1] Zopf, 1884; Atkinson, 1909 i; Cook, 1935.

cell. The thallus remains aseptate during growth, but when growth ceases transverse walls are laid down at irregular intervals, dividing the whole into a number of segments, which may become sporangia, or may be converted into sexual organs. The sporangia bear long exit tubes passing out into the water (fig. 29 *b*); as in *Pythium*, the partly differentiated contents of the sporangium migrate into a vesicle at the end of the tube, where the formation of the zoospores is completed (fig. 29 *c*).

Usually, a host cell is invaded by more than one zoospore of *Lagenidium*. Antheridia and oogonia may occur in the same thallus, but they often develop in separate thalli. The antheridium puts out

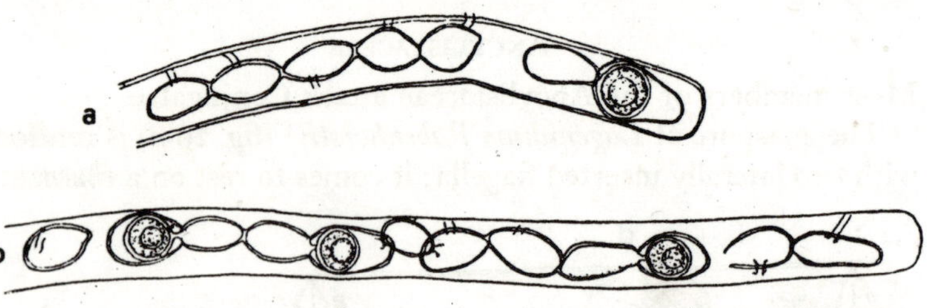

Fig. 30. *Myzocytium proliferum* Zopf; *a*, two plants in *Mesocarpus*, one consists of an antheridium and an oogonium, the other of five sporangia; *b*, large plant bearing sporangia and sexual organs; both ×430; after Zopf.

a conjugation tube, and unites with an oogonium. The contents of the antheridium pass into the oogonium, remain distinct for a short time, and then fuse with the undifferentiated oogonial contents. The mature oospore (fig. 29 *d*) is uninucleate; it is surrounded by a smooth, colourless, two-layered wall.

Lagenidium giganteum[1] is a weak parasite on the larvae of gnats, and on aquatic Crustacea. It is one of the few members of the Archimycetes that have been grown on an artificial medium.

In *Myzocytium proliferum*[2] (fig. 30) the vegetative thallus is divided by a number of deep constrictions, and, later, by septa. The zoosporangia and sexual organs develop as in *Lagenidium*, but the zoospores are nearly mature when they pass into the vesicle; the plants are usually monoecious. The conjugation tube of *Myzocytium* is less obvious than that of *Lagenidium*, and may be replaced by a simple pore between the gametangia.

[1] Couch, 1935 i. [2] Zopf, 1884.

Oospheres are not organised in the oogonia of *Lagenidium* and *Myzocytium* before fertilisation; the oogonial contents contract whilst fertilisation is in progress. When the members of these genera are crowded in the host cells, the thallus may not yield more than a single sporangium, so like that of *Olpidium* that distinction is possible only if the escape of the zoospores is observed.

Ancylistes Closteri[1] is a parasite of *Closterium*. The young thallus is a rather stout, simple or branched filament, becoming septate at maturity and divided into segments (fig. 31 *a*), each equivalent to a reproductive organ. The sexual organs resemble those of *Lagenidium* and *Myzocytium*, but the thalli are usually dioecious, with indications of morphological differences, since antheridia are often formed in thin filaments, oogonia in stouter hyphae. The contents of the antheridium pass into the oogonium through a conjugation tube (fig. 31 *b*), or, if the organs are already in contact, through a simple pore. The wall of the oospore (fig. 31 *c*) has two layers; at germination the endospore appears to be used up, and a long tube (fig. 31 *d*) develops which attacks and infects a cell of *Closterium*. *Ancylistes* does not form zoospores, a circumstance doubtless related to the gregarious habits of its host. A vegetative thallus may become divided into a chain of segments, each segment then putting out a long tubular outgrowth. This lengthens by apical growth, the contents moving forwards, and septa appearing behind them. When

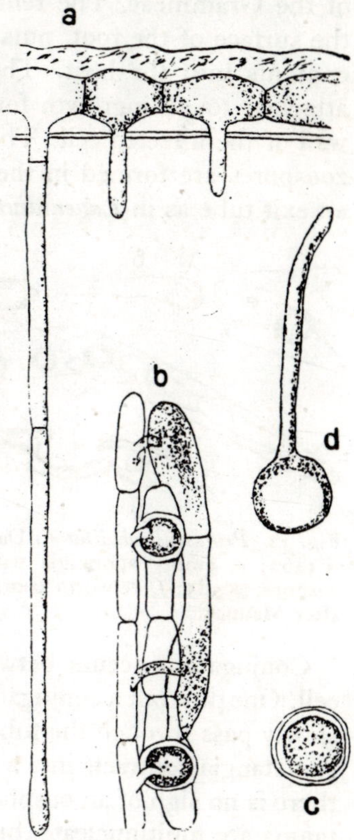

Fig. 31. *Ancylistes Closteri* Pfitzer; *a*, portion of thallus divided into segments which bear infection tubes; *b*, conjugation; *c*, oospore with wall of two layers; *d*, germinating oospore; all ×435; after Dangeard.

[1] Dangeard, 1886; Petersen, 1910.

a host cell is reached, the tube curves and grips it, and a small pore allows the contents of the infection to migrate into the host.

Lagena radicicola[1] is an obligate parasite of the roots of members of the Gramineae. The reniform biflagellate zoospore encysts on the surface of the root, puts in a thin germ tube and so passes its contents into the host. There, a walled thallus develops, and is attached to an ingrowth formed at the point of entry from the wall of the infected cell. The thallus soon becomes multinucleate, zoospores are formed in the usual manner, and are set free from an exit tube as in *Lagenidium*; the vesicle is very ephemeral.

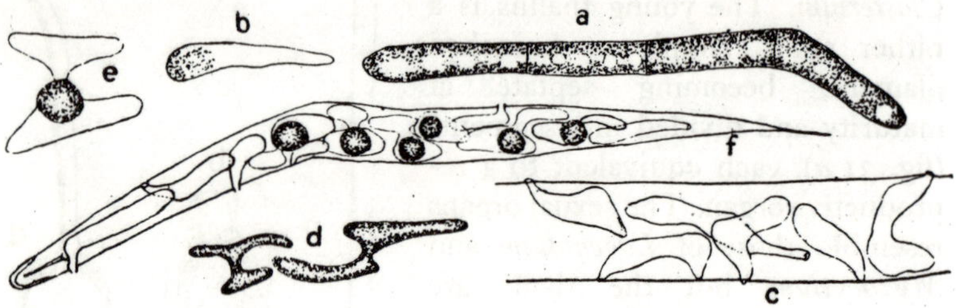

Fig. 32. *Protascus subuliformis* Dang.; *a*, vegetative filament, ×1075; *b*, a spore, ×1460; *c*, empty sporangia within host, ×640; *d*, conjugation, ×270; *e*, oospore, ×385; *f*, eelworm containing oospores and empty sporangia, ×270; after Maupas.

Conjugation occurs between two thalli lying in the same host cell. One puts out a conjugation tube towards the other; its contents slowly pass through the tube and the united material from the two gametangia contracts into a sphere. As is usual in the Ancylistales, there is no sign of an oosphere previous to fertilisation. The gametangia are multinucleate, but the cytological details have not been worked out. Thick-walled oospores form after conjugation. Sometimes, more than one male gametangium may unite with one female gametangium; then more than one oospore may be formed, so that it seems that the contents of the female gametangium must fragment when multiple fusions occur.

Protascus subuliformis[2] occurs as a stout cylindrical filament in the bodies of eelworms; septa form (fig. 32 *a*), and the thallus breaks up into a few multinucleate segments. These increase in size, their

[1] Vanterpool and Ledingham, 1930; Truscott, 1933.
[2] Maupas, 1915.

nuclei divide, and each segment is converted into a sporangium from which non-motile club-shaped spores (fig. 32 *b*) are expelled by an exit canal passing through the skin of the host (fig. 32 *c*).

Sometimes however, after the thallus has separated into segments, two segments of unequal size, probably derived from different thalli, put out papillae (fig. 32 *d*) towards one another. Their contents shrink, a pore opens in the united tips of the papillae, and the contents of the smaller segment pass slowly into the larger. The united contents fuse to form a spherical zygote, surrounded by a thick wall of cellulose (fig. 32 *e*).

As in *Ancylistes*, there seems to be a direct relation between the production of non-motile spores and the gregarious habits of the host; eelworms occur together in large numbers, and are easily reached even by non-motile spores.

PROTOMYCETALES

The Protomycetales include but few species, all parasites in flowering plants. The branching, septate mycelium inhabits the intercellular spaces of the host; haustoria are not formed. Intercalary or terminal resting sporangia develop within the substratum, and afterwards give rise to non-motile spores which tend to fuse in pairs.

There is one family, the Protomycetaceae.

Protomycetaceae

Protomyces macrosporus[1] causes small weals on the stems and veins of the leaves of *Aegopodium Podagraria* and other Umbelliferae. Spores come to rest on the surface of the host, insert germ tubes between the epidermal cells, and give rise to a septate mycelium of multinucleate segments. After a time, some of the segments receive the contents of their neighbours, swell up, and develop a wall of three layers. The rounded resting sporangium so formed remains quiescent during the winter; in spring, the outer layers of the wall split, and the contents emerge in a sac formed from the expanded inner layer of the wall (fig. 33 *a*). A central vacuole enlarges and forces the cytoplasm into a peripheral layer which is cut up into uninucleate portions by the development

[1] von Büren, 1915, 1922.

of radially directed furrows. Nuclear divisions follow, and each portion is converted into four spores; these lie at first in the peripheral layer, but later collect in a ball at the apex of the rounded spc rangium. The sporangial wall splits, the ball of spores is expelled wit i some violence, and the elongated ellipsoid spores scatter a little, put out short processes, and fuse in pairs. Infection of the host follows, beginning a new life cycle. In *Protomyces pachydermus*[1] (fig. 33 *b*, *c*), there is the same sequence of events, but the sporangial sac is elongated.

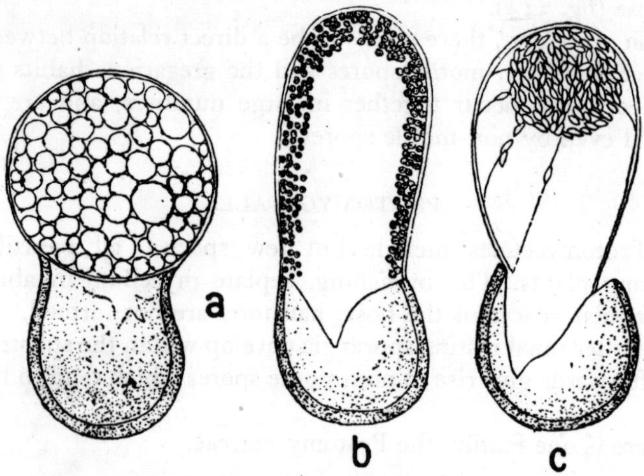

Fig. 33. *Protomyces macrosporus* Unger; *a*, emergence of sporangial sac. *Protomyces pachydermus* Thümen; *b*, spores lying in a peripheral layer; *c*, spores grouped in a ball at the apex of the sac; all × 340; after von Büren.

Protomyces inundatus[2] attacks *Apium inundatum*; some of its sporangia form spores as soon as they are organised, others require a period of rest; the resting sporangia may form spores without any change in their form, or they may put out a sac as a preliminary to spore formation. In *Taphridium umbelliferarum*,[2] the spores are delimited inside the resting sporangium, but they are often set free from a thin-walled sac which develops as they mature; in *T. algeriense*,[2] the spores are cut out from a peripheral layer lining the wall of the resting sporangium, and are set free directly from that structure, no thin-walled sac appearing. In *T. algeriense*,

[1] von Büren, 1915, 1922. [2] Juel, 1921; von Büren, 1922.

although the mycelium ramifies widely in the host tissue, the thin-walled resting sporangia lie just under the epidermis of the host.

The nuclei of the Protomycetaceae are small, and cytological details are scanty. It appears that the spores are uninucleate when they fuse, and that nuclear fusion follows the fusion of the spores. It has been suggested[1] that a reducing division precedes spore formation, but the evidence is not conclusive.

The systematic position of the Protomycetales is doubtful. The spores are formed by a process of cleavage very like that in other Phycomycetes, and the fusion of the spores in pairs recalls phenomena in some Archimycetes; it is possible however that this fusion is pseudapogamous. Until more information is available, it seems best to place the Protomycetales in the Archimycetes, to which they show several obvious resemblances.

OOMYCETES

The Oomycetes contain about 200 species; many live as parasites or saprophytes in water, others inhabit the intercellular spaces of phanerogams, and still other species are common as saprophytes in the soil. The mycelium is well developed and sexual reproduction takes place by means of an antheridium and an oogonium.

In general terms, the lower Oomycetes are aquatic, the higher Oomycetes are terrestrial. In the aquatic species, the sporangia liberate zoospores; in amphibious species zoospores are set free when moisture is abundant, but, when conditions are relatively dry, the sporangium often puts out a germ tube and functions as a conidium; this always occurs in *Pythium ultimum*,[2] a common saprophyte in the soil, and in the species of *Peronospora*. In many aquatic species, the sporangia are renewed by proliferation; in other aquatic species, and in amphibious and terrestrial species, the sporangia may form in basipetal chains, or on sympodial branch systems. As the terrestrial species become less and less dependent on external water for the dispersal of their spores, the sporangia tend more and more to function as conidia, and to ripen in groups on a well-formed sporangiophore, rather than in succession.

In *Allomyces*,[3] the sexual act takes place by the union of motile

[1] von Büren, 1915, 1922. [2] Trow, 1901.
[3] Kniep, 1929–30, 1930; Hatch, 1933–4, 1935.

gametes, and in *Monoblepharis*, an active male gamete unites with a passive or nearly passive[1] female gamete lying in the oogonium. With these exceptions, the contents of the antheridium of the Oomycetes pass through a conjugation tube into the oogonium where one or more oospheres are already present. In most species of Oomycetes the oogonium contains one oosphere, but, rarely in *Pythium*,[2] and commonly in the Saprolegniaceae, several to many are present. In the Saprolegniaceae, the contents of the oogonium are entirely used up in the production of oospheres either rounding up as a whole, or becoming divided into uninucleate portions by cleavage furrows passing out from a central vacuole. In the Monoblepharidaceae, the uninucleate contents of the oogonium shrink somewhat to form one oosphere, without any significant residue. In other Oomycetes, the solitary oosphere lies free in the oogonium, surrounded by some residual material, the **periplasm**; this assists in the formation of the wall of the oospore, after fertilisation has taken place.

The aquatic Oomycetes are very sensitive to the environment[3] in respect to the formation and behaviour of their reproductive organs. The species of *Saprolegnia* show variations in morphology even when growing in their natural habitats; in culture,[4] their behaviour may be profoundly modified by the conditions and by the composition of the medium. *S. mixta*,[5] well supplied with carbohydrates, and frequently transferred to fresh solutions, was maintained in a purely vegetative condition for many years, though portions of the mycelium always yielded reproductive organs soon after transfer to pure water. *S. hypogyna*[6] seldom forms antheridia when grown on dead insects, but does so readily in solutions containing haemoglobin together with inorganic phosphates, nitrates and salts of potassium; indeed, under that treatment, it appears that the material of young oogonia may be diverted to the formation of antheridia. It is probable that the reproductive organs tend to develop in solutions which are nearly but not quite exhausted of their nutritive material,[7] and that they develop at the expense of substances stored in the hyphae. Develop-

[1] Barnes and Melville, 1932.
[2] Sparrow, 1931.
[3] Lechmere, 1910, 1911; Collins, M. I., 1920.
[4] Kauffman, 1908; Coker, 1923.
[5] Klebs, 1899.
[6] Kauffman, 1908.
[7] Obel, 1910.

mental processes are often completely upset if bacteria accumulate in the cultures.

The Oomycetes are divided into the following alliances:

Habitat mostly aquatic or subterranean; sporangia liberating zoospores
- Zoospores uniflagellate; hyphal walls not giving good reactions for cellulose
  - Walls of resting spores pitted: cytoplasm not foamy — BLASTOCLADIALES
  - Walls of resting spores (when formed) bearing hemispherical warts; cytoplasm of hyphae foamy — MONOBLEPHARIDALES
- Zoospores biflagellate; hyphal walls usually giving good reactions for cellulose
  - Hyphae constricted at intervals or at the base; plugs of cellulin present — LEPTOMITALES
  - Hyphae not constricted at intervals; no cellulin plugs — SAPROLEGNIALES

Habitat seldom aquatic, more often subterranean, subaerial or endophytic.
- Zoospores biflagellate; sporangia often functioning as conidia — PERONOSPORALES

BLASTOCLADIALES

The Blastocladiales include two genera, *Allomyces* and *Blastocladia*, with about twenty species occurring on submerged twigs and fruits, on animal material in the water, and in the soil. In both genera, the thallus develops from a basal cell, which is attached to the substratum by rhizoids. The thallus bears sessile zoosporangia liberating zoospores with one posterior flagellum, and rounded resting spores covered by densely pitted walls; in *Allomyces*, gametangia are borne on special thalli.

They form a single family, the Blastocladiaceae.

BLASTOCLADIACEAE

The species of *Allomyces*[1] occur in the soil and on dead animal material in water. The thallus branches subdichotomously from a basal cell attached to the substratum by rhizoids; its branches contain pseudosepta of cellulin at fairly regular intervals. Most often, the thalli bear zoosporangia, either singly or in chains, and

[1] Butler, E. J., 1911; Barrett, 1912 ii; Coker, 1923; Kniep, 1929–30; Hatch, 1933–4, 1935.

a smaller number of resting spores, distinguished by their pitted walls. In *Allomyces javanicus*,[1] which grows well in a weak decoction of peas, uniflagellate zoospores, set free from the sporangia, germinate to give plants bearing zoosporangia and resting spores; this may continue indefinitely. The resting spores, especially after a period of desiccation, are converted into sporangia, without the

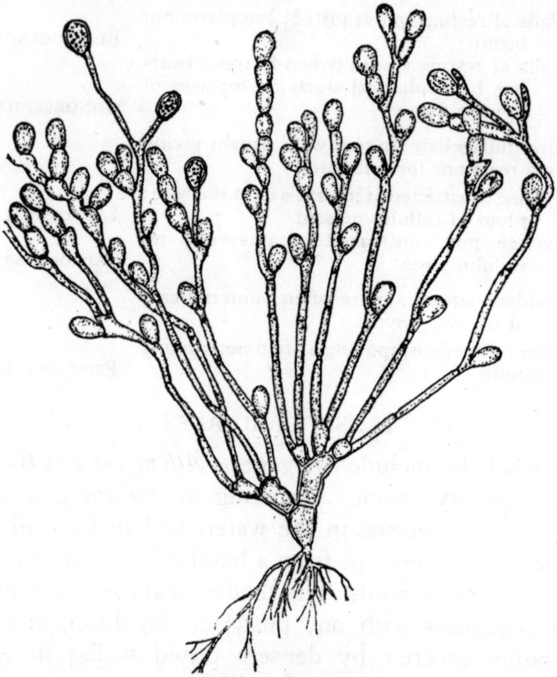

Fig. 34. *Allomyces arbuscula* Butler, E. J.; plant with sporangia, and a few resting spores; after Butler, E. J.

formation of a mycelium; these sporangia emit motile uniflagellate spores which germinate readily giving plants of the same general form as those already mentioned, but bearing chains of gametangia. Some gametangia are relatively small; they liberate small uniflagellate gametes distinguished by a conspicuous orange globule: others are larger; they produce rather larger, colourless uniflagellate gametes. The two sorts of gametes fuse in pairs to give a biflagellate

[1] Kniep, 1929–30, 1930.

motile zygote, which, after a short period of activity, settles down and germinates without delay, yielding a thallus bearing zoosporangia and resting spores. These thalli multiply rapidly by means of their zoospores, and soon crowd out the gametangium-bearing plants. It is not surprising that the latter were overlooked for a long time in *Allomyces arbuscula*[1] (fig. 34), the first species of the genus to be described; its life history resembles that of *A. javanicus*.

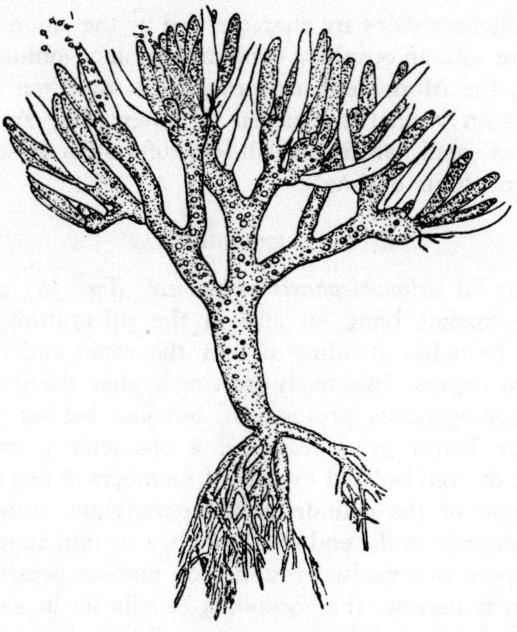

Fig. 35. *Blastocladia Pringsheimii* Reinsch; a plant bearing sporangia; after Thaxter.

Allomyces exhibits a morphological alternation of independent generations. The cytology is not yet quite cleared up. It is probable that zoosporangia form on diploid plants, and that meiosis occurs when the resting sporangia form their motile spores. A life cycle of this kind is so unusual among Phycomycetes that *Allomyces* has been regarded as a colourless alga, though its algal relations are not obvious.

[1] Butler, E. J., 1911; Hatch, 1933–4, 1935.

The thallus of *Blastocladia*[1] (fig. 35) consists of a relatively massive trunk, anchored to the substratum by rhizoids, and crowned by stout branches bearing zoosporangia and resting spores with pitted walls. Up to the present, nothing certain is known of sexuality in *Blastocladia*, but it may well be found to correspond with that of *Allomyces*.

MONOBLEPHARIDALES

The Monoblepharidales are characterised by the union of a motile male gamete with an oosphere contained in an oogonium. There is one family, the Monoblepharidaceae, with some ten species, all saprophytes on vegetable material in water; they appear to the naked eye as small, whitish brush-like tufts or as loose brownish masses of indefinite outline.

MONOBLEPHARIDACEAE

The thallus of *Monoblepharis sphaerica*[2] (fig. 36) consists of branching rhizoids lying on and in the substratum, and stiff, straightish branches standing out in the water and bearing the reproductive organs. In actively growing hyphae, the tip is occupied by fairly homogeneous protoplasm, but just behind the tip the contents are foamy in appearance, a character permitting the recognition of even isolated hyphae of members of this family. In the formation of the cylindrical zoosporangium, cytoplasm and nuclei accumulate in the end of a branch, a septum appears below, and a zoospore is organised round each nucleus present. As the sporangium is narrow, the zoospores usually lie in a row; they escape from the apex of the sporangium. Before the terminal sporangium discharges, a second is formed immediately beneath it; by repetitions of the process, basipetal chains of sporangia are produced.

The zoospores are monoplanetic and uniflagellate; they resemble in appearance and gait the zoospores of many chytrids. They swim for a time, settle on a suitable substratum in well-aerated water and usually put out two germ tubes, one giving rise to the system of rhizoids, the other to the fertile hyphae.

[1] Thaxter, 1896 ii; von Minden, 1916; Kanouse, 1927; Cctner, 1930.

[2] Cornu, 1872; Thaxter, 1895 i; de Lagerheim, 1900; Woronin, 1904; Sparrow, 1933 i.

The oogonium is organised in a terminal swelling; just beneath it, in the unswollen region, an antheridium commonly develops. The fertile hyphae branch sympodially so that groups of paired sexual organs may be formed. The rudiment of the oogonium contains one nucleus and some cytoplasm, delimited by a basal septum. The nucleus does not divide, and, just before the oogonial contents are ripe, there is some loss of water, slight shrinkage, and the formation of a single oosphere. The antheridia contain five or six nuclei; a sperm is organised round each nucleus. A small lateral beak protrudes from the antheridium, its tip opens, and the uniflagellate male gametes emerge. They are smaller than the zoospores, but they resemble them in appearance.

A pore opens in the wall of the mature oogonium, allowing the extrusion of a substance attractive to the male gametes. They alight on the oogonium, where they crawl for a time in amoeboid fashion. Finally, one enters the pore and fuses with the oosphere. After fusion, the zygote may move; it frequently passes towards the base of the oogonium, then towards the pore, and sometimes through it. A thick, warted

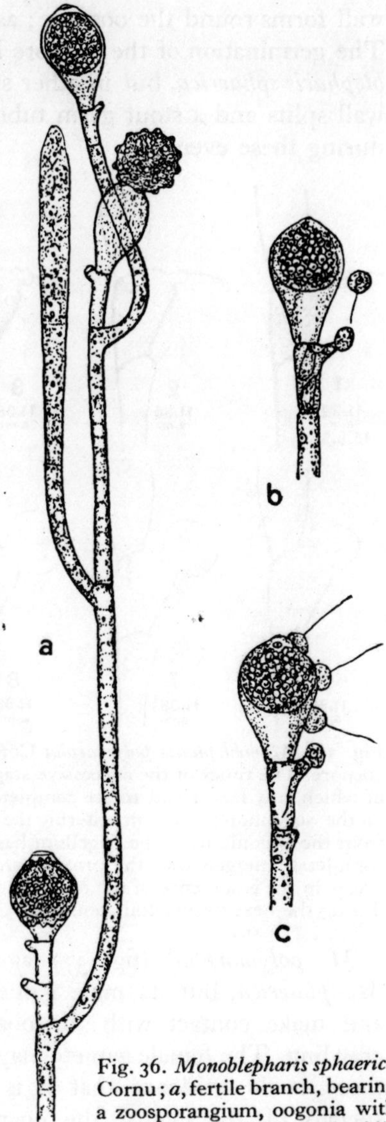

Fig. 36. *Monoblepharis sphaerica* Cornu; *a*, fertile branch, bearing a zoosporangium, oogonia with oospores, and empty antheridia below them, × 520; *b*, emergence of spermatozoids; *c*, spermatozoids creeping on the oogonium; after Woronin.

wall forms round the oospore; as it thickens, the two nuclei fuse. The germination of the oospore has not been observed in *Monoblepharis sphaerica*, but in other species, after a period of rest, the wall splits and a stout germ tube grows out.[1] Meiosis may occur during these events.

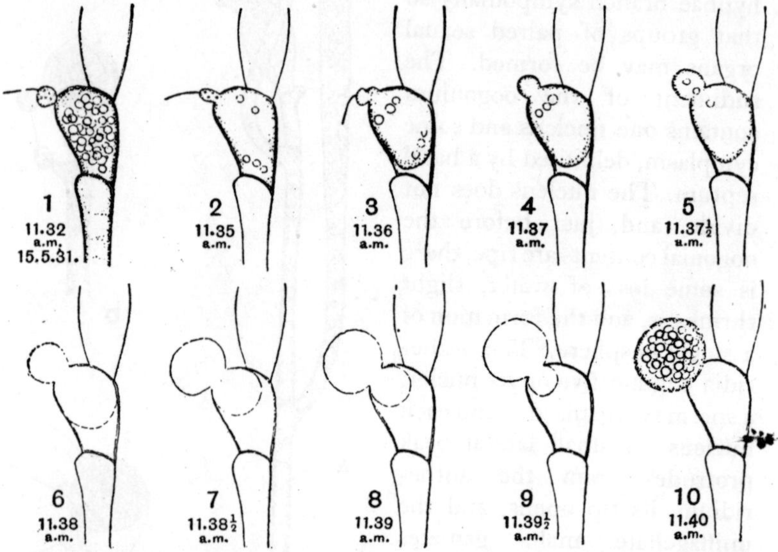

Fig. 37. *Monoblepharis polymorpha* Cornu; fertilisation, and emergence of the oospore. The times of the successive stages appear by the sides of the drawings, of which only nos. 1 and 10 are complete in detail. 1, a sperm has just alighted on the oogonium; 2, sperm entering the oogonium; 3, the oosphere is emerging from the oogonium; 4, the flagellum has disappeared from the sperm, and it is completely merged with the protuberance from the oosphere; 5–10, successive stages in the emergence of the oospore; it gradually formed a thick, warted wall during the next twenty-four hours; × 520. After Barnes and Melville.

M. polymorpha[2] (fig. 37) agrees in general characters with *M. sphaerica*, but its male gametes may swim to the oogonium and make contact with the oosphere without any preliminary crawling. The female gamete may show noticeable activity, though there is no evidence that it is flagellated. In this, and other species of the genus, the oospores usually emerge from the oogonium, and ripen outside it, remaining attached to the edge of the pore.

[1] Laibach, 1927. [2] Barnes and Melville, 1932; Sparrow, 1933 i.

In *Monoblepharis*, the same thallus may bear zoosporangia, oogonia and antheridia; there is no alternation of generations such as occurs in *Allomyces*.

Gonapodya prolifera[1] (fig. 38), like *Monoblepharis*, occurs on submerged twigs, forming vague brownish tufts. The thallus branches freely, each branch having a chaplet-like aspect, since it consists of rounded segments separated by constrictions partly blocked by pseudo-septa. Zoosporangia form at the ends of the branches; they are freely renewed by proliferation. Sexual organs are not certainly known, though oogonia and motile sperms have been ascribed to *Gonapodya*. Despite the difference in the organisation of the thallus, there is a close affinity between *Gonapodya* and *Monoblepharis*; the hyphae of both genera are easily recognised by their characteristic foamy contents, and the zoospores of both have a characteristic internal structure.[2] It is possible that this structure may be the remains of a chloroplast, and that the Blastocladiales and Monoblepharidales are both groups of colourless algae.

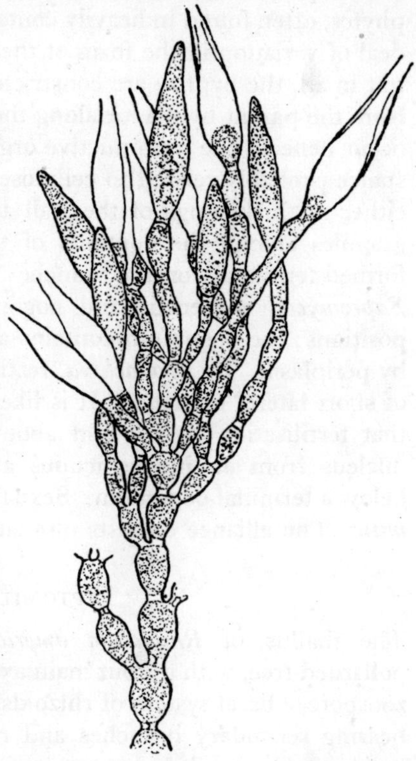

Fig. 38. *Gonapodya prolifera* (Cornu) Fischer; portion of plant, showing constrictions, and renewal of the sporangia by proliferation; after Thaxter.

[1] Thaxter, 1895 ii; Petersen, 1910; von Minden, 1916; Barnes and Melville, 1932; Sparrow, 1933 i.

[2] Sparrow, 1933 i.

LEPTOMITALES

The Leptomitales[1] include about a dozen species; they are saprophytes, often found in heavily contaminated water. There is a good deal of variation in the form of the thallus in the different genera, but in all, the hyphae are constricted, either at the point of origin from the parent hypha, or along their length; similar constrictions occur beneath the reproductive organs. Cellulin, a refractive substance probably related to cellulose, abounds in the Leptomitales, either in thickenings of the wall at the constrictions, or as loose granules among the contents of the hyphae. Zoosporangia are formed terminally on the branches. In *Rhipidium*, *Araiospora* and *Sapromyces*, antheridia and oogonia are developed in terminal positions; the oogonium contains a single oosphere, accompanied by periplasm. In *Apodachlya*, resting structures occur at the ends of short lateral branches. It is likely that these are oospores, and that fertilisation[2] is brought about by the migration of a male nucleus from an inconspicuous antheridium lying immediately below a terminal oogonium. Sexual organs are unknown in *Leptomitus*. The alliance consists of a single family, the Leptomitaceae.

LEPTOMITACEAE

The thallus of *Rhipidium americanum*[3] (fig. 39) resembles a pollarded tree, with a stout main axis derived from the body of the zoospore, a basal system of rhizoids, and an apical group of hyphae bearing secondary branches and reproductive organs. The zoosporangia are developed in succession in sympodia; they are of two kinds. From one kind, spores are liberated immediately; the other kind undergoes a period of rest. At dehiscence, a lid formed from the wall of the sporangium lifts, and the spores emerge in an elongated vesicle, formed from the inner layer of the wall. The vesicle bursts and sets free the biflagellate zoospores. The oogonia are terminal and spherical, the antheridia are small terminal segments of long, sinuous lateral hyphae, winding about among the other branches as they grow towards the oogonia. The contents of the antheridium pass into the oogonium through a conjugation tube.

[1] Kanouse, 1927.
[2] Kevorkian, 1935.
[3] Thaxter, 1896 iii; von Minden, 1916.

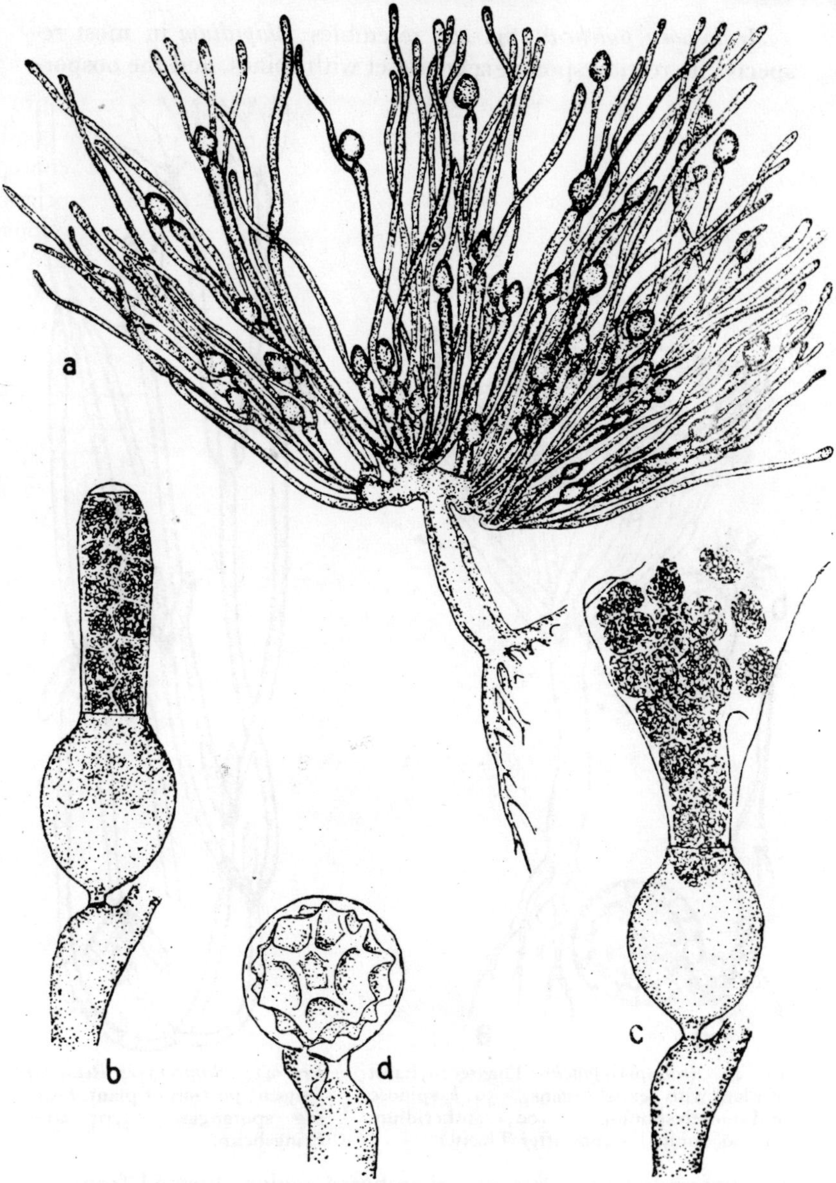

Fig. 39. *Rhipidium americanum* Thaxter; *a*, mature plant bearing sporangia, ×50; *b*, contents of sporangium emerging in a vesicle, ×160; *c*, rupture of vesicle and escape of the zoospores, ×160; *d*, mature oospore in oogonium, with attached antheridium, ×320; all after Thaxter.

Araiospora pulchra[1] (fig. 40) resembles *Rhipidium* in most respects; its resting sporangia are beset with spines, and the oospore

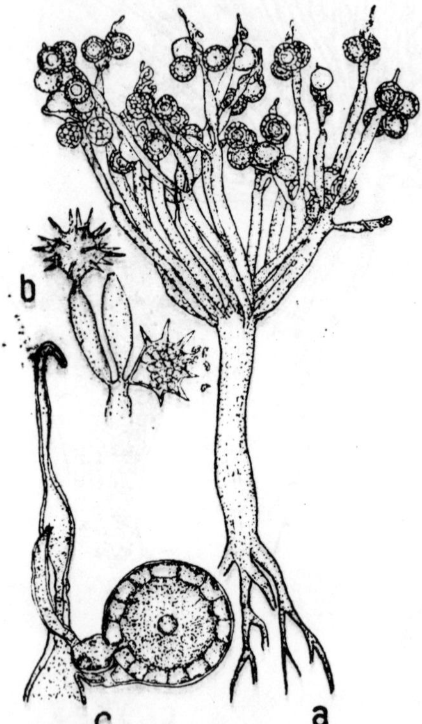

Fig. 40. *Araiospora pulchra* Thaxter; *a*, habit of plant with sexual organs, × 50; *b*, spinose and smooth sporangia, × 100; *c*, antheridium and oogonium, × 400; after Thaxter.

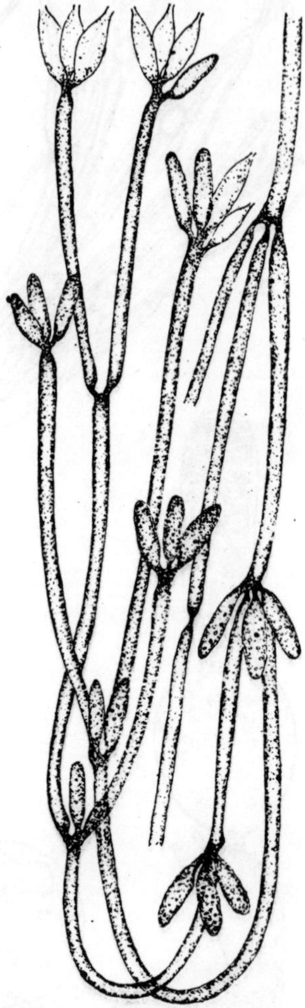

Fig. 41. *Sapromyces Reinschii* Fritsch; portion of plant, bearing sporangia, × 120; after Pringsheim.

is surrounded by a curious, chambered region, formed from the periplasm. *Sapromyces Reinschii*[2] (fig. 41, 42) shows a decline of the

[1] Thaxter, 1896 iii; King, 1903.

[2] Reinsch, 1878; Thaxter, 1896 iii.

tree-like form; this decline is still more pronounced in *Apodachlya pyrifera*[1] (fig. 43 *a*). The thallus of *Apodachlya* consists of thick main hyphae and thinner lateral hyphae, but the thallus has no formal shape. Zoosporangia develop as swollen terminal segments of the branches; their zoospores encyst in a group at the mouth of the sporangium, and finally quit the cysts provided with

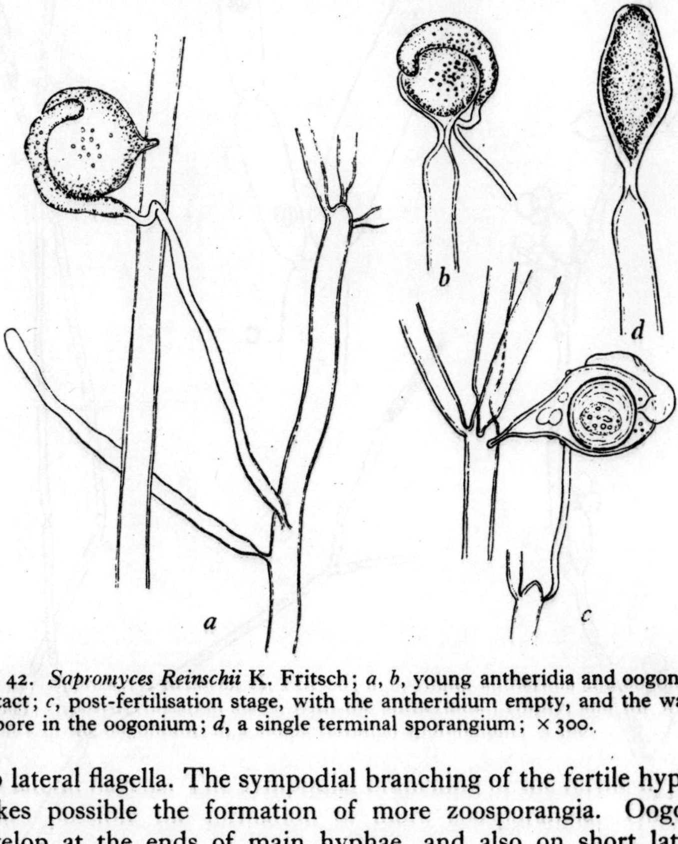

Fig. 42. *Sapromyces Reinschii* K. Fritsch; *a*, *b*, young antheridia and oogonia in contact; *c*, post-fertilisation stage, with the antheridium empty, and the walled oospore in the oogonium; *d*, a single terminal sporangium; × 300.

two lateral flagella. The sympodial branching of the fertile hyphae makes possible the formation of more zoosporangia. Oogonia develop at the ends of main hyphae, and also on short lateral branches. In *Apodachlya brachynema*,[2] they usually occur at the ends of short lateral branches consisting of a few bead-like segments, each of which is probably a potential antheridium. Oospores with thick walls occur inside the oogonia. Their formation may

[1] Zopf, 1888. [2] Kevorkian, 1935.

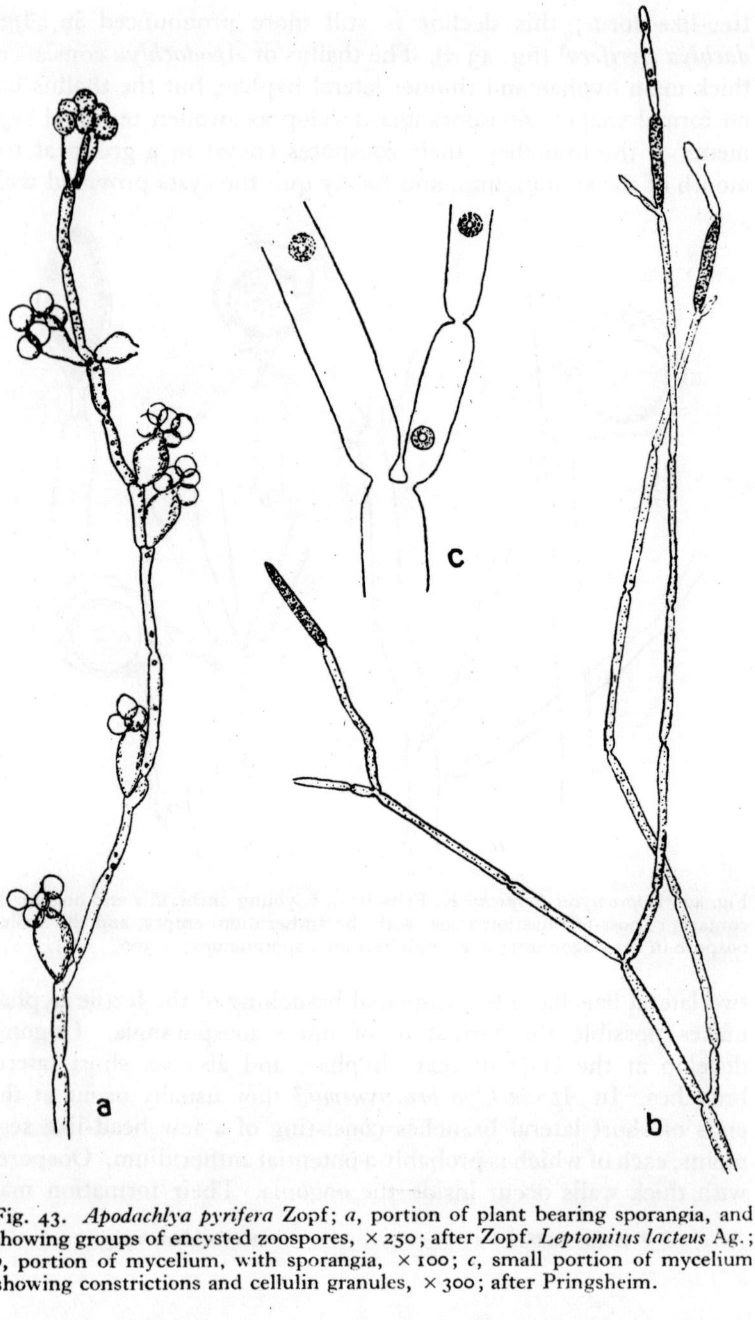

Fig. 43. *Apodachlya pyrifera* Zopf; *a*, portion of plant bearing sporangia, and showing groups of encysted zoospores, ×250; after Zopf. *Leptomitus lacteus* Ag.; *b*, portion of mycelium, with sporangia, ×100; *c*, small portion of mycelium showing constrictions and cellulin granules, ×300; after Pringsheim.

be preceded by the migration of the contents of a hypogynous antheridium into the oogonium through a pore in the septum between the two organs.

In *Leptomitus lacteus*[1] (fig. 43 *b*), a species characteristic of streams contaminated with sewage, the mycelium consists of freely branched hyphae, with no indications of a main axis. The hyphae are deeply constricted, but pores allow of communication between the segments (fig. 43 *c*), in each of which a granule of cellulin is conspicuous. The zoosporangium is formed in the undifferentiated terminal segment of a filament; a granule of cellulin lodges in the pore, shutting off the multinucleate contents of the young sporangium. The zoospores are cut out by a process of cleavage and escape as pyriform cells with two apical flagella. As the sporangium empties, another is organised in the segment next below; in this way, chains of zoosporangia are formed.

SAPROLEGNIALES

The members of the Saprolegniales, while chiefly aquatic, also occur in the soil;[2] a number are parasitic in animals, algae, fungi and higher plants. The mycelium is usually aseptate and branching. The zoosporangia, formed in the tips of hyphae, often do not differ much in diameter from the rest of the filament; when several are formed from the same hypha, they occur in succession. There is little indication within the alliance of a tendency to the production of differentiated sporangiophores.

The zoospores are biflagellate, and often show diplanetism. The hyphal walls give good reactions for cellulose.

Antheridia and oogonia arise at the ends of main or of lateral hyphae, the latter usually in a conspicuous racemose arrangement. Oospheres are formed in the oogonium previous to fertilisation, using up the whole contents of that organ. The antheridium unites with the oogonium by means of a conjugation tube. Although pits often occur in the wall of the oogonium, it does not follow that the conjugation tubes enter through them.

There is one family, the Saprolegniaceae.

[1] Pringsheim, 1858; Guilliermond, 1922; Coker, 1923.

[2] Harvey, 1925, 1927 i, ii; Coker, 1927; Coker and Braxton, 1926; Couch, 1927, 1930–1 i; Barnes and Melville, 1932.

Saprolegniaceae

The Saprolegniaceae[1] include about 100 species, living as saprophytes or parasites in water, as saprophytes in the soil, or as parasites in aquatic organisms and in the roots of higher plants; some species cause disease in crop plants. The mycelium is usually continuous in the vegetative condition; constrictions do not occur in the course of the hyphae or at the insertion of branches. Freely branched vegetative hyphae, creeping on and in the substratum, are distinguished by their narrow diameter and tapering ends from the rather stouter fertile branches on which the reproductive organs are formed. Chlamydospores, often called **gemmae** in this family, arise by the condensation of hyphal contents with subsequent deposition of septa cutting off the evacuated portions of the hypha; they are particularly abundant in *Saprolegnia torulosa*, and occur in other species of the genus.

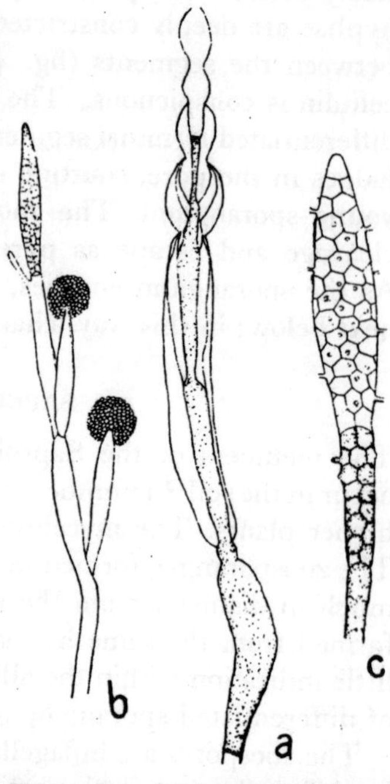

Fig. 44. *Saprolegnia monoica* Prings.; *a*, renewal of the sporangium by proliferation of the fertile hypha, × 225; after Coker. *Achlya racemosa* Hildebr.; *b*, sympodium of sporangia and formation of groups of cysts, × 55; after Pringsheim. *Dictyuchus* sp.; *c*, two successively formed sporangia, the upper showing crowded empty cysts, × 225; after Weston.

In the development of a primary zoosporangium, an accumulation of cytoplasm and nuclei at the end of a hypha is cut off by a wall; the sporangium is cylindrical or club-shaped in most species. It is globose in *Pythiopsis cymosa*, and has complicated lobes in *Plectospira*. In *Saprolegnia* (fig. 44 *a*), secondary sporangia are formed by proliferation; in *Achlya* (fig. 44 *b*) they arise in lateral branches borne at the base of the next older sporangium; in *Dictyuchus* (fig. 44 *c*), chains of sporangia are formed in basipetal succession.

[1] de Bary, 1881 i; Coker, 1923.

Zoospores are differentiated by a process of cleavage,[1] agreeing in its main details with the usual manner of formation of the asexual spores in the Phycomycetes. A centrally placed vacuole in the young sporangium pushes the cytoplasm and nuclei against the wall. Furrows appear on the inner face of this peripheral layer and spread radially outwards, so that many uninucleate rudiments are cut out; they contract somewhat, water is expelled from the sporangium, and shrinkage occurs. This crowds the initials of the zoospores together, so that their outlines are difficult to determine. The final differentiation is completed by further shrinkage of the rudiments; they round off, and are now ready to escape. Nuclear divisions do not accompany the development of the zoospores.

The zoospores of *Saprolegnia* and *Leptolegnia*[2] show diplanetism[3] very clearly. In these genera, the zoospores discharged from the zoosporangium are pyriform, with two apical flagella. They swim for a few minutes and encyst. After a rest of about twenty-four hours, each cyst gives rise to a zoospore shaped like a coffee bean, with two laterally inserted flagella. If contact is made with a suitable substratum, the zoospore produces a germ tube. In *Achlya*[4] and *Aphanomyces*,[4] it is possible that pyriform zoospores exist for a short time inside the sporangium; zoospores moving by means of flagella have been observed in the sporangium, but free-swimming pyriform zoospores have not been seen in the water outside. The zoospores emerge, united by fine threads. They encyst at the mouth of the sporangium, forming a hollow ball of cysts; these cysts in due course set free laterally biflagellate zoospores. The suppression of the first motile stage is carried still farther in *Thraustotheca*[5] (fig. 45); cysts form inside the sporangium, and are liberated as such by the breakdown of the sporangial wall; as in *Achlya*, laterally biflagellate zoospores escape from the cysts. In *Dictyuchus*,[6] the cysts are so crowded in the sporangium that their walls are mutually flattened, giving a network of walls inside the organ; laterally biflagellate zoospores escape separately from the cysts, each through its own pore (fig. 44 *c*) which traverses the walls

[1] Schwarze, 1922. [2] Couch, 1924.
[3] Leitgeb, 1869; Butler, E. J., 1907; Atkinson, 1909 ii; Weston, 1919.
[4] Coker, 1923. [5] Weston, 1918.
[6] Weston, 1919.

of the cyst and the sporangium. In *Dictyuchus*, the sporangia, filled with cysts, often fall away from the hyphae, and may float

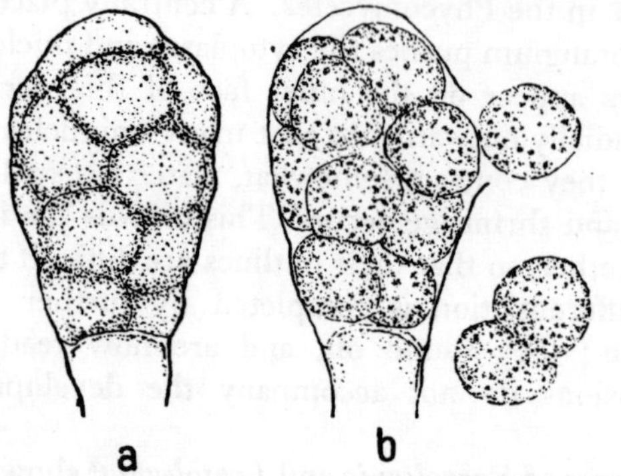

Fig. 45. *Thraustotheca clavata* (de Bary) Humphrey; *a*, mature sporangium; *b*, escape of sporangiospores; × 1120; after Weston.

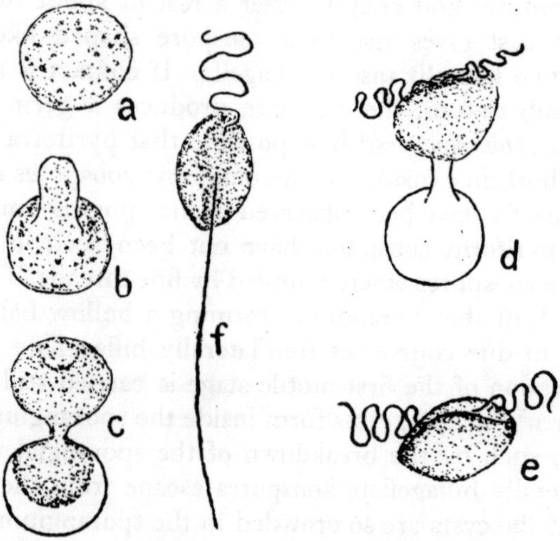

Fig. 46. *Dictyuchus* sp.; *a*, cyst; *b*, *c*, *d*, stages in emergence of the zoospore from the cyst; *e*, *f*, two views of the zoospore; all × 1400; after Weston.

for some time before the zoospores escape; this no doubt facilitates the dispersal of the fungus. The zoospores of *Dictyuchus* may encyst one or more times after they have quitted the sporangium, but however many times this happens, there is no change in the

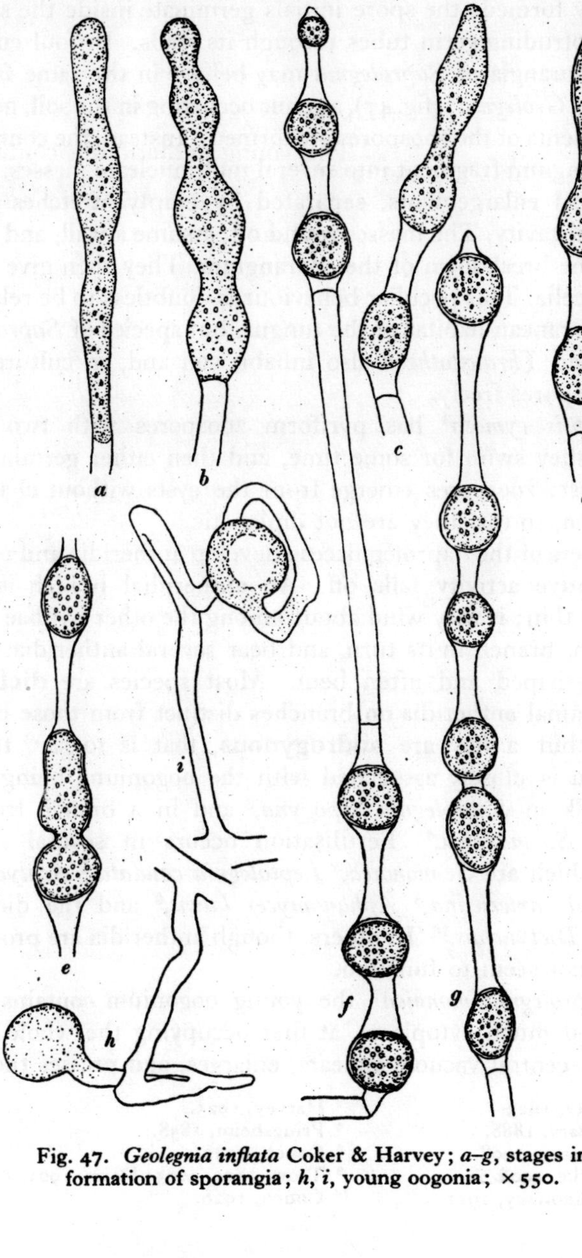

Fig. 47. *Geolegnia inflata* Coker & Harvey; *a–g*, stages in formation of sporangia; *h*, *i*, young oogonia; × 550.

morphology of the zoospore (fig. 46). In *Aplanes*,[1] active zoospores are rarely formed; the spore initials germinate inside the sporangium, protruding germ tubes through its walls. In foul cultures, the zoosporangia of *Saprolegnia* may behave in the same fashion. Finally, in *Geolegnia*[2] (fig. 47), a genus occurring in the soil, not even the rudiments of the zoospores are formed. Instead, the contents of the sporangium fragment into several multinucleate masses. These lie in local enlargements, separated by empty stretches of the sporangial cavity. The masses round off, assume a wall, and are set free by the breakdown of the sporangium. They then give rise to fresh mycelia. This peculiar behaviour is doubtless to be related to the subterranean habitat of the fungus, but species of *Saprolegnia*, *Achlya* and *Thraustotheca* also inhabit soil and, in culture, they form zoospores freely.

Pythiopsis cymosa[3] has pyriform zoospores with two apical flagella; they swim for some time, and then either germinate, or else encyst; zoospores emerge from the cysts without change of flagellation, so that they are not diplanetic.

Members of the Saprolegniaceae develop antheridia and oogonia as vegetative activity falls off. An antheridial branch is often long and thin; it may wind about among the other hyphae of the mycelium, branch in its turn, and bear several antheridia. These are club-shaped and often bent. Most species are **diclinous**, with terminal antheridia on branches distinct from those bearing oogonia, but a few are **androgynous**, that is to say, the antheridium is closely associated with the oogonium, being found in its stalk in *Saprolegnia hypogyna*,[3] and in a branch from the stalk in *S. monoica*.[4] Fertilisation occurs in several species, among which are *S. monoica*,[5] *Leptolegnia caudata*,[6] *Achlya polyandra*,[7] *A. americana*,[8] *Aphanomyces laevis*,[9] and the dioecious forms of *Dictyuchus*.[10] In others, though antheridia are produced, they do not seem to function.

In *Saprolegnia monoica*[5] the young oogonium contains many nuclei and much cytoplasm, at first occupying the whole of the space. A central vacuole appears, enlarges, and presses the cyto-

[1] Coker, 1923.
[2] Harvey, 1925.
[3] de Bary, 1888.
[4] Pringsheim, 1858.
[5] Claussen, 1908.
[6] Couch, 1932.
[7] Mücke, 1908.
[8] Trow, 1899, 1905; Davis, 1903, 1905.
[9] Kasanowsky, 1911.
[10] Couch, 1926.

plasm and nuclei into a peripheral layer; it seems that some degeneration of the oogonial contents occurs at this stage. The nuclei divide in the peripheral layer, where some of the daughter nuclei break down. Cleavage furrows cut out a number of uninucleate oospheres (fig. 48 *a*); these round off, and lie free in the oogonium, without any residual periplasm.

The young antheridium contains dense cytoplasm with a few nuclei; these divide at the same time as the nuclei in the oogonium.

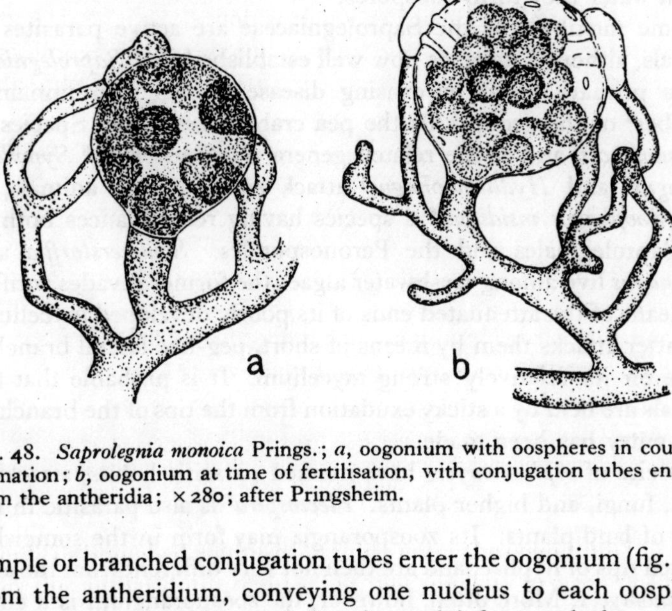

Fig. 48. *Saprolegnia monoica* Prings.; *a*, oogonium with oospheres in course of formation; *b*, oogonium at time of fertilisation, with conjugation tubes entering from the antheridia; ×280; after Pringsheim.

Simple or branched conjugation tubes enter the oogonium (fig. 48 *b*) from the antheridium, conveying one nucleus to each oosphere. Subsequently the male and female nuclei fuse and the oospore develops a smooth colourless wall, rather thick, and somewhat refractive. A thick epispore and a thinner endospore can usually be detected.

The number of oospores formed from one oogonium is not alike in all members of the Saprolegniaceae; in *Achlya*, *Saprolegnia* and *Thraustotheca*, the oogonia contain from few to numerous oospores, but in *Aphanomyces*, *Dictyuchus*, *Geolegnia* and other genera, the oospore is usually solitary. The contents of the oospore are often brown or yellow, owing to the accumulation of fatty material. The ripe oospore contains a large oil drop, sometimes centrally placed, sometimes lying to one side; the position of this globule appears to be a character of systematic importance. Oospores may germinate as soon as they are ripe, or they may do so only after a resting period. It is probable that meiosis occurs at this stage. In germination, the outer layer of the wall splits, a germ tube emerges, and, either directly invades a substratum, or, with or without branching, gives rise to one or more zoosporangia. The oospores of *Aphanomyces euteiches*,[1] a cause of root rot in peas, germinate as soon as they are ripe; they emit germ tubes when placed in nutrient solutions, but in water they form zoospores.

Some members of the Saprolegniaceae are active parasites of animals, although it seems now well established that *Saprolegnia* is not a primary agent in causing disease in fish. An unnamed member[2] of the group kills the pea crab, *Pinnotheres*. Species of *Aphanomyces*, and of the related genera *Sommerstorffia*,[3] *Synchaetophagus*[4] and *Hydatinophagus*[5] attack small aquatic animals, as does *Zoophagus insidians*,[6] a species having resemblances both to the Saprolegniales and the Peronosporales. *Sommerstorffia* and *Zoophagus* live among freshwater algae; the former invades rotifers by means of the attenuated ends of its poorly developed mycelium, the latter attacks them by means of short, peg-like lateral branches borne on its relatively strong mycelium. It is probable that the animals are held by a sticky exudation from the tips of the branches, until entry has been made.

Species of *Aphanomyces* have marked parasitic habits, attacking algae, fungi, and higher plants. *Plectospira*[7] is also parasitic in the roots of land plants. Its zoosporangia may form in the somewhat inflated tips of hyphae, and are then not very different from those of *Aphanomyces*. More often, however, the zoosporangium is a complicated structure, consisting of several swollen hyphae bearing short lateral branches in their turn, the whole tangled together. This sporangium forms by branching from the tip of an ordinary hypha, and it is cut off below by a septum. It recalls the lobulate sporangia formed by some species of *Pythium*. The zoospores

[1] Jones and Drechsler, 1925.
[2] Atkins, 1929.
[3] Arnaudow, 1923.
[4] Apstein, 1911; Valkanov, 1932.
[5] Valkanov, 1932.
[6] Arnaudow, 1923; Sparrow, 1929; Barnes and Melville, 1932.
[7] Drechsler, 1929.

escape from one or more elongated exit tubes put out from the lobes of the sporangium. They are diplanetic in the same manner as the zoospores of *Achlya* and *Aphanomyces*. The sexual organs are of saprolegniaceous type, but the antheridia develop in large numbers so that many are applied to each oogonium. Since the oogonium contains only one oosphere, the wastage must be great.

PERONOSPORALES

Most of the 200 species of the Peronosporales are parasitic in land plants, the mycelium usually inhabiting the intercellular spaces; host cells are entered by means of haustoria. Zoosporangia, or conidia which are modified zoosporangia, provide for asexual reproduction; in many species, the same organ may function as a zoosporangium or as a conidium according to external conditions. Low temperatures and humidity favour the liberation of zoospores, higher temperatures and relative dryness promote the direct germination of the sporangium. In most of the Peronosporales, the sporangiophores emerge from the host plant, while antheridia and oogonia are formed within its tissues. The oogonium contains a single oospore, and periplasm is present.

There are three families, as follows:

Species living as parasites or as saprophytes; sporangiophores differing little from vegetative hyphae	PYTHIACEAE
Species living as parasites	
Sporangia formed in chains on club-shaped sporangiophores crowded beneath the epidermis of the host	ALBUGINACEAE
Sporangia formed singly at the ends of branched sporangiophores which emerge from the host in early stages of development	PERONOSPORACEAE

PYTHIACEAE

The Pythiaceae are characterised by their saprophytic tendencies, the general absence of well-defined sporangiophores, the development of sporangia directly from the mycelium or in succession on a sympodially branched system of hyphae, and the capacity of the sporangia to form germ tubes or liberate zoospores while still attached to the fertile hyphae.

Pythium de Baryanum[1] causes damping off in seedlings; these are attacked at ground level, the tissues are softened and the seedling falls over. The disease is favoured by excessive moisture, crowding and insufficient ventilation. In seedlings, the mycelium inhabits the intercellular spaces of the host, but *P. de Baryanum* also grows readily on animal and vegetable material floating in water, and then forms a strong mycelium in the water surrounding the substratum. The mycelium consists of branched hyphae with thick main strands and thinner lateral branches; the hyphae taper evenly towards their tips, in this respect recalling the hyphae of the higher Chytridiales. Septa appear here and there in old mycelia inside seedlings, and are abundant when mycelium floating in water is forming reproductive organs.

Globular sporangia develop terminally on hyphae projecting from the substratum. If this is of vegetable origin and exposed to the air, the sporangium often functions as a conidium, putting out a lateral process which branches freely and gives rise to a mycelium. If however conditions are wet, and particularly when the fungus is growing in undisturbed cultures on insects in shallow water, zoospores may be produced in abundance.[2] The sporangium puts out a lateral tube (fig. 49 *a*), its end swells (fig. 49 *b*) and the contents of the sporangium show changes indicating that rudiments of zoospores have been formed by cleavage. The contents migrate into a vesicle formed at the end of the tube, and here the organisation of the zoospores is completed. The vesicle bursts, setting free bean-shaped zoospores with two flagella inserted in a lateral depression (fig. 49 *c*). The zoospore germinates by means of a germ tube.

Sexual organs appear when the host has been killed and the fungus is living saprophytically. Usually an oogonium develops terminally on a lateral branch, and an antheridium is cut off at the end of a branch from the stalk of the oogonium (fig. 49 *d–g*). The organs are brought together by the curvature of the hyphae, so that the tip of the antheridium is applied to the side of the oogonium; such an antheridium is described as **paragynous**.

Young gametangia are separated from their stalks by septa (fig. 49 *j, k*); they contain dense cytoplasm and many nuclei. As the oogonium ripens, its contents contract to form a globose oosphere,

[1] de Bary, 1881 i, ii; Butler, E. J., 1907, 1913. [2] Butler, E. J., 1913.

surrounded by scanty periplasm. The nuclei pass to the periphery of the ooplasm, and divide once; all but one of the daughter nuclei degenerate, the survivor being the functional female nucleus. The nuclei in the antheridium also divide once, a single male nucleus surviving the subsequent degeneration. A receptive papilla forms where the antheridium touches the wall of the oogonium. A con-

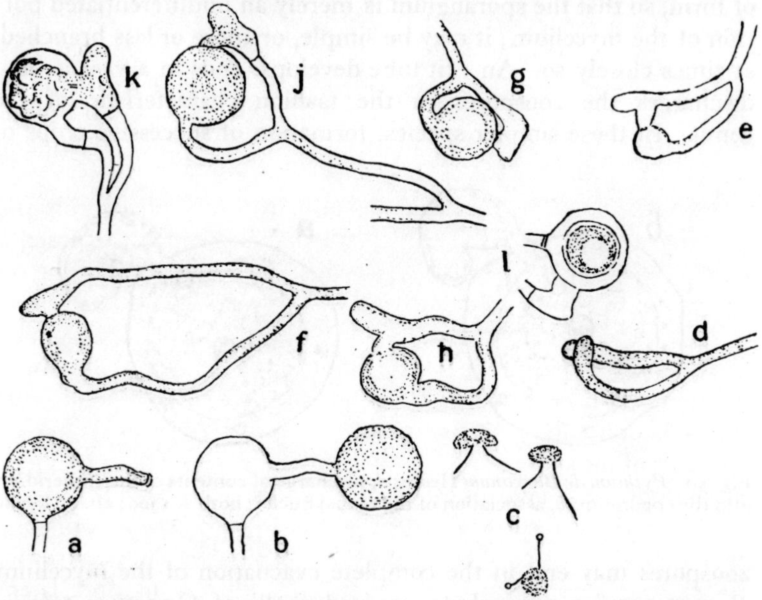

Fig. 49. *Pythium de Baryanum* Hesse; *a*, sporangium with lateral hypha; *b*, formation of the vesicle; *c*, free zoospores, one withdrawing its flagella; *d*, early stage in formation of the sexual organs; *e–k*, successive stages in the development of the sexual organs; *l*, oospore within the oogonium; all ×512; after Butler, E. J.

jugation tube passes from the antheridium into the ooplasm, opens, and emits the male nucleus (fig. 50 *a*), together with some cytoplasm, which apparently mingles with the periplasm.[1] The male and female nuclei associate (fig. 50 *b*) and soon fuse. The oospore then forms a thick, stratified wall, and, after a rest of some months, germinates by means of a germ tube.

Pythium de Baryanum may be taken as a central type of its genus,

[1] Miyake, 1901.

but other species need comment. The sporangia of *Pythium* are not always formed at the end of a hypha. *P. gracile*[1] and other aquatic species are parasites of algae, the mycelium lying in the host cell. Any part of the contents of the mycelium may be converted into zoospores, without any formation of a septum between the fertile and vegetative regions. The hypha undergoes no change of form, so that the sporangium is merely an undifferentiated portion of the mycelium; it may be simple, or more or less branched, at times closely so. An exit tube develops, ends in a vesicle, and discharges the zoospores in the fashion characteristic of the genus. In these simpler species, formation of successive crops of

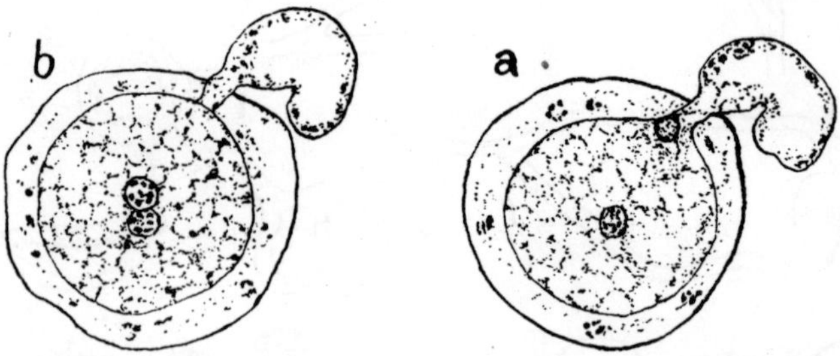

Fig. 50. *Pythium de Baryanum* Hesse; *a*, discharge of contents of the antheridium into the oogonium; *b*, association of the sexual nuclei; both × 1300; after Miyake.

zoospores may end in the complete evacuation of the mycelium. *P. marinum*,[2] a saprophyte in dead thalli of *Ceramium rubrum*, is one of the simpler species of the genus; it is the only well-known member of its family occurring in a marine habitat.

Pythium aphanidermatum[3] and *P. myriotylum*[4] cause rots in the fruits of Cucurbitaceae. Their sporangia form in the ends of more or less closely packed branch systems, the whole complex being cut off by a basal septum; the condition is thus a slight advance on *P. gracile* and its relatives, as there are some indications of the assignment of a special part of the mycelium to the function of zoospore formation. *P. proliferum*[5] (fig. 51*a*) is an aquatic species, forming globular sporangia renewed by

[1] Butler, E. J., 1907.
[2] Sparrow, 1934.
[3] Drechsler, 1925, 1929; Ayyar, 1928.
[4] Drechsler, 1930 ii.
[5] Butler, E. J., 1907.

proliferation. *P. intermedium*[1] (fig. 51 *b*) bears its rounded sporangia in basipetal chains, an unusual feature in the genus; this species may be a link with the Albuginaceae. It occurs as a saprophyte in the soil, and as a parasite in the prothalli of Pteridophyta.

The capacity for direct germination possessed by the sporangia of *Pythium* seems to be an adaptation to a subaerial habitat where there may not be sufficient water for the successful distribution of zoospores. *P. ultimum*,[2] perhaps the commonest species of the genus occurring in soil, is highly specialised in this respect, since its sporangia always behave as conidia.

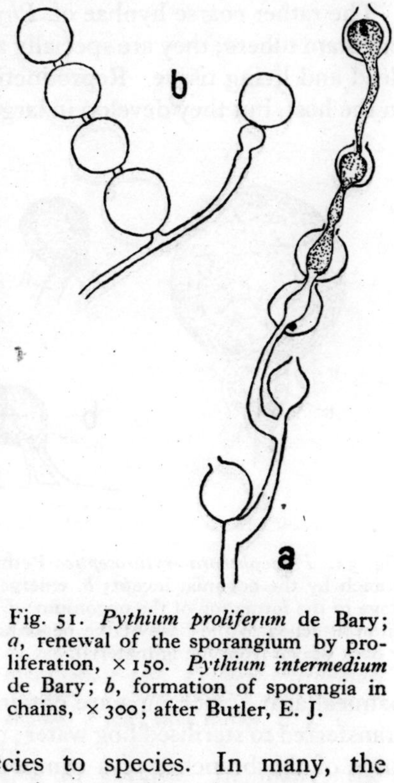

Fig. 51. *Pythium proliferum* de Bary; *a*, renewal of the sporangium by proliferation, × 150. *Pythium intermedium* de Bary; *b*, formation of sporangia in chains, × 300; after Butler, E. J.

The zoospore of *Pythium* is bean-shaped, with two lateral flagella;[3] in species parasitic in algae, it germinates as in *Lagenidium*, by the passage of its contents into the host cell; in other species, a germ tube is produced.

The sexual organs of *Pythium* seem to be fairly uniform in their characters, though there are differences in detail from species to species. In many, the agreement with *P. de Baryanum* is very close; in some, as in *P. angustatum*, several antheridia of diclinous origin are applied to one oogonium, which, moreover, may contain more than one oospore; in others, as in *P. Artotrogus*, the oogonia bear spinous outgrowths.

The genus *Phytophthora*[4] is closely allied to *Pythium*; it differs

[1] Butler, E. J., 1907. [2] Trow, 1901.

[3] Atkinson, 1894 i; Butler, E. J., 1913.

[4] de Bary, 1876; Butler, E. J., 1907; Butler, E. J. and Kulkarni, 1913; Dastur, 1913; Pethybridge, 1913, 1922; Rosenbaum, 1917; Sherbakoff, 1917; McRae, 1918; Fitzpatrick, 1923; Godfrey, 1923; Leonian, 1925–34; Drechsler, 1931.

in the more substantial appearance of the contents of the hyphae, in the looser branching of the mycelium, and in the preference for terrestrial flowering plants as hosts. Old hyphae of *Phytophthora* may be irregularly septate. Like those of *Pythium*, the species of *Phytophthora* are able to live as saprophytes.

The rather coarse hyphae of *Phytophthora erythroseptica*[1] occur in potato tubers; they are specially abundant at the margin between dead and living tissue. Reproductive organs do not seem to form in the host, but they develop in large numbers in cultures grown on

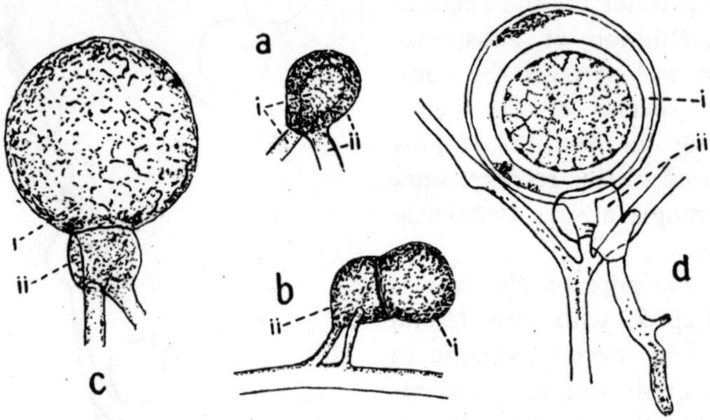

Fig. 52. *Phytophthora erythroseptica* Pethyb.; *a*, penetration of the antheridial branch by the oogonial incept; *b*, emergence of the young oogonium; *c*, later stage in the formation of the oogonium; *d*, ripe oospore, with the empty antheridium attached to the base of the oogonium; i, oogonium; ii, antheridium; all ×860; drawn from living material.

oatmeal agar. Sporangia are obtained when vigorous mycelium is transferred to sterilised bog water; they develop in succession at the ends of the branches of a sympodial system. The sporangia are pyriform; they usually germinate by means of a germ tube, thus acting as conidia.

Antheridia and oogonia arise on short lateral branches, which in early stages of development cannot be distinguished from ordinary vegetative hyphae. They grow towards one another, and the tip of one, the **oogonial incept**, penetrates the other; in the latter, dense material accumulates and an antheridium is cut off by one or two septa. The oogonial incept grows through the antheridium (fig. 52 *a*,

[1] Pethybridge, 1913; Murphy, 1918.

b); it emerges on the other side to form a rounded swelling (fig. 52 *c*), containing cytoplasm and nuclei, which is the young oogonium. Its funnel-shaped base lies within the antheridium which is described as **amphigynous**, and it is separated from its stalk by a thick irregular plug. Many of the nuclei degenerate in the young oogonium; the survivors divide once, and, after further disintegration, a single nucleus remains. The contents of the oogonium are now organised into a solitary oosphere, surrounded by a little periplasm.

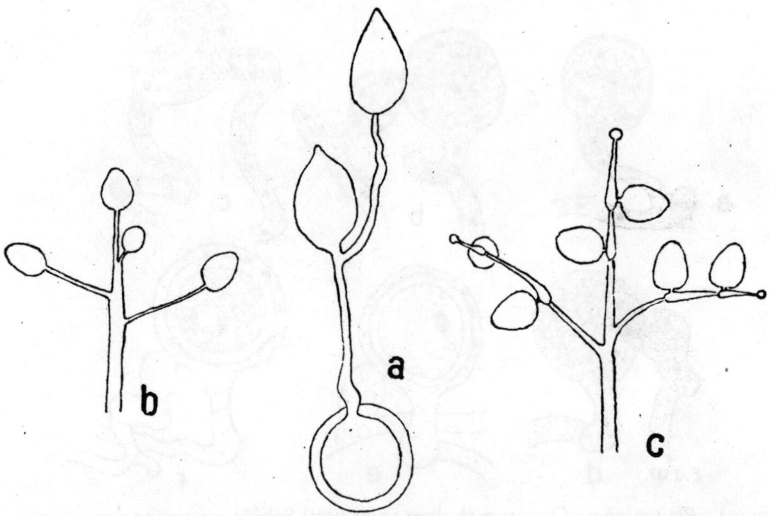

Fig. 53. *Phytophthora palmivorum* (Butl.) Butl.; *a*, germination of oospore, × 400; after Butler, E. J. *Phytophthora infestans* de Bary; *b*, *c*, stages in development of the sporangiophore, × 200; after de Bary.

The antheridial nuclei also divide once, and then all but one perish. The wall of the oogonium bulges out inside the antheridium to form a large receptive papilla; as this is withdrawn, a conjugation tube follows it, and delivers the male nucleus and some cytoplasm into the oosphere; by this time the periplasm has nearly disappeared. A wall forms round the oospore, the male and female nuclei fuse, and the oospore (fig. 52 *d*) rests in a uninucleate condition. Its further history is unknown in *Phytophthora erythroseptica*, but in *P. palmivorum*[1] and other species, the germinating oospore

[1] Butler, E. J., 1907; Fitzpatrick, 1923.

gives rise to a short hyphal system on which sporangia are borne (fig. 53 *a*).

Phytophthora infestans[1] is the cause of the well-known potato blight. It first became notorious in 1845–6, when it caused a potato famine in Ireland, and so contributed to the repeal of the Corn Laws. The mycelium is widely distributed in the host; in damp air, sporangia borne on sympodially branched sporangiophores (fig. 53 *b*, *c*) appear in abundance from the leaves; there is a charac-

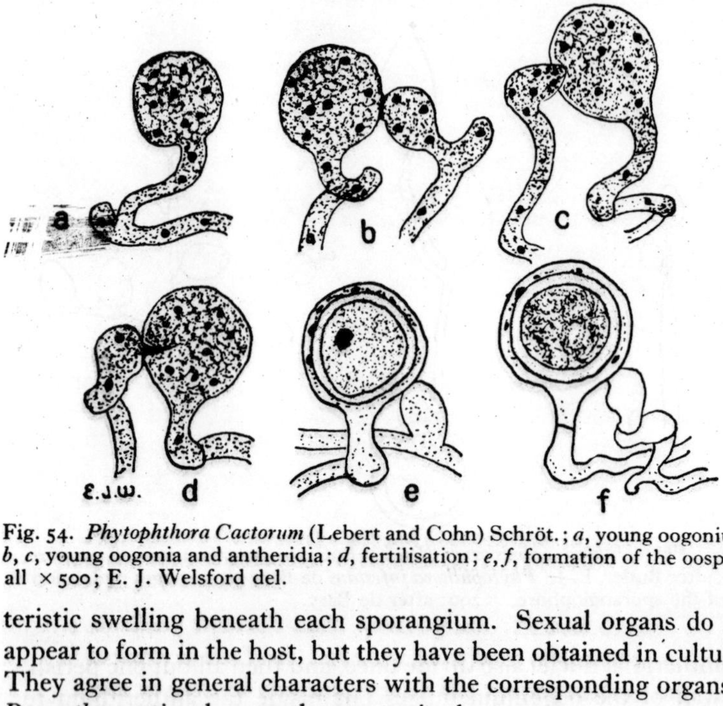

Fig. 54. *Phytophthora Cactorum* (Lebert and Cohn) Schröt.; *a*, young oogonium; *b*, *c*, young oogonia and antheridia; *d*, fertilisation; *e*, *f*, formation of the oospore; all × 500; E. J. Welsford del.

teristic swelling beneath each sporangium. Sexual organs do not appear to form in the host, but they have been obtained in culture.[2] They agree in general characters with the corresponding organs of *P. erythroseptica*, but parthenogenesis also occurs.

In *Phytophthora Cactorum*[3] (fig. 54) the sexual organs usually develop and function like those of *Pythium de Baryanum*; occasionally however they behave like those of *Phytophthora erythroseptica*. As in *Pythium*, the partly differentiated zoospores may pass into a vesicle before they escape, but the vesicle may be evanescent. There is a similar method of discharge in *Phytophthora cryptogea*,[4]

[1] de Bary, 1876. [2] Clinton, 1911; Pethybridge and Murphy, 1913.
[3] Lafferty and Pethybridge, 1922–4. [4] Pethybridge and Lafferty, 1919.

a species often renewing its sporangia by proliferation. *Phytophthora Faberi*,[1] a tropical species parasitic on *Theobroma Cacao* is dioecious; the emergence of its zoospores is favoured by light, by a supply of free oxygen, and by moderate temperatures.

Trachysphaera fructigena[2] is also a tropical species, attacking coffee and cocoa. The gametangia unite in the same manner as those of *Phytophthora erythroseptica*; the oogonia bear numerous, irregular, rounded outgrowths. The sporangia are spherical, with many spines projecting from their walls, and quite distinct from the sporangia of other members of the family. Sometimes, a sporangiophore ends in a solitary sporangium, but more often it ends in a swelling bearing either a whorl of sporangia or a number of secondary fertile branches. Sporangia seem always to function as conidia. They are produced in great numbers, forming thick, pinkish-brown incrustations on the surface of the fruit of the host.

The species of *Pythiogeton*[3] occur on vegetable material submerged in fresh water. The mycelium is delicate, but fairly copious. Although a sporangium resembling that of *Pythium* may be produced, the characteristic sporangium of *Pythiogeton* is an elongated sac attached to the supporting hypha so that its longest axis is transverse to the general direction of the stalk. When the sporangium is ripe, its contents pass, without any apparent differentiation, into a thin-walled vesicle; they are shot out from this, and whilst lying free in the water, break up into zoospores like those of *Pythium*. When *Pythiogeton* is grown on agar, the contents of the sporangium are shot into the air, suddenly, and with sufficient force to carry them several millimetres. Little is known of the sexual organs; the oogonium is globose or somewhat angular, and contains a single, thick-walled oospore. The stalk of the oogonium coils round the antheridial branch, a condition present in some species of *Pythium*.

Pythiomorpha gonapodioides[4] is another aquatic species; its hyphae show irregular variations in calibre, giving the mycelium a characteristic aspect. The sporangia are very like those of *Phytophthora*; they are renewed freely by proliferation, and liberate zoo-

[1] Ashby, 1922; Gadd, 1924.
[2] Tabor and Bunting, 1923.
[3] von Minden, 1916; Drechsler, 1932; Sparrow, 1936 i.
[4] Petersen, 1910; Kanouse, 1925, 1927; Drechsler, 1932; Forbes, 1935.

spores of the usual kind. It seems probable that this fungus is an aquatic species of *Phytophthora*.

Stigeosporium Marattiacearum,[1] the fungal member of the mycorrhiza of the Marattiaceae, has some resemblance to *Phytophthora*, and may be a member of the Pythiaceae.

ALBUGINACEAE

The Albuginaceae include some six species, all parasitic in flowering plants. They are characterised by the production of sporangia in chains.

Cystopus candidus[2] attacks *Capsella Bursa pastoris* and other Cruciferae, forming shining white patches on the stems and leaves, and causing deformation of the flowering shoots and fruits. The mycelium consists of rather coarse hyphae living in the intercellular spaces of the host and bearing numerous small, rounded haustoria (fig. 55 *a*). Hyphae collect beneath the epidermis, branch profusely, and give rise to a closely packed series of club-shaped sporangiophores, arranged in palisade fashion, perpendicular to the surface of the leaf (fig. 55 *b*, *c*). Dense cytoplasm, containing five or six nuclei, accumulates in the apex of the sporangiophore, delimited below by a broad septum. This is the first sporangium of a series each formed in the same way. The septum is gelatinous in consistency, shrinking as the sporangia ripen. As the chains of sporangia elongate, they press on the epidermis above until it gives way. The gelatinous pads between the sporangia disintegrate in moist air, so that the sporangia separate; they are blown away by the wind or washed off by water. The oldest sporangium of a chain develops a thicker wall than its successors. It takes most of the pressure set up by the resistance of the epidermis. Sometimes in *Cystopus candidus*, and frequently in related species, this sporangium fails to develop further.

In the presence of water, at low temperatures, the sporangia set free a few biflagellate zoospores; when conditions are relatively dry, germ tubes are produced.

Antheridia and oogonia appear after sporangia have developed. They are buried in the host tissue, and tend to form in the flowers and in the axes of the inflorescences rather than in foliage leaves.

[1] West, 1917.

[2] *Cystopus candidus* (Pers.) Lév. = *Albugo candida* (Pers.) O. Ktze.

The details of the sexual process[1] differ little from those in *Pythium de Baryanum*; periplasm is more abundant, and the functional female nucleus lies in a deeply staining mass of cytoplasm occupying the centre of the ooplasm. Divisions take place in the oospore soon after the male and female nuclei have fused, so that there are about

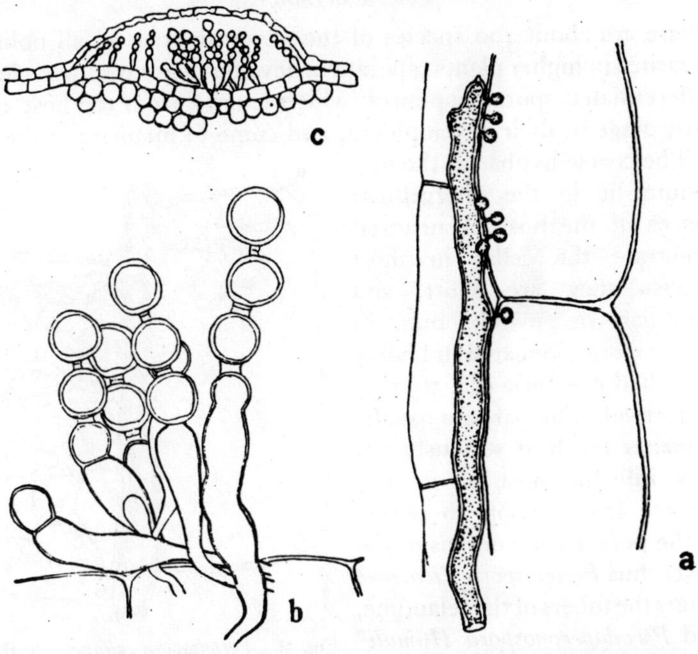

Fig. 55. *Cystopus candidus* (Pers.) Lév.; *a*, portion of mycelium bearing haustoria, ×400; after de Bary; *b*, group of sporangiophores with chains of sporangia, ×450; *c*, diagram showing position of sporangiophores beneath the raised epidermis of the host, ×75.

thirty-two nuclei in the resting oospore. There is evidence that meiosis occurs when the fusion nucleus divides. Zoospores are set free when the oospore germinates.

In allied species, the cytological details of the sexual process differ from those of *Cystopus candidus*. In *Albugo Bliti*[2] and *A. Portulacae*,[2] the oogonia contain about 250 nuclei, the antheridia about thirty-five. Ooplasm is formed in the oogonium,

[1] Wager, 1896; Davis, 1900; Krüger, 1910.
[2] Stevens, 1899, 1901, 1904.

and the nuclei pass into the periplasm, where they divide. About 50 nuclei pass back into the ooplasm, and, after a second nuclear division, many of the daughter nuclei degenerate. A number of nuclei enter from the antheridium, and the male and female nuclei then fuse in pairs.

PERONOSPORACEAE

There are about 100 species of the Peronosporaceae, all obligate parasites in higher plants; sporangia develop in groups upon clearly differentiated sporangiophores, which emerge from the host at an early stage in their development, and come to maturity in the air.

The coarse hyphae of the mycelium lie in the intercellular spaces of the host. Haustoria[1] penetrate the cells; in most species they are short, and vesicular or ovoid, but, in *Peronospora*, elongated and richly branched haustoria (fig. 56) may be formed. The parasite usually damages the host seriously. It may kill the host or it may survive from season to season in the perennating organs of the host; thus *Peronospora Ficariae*[2] enters the tubers of the celandine, and *Pseudoperonospora Humuli*[3] passes the winter in the stools of the hop.

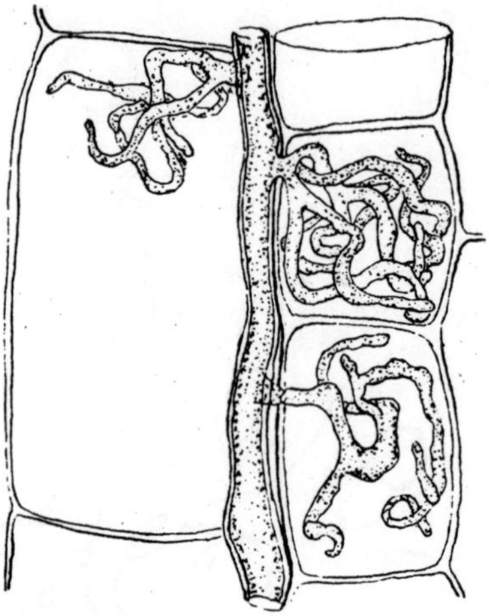

Fig. 56. *Peronospora calotheca* de Bary; richly branched haustoria, ×400; after de Bary.

Hyphae accumulate beneath the stomata, branch, and give rise to the rudiments of the sporangiophores; these emerge in groups from the stomata, elongate, usually branch, and develop sterigmata on the ends of the branches. The tip of the sterigma swells, nuclei and cytoplasm pass into the swelling, and, when the sporangium has reached full size, communication with the sporangiophore is stopped by the blocking of the lumen of the sterigma.[4] The sporangiophores, with their sporangia, form conspicuous whitish, greyish

[1] de Bary, 1863 i.
[2] Krüger, 1910.
[3] Ware, 1926.
[4] Rostowzew, 1903.

or bluish patches on the host, differing in their downy appearance from the shining pustules caused by the Albuginaceae.

The sporangium falls when it is ripe; it may liberate zoospores resembling those of *Pythium* or it may germinate by means of a germ tube.

The sexual process[1] in the Peronosporaceae resembles closely that of *Pythium*. The wall of the oogonium may disappear soon after the oospore is ripe; in *Plasmopara* it persists after the tissue of the host has rotted; in *Sclerospora* the thick wall of the oogonium and that of the oospore unite to form a common envelope.

Basidiophora entospora[2] (fig. 57 *a*) attacks *Erigeron canadense*; the elongated club-shaped sporangiophores bear a few sterigmata at the apex; each supports a broadly ovoid sporangium from which the zoospores escape, one by one.

Plasmopara viticola[3] causes downy mildew of the vine. The sporangiophore consists of a main axis, bearing four to eight lateral branches; sporangia develop singly on short sterigmata. They soon liberate zoospores if placed in water, but these zoospores only swim for about twenty minutes before they settle down and produce a germ tube; this short period of activity suggests that the tendency to produce zoospores is weakening. *P. densa*[4] does not form zoospores; the contents of the sporangium emerge as a whole, form a wall, and develop a germ tube.

Bremia lactucae[5] (fig. 57 *b*) occurs on lettuce and other members of the Compositae. The sporangiophore has a long main axis with dichotomously or trichotomously arranged branches which may branch again. The fertile axes end in saucer-like expansions (fig. 57 *c*), with a few sterigmata round the margin; each sterigma bears one sporangium. In darkness, at low temperatures, and with sufficient moisture, the sporangia may produce zoospores, but usually they germinate directly by means of a germ tube arising from the apex of the sporangium.

Pseudoperonospora Cubensis[6] attacks and kills cucumber plants. The sporangiophores are like those of *Peronospora*, but the grey or smoky violet sporangia may give rise to zoospores, or may act as

[1] Wager, 1900; Stevens, 1902; Rosenberg, 1903; Ruhland, 1904.
[2] Roze and Cornu, 1869.
[3] Farlow, 1876.
[4] Ruhland, 1904.
[5] Schweizer, J., 1918; Milbraith, 1923.
[6] Rostowzew, 1903.

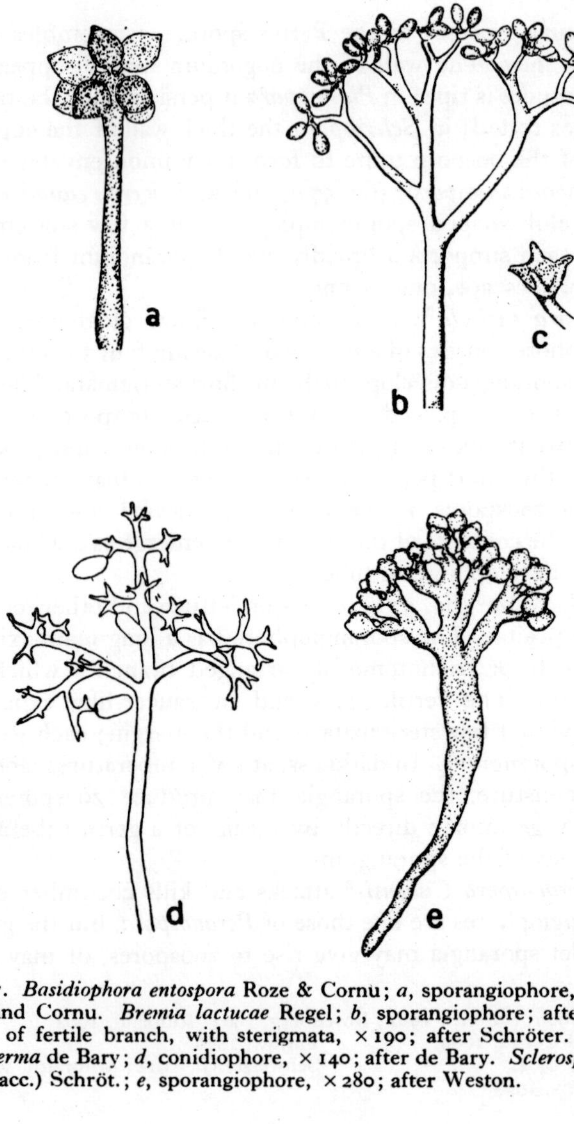

Fig. 57. *Basidiophora entospora* Roze & Cornu; *a*, sporangiophore, × 225; after Roze and Cornu. *Bremia lactucae* Regel; *b*, sporangiophore; after Milbraith; *c*, end of fertile branch, with sterigmata, × 190; after Schröter. *Peronospora leptosperma* de Bary; *d*, conidiophore, × 140; after de Bary. *Sclerospora graminicola* (Sacc.) Schröt.; *e*, sporangiophore, × 280; after Weston.

conidia, the germ tube proceeding from the apex of the sporangium, as in *Bremia*, or from some other point of the surface, as in *Peronospora*. The sporangia of *Sclerospora graminicola*[1] (fig. 57 *e*) show similar behaviour. The sporangiophores of this species develop and mature at night, possibly in relation to the better supply of moisture on the surface of the host at that time.

The tendency for the sporangium to become a conidium culminates in *Peronospora*, in which zoospores are unknown. The conidiophores (fig. 57 *d*) frequently show forked branching, and a main axis is not clearly recognisable once branching has begun. The conidia are often bluish or violet; they develop in the same way as the sporangia of other members of the family. They fall from their supports, and germinate by the protrusion of a germ tube from any point of the surface.

ZYGOMYCETES

The outstanding characteristic of the Zygomycetes is their method of sexual reproduction by the union of similar or slightly dissimilar gametangia. Asexual reproduction is by non-motile spores developed in sporangia; usually the sporangium contains many spores, but small sporangia, known as **sporangiola**, also occur, and contain one or a few spores each, or the sporangia may be further modified to form conidia.

The Zygomycetes include about 300 species, and may be divided as follows:

Accessory multiplication by sporangiospores	MUCORALES
Accessory multiplication by conidia.	ENTOMOPHTHORALES

MUCORALES

The Mucorales[2] comprise about 200 species; they are saprophytes on many kinds of substrata, or, less often, they are parasites, chiefly on other members of the same alliance.

The mycelium is usually well developed, consisting of richly branched aseptate hyphae lying on and in the substratum. A young mycelium shows a very regular pattern, composed of radiating main hyphae with evenly disposed lateral branches. As the myce-

[1] Stevens, 1902; Kulkarni, 1913; Weston, 1923, 1924.

[2] Lendner, 1908; Moreau, 1913 i; Namyslowski, 1920; Naoumoff, 1924; Nadson and Phillipov, 1925 i; Satina and Blakeslee, 1925–30; Zycha, 1935.

lium ages, the regular pattern is lost, and septa appear here and there in the hyphae. Except in the Mortierellaceae, hyphal anastomoses are uncommon. In old hyphae, the contents may contract to form intercalary chlamydospores; these are common in *Mucor hiemalis*, and particularly so in *M. racemosus*, where they also form in the sporangiophores. A mycelium immersed in a fluid, and poorly supplied with elementary oxygen, may break up into yeast-like bodies, multiplying by budding and causing energetic alcoholic fermentation. The yeast-forms grow into ordinary mycelia if they are transferred to a solid medium.

Multiplication is very largely by means of non-motile spores produced in stalked sporangia; they are carried by the air, and form germ tubes on germination. The young sporangiophore arises as a branch of the mycelium. Its apex swells, receives nuclei, cytoplasm and food material, and is cut off by a wall, forming the sporangium. In the sporangia of the Mucoraceae, and in the large sporangia of the Choanephoraceae, the septum is dome-shaped from the first, and is called the **columella**. It remains as a conspicuous structure after the outer wall has broken away, and the spores are shed. The septum is flat and does not form a columella in the sporangia of the Mortierellaceae, nor in sporangiola and their conidium-like derivatives.

Sexual reproduction is by the fusion of multinucleate gametangia, formed in contact where the parent hyphae have met. Each gametangium is seated on a short branch known as a **suspensor**. Union of the gametangia results in the formation of a **zygospore**; this produces a thick wall which may be smooth or sculptured, and is sometimes surrounded by a weft of vegetative hyphae. Occasionally gametangia develop without fusion as **azygospores**.

The sexual method of reproduction was first observed in *Sporodinia grandis*,[1] a form occurring on the fructifications of Autobasidiomycetes, and zygospores were found to be readily obtained in this species. In others, however, production of zygospores, though occasionally observed, proved exceedingly erratic; particular chemical substances or degrees of concentration were stated to be favourable to zygospore formation, but none of them proved reliable or stood the test of continued experiment.

The necessary condition was discovered by Blakeslee.[2] In the

[1] Ehrenberg, 1820; de Bary, 1864. [2] Blakeslee, 1904.

course of his investigations at Harvard he observed that, while zygospores of certain species, such as *Sporodinia grandis*, developed readily in pure culture from a single spore, yet in *Rhizopus nigricans* and many other forms, zygospores were never obtained under such conditions, but only when a mass of spores from a culture which had borne gametangia was used for inoculation. In due course it was demonstrated that such species consist of two strains, or kinds of thalli, which, when grown apart, produce only sporangia, but which form zygospores in contact. In some species there are visible differences between the two strains, one being rather more luxuriant than the other; to the relatively well-developed strain the term (+) was applied, and the term (−) to the strain with which it produced zygospores. In other species, though a physiological difference evidently exists, since neither strain can form zygospores alone, yet any morphological distinction is lacking, and a (+) or a (−) strain can only be defined as one which produces zygospores in conjunction with the other.

Species in which conjugation depends on the interaction of two thalli are said to be heterothallic, while those which give rise to zygotes in mycelia from a single sporangiospore, and consist, therefore, of only one kind of thallus, are described as homothallic.

Though a comparison with incompatibility in other fungi merits consideration, it seems probable that the physiological difference between (+) and (−) strains in this alliance is equivalent to the difference in sex in other plants and animals, the homothallic mycelium being bisexual, like a fern prothallus, and the heterothallic, unisexual. Since there is no difference between the sexual organs in the heterothallic Mucorales, the distinction of sex, if such it be, is expressed in them in the simplest possible form, as the capacity to produce zygotes with a different, but not with a similar individual. Conjugation does not invariably occur when the appropriate strains are brought together; poor nutrition and unsuitable temperature may prevent the formation and union of the gametangia. Moreover, in *Mucor Mucedo* and other species, prolonged cultivation under unfavourable conditions may cause the loss of conjugation, so that **neutral** races are developed, unable to respond either to (+) or to (−) strains.

A response may be observed not only between the (+) and (−) strains of the same species, but between (+) and (−) strains of different species and even of different genera. Between closely related species, hybrid zygospores have been obtained, but they have not given rise to viable mycelia. When the relationship is more remote, gametangia are seldom cut off, and conjugation does not take place. The reaction has been used to discriminate (+) and (−) strains in forms in which no morphological distinction exists. Imperfect attempts at response may occur between a homothallic species and the (+) and (−) strains of a heterothallic species. Thus, when the homothallic *Absidia spinosa* was hybridised with a heterothallic form known as *Mucor V*, the smaller of the gametangia of *Absidia* reacted exclusively with those of the (+) strain, and the larger with those of the (−) strain of *Mucor V*. Differences of size indicate the sex of the gametes, not of the organs which bear them; here no gametes are differentiated, but it is perhaps justifiable, in the absence of other evidence, to infer that the larger gametangium, like the larger gamete, is female. If so, the above observations connect the heterothallism of *Mucor V* and its allies with the ordinary, morphological differences of sex, and suggest that the (−) strain may properly be described as male, and the (+) strain as female. The results of biochemical experiments,[1] though not decisive, point to a similar conclusion.

On the germination of the zygospore a sporangium is usually formed at the end of a short germ tube; the spores produced therein are the first stage of the new gametophyte.

The Mucorales may be divided as follows:

- Sporangia globose to ovoid, usually containing numerous spores, sometimes one or a few; zygospores formed from the whole of the two gametangia
  - Columella present, zygospore naked or invested by outgrowths from its own wall or from those of the suspensors
    - Principal sporangia containing numerous spores
      - Sporangiola, if any, developed on lateral branches of principal sporangiophores — MUCORACEAE
      - Sporangiola developed independently — CHOANEPHORACEAE
    - Monosporous sporangiola (conidia) only — CHAETOCLADIACEAE

[1] Satina and Blakeslee, 1925–30; Burgeff and Seybold, 1927.

Columella absent, zygospore surrounded by a dense weft of vegetative hyphae	
Sporangia or zygospores formed singly	**MORTIERELLACEAE**
Sporangia or zygospores formed in groups	**ENDOGONACEAE**
Sporangia elongated, springing from a swollen head, each containing a single row of spores; zygospore an outgrowth from one or both gametangia	**CEPHALIDACEAE**
Sporangia absent, replaced by a simple conidial apparatus; zygospore wall may be free from the walls of the gametangia	**ZOOPAGACEAE**

MUCORACEAE

The sporangiophores arise as aerial branches from the mycelium; they are negatively hydrotropic, thus emerging from a wet substratum, negatively geotropic, and, sometimes, positively phototropic to a marked degree. The simplest sporangiophores are erect hyphae each bearing one terminal sporangium; this condition is found in the longer fertile axes of *Mucor Mucedo*, and especially in *Phycomyces Blakesleeanus*, where the robust sporangiophore may be thirty centimetres in height. In most Mucoraceae, however, the sporangiophore branches. Cymose and racemose branch systems occur in many species of *Mucor*; in *Thamnidium elegans* the erect main hypha bears lateral tufts of forked branches, each ultimate branchlet ending in a small sporangium or in a sporangiolum; in *Sporodinia grandis* the fertile apparatus consists of a repeatedly dichotomising branch system.

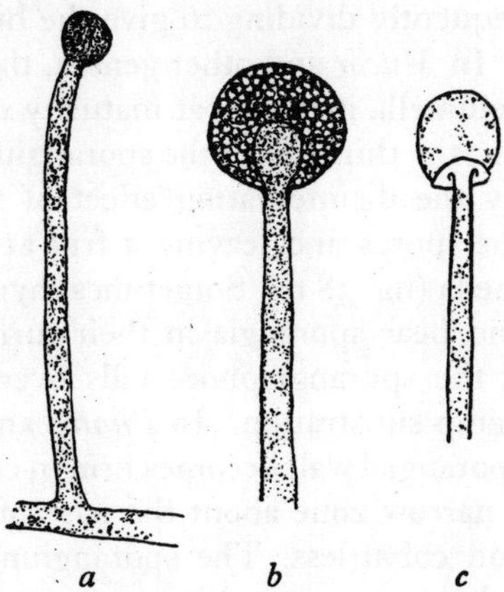

Fig. 58. *Mucor hiemalis* Wehm.; *a*, young sporangiophore, before the sporangium is cut off; *b*, mature sporangium; *c*, sporangiophore and columella after the spores are shed; ×280; H. S. Williamson del.

The apex of a young fertile branch enlarges to form the sporangium (fig. 58 *a*), and becomes crowded with cytoplasm, nuclei and food material. It is cut off by the dome-shaped columella (fig. 58 *b*), the position of which is at first indicated by a series of vacuoles; a plane wall is never present. Very soon, sometimes before the formation of a columella is complete, furrows from the periphery cut centripetally into the cytoplasm. They meet and fuse with furrows starting from the columella, thus cutting the spore mass into blocks of variable size. These divide further and grow, nuclear division sometimes occurring inside them. Contraction follows, the spores round up, and each develops a wall. Some residual mucilaginous material may lie between the spores.

In *Sporodinia grandis* the process of spore formation is abbreviated, the protoplasm being cut into relatively large blocks which at once round up to form spores; they contain many nuclei, with little reserve material, and are short-lived. In *Mucor Mucedo*, *Rhizopus nigricans* and others, segmentation is carried further, forming smaller, but still multinucleate spores. In *Pilobolus crystallinus* cleavage continues until the protoplasm is divided into uninucleate masses; these grow and become multinucleate, subsequently dividing to give the binucleate spores.[1]

In *Mucor* and other genera, the quantity of fluid in the stalk and columella increases at maturity so that considerable pressure is set up; the thin wall of the sporangium bursts, a process often favoured by the disintegrating effect of moist air around it, setting free the spores and leaving a frill at the base of the distended columella (fig. 58 *c*). Sometimes, hyphae may grow from the columella and bear sporangia in their turn; this is specially liable to occur if the sporangiophore falls over so that the columella touches a damp substratum. In *Pilaira* and *Pilobolus*, the upper part of the sporangial wall becomes firm in consistency and dark-coloured, only a narrow zone about the base of the columella remaining delicate and colourless. The sporangium of *Pilobolus* contains a layer of gelatinous material between spores and the wall, especially at the base of the latter; this material swells as water is absorbed from the substratum. The delicate portion of the wall ultimately gives way under the pressure, and the upper firm part of the wall, together with the mass of spores, is violently shot off; this will take place even in very dry air.[2] Large vesicles develop on the sporangiophore, one at the base, and one just below the sporangium; the upper appears

[1] Harper, 1899 ii; Schwarze, 1922.

[2] We are indebted to Miss E. Green, M.Sc., for the observation that both *Mucor* and *Pilobolus* will discharge their spores in a desiccator.

to act as an ocellus,[1] assisting in the orientation of the positively phototropic sporangiophore, ensuring that the mass of spores is aimed towards the light.[2] Dehiscence occurs in similar fashion in *Pilaira*, but it is less vigorous, so that the hard upper part of the wall is simply lifted, exposing the spores.

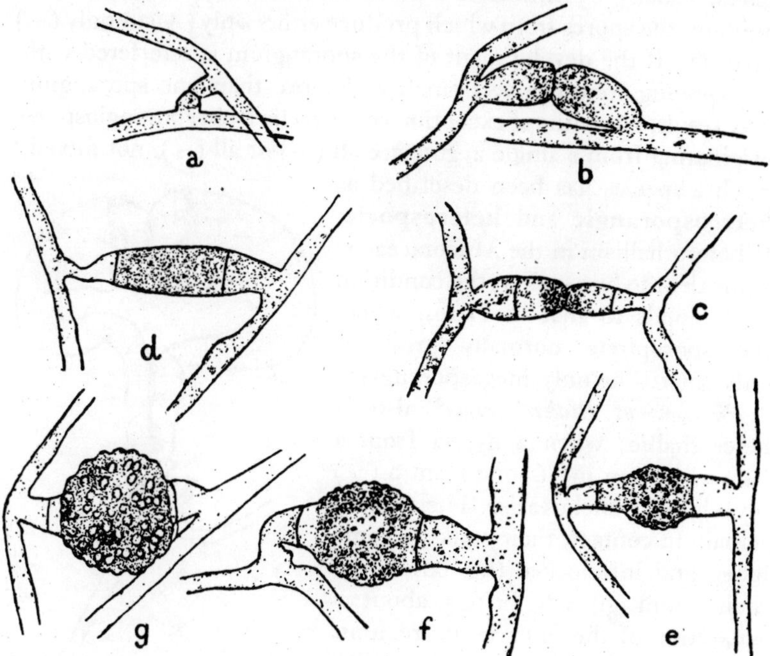

Fig. 59. *Mucor hiemalis* Wehm.; *a*, *b*, contact and growth of progametangia; *c*, progametangia divided to form gametangia and suspensors; *d*, conjugation; *e*, *f*, *g*, development of zygospore; ×280; H. S. Williamson del.

In *Mucor hiemalis*[3] and *M. Mucedo* when the hyphae of opposite strains cross or touch one another, branches develop on contact and, as they grow, push the parent filaments apart. Each branch enlarges to form a club-shaped **progametangium** (fig. 59 *a*, *b*). This is divided by a transverse septum into a terminal gametangium and a basal suspensor (fig. 59 *c*). The gametangia are more or less equal in size; in due course open communication between them is

[1] Buller, 1921; Ingold, 1932 [2] Parr, 1918. [3] Blakeslee, 1904.

afforded by the dissolution of the intervening wall (fig. 59 *d*). The contents mingle, the nuclei fuse in pairs, the wall becomes greatly thickened and differentiated into layers, and the formation of the zygospore is complete (fig. 59 *e*, *f*, *g*). After a period of rest lasting from five to nine months, the zygospore of *Mucor Mucedo* germinates, usually by means of a germ tube bearing a terminal sporangium, the spores from which produce either only (+) or only (−) mycelia. If the development of the sporangium is interfered with, the sporangiophore may branch and more than one sporangium may result, but without exception the mycelia from sporangiospores originating from a single zygote are all (+) or all (−), not mixed.[1] Such a species has been described as **heterosporangic** and **heterosporic**. If heterothallism in the Mucoraceae is equivalent to bisexuality, the condition corresponds to that in *Taxus*, where the sporophyte normally produces only micro- or only megasporangia.

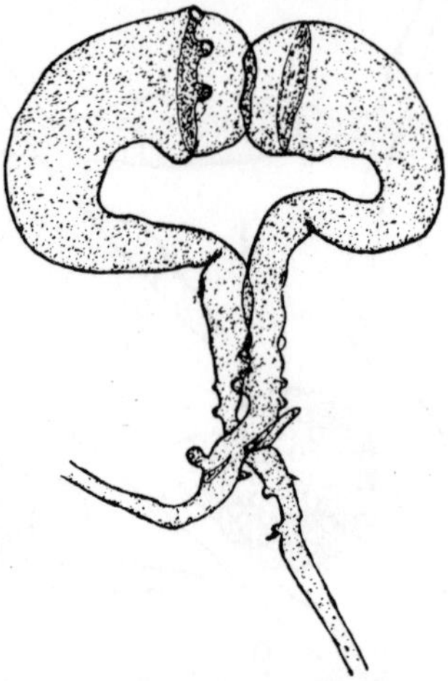

Fig. 60. *Phycomyces Blakesleeanus* Burgeff; interlocked branches bearing suspensors and gametangia, ×60; H. S. Williamson del.

Phycomyces Blakesleeanus[2] also is heterothallic; when a hypha from a (+) mycelium meets one from a (−) mycelium, branches arise. Their apices remain in contact, their stalks become lobed and interlocked (fig. 60), while their rapid growth brings about a separation of the intervening regions. Gametangia are cut off and fuse. The zygospores (fig. 61) are protected by dichotomously branched outgrowths arising from the suspensors, one suspensor being more energetic in this respect than the other. Germination takes place after some months, and the germ tube gives rise to a sporangium in which both (+) and (−) spores are formed, as well as spores which produce homothallic mycelia distinguished by weak growth, contorted aerial branches, and the occasional production of homothallic zygospores.[1]

[1] Blakeslee, 1906 i.

[2] Blakeslee, 1906 i; Burgeff, 1914, 1915, 1925, 1928; Orban, 1919; Walter, 1921; Baird, 1924.

Both the sporangia of the homothallic mycelium and those derived from its zygospores produce spores which give rise to (+) or (−) or homothallic mycelia.

Pilobolus crystallinus[1] is another heterothallic species. Conjugation occurs as in *Phycomyces* but it is not accompanied by the development of outgrowths from the suspensors. The zygospore has a smooth wall, and, unlike the zygospores of most members of the Mucorales, germinates without a resting period.

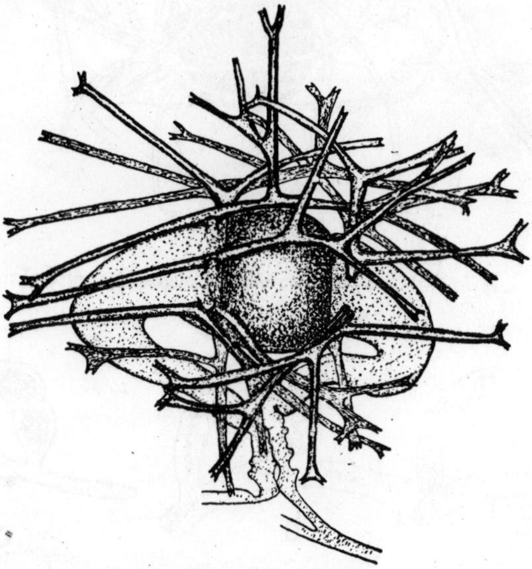

Fig. 61. *Phycomyces Blakesleeanus* Burgeff; mature zygospore, ×60; H. S. Williamson del.

The genus *Absidia*[2] contains both homothallic and heterothallic species. The zygospores (fig. 62) are surrounded loosely by curved appendages arising from the suspensors, or occasionally from one suspensor only. In *Absidia*, and in *Rhizopus*, the fungus spreads over the substratum by means of prostrate hyphae, which bear tufts of rhizoids at intervals, and also carry the short sporangiophores. The latter develop in tufts in *Rhizopus*, but in *Absidia* they are usually scattered in distribution.

Parasitella simplex[3] resembles closely in morphology the smaller

[1] Krafczyk, 1931. [2] Blakeslee, 1904, 1915, 1920; Lendner, 1908, 1910, 1923.
[3] Bainier, 1903; Burgeff, 1920 ii, 1924, 1930.

species of *Mucor*, but differs in its manner of nutrition, since it is a parasite on other members of the family. After a hypha of *Parasi-*

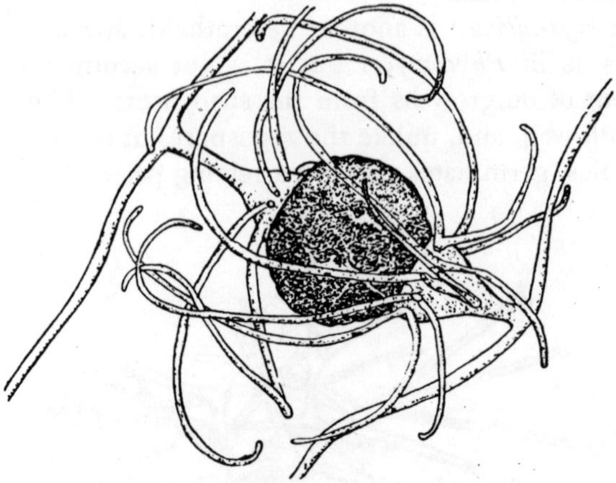

Fig. 62. *Absidia glauca* Hagem; mature zygospore; after Lindner.

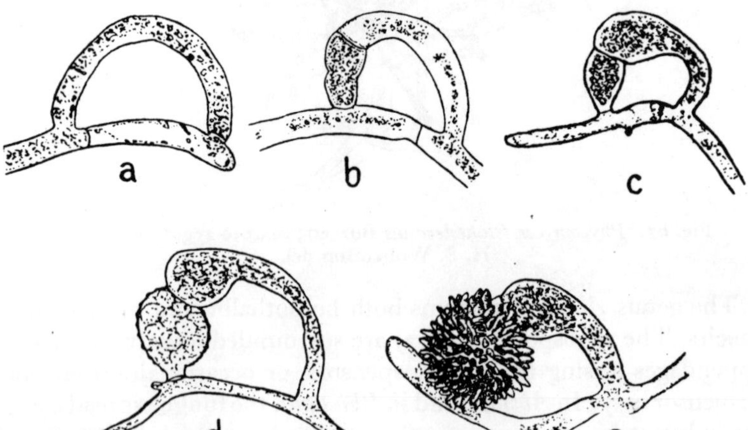

Fig. 63. *Zygorhynchus Moelleri* Vuill.; *a*, young progametangia; *b*, conjugation; *c*, *d*, *e*, development of zygospore; × 533; after Green.

tella has come into contact with one of the host, its tip is cut off as a small multinucleate segment; this soon unites with the hypha of the host by the dissolution of the intervening walls. The multinucleate segment is called the **cupping cell**; it receives cytoplasm and nuclei from the host, and then branches. Behind it, in the original invading hypha, a second segment is cut off to form a storage organ. *Parasitella simplex* is heterothallic, and since its (+) strain attacks only the (–) strain of *Absidia glauca*, and vice versa, it is possible that the parasitic relation may have arisen from abortive attempts at conjugation.

The gametangia of *Rhizopus nigricans*[1] often differ in size. The larger gametangia occur indiscriminately on (+) and (–) mycelia, and the difference may be due to nutrition. Among homothallic moulds, *Sporodinia grandis*[2] shows no distinction between the gametangia, but in many species a regular difference in size exists. In *Zygorhynchus*[3] and *Dicranophora*,[4] union always occurs between a large and a small gametangium.

Zygorhynchus Moelleri occurs in soil; conjugation may take place between closely related branches. A septum is formed near the tip of an erect hypha; below the septum a branch grows out (fig. 63*a*); it bends over and comes into contact with the side of the terminal segment, where a second branch is formed. Both grow, the curved branch especially receiving abundant protoplasm; gametangia of unequal size are cut off, their contents mingle (fig. 63 *b*), the curved suspensor becomes greatly swollen (fig. 63 *c*, *d*), and supplies food to the zygote. This develops a spiny coat (fig. 63 *e*), and germinates after some months, giving rise to a mycelium.

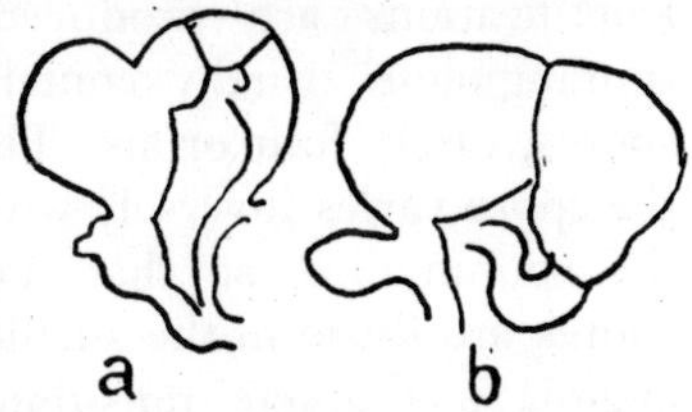

Fig. 64. *Dicranophora* sp.; *a*, formation of a progametangium from the larger and a gametangium from the smaller branch; *b*, young zygospore; after Blakeslee.

In *Dicranophora*[4] also, the gametangia are close together, and, while one remains scarcely wider than the hypha on which it is borne, the other swells and a characteristic bulge develops on its stalk (fig. 64). Conditions are much like those in *Zygorhynchus*, but the disproportion between the gametangia is greater.

[1] Blakeslee, 1907; McCormick, 1911.
[2] Dangeard, 1906; Keene, 1914; Robinson, 1926 i,
[3] Blakeslee, 1904, 1913; Dangeard, 1906; Moreau, 1911, 1912, 1913 ii; Kominami, 1914; Green, 1927.
[4] Blakeslee, 1904.

Choanephoraceae

The Choanephoraceae is a small family, consisting chiefly of tropical species; its members show a gradual transition from the sporangium to the conidium as a unit of asexual reproduction. Sporangia may be produced singly or in groups on rounded heads. In *Blakeslea*,[1] the single sporangia are spherical; some possess a well-marked columella and many spores, while in others the spores are few, and the columellas small or obsolete (fig. 65). The rounded heads bear from twelve to forty sporangiola, often arranged in groups of ten or more, so that conspicuous fructifications are produced. Each sporangiolum usually contains three spores, rarely four or six. The size of the spore varies inversely with that of the sporangium, so that the largest spores are those in the smallest sporangiola; a complete transition series is available from the largest to the smallest size.

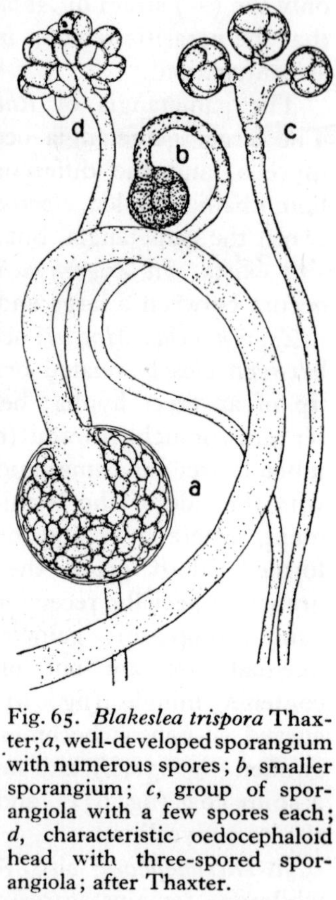

Fig. 65. *Blakeslea trispora* Thaxter; *a*, well-developed sporangium with numerous spores; *b*, smaller sporangium; *c*, group of sporangiola with a few spores each; *d*, characteristic oedocephaloid head with three-spored sporangiola; after Thaxter.

Similar heads are produced by *Cunninghamella*[2] and *Choanephora*,[3] but the sporangiolum containing three or four spores is replaced by a single cell. Careful examination in *Choanephora* shows that this cell is surrounded by a partly separable membrane, so that it may be regarded as the single sporangiospore of a sporangium which falls off as a whole, and has, in effect, become a conidium. When cultures are kept very moist, sporangia

[1] Thaxter, 1914 ii; Jochems, 1927.
[2] Matruchot, 1903; Blakeslee, 1904, 1906 ii; Thaxter, 1914 ii; Burger, 1919; Blakeslee, Cartledge and Welch, 1921; Torrey, 1921.
[3] Currey, 1873; Cunningham, D. D., 1879, 1895; Thaxter, 1903; Dastur, 1920; Couch, 1925.

resembling those of *Mucor* may develop freely. Conjugation occurs in the usual way, the gametangia often differing considerably in size. The wall of the zygospore of *Choanephora infundibulifera* may be free from those of the united gametangia, an unusual condition in the Mucorales.

When *Cunninghamella echinulata* was first discovered, it was placed in the form genus *Oedocephalum* in the *Fungi imperfecti*; it was later transferred [1] to the Mucorales on the ground of its non-septate mycelium and its liability to be attacked by *Piptocephalis*, a member of the Cephalidaceae and an obligate parasite on members of the Mucorales. The discovery [2] of zygospores confirmed this diagnosis. *Cunninghamella echinulata* is heterothallic; its zygospores are without appendages.

Chaetocladiaceae

The family Chaetocladiaceae[3] includes a single genus, *Chaetocladium*: the species may live as saprophytes, or parasitically on other members of the Mucorales. They occur frequently on *Mucor Mucedo* growing on horse dung. The fertile hyphae have sterile tips, and bear laterally tufts or whorls of branches, at the ends of which the monosporous sporangiola appear. These function as conidia, being dispersed without dehiscence; at germination the empty wall of the sporangiolum is left behind. Zygospores are formed as in *Mucor*, and heterothallism occurs in *Chaetocladium Brefeldii*.

Chaetocladium, like *Parasitella*, forms a cupping cell where a hypha meets a branch of the host; it contains cytoplasm and nuclei derived from both host and parasite and puts out a tangle of short branches from which the fertile hyphae of the parasite ultimately develop.

Mortierellaceae

The mycelium of the Mortierellaceae,[4] a family including about twenty species, lives in the soil, or less often, attacks members of the Autobasidiomycetes. The hyphae branch dichotomously and anastomose freely, so that dense felted masses may result. The sporangiophores are usually branched, and, as a rule, develop only on old mycelia. The sporangia are small, terminal and colourless, with a

[1] Matruchot, 1903. [2] Blakeslee, 1904. [3] Burgeff, 1920 i, 1924.
[4] Dauphin, 1908; Thaxter, 1914 ii; Dixon-Stewart, 1932.

deliquescent wall; they may contain many spores, but there is no columella. In addition, special hyphae may bear rounded spores of somewhat doubtful status; they are termed **stylospores**, and are probably monosporous sporangiola, since they germinate after the thick outer wall splits. *Mortierella stylospora* appears to form stylospores only; sporangia are not known.

The zygospores are presumably as in the Mucoraceae, but observation is hindered by the presence of a dense envelope formed of outgrowths from the suspensors, and probably from neighbouring hyphae. The wall may be smooth or rough; germination has not been observed.

Haplosporangium[1] and *Dissophora*[1] are imperfectly known genera which appear to belong to the Mortierellaceae; their zygospores have not been discovered. In *Haplosporangium*, a dense weft of branches gives rise to long hyphae; these become irregularly septate, and put out short sporangiophores as lateral branches, each ending in a sporangiolum. In *Haplosporangium bisporale*, the sporangiolum contains one or two spores; in *Haplosporangium decipiens*, the sporangiolum is monosporous, and functions as a conidium. The fertile hyphae of *Dissophora* do not develop septa until they are old; they always end in a sterile tip, a feature distinguishing the genus from *Mortierella*. The multisporous sporangia form on lateral branches from the main fertile hyphae; they agree in structure with those of *Mortierella*.

Endogonaceae

The Endogonaceae[2] include about twenty-five species. The position of the family is somewhat doubtful, though the characters of the mycelium and of the reproductive structures resemble those of the Mortierellaceae. The species live in the soil or on rotting vegetable material. They form more or less compact fruit bodies containing groups of sporangia, chlamydospores or zygospores. In *Endogone lactiflua*[3] the young fructification is a tangle of interwoven hyphae, the outermost of which combine to form a peridium or protective layer. From the inner filaments, gametangia develop as multinucleate outgrowths of unequal size. In each, a single large

[1] Thaxter, 1914 ii.
[2] Bucholtz, 1912; Thaxter, 1922; Fischer, E., 1923 ii.
[3] Bucholtz, 1912.

nucleus appears, and a terminal uninucleate portion is cut off by a wall. The gametangia fuse (fig. 66 *a*, *b*), the zygote grows out from the larger of the two, and into it both nuclei pass (fig. 66 *c*); their union has not been seen. Each zygote is surrounded by a sheath of vegetative hyphae, lying inside the general peridium. Sporangia

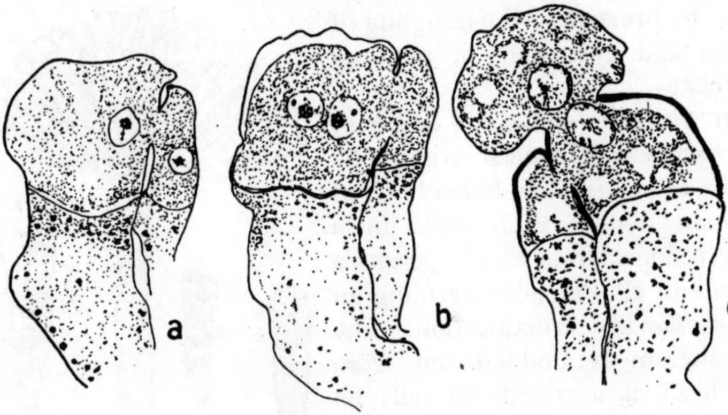

Fig. 66. *Endogone lactiflua* Berk.; *a*, *b*, conjugation; *c*, development of zygospore; after Bucholtz.

have been found in two species only, of which *Endogone malleola*[1] is the better known. They form on the surface of compact masses of hyphae lying on and in the soil; in structure they recall the sporangia of *Mortierella*. Thick-walled chlamydospores occur in many species; in *Endogone fasciculata*[1] alone have chlamydospores and zygospores been found in the same fructification.

CEPHALIDACEAE

The members of the Cephalidaceae,[2] about forty in number, are distinguished by the narrow sporangia associated in groups on the swollen ends of fertile hyphae, each containing a single row of spores. Some species are saprophytes on dung and other substances, but many occur as parasites on members of the Mucorales. They are usually regarded as obligate parasites, but since one of their number, *Dispira cornuta*,[3] has been grown saprophytically on a substratum rich in proteins, it is not unlikely that others may

[1] Thaxter, 1922. [2] Vuillemin, 1902.
[3] Thaxter, 1895 iii; Ayers, 1933, 1935.

prove capable of similar development when suitable media have been found.

The parasitic species of the Cephalidaceae invade their hosts by means of haustoria, and not by the formation of cupping cells. This difference in the method of attack may be an indication of a more remote relationship between host and parasite, preventing the mingling of nuclei and cytoplasm of the two associates.

In *Syncephalastrum*,[1] a genus of saprophytes of tropical origin, the fertile branches arise laterally from a main hypha; each ends in a spherical head from which sporangia radiate in all directions, forming an *Aspergillus*-like fructification. The sporangium is budded out from the head as a cylindrical cell; its contents divide transversely to form a row of walled spores (fig. 67 *b*, *c*, *d*). These are set free by the disappearance of the sporangial wall, which shrivels and breaks up. By careful manipulation of immature material, however, the spores can be squeezed out and the sporangial wall left intact, showing that the units of multiplication are sporangiospores, not exogenously formed conidia. In *Piptocephalis*, a frequent parasite of species of *Mucor* growing on horse dung, the main fertile hypha forks several times, each ultimate branchlet ending in a swollen head; transverse septa are often present. In *Syncephalis* (fig. 67 *a*, *e*, *f*) the fertile hypha is usually stout, and either unbranched or crowned

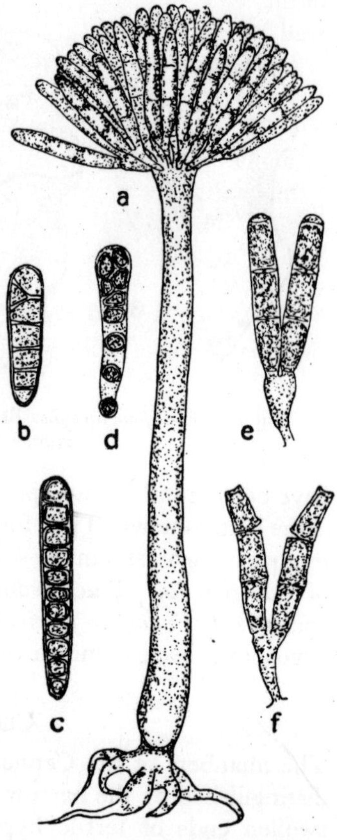

Fig. 67. *a*, *Syncephalis pycnosperma* Thaxter; nearly mature sporangiophore with numerous sporangia. *b*, *c*, *d*, *Syncephalastrum racemosum* Cohn; stages in the separation of spores. *e*, *f*, *Syncephalis pycnosperma* Thaxter; stages in the separation of the spores. After Thaxter.

[1] Thaxter, 1897; Vuillemin, 1922.

by a whorl of short branches. In both genera, the sporangia develop much as in *Syncephalastrum*; the early dissolution of the wall gives to the spores the appearance of a chain of conidia.

As in most of the Mucorales, the gametangia of *Piptocephalis* are similar in form and size; the zygote, however, arises as an outgrowth from the point of fusion; it develops a thick and spiny wall.

In *Syncephalis nodosa*[1] (fig. 68) sexual reproduction is initiated by the association of a pair of hyphae, one of which is twisted about the other; the end of the relatively straight hypha swells to form a gametangium and is cut off by a wall. The tip of the enveloping hypha winds about this swollen extremity, and a septum cuts off the last turn of the spiral from the rest, so that the second gametangium is produced. The wall between the two gametangia now breaks down and conjugation takes place; the contents of the swollen head pass into the spiral, and a zygospore is budded out from the latter in the neighbourhood of the septum. The stalk below the septum gives rise to bladder-like outgrowths. In this, and apparently in other species of the genus, there is well-marked differentiation in the gametangia, the spiral branch, since it gives rise to the zygote, being presumably female.

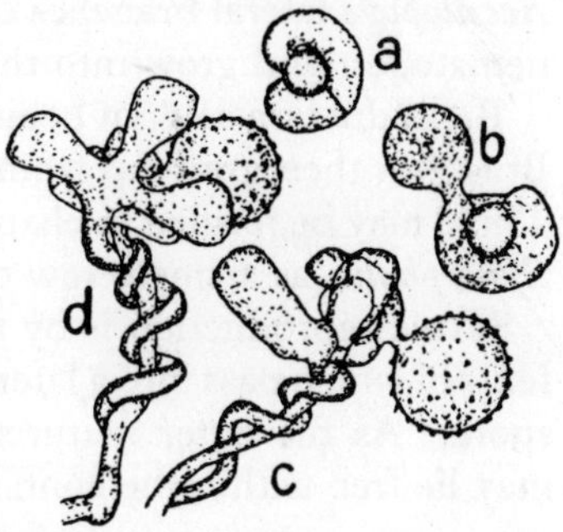

Fig. 68. *Syncephalis nodosa* Van Tieghem; *a*, gametangia seen from above; *b*, the same a little later, with a zygospore growing out from the coiled gametangium; *c*, *d*, later stages, viewed laterally; after Thaxter.

Zoopagaceae

The Zoopagaceae[2] is a family of rather uncertain position and of peculiar characters; the species are parasites on amoebae and nematodes in soil and rotting vegetable material; less than twenty species are known, but it seems probable that more will be discovered.

The thallus is always weakly developed; sometimes it is not a mycelium. In *Endocochlus asteroides*[3] it is a twisted sac lying free in

[1] Thaxter, 1897. [2] Drechsler, 1933–5.
[3] Drechsler, 1935 i.

the protoplasm of the host, and in *Bdellospora helicoides*[1] it recalls the thallus of a member of the Rhizidiaceae. A weak mycelium is present in *Zoopage phanera*[1] and in the species of *Acaulopage*.[2]

The prey is attacked in a variety of ways. The spores of *Endocochlus* are ingested by amoebae and germinate inside the host; the amoebae are unable to digest these spores, although they readily destroy those of many other fungi. Conidia of *Bdellospora* adhere firmly to the pellicle of an amoeba, form haustoria, and enlarge on the outside of the animal to form ovoid thalli. In *Zoopage* and *Acculopage* lateral branches of the mycelium stick to amoebae and nematodes, and grow into them.

Reproduction seldom begins until the host is weakened or dead. Branches then grow up from the thallus, and give rise to conidia. These may be formed in chains, and probably develop, like those of *Syncephalis*, as a single row of spores in a sporangium.

Sexual reproduction is by the union of similar hyphae, the contents of which pass into a lateral outgrowth and there form the zygospore. As the latter matures, it develops a thick membrane and may lie free within the containing walls.

ENTOMOPHTHORALES

The Entomophthorales include some fifty species which have many characters in common, and are classified in a single family, the Entomophthoraceae. Most of the species are parasites on insects, some attack plants, a few occur as saprophytes on dung and other substrata. The mycelium is often feebly developed, and generally shows signs of loose organisation; it produces stout conidiophores from the ends of which conidia are violently discharged. The conidia are large, colourless, and often multinucleate; they may be regarded here, as elsewhere in the Zygomycetes, as modified sporangia.

In sexual reproduction, fusion takes place between segments of the mycelium. The zygospore is formed by the enlargement of one or both segments, or as a lateral outgrowth into which their contents pass. It germinates after a period of rest and gives rise to a mycelium bearing conidia. Fusion often fails to occur, and azygospores are formed.

[1] Drechsler, 1935 i. [2] Drechsler, 1935 ii.

Entomophthoraceae

The species of *Entomophthora*[1] and *Empusa*[2] are responsible for serious epidemics among insects, often among those injurious to plants or animals. At first, the fatty reserves of the host are attacked, but, eventually, all organs are destroyed, and the insect is reduced to an empty skin.[3] Some species will grow as saprophytes[4] on media rich in proteins. The germ tube from the conidium penetrates the skin of the host and gives rise inside its body to numerous irregular segments which multiply by budding, or to a branched, anastomosing, often septate mycelium. At about the time of the death of the host, the mycelium, if present, breaks into short, thin-walled, multinucleate portions known as **hyphal bodies**. These may continue for a time to increase by division or budding, or they may become chlamydospores, the walls thickening, and the bodies passing into a resting stage; warm, moist conditions cause the rapid resumption of activity. The germinating chlamydospore, or the unaltered hyphal body, sends to the surface of the host a long conidiophore, which, in *Entomophthora*, sometimes undergoes septation, forming a series of usually binucleate segments. When conditions for growth are favourable, the primary conidiophore may branch again and again, so that a palisade-like group of crowded conidiophores is formed, appearing as a conspicuous tuft on the surface of the host. At the end of each conidiophore, a large uninucleate conidium develops; it is shot off to a distance of two or three centimetres, and, in some species at least, bears an adhesive pad by means of which it sticks to anything it touches; this character facilitates attack on active insects.

In a few species, sexual reproduction occurs; in *Entomophthora americana*,[5] two hyphal bodies may fuse near their tips (fig. 69 *a*), and the young zygote buds out from the point of fusion as in *Piptocephalis*; alternatively, if the hyphal bodies fuse side by side, an H-shaped figure is produced (fig. 69 *b*), and the nuclei and most of the cytoplasm from both bodies pass into an outgrowth from one of them at a point remote from the point of fusion. The young zygospore is cut off by a wall, and passes into a resting condition

[1] Thaxter, 1888; Riddle, 1906; Molliard, 1918.
[2] Thaxter, 1888; Riddle, 1906; Majmone, 1914.
[3] Sawyer, 1933. [4] Sawyer, 1929.
[5] Riddle, 1906.

(fig. 70 *b*) which lasts until the succeeding summer. Nuclear fusion has not been observed, but may occur on germination.

The conidiophores of *Empusa* differ from those of *Entomophthora* in being unbranched (fig. 71); the nuclei are smaller and more

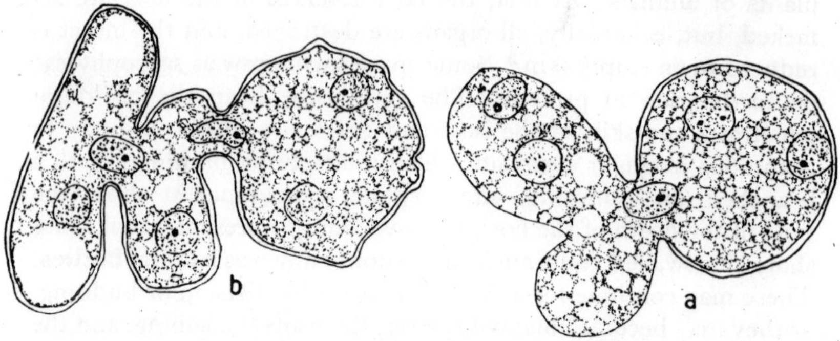

Fig. 69. *Entomophthora americana* Thaxter; *a*, fused hyphal bodies with round zygospore growing from point of contact; *b*, same, round zygospore arising from the side of one of the hyphal bodies; ×650; after Riddle.

Fig. 70. *Entomophthora rhizospora* Thaxter; *a*, young zygospores; *b*, resting zygospores; ×170; after Thaxter.

numerous, each conidium containing twelve to fifteen. Fusion between hyphal bodies has not been observed, but single hyphal bodies bud out azygospores into which the contents pass. A thick coat is formed, and a resting stage entered, as in the zygospores of *Entomophthora*. *Empusa Muscae* is sometimes common in autumn on house flies; these insects may be found attached to windows and curtains by a halo of sterile hyphae projecting from their bodies.

Conidiobolus utriculosus[1] is a parasite on the sporophores of *Auricularia* and its allies. *C. villosus*[2] occurs in fern prothalli, and in rotting wood attacked by *Hypochnus*; it grows readily on nutrient media. The young mycelium is well developed and septate; as it ages, it breaks into hyphal bodies. The simple conidiophores (fig. 72) develop from these; each bears one large conidium. The conidia may germinate at once, or may be trans-

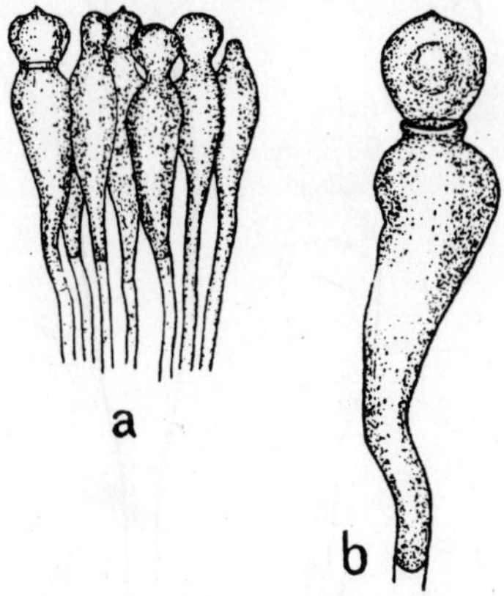

Fig. 71. *Empusa Muscae* Cohn; *a*, group of conidiophores, × 185; *b*, single conidiophore; × 350; after Thaxter.

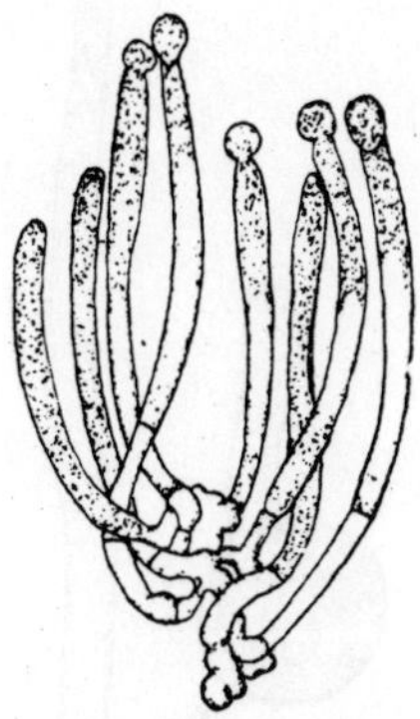

Fig. 72. *Conidiobolus utriculosus* Brefeld; group of conidiophores, × 80; after Brefeld.

formed, by thickening of their walls, into resting cells or into hyphal bodies. In *C. utriculosus*, zygospores are formed after the union of two outgrowths from hyphal bodies, a large, thick-walled spore forming in the swollen end of one of the conjugating hyphae.

Completoria complens,[3] the only species of the genus, forms brown spots on the prothalli of ferns. The mycelium exists inside the cells as a cluster of short, thick hyphae, confined to one cell, or spreading into several; hyphal bodies may also form. Resting spores have been observed as well as conidiophores and conidia like those of other members of the group.

[1] Brefeld, 1884. [2] Martin, G., 1925.
[3] Leitgeb, 1882; Atkinson, 1894 ii.

Massospora cicadina[1] attacks the insect *Magicicada septendecim*, a species which appears above ground at infrequent intervals. It is probable that the host is infected during its subterranean existence. The fungus attacks the posterior segments of the host, causing their progressive destruction, but not necessarily killing the insect. The spores are not projected, as in other members of the Entomophthorales; they are formed inside the host. The ruined segments are sloughed off as they disintegrate, liberating the spores among healthy insects, and ensuring infection. Hyphal bodies and resting spores have been recognised; the latter are probably azygospores.

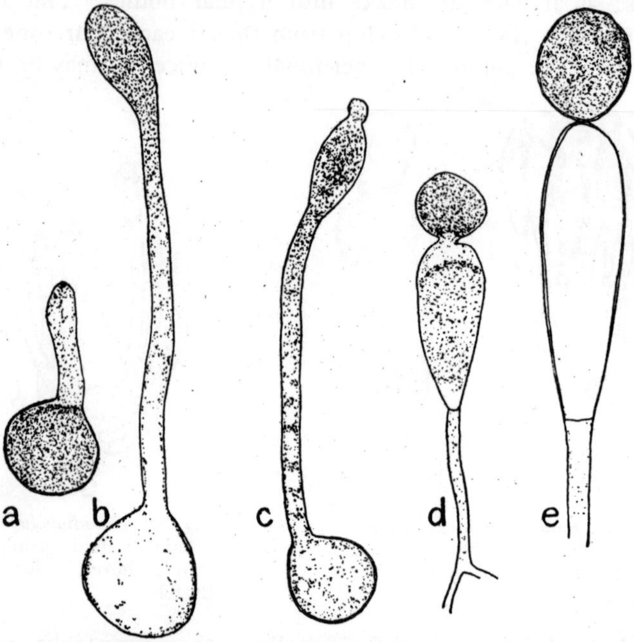

Fig. 73. *Basidiobolus ranarum* Eidam; *a*, germinating conidium; *b*, *c*, same preparing to give rise to a new conidium at the end of the germ tube; *d*, *e*, later stages in the development of the conidium; ×210; H. S. Williamson del.

The best-known species of *Basidiobolus*,[2] *B. ranarum*, occurs on the excrement of frogs; the mycelium consists of branched septate hyphae, each compartment normally containing one nucleus.

[1] Speare, 1921; Goldstein, 1929.

[2] Eidam, 1886; Raciborski, 1896; Fairchild, 1897; Fries, 1899; Olive, 1907; Lakon, 1926.

Large, spherical conidia are discharged[1] from slender conidiophores (fig. 73); these are positively phototropic. The club-shaped swelling just below the conidium probably acts as an ocellus, like the subsporangial swelling in *Pilobolus*. The conidia are swallowed by beetles, these in turn being eaten by frogs.[2] The conidia are set free in the alimentary canal of the frog, where, after nuclear divisions have occurred, spores develop in each; the conidium thus becomes a sporangium. The sporangiospores pass into the gut, multiply by fission, and finally emerge with the dung, then germinating to give the ordinary mycelium. When conidia are sown

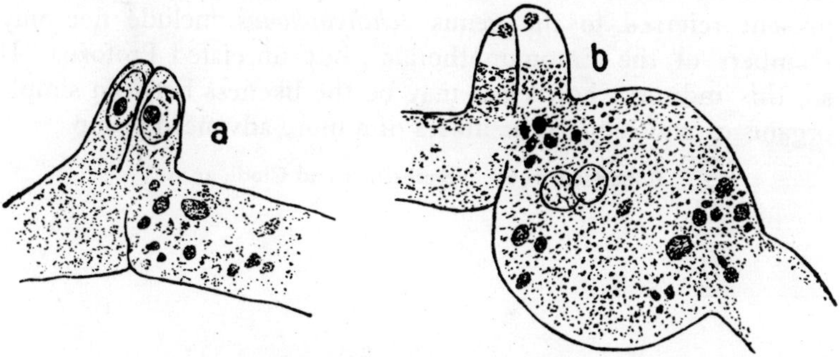

Fig. 74. *Basidiobolus ranarum* Eidam; *a*, fused gametangia; *b*, association of nuclei in swollen megagametangium; ×1110; after Fairchild.

on very rich media they may give rise to spores as they do inside the frog.

In sexual reproduction, two adjacent uninucleate cells of a hypha become gametangia, each putting out a short branch into which the nucleus passes. Meanwhile, the wall between the cells breaks down (fig. 74 *a*), and one of them enlarges considerably. The nuclei divide, one daughter nucleus remaining in each branch, where it is retained by a septum and plays no further part in development. The others pass into their respective gametangia, and the nucleus of the smaller now moves into the larger. The nuclei lie side by side (fig. 74 *b*) for a time, and then fuse. The zygospore contracts, forms a wall, and, after a period of rest, puts out a germ tube; this may end in a conidium, or may branch to form the ordinary mycelium.

Ichthyophonus intestinalis[3] occurs in the alimentary canal of the

[1] Barnes, 1934 i; Ingold, 1934. [2] Levisohn, 1927.
[3] Leger and Hesse, 1923; Leger, 1927.

trout. In the oesophagus and stomach, it has the form of rounded, multicellular bodies, comparable with the hyphal bodies of *Entomophthora*; in the intestine and faeces, small uninuclear structures occur; they arise by the division of the larger bodies, and may be comparable with the sporangiospores of *Basidiobolus*. They are voided with the faeces, and appear to provide for the dissemination of the fungus. On trout agar, they give rise to a septate mycelium, and, ultimately, to zygospores, formed as in *Basidiobolus*. *Ichthyophonus intestinalis* does not seem to harm the fish, but other species resembling it cause fatal disease.[1] It is possible that the forms at present referred to the genus *Ichthyophonus* include not only members of the Entomophthorales, but unrelated Protozoa. If so, this indicates how close may be the likeness between simple organisms and reduced members of a more advanced group.

[1] Robertson, 1909; Neresheimer and Clodi, 1914.

ASCOMYCETES

The Ascomycetes include over 16,000 species, all of which, excepting only the yeasts, possess a well-developed mycelium of branched and septate hyphae. The cells of the mycelium may be uninucleate, as in the Erysiphaceae and in species of *Chaetomium* and *Sordaria*, or they may contain a few or several nuclei.

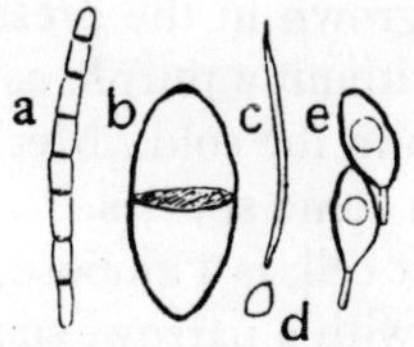

Fig. 75. Spores of *a*, *Geoglossum difforme* Fr.; *b*, *Delitschia furfuracea* Niessl.; *c*, *Rhytisma Acerinum* (Pers.) Fr.; *d*, *Chaetomium Kunzeanum* Zopf; *e*, *Podospora minuta* (Fuck.) Wint.; × 250.

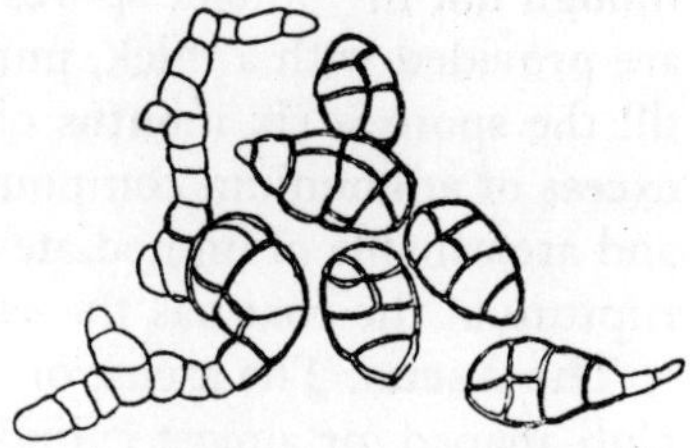

Fig. 76. *Pleospora* sp.; germinating spores, × 500.

Multiplication may take place by means of conidia[1] and chlamydospores, but the characteristic method of reproduction is by ascospores, that is, by spores produced in the interior of a specialised mother cell, or ascus.

The Ascospores. The ascospores are usually elliptical in outline, but in some species spherical, and in some long and narrow (fig. 75); they contain densely granular cytoplasm and frequently oil drops. The epispore may be smooth or sculptured; hyaline or opaque, colourless or variously coloured. The young spore is ordinarily unicellular and uninucleate, though in some species it contains two nuclei. Nuclear division and wall formation may take place during development, the mature spore is then a row or a mass of cells (fig. 76). Each cell of a multicellular spore is usually capable of putting out a separate germ tube.

Most spores are able to germinate as soon as they leave the ascus, but some require a period of maturation during which changes take

[1] Klebahn, 1918.

place in the contents or in the wall. To some extent these changes may be accelerated by heat. At ordinary temperatures the spores of *Lachnea cretea* will not germinate till eight weeks after they are shed, but, if exposed for some hours to hot sunshine, even newly evacuated spores will germinate at once. The same result is obtained artificially by incubation at 60° or 70° C. for a few minutes or at more moderate temperatures for a longer time. As maturation proceeds, progressively less heat is needed to induce development. Similarly, in *Humaria granulata*, germination does not normally take place till seven months after the spore leaves the ascus, but the application of heat causes germination in spores five months old, though not in younger spores. The spores of *Dasyobolus immersus*[1] are provided with a thick, purple coat which prevents germination till the spore is six months old. Spores grown in the presence of excess of ammonium compounds ripen without a purple covering, and are capable of immediate germination in the cold. Mechanical rupture of the coat has the same effect in some species.

The Ascus. The ascus, or spore mother cell, is a globose, ovoid, club-shaped, or almost cylindrical organ with a narrow, somewhat elongated base. Before reaching maturity it contains a single nucleus, often termed the **definitive** nucleus, which undergoes three karyokinetic divisions, in which reduction takes place, giving rise to eight daughter nuclei (fig. 77), around which the spores are delimited by a process of free cell formation. All the cytoplasm of the ascus is not included in the spores; the remainder constitutes the **epiplasm**, and becomes rich in food substances and glycogen. The glycogen in slender asci, such as those of *Peziza vesiculosa*, is usually confined to the region below the spores, while in larger asci it surrounds the spores also;[2] it exerts little or no osmotic pressure, but is capable of being rapidly transformed into sugars of high osmotic value. Such a transformation and the consequent absorption of water may be responsible for the bursting of the ascus and the violent ejection of the spores. There is evidence that the amount of glycogen present may be a factor in determining the distance to which the spores are thrown. Glycogen, or similar substances, the alteration of which results in a change of osmotic pressure, play an important part in many of the rapid cell expansions characteristic of fungi.

[1] Gwynne-Vaughan and Williamson, 1933 ii.
[2] Walker and Andersen, 1925.

In most of the Ascomycetes the mature ascus contains eight spores, but, in some, though eight nuclei are produced, certain of the spores initiated may prove abortive, or spores may be formed round only one or a few nuclei. In some asci the three original nuclear divisions are succeeded by others, spores being formed around sixteen, thirty-two, or a larger number of nuclei. Thus a

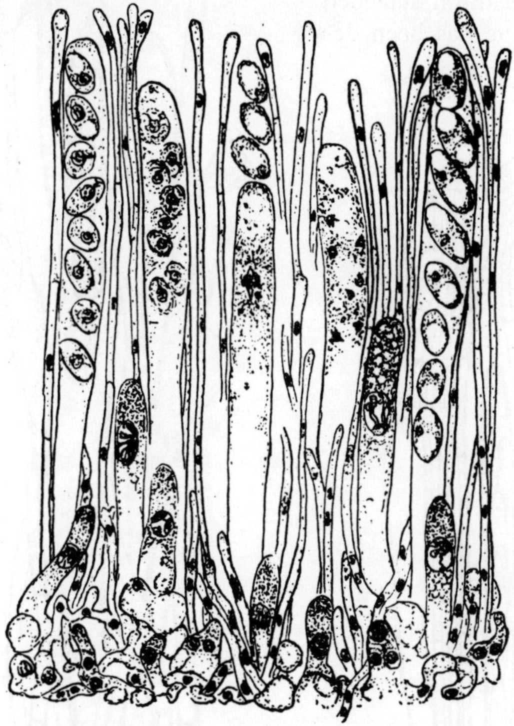

Fig. 77. *Humaria rutilans* (Fr.) Sacc.; hymenial layer showing asci and paraphyses in various stages of development, ×400.

single spore is sometimes developed in the ascus of *Tuber*, two is the regular number in *Phyllactinia*, four in many of the Laboulbeniales, and in *Sordaria maxima*; there are sixteen or thirty-two in species of *Rhyparobius*, sixteen to sixty-four in *Philocopra pleiospora*, and in *P. curvicolla* one hundred and twenty-eight. In species of *Neurospora*, *Sordaria* and other genera, four binucleate spores are found or mixed groups of binucleate and uninucleate spores

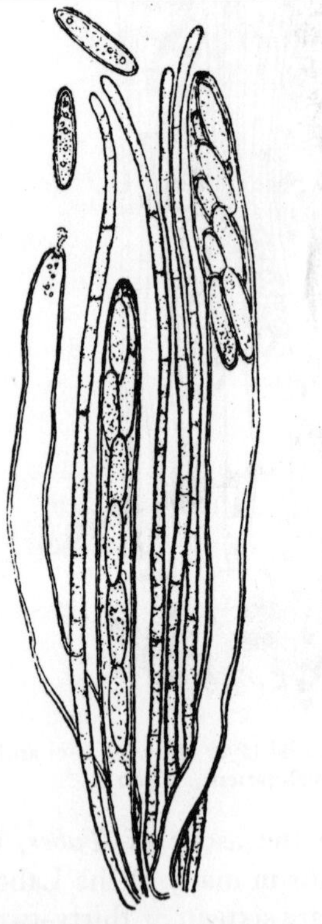

Fig. 78. *Mitrula laricina* Mass.; development and ejection of biseriate spores, ascus opening by a plug; ×600.

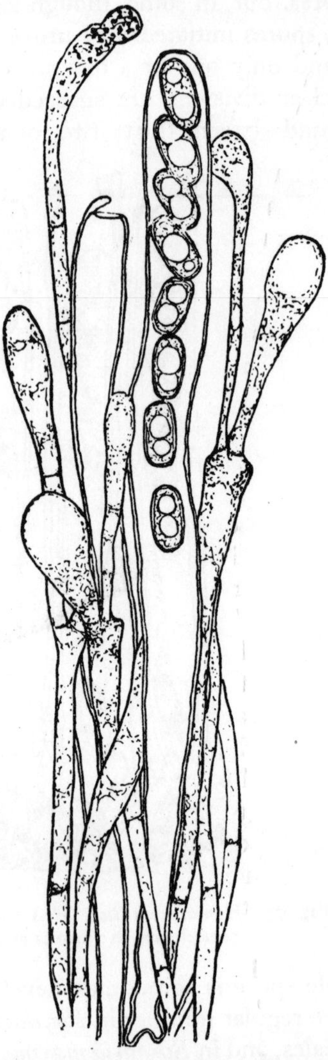

Fig. 79. *Sepultaria coronaria* Mass.; uniseriate spores; ascus opening by a lid; branched, septate, clavate paraphyses; ×600.

in any proportion that utilises the eight nuclei. In the Plectomycetes spores are liberated by the disintegration of the ascus wall, and, if a definite sheath surrounds the asci, remain for a time enclosed by its outer layers; in the Discomycetes and Pyrenomycetes, however, the ascus opens explosively, by an irregular tear, or by dehiscence along a definite line (figs. 78, 79), and the spores are shot out in a jet of liquid, while the deflated ascus sinks back to about half its size. Where the asci are enclosed in a fructification with a long neck, they may deliquesce to form a mucilaginous mass which readily absorbs water and expands, being squeezed up the neck and exuded at the ostiole. In other flask-shaped fructifications the ascus elongates to reach the ostiole and the spores either float near the tip of the ascus in its fluid contents or are attached to the apex and to one another by cell prolongations or strands of wall material. Elongation of the ascus before the discharge of the spores is also found in the cup-shaped fructifications of the Ascobolaceae.

In forms with an exposed hymenium the asci, under damp conditions, discharge their spores in succession; but, on a quiet, dry day, any disturbance causes large numbers of ripe asci to eject their contents together, so that a cloud of spores is visible to the naked eye. This phenomenon, which is known as puffing, may be brought about by shaking the fructifications, or even by the currents of air set up in walking past them. It can be induced, when ripe asci are lying in water, by exposing them suddenly to the action of glycerine or alcohol, and is due to alterations of pressure affecting all asci at the same stage of development.

The Ascocarp. In the Endomycetaceae and Saccharomycetaceae the asci develop singly; in most families they arise in associated groups, and are protected by a common wall of sterile filaments known as the **peridium**; in this way a definite fructification, the **ascocarp** or **ascophore**, is formed. The young ascocarp is globose and usually completely closed; it may retain this form at maturity, opening only by the decay or irregular splitting of its walls; it may assume a flask-shaped outline, opening by a terminal pore, the **ostiole** (fig. 80); or it may spread out to form a cup in the concavity of which the asci are fully exposed. The cup-shaped ascocarp is known as an **apothecium**; to the other forms the term **perithecium** is applied. In some of the simpler ascocarps the asci are irregularly disposed, in the rest they are arranged in parallel series (fig. 77) and are often intermingled with paraphyses.

The paraphyses are slender filaments of about the same length as the asci; they develop earlier than the latter (fig. 81); in

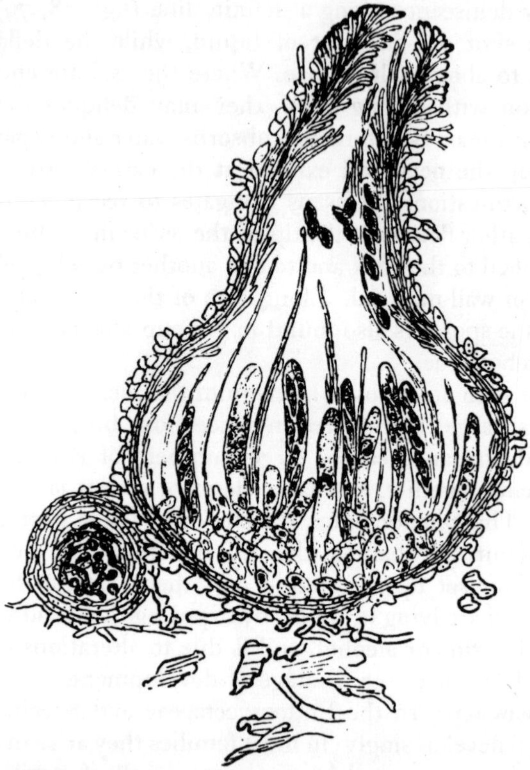

Fig. 80. *Sordaria* sp.; two ascocarps in longitudinal section, left very young, right mature, showing asci and paraphyses, × 400.

apothecia they usually form a wedge-shaped group, and their growth forces open the sheath. They are cylindrical, pointed or club-shaped, and often contain bright coloured granules. The peridium varies greatly in thickness; hairs and other appendages are in some species developed from its outer cells. Filaments known as **secondary mycelium**[1] grow down into the substratum and supply food.

[1] This meaning of the term must not be confused with the significance more recently given to it in Basidiomycetes, cf. p. 287.

Sexual Reproduction. The Ascomycetes show a well-marked alternation of generations modified by the wide occurrence of apogamy. In most of the species which possess male and female organs, these are produced on the same mycelium, but in *Pericystis apis* and in certain of the Laboulbeniaceae they occur on separate individuals so that dioecism is established. Among monoecious forms, self-sterility is associated in *Ascobolus magnificus* with normal fertilisation, while in self-incompatible species which lack

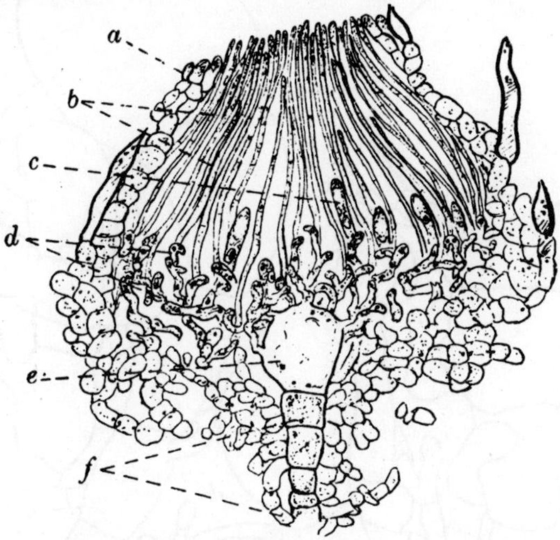

Fig. 81. *Lachnea stercorea* (Pers.) Gill.; ascocarp in longitudinal section showing young asci and paraphyses; the oogonium is still recognisable; × 160. *a*, sheath; *b*, paraphyses; *c*, ascus; *d*, ascogenous hyphae; *e*, oogonium; *f*, stalk of female branch.

antheridia, the complementary strains are brought together by mycelial fusion. In *Glomerella cingulata*[1] perithecia are produced on both *A* and *B* strains growing alone, but they are more abundant when these meet, indicating that some stimulus is provided by their union. Similarly two saltants of *Cytospora ludibunda*[2] are infertile when grown separately, but form pycnidia when they meet.

In *Diaporthe perniciosa*[3] and in *Aspergillus nidulans*[4] other types of heterothallism are known, strains with different vegetative characters having been bred in culture.

[1] Edgerton, 1914.
[2] Das Gupta, 1933.
[3] Cayley, 1923 ii.
[4] Henrard, 1934.

In forms with normal fertilisation the mycelia produced on germination of the spores give rise to sexual organs. These may be simple in form and almost identical, or clearly differentiated into male and female structures. The male branch consists of an antheridial hypha, or stalk, and a terminal antheridium. The female branch, sometimes called the **archicarp**, possesses a stalk of one

Fig. 82. Sexual branches for comparison of size; *a*, *Humaria granulata* Quél.; *b*, *Ascobolus magnificus* Dodge; *c*, *Pyronema confluens* Tul.; *d*, *Ascophanus Aurora* (Crouan) Boud.; all × 1000.

or more cells bearing the oogonium (fig. 81). In many species this branch is extended as a **trichogyne** by means of which communication is established between the male and female organs. The oogonium may be a single, uninucleate cell (*Erysiphe*, *Stigmatomyces*), the multinucleate segment of a narrow hypha (*Eurotium*), or a pyriform or globose structure containing, according to its size, as many as 1400 nuclei (*Humaria granulata*) or as few as four (*Monascus purpureus*). The trichogyne may consist of one, two or several cells, and may be the distal region of the female branch (*Ascobolus*) or a secondary outgrowth from the oogonium (*Pyronema*, *Ascophanus*). The antheridium is usually somewhat smaller than the female organ; it may contain one or a few nuclei or as many as 400 (*Ascobolus*). In *Ascophanus Aurora* it is broken away from its stalk after fertilisation and remains attached to the tip of the trichogyne. Elsewhere (*Sclerotinia*, *Podospora*) an antheridium detached before fertilisation and carried by wind or insects has been described, but detailed cytological study is needed to ascertain whether the small, round bodies, which are known to stimulate fruit production, do so as antheridia fertilising female organs or as germinating conidia of a complementary strain.

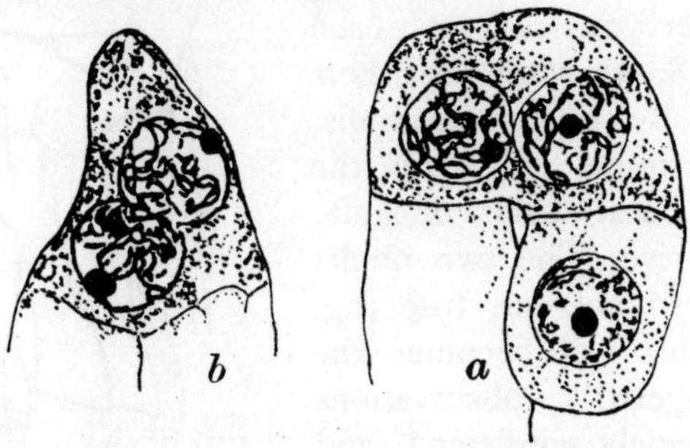

Fig. 83. *Humaria rutilans* (Fr.) Sacc.; *a*, crozier with ascogenous cell containing two nuclei, uninucleate terminal cell and stalk; *b*, fusion in the ascus, the nuclei are just passing into synizesis; both × 1875.

Often the oogonium becomes septate after the fertilisation stage (*Erysiphe*, *Ascodesmis*); sometimes several cells of the female branch communicate by wide pores (*Lachnea cretea*, *L. melaloma*) and together form an **oogonial region**. From the oogonium, or from one or more cells of the oogonial region, a number of filaments, the **ascogenous hyphae**, bud out. These receive the oogonial contents and ultimately give rise to asci in the first two nuclear divisions in which meiosis takes place. The ascogenous hyphae thus form the diplophase or sporophyte, while the vegetative filaments on which the sexual organs are borne are gametophytic, and may also give rise to conidia and chlamydospores. These asexual or accessory spores play no part in the alternation of generations, but serve for the rapid spread of the gametophyte or haplophase; it is the ascospores alone which correspond to the spores of vascular plants.

Cytology. Thanks to the researches of de Bary and others, sexual organs had been reported in a number of Ascomycetes during the latter half of the nineteenth century, before the study of nuclear detail was undertaken. Nuclear fusion, however, was first observed by Dangeard in 1894, not in the oogonium, but in the ascus. He found[1] that, in *Peziza vesiculosa* and other species with a well-developed fructification, the ascogenous hypha bends over, forming a **crozier**, its two nuclei divide simultaneously (fig. 87), the curved, subterminal segment of the branch retaining one daughter nucleus of each division (fig. 83 *a*), while the others are cut off by transverse walls, one in the terminal cell or tip, the other in the basal region or stalk, below the curve. The two nuclei in the subterminal cell fuse (fig. 83 *b*), and this cell becomes the ascus. Dangeard's observations have been widely confirmed, and it would appear that, apart from occasional trinucleate and quadrinucleate examples, all asci borne on ascogenous hyphae are binucleate at their first formation; in many species, however, a crozier is not formed, but the binucleate segment of the ascogenous hypha at once becomes an ascus. Among Ascomycetes anastomoses are common, not only between vegetative filaments, but between the tip and stalk cells of ascogenous hyphae when the bending over of the ascus occurs. Often the nucleus of the stalk passes into the tip, which, having thus acquired two nuclei, grows on to form a second ascus;[2] a third (fig. 84) and a fourth may arise in the same way. The method of ascus formation has been compared by some

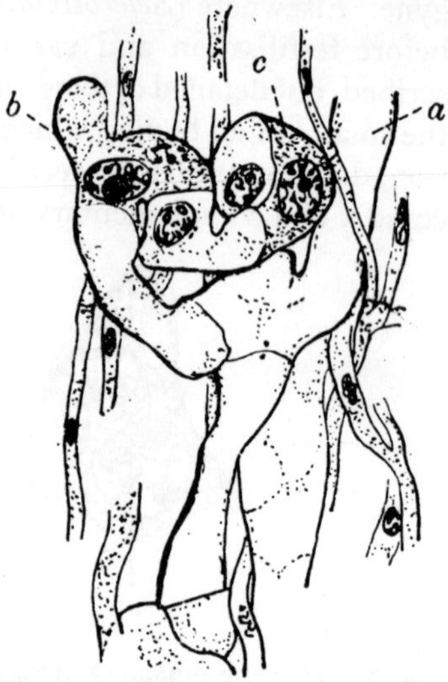

Fig. 84. *Humaria rutilans* (Fr.) Sacc.; an ascus (*a*) the terminal cell connected with which has continued its growth and given rise to another ascus (*b*) from the terminal cell of which a third ascus (*c*) has arisen, × 1250.

[1] Dangeard, 1894 ii.

[2] Fraser, H. C. I., 1908; McCubbin, 1910; Brown, W. H., 1910, 1911; Claussen, 1912.

authors[1] with the production of clamp connections in the secondary mycelium of the Basidiomycetes, but a study of development shows that, whereas the ascus is formed by the curling over of a branch of which the beak is the terminal cell, the clamp connection is a lateral outgrowth. Clamp connections have been reported in the ascogenous hyphae of lichens[2] but only near the region of ascus formation, where the proliferation of stalk and terminal cell may produce a similar appearance.

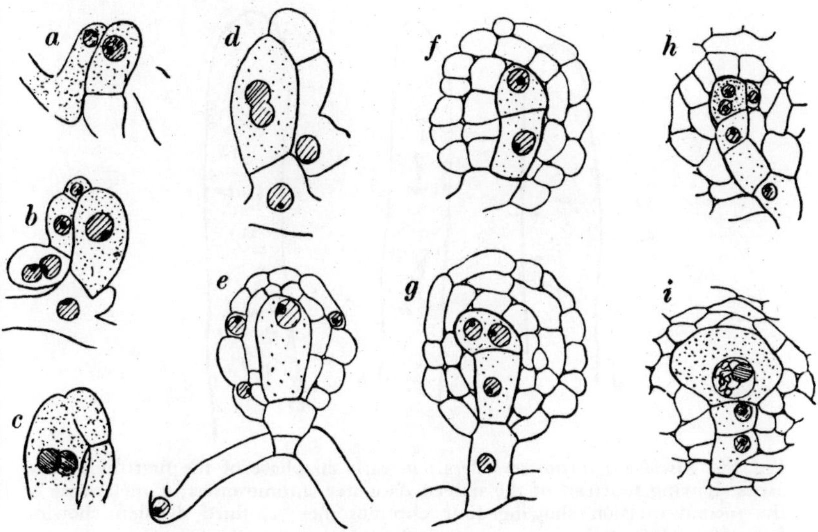

Fig. 85. *Sphaerotheca Humuli* (DC.) Burr.; *a* and *b*, sexual branches; *c*, male nucleus in oogonium; *d*, fusion in oogonium, antheridium without nucleus; *e*, fusion nucleus in oogonium; *f* and *g*, septation of oogonium; *h*, two nuclei in ascus; *i*, ascus after nuclear fusion; after Harper.

The year after Dangeard's discovery of the binucleate ascus, his observations and those of de Bary were confirmed by Harper for the hop mildew, *Sphaerotheca Humuli* (fig. 85). Harper[3] saw the development of a uninucleate antheridium and a uninucleate oogonium; he observed the passage of the male nucleus into the oogonium and its fusion with the female nucleus. After fertilisation the oogonium underwent septation, giving rise to a row of cells,

[1] Claussen, 1912; Kniep, 1916; Bensaude, 1918; Martens, 1932; cf. p. 289.
[2] Moreau and Moreau, 1922; Moreau, 1925. [3] Harper, 1895.

the penultimate of which contained two nuclei and, when these had fused, developed into the ascus. In the life history of this fungus there occurs, then, not only the fusion of sexual nuclei usual in other plants and animals, but a second fusion, that in the young ascus, so that the ascus nucleus is regularly tetraploid. These observations have been confirmed in many other Ascomycetes, and it has moreover been shown that, while the first two divisions in the ascus constitute a meiotic phase (figs. 87 *b*, *c*, 155 *c*, *d*),

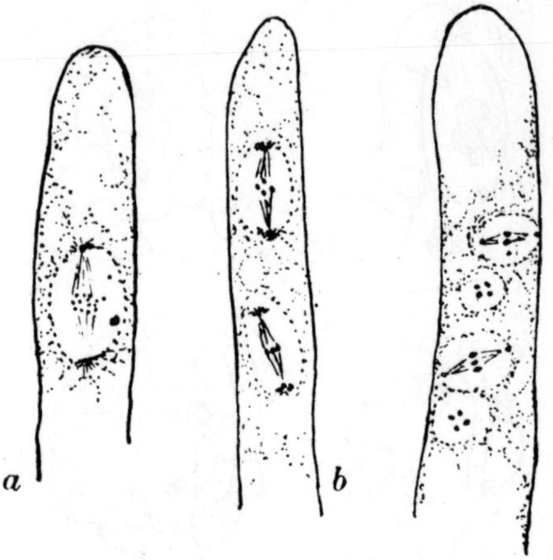

Fig. 86. *Ascobolus furfuraceus* Pers.; *a*, early anaphase of the first division in ascus showing fourteen of the sixteen daughter chromosomes; *b*, metaphase of the second division showing four chromosomes; *c*, third division showing four; after Dangeard.

a second reduction,[1] known as **brachymeiosis**, takes place in the third division (figs. 86, 155 *e*). Thus the nuclei of the ascogenous hyphae after fertilisation are diploid, the ascus nucleus before division is tetraploid, after the first two mitoses the ascus contains four diploid nuclei, and, after the third mitosis, there are eight haploid nuclei in the ascus about which the spores are formed. The allelomorphs which are to be separated in brachymeiosis may become associated during the second division (fig. 86 *b*).

In order to demonstrate the occurrence or otherwise of these reductions, chromosome counts are required in the prophases and

[1] Maire, 1905; Dangeard, 1907; Fraser, H. C. I., 1908; Fraser, H. C. I., and Welsford, 1908; Fraser, H. C. I., and Brooks, 1909; Tandy, 1927; Gwynne-Vaughan and Williamson, 1930, 1931, 1932, 1933 i, 1934.

telophases of each division; the massing of the chromatin in synizesis and the form of the gemini are valuable evidence of meiosis.

Similarly the occurrence or otherwise of nuclear fusion in the oogonium can be determined by chromosome counts, the chromosomes in the nuclei of the ascogenous hyphae, if fusion has taken place, being twice as numerous as those in the vegetative cells and virgin oogonia. Equally significant is the observation of stages when

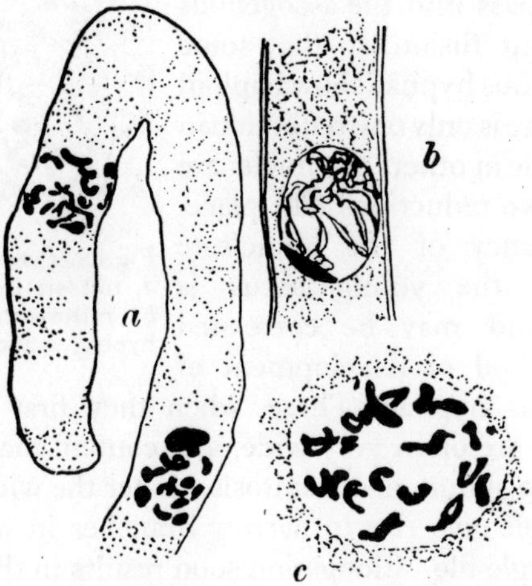

Fig. 87. *Humaria rutilans* (Fr.) Sacc.; *a*, ascogenous hypha showing sixteen chromosomes in each nucleus, × 1950; *b*, fusion nucleus of ascus with double spireme, × 1300; *c*, fusion nucleus of ascus showing sixteen gemini, × 1950.

the male nuclei are entering the female organ (fig. 89 *a*) and of stages when nuclei are in process of fusion (fig. 89 *b*, *c*) but negative evidence is of little value unless the material can be shown by other criteria, such as nuclear division, to be in an active state. The number of nuclei in a series of oogonia may also be counted. This number is increased by the entrance of male nuclei and, if fusion occurs, it is later diminished, while the individual nuclei are of larger size.

The occurrence of fertilisation is by no means universal among Ascomycetes. Often the sexual apparatus has wholly disappeared, the ascogenous hyphae arising from vegetative cells and containing haploid nuclei. The only nuclear fusion is that in the ascus, and brachymeiosis no longer occurs.

Between such apogamous species and those with normal fertilisation is a series of pseudapogamous forms, the female or vegetative nuclei fusing in pairs before they pass into the ascogenous hyphae. Of a rather different character is the intermediate condition observed in *Pyronema domesticum*. Here the sexual organs are normal and male nuclei enter the oogonium, but, while some of the sexual nuclei unite in pairs, others pass into the ascogenous hyphae without fusion, so that some of the ascogenous hyphae show haploid nuclei and there is only one reduction in their asci, while in others the nuclei are diploid and two reductions take place.

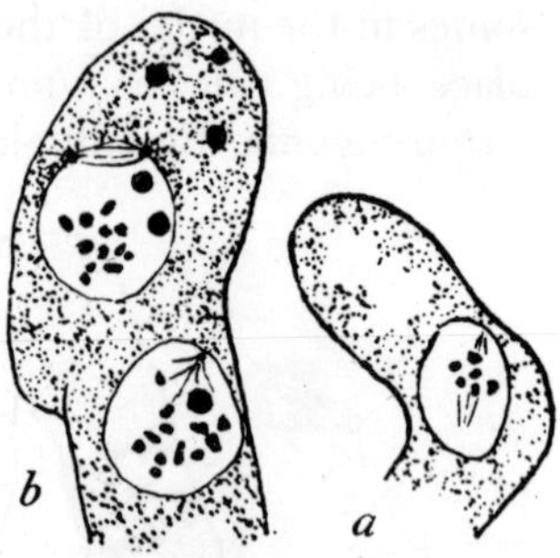

Fig. 88. *Pyronema confluens* Tul.; *a*, mitosis in a vegetative cell; *b*, mitosis in an ascogenous hypha; × 2600.

The constancy of the binucleate condition in the young ascus is remarkable and may be correlated with the method of development of the ascogenous hyphae. These, when they first grow out from an oogonium, are often very wide, and contain many nuclei, all of which undergo simultaneous mitosis. Later the wide hyphae or the oogonium itself give rise to narrow branches in which the nuclei must lie in single file. Elongation soon results in the appearance of tracts devoid of cytoplasm separating the branches from the oogonium and from one another, so that, henceforth, divisions in each proceed independently. In these narrow filaments the spindles in mitosis lie more or less parallel to the long axis of the hypha (fig. 90 *a*), and walls are formed at right angles to them, separating the sister nuclei.[1] Consequently the proximal and distal cells of each hypha contain one nucleus and each of the intervening cells contains two (fig. 90 *b*, *c*), which are daughter nuclei of different divisions. The uninucleate cells do not, as a rule, function further, but the binucleate cells grow out (fig. 90 *d*) and, when their nuclei divide (fig. 90 *e*), septa appear at right angles to the spindles, so that they also form uninucleate cells with a binucleate cell between (fig. 90 *f*). Thus all active cells are provided with two nuclei. The arrangement is inevitable if mitosis accompanied by septation takes place in a narrow cell containing two or more nuclei in single file,

[1] Gwynne-Vaughan and Williamson, 1930, 1932, 1934.

just as mitosis, with septation at right angles to the spindle, in a uninucleate cell, necessarily gives cells with one nucleus each. Any binucleate cell may grow out laterally, curling over and forming a

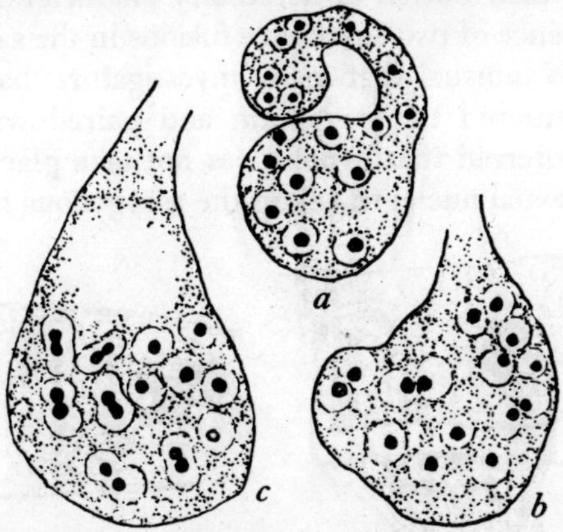

Fig. 89. *Ascophanus Aurora* (Crouan) Boud.; *a*, sexual organs in transverse section with male nuclei passing through trichogyne from small antheridium to larger oogonium; *b*, oogonium in longitudinal section with paired male and female nuclei; *c*, fusion of sexual nuclei; unpaired female nuclei are present in *b* and *c*; × 2600.

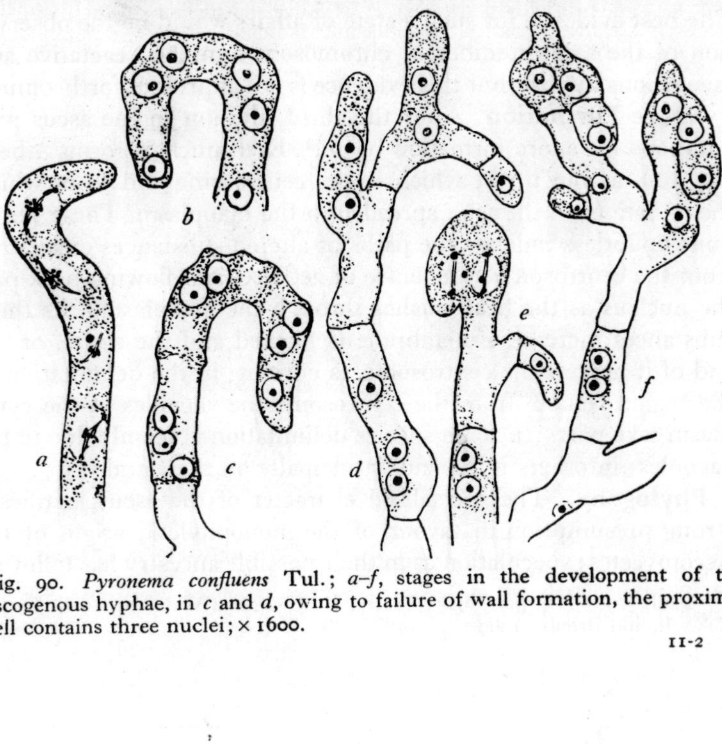

Fig. 90. *Pyronema confluens* Tul.; *a–f*, stages in the development of the ascogenous hyphae, in *c* and *d*, owing to failure of wall formation, the proximal cell contains three nuclei; × 1600.

crozier, the divisions in which correspond exactly to those in other ascogenous branches. The fusion of nuclei in the young ascus facilitates the distribution of hereditary characters.

The occurrence of two successive fusions in the same life history is, however, so unusual that many investigators, having seen that male nuclei entered the oogonium and paired with the female nuclei, have inferred that fusion does not take place at this stage, but that the sexual nuclei travel up the ascogenous hyphae in pairs

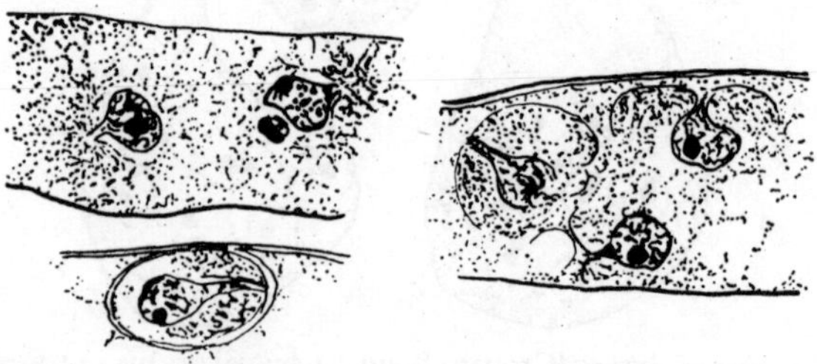

Fig. 91. *Humaria rutilans* (Fr.) Sacc.; stages of spore formation, × 1875.

and that the fusion in the ascus represents their ultimate union. The best evidence for such a state of affairs would be the observation of the same number of chromosomes in the vegetative and ascogenous hyphae, but this evidence is not at present forthcoming.

Spore Formation. After the third division in the ascus preparations for spore formation begin.[1] Each nucleus forms a beak (fig. 91), at the tip of which is the centrosome and from which the radiations of the aster spread into the cytoplasm. These radiations doubtless indicate the paths of altered substances emanating from the centrosome as a centre of activity, and flowing back past the nucleus as the beak pushes through the cytoplasm. As these substances increase a membrane is formed and the spore, or the end of it nearest the centrosome, is cut out; in the delimitation of the region remote from the centrosome the vacuoles of the cytoplasm take part. In some species delimitation is mainly due to the vacuoles; in others it depends principally on the astral rays.

Phylogeny. The specialised character of the ascus justifies a strong presumption in favour of the monophyletic origin of the Ascomycetes; speculation as to their possible ancestry has followed

[1] Faull, 1905; Harper, 1905; Fraser, H. C. I., and Welsford, 1908; Fraser, H. C. I., and Brooks, 1909.

two main lines; they have been regarded as derived either from the Phycomycetes,[1] or from the red algae or the ancestors of the latter.[2] There is an interesting resemblance to be traced between the sporogenous filaments and carpospore fructifications of the Florideae and the ascogenous hyphae and ascocarps of the Ascomycetes, and also between the male and female organs of the two groups; but any suggestion of relationship between them is concerned with the higher Ascomycetes and involves the assumption that the simpler forms are reduced from the more complex.

The structure of these simpler forms constitutes a similar objection to the derivation of the Ascomycetes from any but the more primitive of existing Phycomycetes, though the evolution of the conidium both in the Peronosporaceae and in the Choanephoraceae and Mortierellaceae suggests a possible origin for the accessory spores, and the sexual apparatus of *Syncephalastrum* offers at least an example of parallel development. There is not, however, sufficient evidence to exclude the possibility of a separate flagellate ancestry, and effective speculation as to the origin of the Ascomycetes must await knowledge of the development of a larger number of forms.

Within the group it seems probable that the most primitive species are to be sought among the relatively simple forms included in the Plectascales. In most of the Endomycetaceae the oogonium becomes the ascus directly, so that there are no ascogenous hyphae and the first nuclear divisions after fertilisation bring about reduction. Elsewhere within the family a slight development of ascogenous hyphae is found. It is possible that we have here the origin of the sporophyte of the Ascomycetes; the neighbouring family, Gymnoascaceae, with its loose network of protective hyphae around the asci, certainly suggests the origin of the peridium. Attempts at a natural classification, however, are limited by the scarcity of detailed knowledge, so that the efficient study of additional forms is greatly to be desired.

The Ascomycetes may be subdivided as follows:

Ascocarp, if present, either with no definite ostiole, or shield-shaped, or with asci irregularly arranged	PLECTOMYCETES
Ascocarp wide open when ripe; asci in parallel series	DISCOMYCETES
Ascocarp flask-shaped when ripe; asci in parallel series	PYRENOMYCETES

[1] Atkinson, 1915.

[2] Dodge, 1914.

PLECTOMYCETES

The Plectomycetes[1] include those relatively simple forms which possess neither the cup-shaped apothecium of the Discomycetes nor the flask-shaped perithecium of the Pyrenomycetes. In most of the species a rounded ascocarp is produced, opening either by the decay of its wall, or by an irregular tear or split. The asci may arise from the floor of this fructification and stand parallel one to another, or they may be irregularly disposed, the fertile hyphae forming a tangled weft. In many species a definite fructification is lacking.

The asci are usually pyriform, club-shaped or ovoid, in contrast to the nearly cylindrical asci of most of the Discomycetes and Pyrenomycetes. They arise indifferently from terminal or intercalary cells of the ascogenous hyphae. Except in the Perisporiaceae, the ascospores are usually continuous and hyaline. In most species the gametophytic mycelium gives rise to conidia.

The sexual apparatus is simple, and the oogonium after fertilisation may become an ascus directly, or may give rise to ascogenous hyphae.

The Plectomycetes include some 1200 species and may be subdivided as follows:

Asci irregularly arranged	**Plectascales**
Asci parallel	
Ascocarp present	**Erysiphales**
Ascocarp lacking	**Exoascales**

PLECTASCALES

The Plectascales include all those Ascomycetes in which the asci are irregularly arranged. In the simplest forms the asci are produced singly, one from each oogonium, and are without protective filaments; in *Gymnoascus* and its allies ascogenous hyphae are formed after fertilisation, and they, and the asci which they bear, are surrounded by an open weft of thick-walled hyphae; in the higher families a well-developed ascocarp is present.

[1] Gwynne-Vaughan, 1922.

The alliance includes the following families:

Asci naked	
Cells forming a mycelium; asci distinct from mycelial cells	ENDOMYCETACEAE
Cells single or loosely attached; asci not differentiated from other cells	SACCHAROMYCETACEAE
Asci surrounded by loosely interwoven hyphae	GYMNOASCACEAE
Asci surrounded by a definite peridium	
Ascocarp subaerial	
Sessile	ASPERGILLACEAE
Stalked	ONYGENACEAE
Ascocarp subterranean	
Peridium distinct from contents of ascocarp; spore mass powdery at maturity	ELAPHOMYCETACEAE
Peridium continuous with contents of ascocarp; spore mass never powdery	TERFEZIACEAE

ENDOMYCETACEAE

In the Endomycetaceae[1] the mycelium bears numerous asci, each of which is either the product of a separate sexual fusion, or is developed parthenogenetically. Oidia, chlamydospores, and yeast-like conidia may be formed. Most of the Endomycetaceae are saprophytic on sugary substances or on exudations from plants; *Endomyces Mali* is an active parasite on apples, some other species are parasitic on fungi, and *Endomyces albicans* causes the human disease known as thrush. The principal genera of the family are *Eremascus* and *Endomyces*

Only two species of *Eremascus* are known. *E. albus*[2] was discovered by Eidam in 1881, in a bottle of malt extract. The contents had gone bad and were covered with a growth of fungi, among which was the new form. It produced a fine, snowy white, septate mycelium from which pairs of fertile hyphae grew out, curled round one another, and fused at their tips (fig. 92). The fused portion was cut off by a wall and eventually produced eight spores. Unfortunately the species was lost and has never reappeared.

In 1907, however, Stoppel, on opening some pots of currant and apple jelly, discovered a similar form which she named *Eremascus fertilis*. This species,[3] like *E. albus*, possesses a branching, septate mycelium. Pairs of uninucleate cells grow up from the

[1] Kløcker, 1909 i; van der Wolk, 1913; Ramsbottom, 1914; Juel, 1921; Mangenot, 1922.

[2] Eidam, 1883.

[3] Stoppel, 1907; Guilliermond, 1909.

same or different hyphae; they unite (fig. 93), their nuclei fuse, and after three karyokinetic divisions eight spores are formed. Sometimes, especially in old cultures, the fertile hyphae may produce asci without fusion; these are usually smaller than the ordinary asci and contain four or fewer spores.

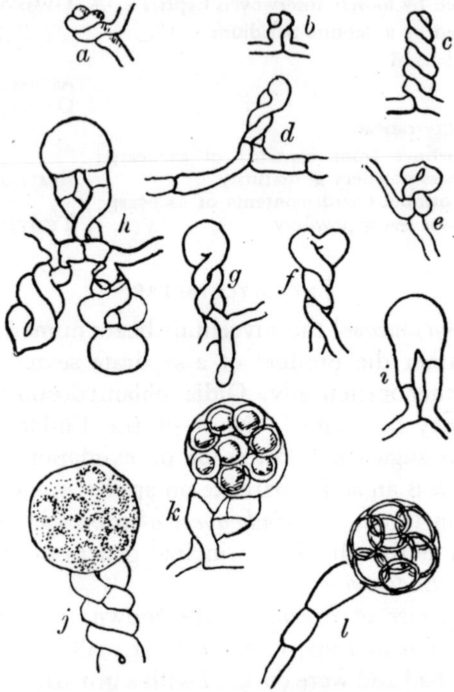

Fig. 92. *Eremascus albus* Eidam; *a, b, c, d*, sexual apparatus; *e, f, g, h*, fusion of gametangia; *i, j, k*, development of asci; *l*, parthenogenetic ascus; × 900–1000; after Eidam.

The species of *Endomyces*[1] possess a branched, septate mycelium. This may break up into oidia, which sometimes become surrounded by thick walls and form cysts, or may produce yeast-like conidia which themselves multiply by budding. The mycelium also bears naked, four-spored asci, the spores of which are often of a characteristic bowler-hat shape.

In *Endomyces Magnusii* vegetative multiplication is by means

[1] Guilliermond, 1909, 1913; Juel, 1921.

of oidia. The ascus is the product of a sexual process in which an elongated, swollen cell and a relatively narrow cell unite. The single nucleus of the smaller cell passes into the larger and fuses with its nucleus; the fusion nucleus divides twice, and four spores are formed (fig. 94). Fusion appears to take place indifferently between related or unrelated filaments, and parthenogenesis is frequent.

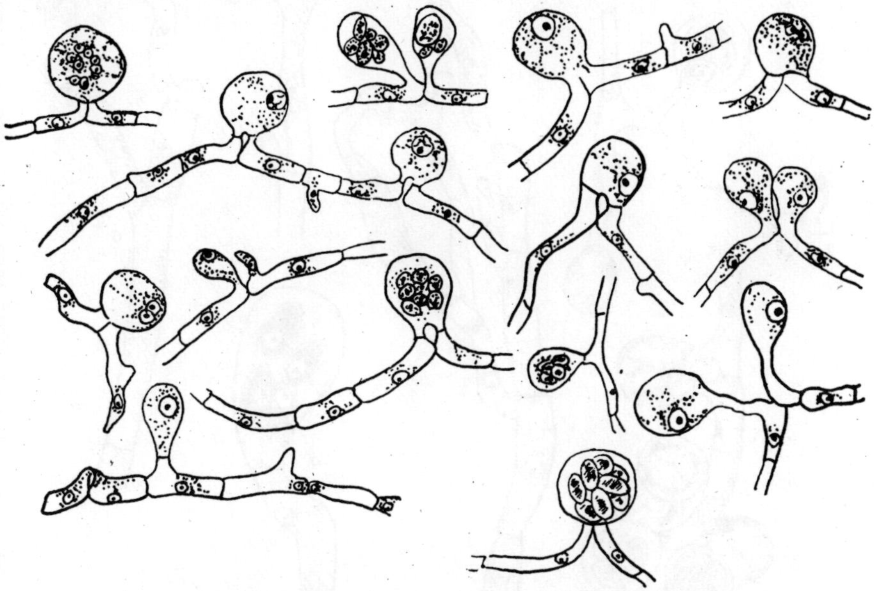

Fig. 93. *Eremascus fertilis* Stoppel; stages in the formation of the ascus, both by fusion of two cells and parthenogenetically; after Guilliermond.

In *Endomyces fibuliger* about half the asci result from the union of two filaments while the remainder are parthenogenetic. The fusing hyphae are usually related; both are similar at the time of fusion, but afterwards the growth of one ceases and the other swells to form the ascus. Vegetative multiplication (fig. 95) is by yeast-like conidia.

Endomyces Lindneri,[1] a Chinese species, resembles *E. fibuliger* in its vegetative characters, but the gametangium, instead of functioning as an ascus, sometimes gives rise to short branches

[1] Mangenot, 1919.

bearing two or more asci (fig. 96). These may be regarded as constituting a brief sporophytic stage and as foreshadowing the ascogenous hyphae of higher forms.

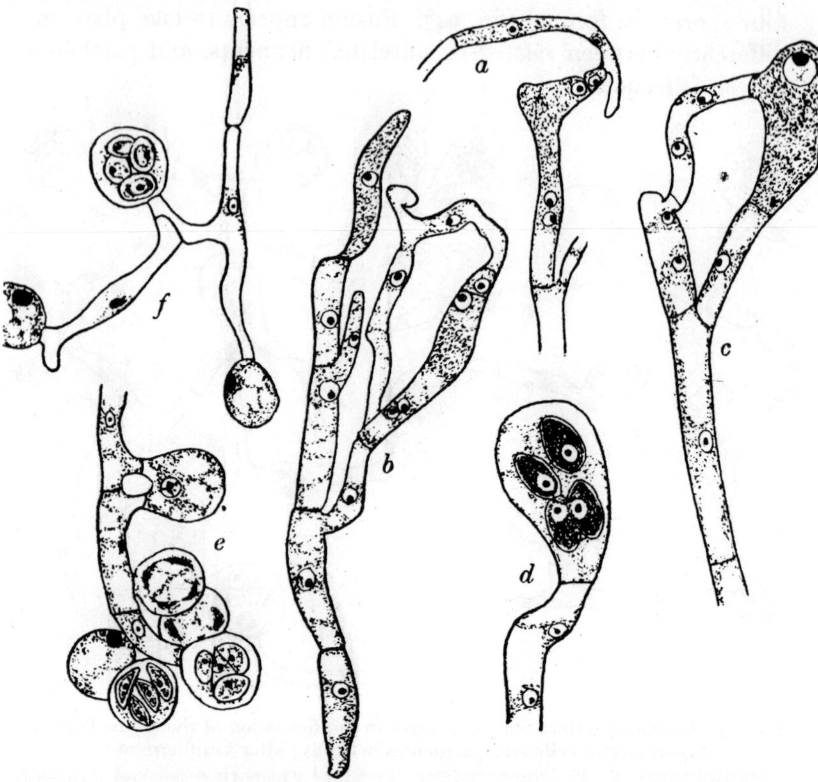

Fig. 94. *Endomyces Magnusii* Ludw.; *a*, *b*, antheridium and oogonium in contact; *c*, oogonium after fusion of sexual nuclei; *d*, parthenogenetic ascus. *Endomyces fibuliger* Lindner; *e*, conjugation between two neighbouring cells, at the end of the hypha is a group of young asci; *f*, ascus containing spores and two young parthenogenetic asci; after Guilliermond.

Dipodascus[1] (fig. 97) differs from *Endomyces* and *Eremascus* in its multinucleate cells and in the presence of numerous spores in its ascus. In the formation of the ascus two branches grow up from the same or different hyphae; they fuse at their tips, one outgrows the other and, after fertilisation, becomes the single ascus. When the gametangia have fused, one nucleus in each is seen to be larger

[1] de Lagerheim, 1892; Juel, 1902, 1921; Dangeard, 1907

than its neighbours. While the remaining nuclei degenerate, these two unite and, after a number of divisions, give rise to the nuclei of the spores (fig. 97 *c*).

In *Pericystis*, also, the cells are multinucleate and the spores numerous. *P. apis*,[1] the cause of chalk brood, a serious disease of

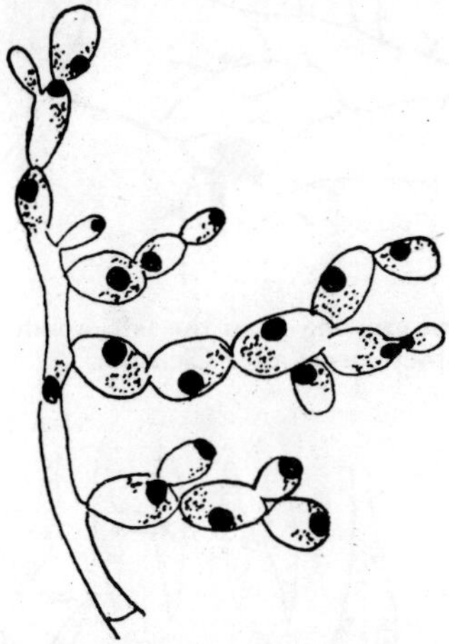

Fig. 95. *Endomyces fibuliger* Lindner; formation of conidia; after Guilliermond.

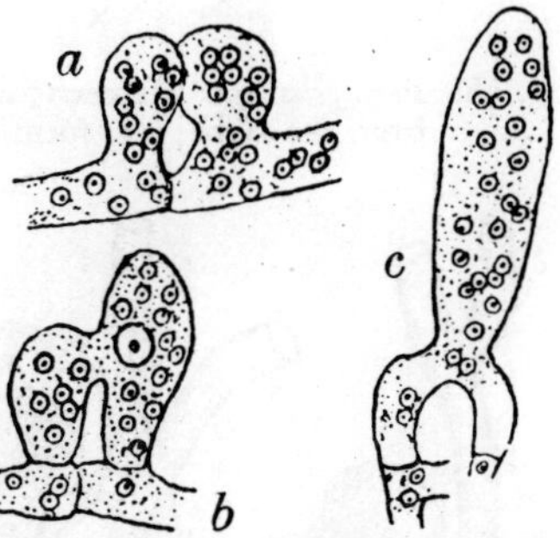

Fig. 97. *Dipodascus albidus* Lagerh.; *a*, fusion of gametangia; *b*, gametangia with fusion nucleus in the larger, which will become the ascus; *c*, young ascus with nuclei due to division of fusion nucleus; after Juel.

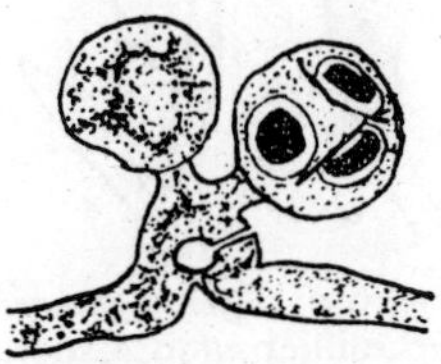

Fig. 96. *Endomyces Lindneri* (Saïto) Mang.; two asci on short ascogenous hyphae developed from a pair of fused gametangia; after Mangenot.

bees, is dioecious with male and female thalli which differ in their vegetative development, the female being denser but less extensive. Where they meet (fig. 98) gametangia are formed, and communicate by means of a conjugation tube from the male organ (fig. 99 *a*). Male nuclei enter the oogonium and fuse in pairs with the female nuclei, while any superfluous nuclei degenerate. Nuclear divisions

[1] Claussen, 1921; Varitchak, 1933.

follow, the first being probably meiotic, the oogonium enlarges considerably and becomes filled (fig. 99 *b*) with balls of spores.

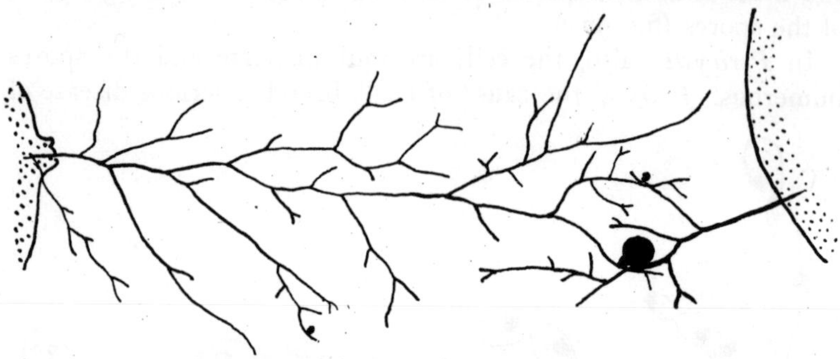

Fig. 98. *Pericystis apis* Maassen; a male plant growing from the left, a female from the right; asci forming where they meet; after Claussen.

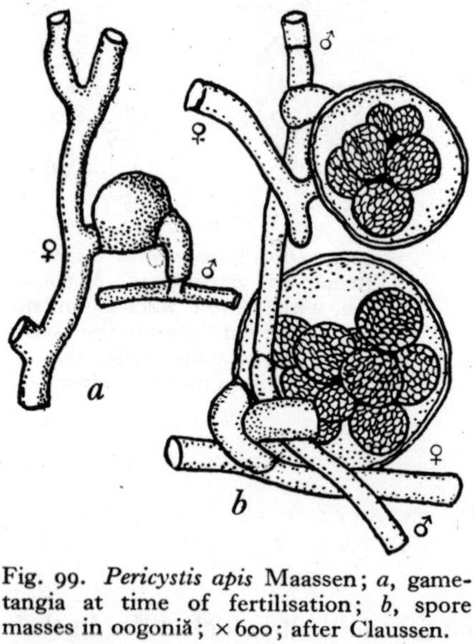

Fig. 99. *Pericystis apis* Maassen; *a*, gametangia at time of fertilisation; *b*, spore masses in oogonia; ×600; after Claussen.

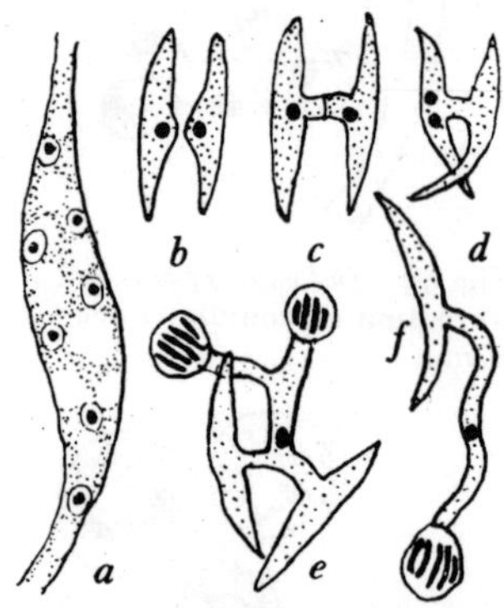

Fig. 100. *Spermophthora Gossypii* Guilliermond; *a*, sporangium; *b*, *c*, *d*, spores from sporangia in process of fusion; *e*, ascogenous hypha and asci developed from fused spores; *f*, ascogenous hypha and ascus from single spore; after Guilliermond.

Spermophthora Gossypii[1] has been isolated from seeds of cotton and grows readily on solid media. The mycelium is aseptate and branches dichotomously; chlamydospores are cut off by walls in the older parts. Terminal or intercalary swellings are formed (fig. 100 *a*) and in these, after two mitoses, fusiform spores are differentiated

[1] Ashby and Nowell, 1926; Guilliermond, 1927 i, 1928 i.

about the nuclei. The spores are liberated and develop independently or unite in pairs (fig. 100 *b*, *c*, *d*) by means of a canal. Both nuclei pass into the canal, they fuse, a hypha grows out and gives rise to one or more asci (fig. 100 *e*) containing eight spores, which germinate to form the aseptate mycelium. The aseptate mycelium thus constitutes the haplophase, and the paired spores give rise to the ascogenous hyphae or diplophase, while the union of spores constitutes a syngamy which may be primitive or reduced. Possibly we have here one of the early experiments in the formation of a sporophyte. A point of special interest is the development of the asci from uninucleate cells.

SACCHAROMYCETACEAE

The Saccharomycetaceae, or yeasts,[1] occur mainly as separate cells, which are only exceptionally united to form a mycelium; they are widely distributed in or on sugary media, and some are found as parasites in plants and animals, including man. As a rule they do little damage, those in healthy animals being soon destroyed by leucocytes, but the needle-like ascospores of *Monospora bicuspidata*[2] penetrate the gut wall of *Daphnia*, multiply by budding, and may bring about the death of the host, while *Nematospora*[3] injures or kills the seeds of legumes.

The individual cell is round or elliptical, it is bounded by a delicate membrane and possesses a relatively large nucleus with which is associated a vacuole containing metachromatic granules. Division is usually of a simple type, but karyokinesis occurs in the cells of *Schizosaccharomyces octosporus*,[4] *Nematospora Phaseoli*[5] and other species. The cytoplasm contains glycogen, oil and refractive granules of volutin.

Multiplication is brought about by transverse division and separation of the daughter cells, as in *Schizosaccharomyces*, or, more usually, by budding, that is, by the formation of successive outgrowths which attain the form and size of the parent cell; each bud receives a nucleus and cytoplasm and is cut off by a wall; before its separation it may bud again, and in this way considerable colonies are formed. Budding occurs in *Saccharomyces*, *Zygo-*

[1] Guilliermond, 1901, 1903, 1905 ii, 1910 i, ii, 1920 i, ii, 1928 iii, 1931; Hansen, 1901; Kløcker, 1909 ii; Nadson and Konokotine, 1910; Wager and Peniston, 1910; Marchand, 1913; Satava, 1918 i, ii; Guilliermond and Péju, 1920.

[2] Keilin, 1921.

[3] Wingard, 1922, 1925; Guilliermond, 1927 ii.

[4] Guilliermond, 1920 ii.

[5] Wingard, 1925.

saccharomyces and *Saccharomycopsis*. Under suitable conditions, the cell contents may round themselves up and form one to eight spores. These **endospores** are the ascospores of the yeast, the cell in which they develop functioning as an ascus. It has been shown in several species that the ascospores may conjugate in pairs,[1] and that they may form colonies (fig. 101) of cells with diploid nuclei. Any of these cells may itself become an ascus.[2] Alternatively

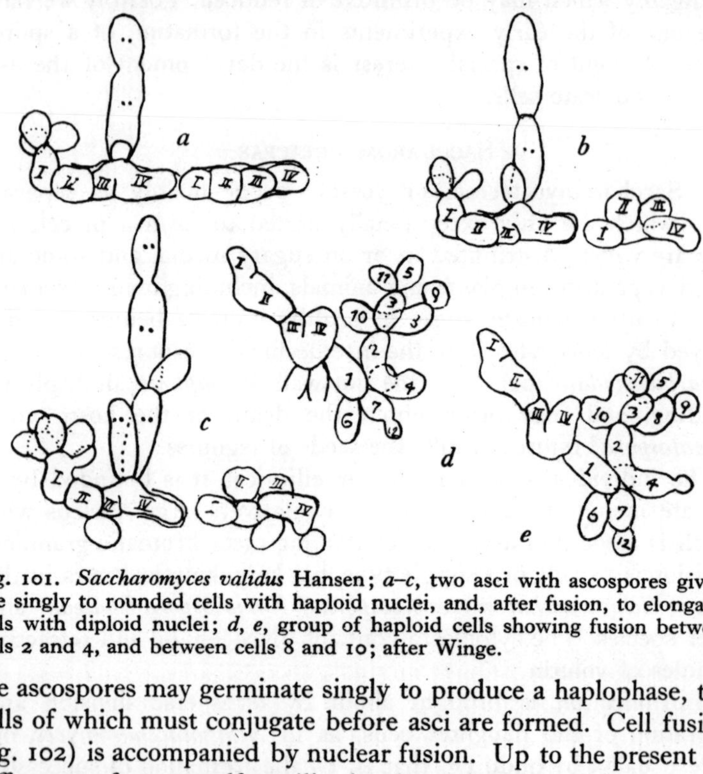

Fig. 101. *Saccharomyces validus* Hansen; *a–c*, two asci with ascospores giving rise singly to rounded cells with haploid nuclei, and, after fusion, to elongated cells with diploid nuclei; *d*, *e*, group of haploid cells showing fusion between cells 2 and 4, and between cells 8 and 10; after Winge.

the ascospores may germinate singly to produce a haplophase, the cells of which must conjugate before asci are formed. Cell fusion (fig. 102) is accompanied by nuclear fusion. Up to the present no difference of sex or self-sterility has been demonstrated.

One of the most striking features of the yeasts, which gives them considerable economic importance, is the power possessed by many species of producing alcoholic fermentation[3] in solutions of suitable sugars. Under conditions of plentiful aeration the yeast grows and multiplies rapidly, little alcohol is formed and the sugar

[1] Schiönning, 1895; Barker, 1901; Guilliermond, 1901.

[2] Satava, 1918 i, ii, 1934; Winge, 1935.

[3] Cf. p. 17.

is used as food. When free oxygen is insufficient, the main part of the sugar metabolised is used in anaerobic respiration, and the

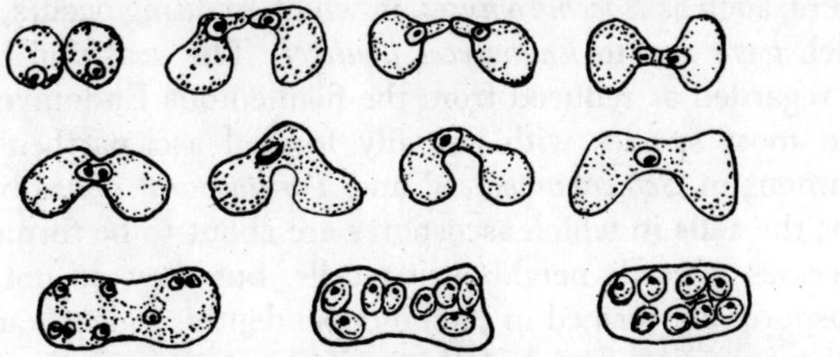

Fig. 102. *Schizosaccharomyces octosporus* Beyerinck; conjugation and formation of ascospores; after Guilliermond.

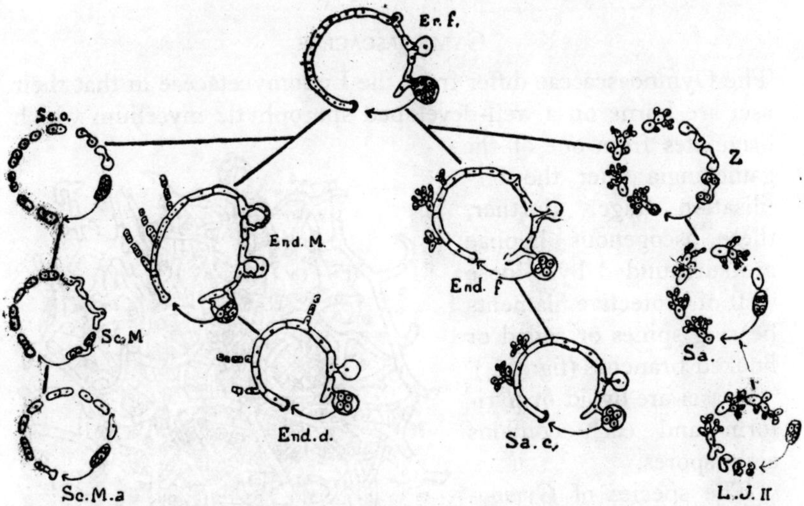

Fig. 103. Diagram of the phylogeny of the Yeasts; after Guilliermond. Er. f., *Eremascus fertilis*. End. f., *Endomyces fibuliger*. Sa. c., *Saccharomycopsis capsularis*. Z., *Zygosaccharomyces*. Sa., *Saccharomyces*. L. J. II, Johannisberg yeast II. End. M., *Endomyces Magnusii*. End. d., *Endomyces decipiens*. Sc. o., *Schizosaccharomyces octosporus*. Sc. M., *Schizosaccharomyces mellacei*. Sc. M. a., *Schizosaccharomyces mellacei*, apogamous variety.

quantity of alcohol produced is greater in proportion to the number of cells concerned.

On the ground that their daughter cells are produced by septation, and not, as in other genera, by budding, Guilliermond[1] (fig. 103)

[1] Guilliermond, 1909.

has postulated for the species of *Schizosaccharomyces* a derivation from the neighbourhood of *Endomyces Magnusii*; he refers the genera, such as *Saccharomyces*, in which budding occurs, to the line which gave rise to *Endomyces fibuliger*. The yeasts in both series are regarded as reduced from the filamentous Endomycetaceae.

In most species with sexually formed asci parthenogenesis is common; in *Schwanniomyces*[1] and *Torulaspora*[2] it has become the rule; the cells in which ascospores are about to be formed put out processes towards neighbouring cells, but they do not fuse, and ascospores are formed in each independently. Such examples lead to the condition found in the majority of the yeasts where ascospore formation takes place, not only without conjugation, but without any vestige of that process.

GYMNOASCACEAE

The Gymnoascaceae differ from the Endomycetaceae in that their asci are borne on a well-developed sporophytic mycelium which originates from one of the gametangia after the fertilisation stage. Further, these ascogenous hyphae are surrounded by a loose weft of protective filaments bearing spines or coiled or hooked branches (fig. 104). The asci are ovoid or pyriform and each contains eight spores.

Fig. 104. *Gymnoascus* sp.; *a*, ascocarp, ×265; *b*, ascus and free ascospores, ×1040.

The species of *Gymnoascus* occur as saprophytes on dung, dead grass, bees' nests, and other habitats. In *G. Reesii*[3] two branches grow up from the same hypha, one on each side of a septum, and become twisted round one another. Their free ends swell and are cut off by transverse septa forming the gametangia; the walls be-

[1] Guilliermond, 1910 ii. [2] Rose, 1910. [3] Dale, 1903.

tween them break down, the contents of the antheridium pass into the oogonium (fig. 105 *a*, *b*), and the oogonium puts out a branch which undergoes septation and gives rise to the ascogenous hyphae. The sexual organs are at first uninucleate but later coenocytic; the fusion of the nuclei has not been recognised. In *G. candidus*[1] (fig. 105 *d*), the gametangia usually arise from different filaments; they differ in form at the time of their union, the antheridium is

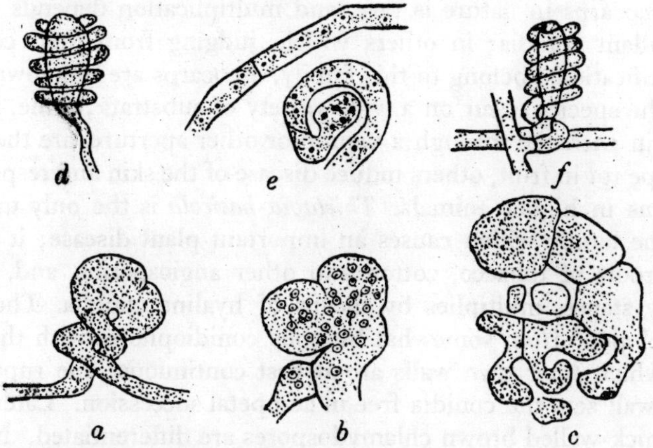

Fig. 105. *Gymnoascus Reessii* Baran.; *a*, surface view of conjugating cells; *b*, the same in longitudinal section; *c*, a later stage, septate branch from oogonium giving rise to ascogenous hyphae. *Gymnoascus candidus* Eidam; *d*, surface view of conjugating cells; *e*, same in longitudinal section; all after Dale. *Ctenomyces serratus* Eidam; *f*, surface view of conjugating cells; ×400; after Eidam.

straight and somewhat swollen, the oogonium grows spirally round it till the apices meet and fuse (fig. 105 *e*); afterwards the oogonium undergoes septation and gives rise to ascogenous hyphae; the sheath of protective branches is very scanty.

Ctenomyces serratus[2] (fig. 105 *f*), which occurs saprophytically on feathers and other debris, and *Amauroascus verrucosus*,[3] on leather, form sexual organs which, in the early stages, closely resemble those of *Gymnoascus*. In *Amauroascus*, however, the oogonium is stated to give rise to ascogenous hyphae without septation.

Several of the fungi responsible for skin diseases[4] have been shown to be degenerate members of this family; traces of the

[1] Dale, 1903.
[2] Eidam, 1880; Marsh, 1926.
[3] Dangeard, 1907.
[4] Grigoraki, 1925.

characteristic outgrowths from the weft of hyphae which surrounds the asci of normal forms are still found on the mycelium.

Aspergillaceae

The Aspergillaceae[1] are distinguished from the earlier families of the Plectascales by the fact that their ascogenous hyphae are surrounded by a continuous sheath of gametophytic filaments, so that a closed fructification is formed. In many species the development of ascocarps in nature is rare, and multiplication depends on the abundant conidia; in others which, judging from their conidial fructifications, belong to this family, ascocarps are unknown.

The species occur on a wide variety of substrata; some, if they obtain entrance through a wound or other aperture, are the cause of ripe rot in fruit, others induce disease of the skin and respiratory organs in higher animals. *Thielavia basicola* is the only member of the family which causes an important plant disease; it infects the roots of tobacco, cotton and other angiosperms, and, in the early stages, multiplies by means of hyaline conidia. These are developed[2] from somewhat bulbous conidiophores with the walls of which their own walls are at first continuous; the rupture of the wall sets the conidia free in acropetal succession. Later, rows of thick-walled brown chlamydospores are differentiated. Normal development of the infected root is prevented and the host is killed or stunted; if death ensues, ascocarps develop on the dead plant.

The formation of an ascocarp in the Aspergillaceae is initiated by the appearance of sexual organs, from the stalks of which the cells of the sheath arise. In *Microascus* and *Emericella* the sheath opens by a pore, in other cases it remains closed, and the ascospores are liberated by its decay. The asci are spherical or pyriform and contain two to eight spores. Both in *Penicillium* and *Eurotium* sclerotia sometimes appear which resemble the ascocarps, but possess an exceptionally thick wall and contain no asci; they may prove to be immature fructifications which have passed into a resting state.

In the investigated species of *Eurotium* the cells of the mycelium are multinucleate; conidiophores appear early, as thick aseptate hyphae, the free end becomes swollen and buds out numerous

[1] Thom and Church, 1926.
[2] Brierley, 1915.

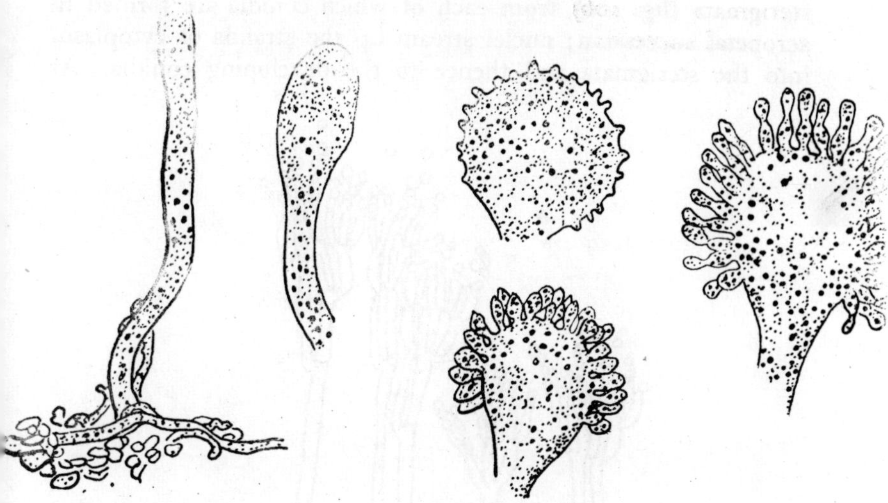

Fig. 106. *Eurotium herbariorum* (Wigg.) Link; development of conidiophores and conidia, ×625.

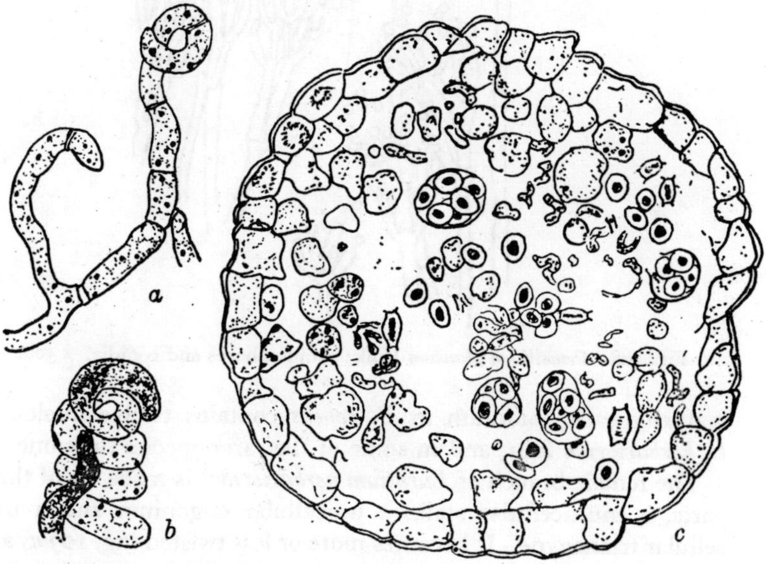

Fig. 107. *Eurotium herbariorum* (Wigg.) Link; *a*, young archicarp; *b*, archicarp and abortive antheridium; *c*, ascocarp containing asci and spores; ×625.

sterigmata (fig. 106), from each of which conidia are formed in acropetal succession; nuclei stream up the strands of cytoplasm into the sterigmata and thence to the developing conidia. At

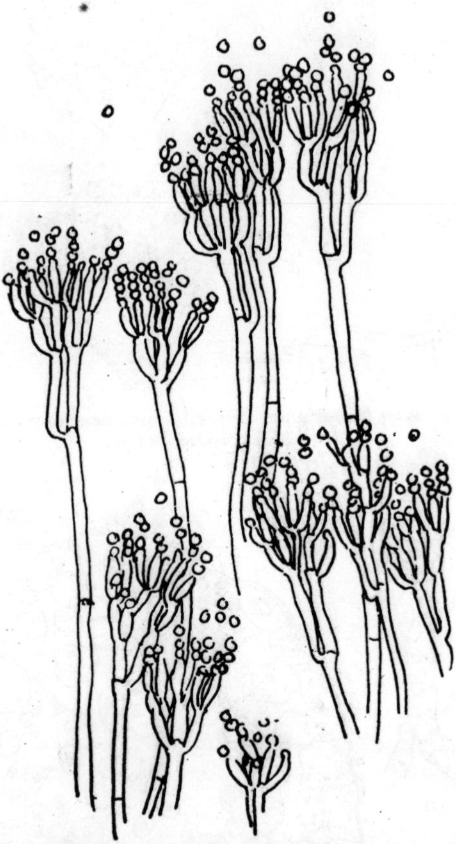

Fig. 108. *Penicillium glaucum* Link; conidiophores and conidia, × 500.

maturity each conidium in *E. repens* contains twelve nuclei, in *E. herbariorum* four, and in some of the rarer species only one.

The female branch of *Eurotium herbariorum*[1] is made up of three parts, a multicellular stalk, a unicellular oogonium and a unicellular trichogyne. It becomes more or less twisted (fig. 107 *a*) and

[1] de Bary, 1870; Fraser, H. C. I., and Chambers, 1907.

near it another septate hypha may appear from the end of which the unicellular antheridium is cut off. Like the cells of the mycelium all parts of the sexual apparatus are coenocytic. Fusion may take place between the antheridium and trichogyne, but the contents of the male organ have not been seen to enter the oogonium. Even if normal fertilisation still sometimes occurs, it is certainly not general, for the antheridium often fails to reach the trichogyne (fig. 107 *b*) and is frequently absent altogether. Both in *E. repens*[1] and *E. Fischeri*[2] fusions of nuclei in the oogonium have been recorded, but they are probably fusions of female nuclei in pairs, and not of male and female nuclei.

Nevertheless the oogonium becomes septate and from its cells branches develop and give rise to asci. The ascus is formed from the terminal or penultimate cell of a hypha and in it fusion of two nuclei takes place. The spores have a sculptured epispore and are of characteristic form, so that, when seen laterally, they somewhat resemble a butcher's tray.

Shortly before the septation of the oogonium vegetative hyphae begin to grow up about the sexual organs; some of them extend inwards forming a nutritive layer, while others constitute the outer wall of the sheath and secrete a thick, brittle pellicle of yellow substance, readily soluble in alcohol. Later the ascogenous hyphae, the nutritive cells and the walls of the asci are absorbed, supplying food to the developing spores, so that the spores at last lie free within the outer layers of the sheath and are liberated by its decay. *Eurotium* was known for some time before its ascocarps and conidia were recognised as belonging to the same form; the generic name *Aspergillus* was given to the conidial stage and still survives in the name of the family.

The fertile branch in *Microeurotium albidum*[3] consists, as in *Eurotium herbariorum*, of a trichogyne, oogonium and stalk. The oogonium, however, is at first uninucleate and later contains two nuclei probably as a result of mitosis. The nuclei fuse, the oogonium becomes the single ascus, it is surrounded by a perithecial wall and later forms from two to over twenty spores. This fungus appears to be a reduced form, comparable, in respect of its single ascus, to *Sphaerotheca* among the Erysiphaceae.

[1] Dale, 1909. [2] Domaradsky, 1908.
[3] Ghatak, 1936.

In *Penicillium*[1] multiplication takes place chiefly by means of the abundant conidia borne in chains on branched brush-like conidiophores (fig. 108). In most species ascocarps are rare. In *P. glaucum*,[2] *P. luteum*[3] and *P. bacillosporum*[3] they are initiated by the development of two short, intertwined branches (fig. 109), while in others there is a coil similar to that found in *Eurotium*. Most of the species are self-fertile, but in *P. luteum*[4] self-incompatibility has been recorded and the strains also differ in the feebleness or vigour of their develop-

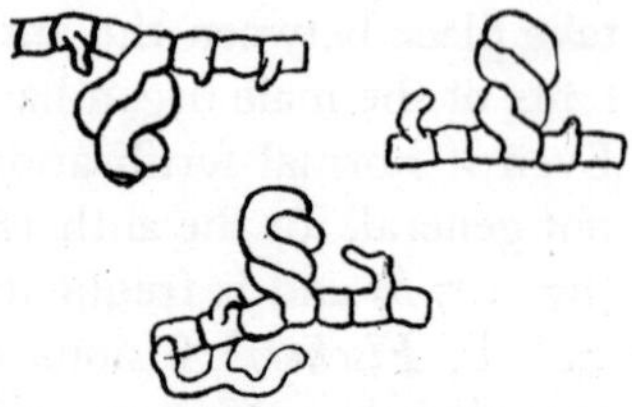

Fig. 109. *Penicillium glaucum* Link; conjugating çells, ×630; after Brefeld.

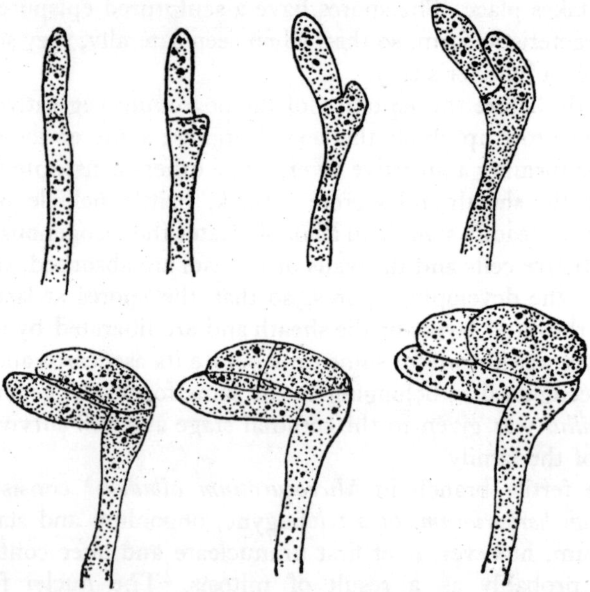

Fig. 110. *Monascus Barkeri* Dang.; development of oogonium, trichogyne and antheridium, ×900; after Barker.

ment. On the vigorous strains ascocarps, though without asci, appear in single spore culture, numerous fertile ascocarps result when two

[1] Brefeld, 1874; Zukal, 1887; Kløcker, 1903; Dangeard, 1907; Thom, 1910; Bezssonof, 1918 i; Dodge, 1933.
[2] Brefeld, 1874.
[3] Emmons, 1935.
[4] Derx, 1926.

vigorous strains meet, few when two feeble strains come into contact and an intermediate crop when a feeble and a vigorous strain are brought together.

In some species asci are borne on lateral outgrowths of the ascogenous hyphae, in others they are formed in chains, presumably

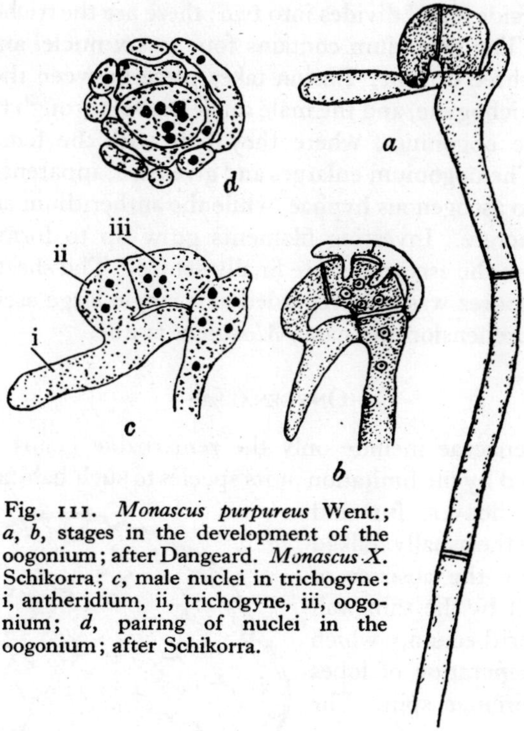

Fig. 111. *Monascus purpureus* Went.; *a*, *b*, stages in the development of the oogonium; after Dangeard. *Monascus X.* Schikorra; *c*, male nuclei in trichogyne: i, antheridium, ii, trichogyne, iii, oogonium; *d*, pairing of nuclei in the oogonium; after Schikorra.

because the binucleate cells of the ascogenous filaments do not branch out.

A further study of the genus is much to be desired, especially in view of the resemblance of the gametangia to those of *Gymnoascus* and *Eurotium*.

In the genus *Monascus*[1] there is an abundant mycelium bearing chains of ovoid conidia. *M. Barkeri* is used by the Chinese for the

[1] Barker, 1903 i; Ikeno, 1903 ii; Kuyper, 1905; Olive, 1905; Schikorra, 1909; Young, 1931.

manufacture of an alcoholic liquor known as Samsu. The development of the sexual organs has been seen in this species, in *M. purpureus* and in *M. ruber*. A branch of the mycelium cuts off a small, terminal cell which elongates and bends sideways to form the antheridium (fig. 110). A prolongation of the penultimate cell grows alongside it and divides into two; these are the trichogyne and oogonium. The oogonium contains four to six nuclei and the antheridium three or four. Fusion takes place between the antheridium and trichogyne, and the male nuclei travel through the trichogyne to the oogonium, where they pair with the female nuclei (fig. 111). The oogonium enlarges and gives rise, apparently without septation, to ascogenous hyphae, while the antheridium and trichogyne degenerate. Investing filaments grow up to form a sheath within which the ascospores are finally set free. The sheath with its contained spores was long regarded as a single large ascus, and to this misapprehension the name *Monascus* is due.

Onygenaceae

The Onygenaceae include only the remarkable genus *Onygena*,[1] characterised by the limitation of its species to such habitats as derelict horns, hooves, fur and feathers, by the usually stalked ascocarp, by the absence of conidia, and by the thin wall of the perithecium, which opens by separation of lobes or by a circular split. The best known species is *Onygena equina*, in which the ascocarp first appears as a dome-shaped mass of white hyphae; the outer of these divide into short segments which are liberated as chlamydospores. Asci are produced internally and give rise to eight spores each; the ascus walls soon disappear and the ascospores lie free among the vegetative hyphae. This mature stage, in which there is no trace of

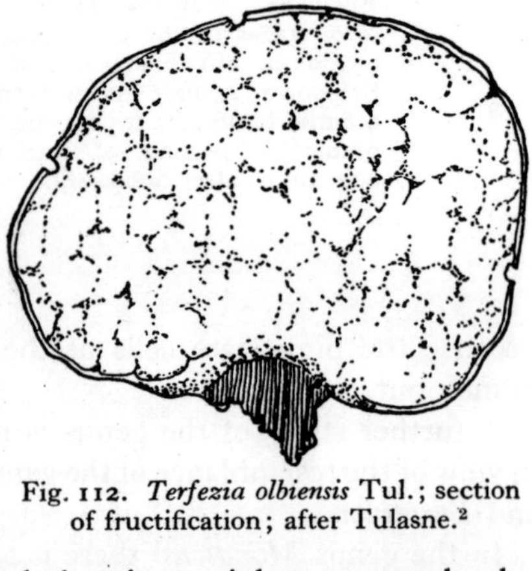

Fig. 112. *Terfezia olbiensis* Tul.; section of fructification; after Tulasne.[2]

[1] Ward, 1899; Brierley, 1917.

[2] Tulasne, 1851.

asci, caused *Onygena* to be classified with the Myxomycetes and also with the Lycoperdaceae before its true position was discovered. There is no sign of sexual organs.

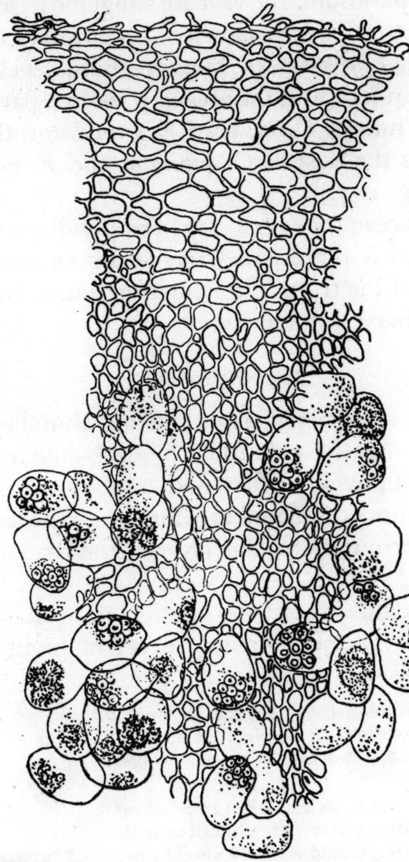

Fig. 113. *Terfezia olbiensis* Tul.; section through hymenium, showing asci irregularly arranged; after Tulasne.

ELAPHOMYCETACEAE AND TERFEZIACEAE[1]

In these two families the fruit is subterranean; it differs from that of other hypogeal members of the Ascomycetes, and resembles that of the subaerial Plectascales in the irregular arrangement of the asci, which are scattered or grouped in nests surrounded by

[1] Tulasne, 1851.

sterile branches; the fertile tissue is not exposed at any stage of development.

In the Elaphomycetaceae the ascocarp[1] is surrounded by a thick, yellow or brown peridium, the asci are subglobose, and the interior of the perithecium is filled at maturity by a powdery mass of spores. The chief genus is *Elaphomyces*. The mycelium develops in relation to the roots of conifers, and the ascocarp is often parasitised by the pyrenomycetous fungus *Cordyceps*;[2] *E. granulatus*, the commonest British species, is the host of *C. capitata*, and *E. variegatus* of *C. ophioglossoides* (fig. 184 *b*).

In the Terfeziaceae (figs. 112, 113) the peridium is less distinct, and in some cases is represented merely by an ascus-free region in the periphery of the fruit. The spores do not, as in *Elaphomyces*, form a powdery mass at maturity.

ERYSIPHALES

The Erysiphales are characterised by an abundant, superficial mycelium which may be colourless or dark-coloured. The perithecia are spherical, ovoid or flattened, and are usually without an ostiole; the peridium is thin and membranous; the asci are parallel one to another, forming a regular layer at the base of the fructification.

The alliance includes some 600 species, most of which are epiphytes or external parasites on the leaves of higher plants, and which may be classified as follows:

Aerial mycelium colourless. Perithecia more or less globose without an ostiole, furnished with conspicuous appendages. Conidia in single rows	ERYSIPHACEAE
Aerial mycelium dark in colour, rarely absent. Perithecia globose or ovoid, without an ostiole or appendages. Conidia not in single rows	PERISPORIACEAE
Aerial mycelium dark in colour or absent. Perithecia flattened or shield-shaped, with an ostiole, without appendages. Conidia absent	MICROTHYRIACEAE

The Erysiphaceae are apparently a primitive family and approach the Plectascales in the characters of their sexual organs, in their type of fructification, and especially in their globose asci with colourless, continuous spores. In the Perisporiaceae, on the other hand, the spores are commonly two or more celled and often dark-coloured,

[1] Clémencet, 1932. [2] Cf. p. 252.

the asci are usually cylindrical, as in the Pyrenomycetes, and some of the characters of the perithecium also recall the lower members of that group. The Microthyriaceae approach the Perisporiaceae in the characters of the ascus and spore, but are clearly distinguished by their curious, flattened perithecium.

Erysiphaceae

The members of the Erysiphaceae[1] are popularly known as white or powdery mildew or blight. They have been recorded especially in Europe and North America, but their distribution is almost cosmopolitan.

They are obligate parasites on the leaves, young shoots and inflorescences of flowering plants. Both the conidium and the ascospore give rise on germination to an abundant, superficial mycelium of uninucleate cells, which forms a white, web-like coating over the leaf and sends haustoria into the epidermal cells of the host. In the simplest examples the haustorium is a slender tube which swells inside the host cell; in others it is branched, sometimes forming finger-like processes, and frequently provided with an external disc or appressorium, from which, or from the mycelium in the neighbourhood of which, the haustorium proper arises and pushes into the epidermal cell. As a rule the fungus does not penetrate farther, but in *Erysiphe Graminis* endophytic growth is induced if the conidia germinate on wounded leaves; in *Phyllactinia Corylea* the branches of the aerial mycelium enter through the stomata, extend through the intercellular spaces and send haustoria into the neighbouring cells; in *Erysiphe taurica*, a species occurring in hot, dry climates, the whole mycelium during the conidial stage is located in the tissues,[2] and conidiophores grow out through the stomata. When perithecia are about to be produced, the mycelium emerges and spreads over the surface of the host; both in *Phyllactinia* and in *E. taurica* the aerial hyphae show appressoria, although no haustoria are formed. It may be inferred that they are derived from forms which possessed a purely sub-aerial mycelium with haustoria, and that the ectophytic condition is primitive in the family. An alternative descent from endophytic Pyrenomycetes with stromata in the tissues of the host has, however, been proposed.[3]

[1] Salmon, 1900–7; Sands, 1907. [2] Salmon, 1905–6. [3] Arnaud, 1920.

The Erysiphaceae are propagated in summer by rather large, oval, uninucleate conidia[1] (fig. 114), produced in rows on unbranched conidiophores. Before the connection of the conidial with the ascigerous stage was recognised, the generic name *Oidium* was used for the former and it still survives for species in which the perithecia have not been seen.

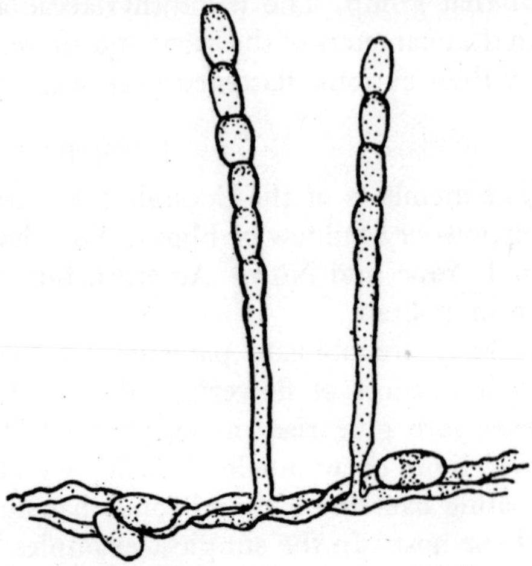

Fig. 114. *Sphaerotheca pannosa* Wallr.; conidiophores and conidia, × 240.

The perithecia of the Erysiphaceae appear in the late summer or autumn; they are spherical or subglobose, 0·05 to 0·3 mm. in diameter, and furnished with simple or branched appendages; a secondary mycelium fixes them to the substratum. Young ascocarps are white and glistening like the vegetative hyphae, later they be-

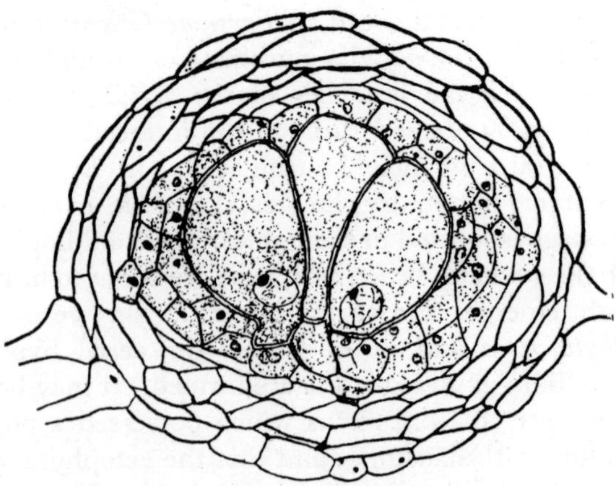

Fig. 115. *Erysiphe Polygoni*; young perithecium containing uninucleate asci; after Harper.

[1] Foëx, 1912.

come clear yellow, and finally brown; a considerable range of stages can often be seen within the field of a hand lens.

During development the wall of the ascocarp is differentiated into inner and outer layers (fig. 115). The inner layer is made up

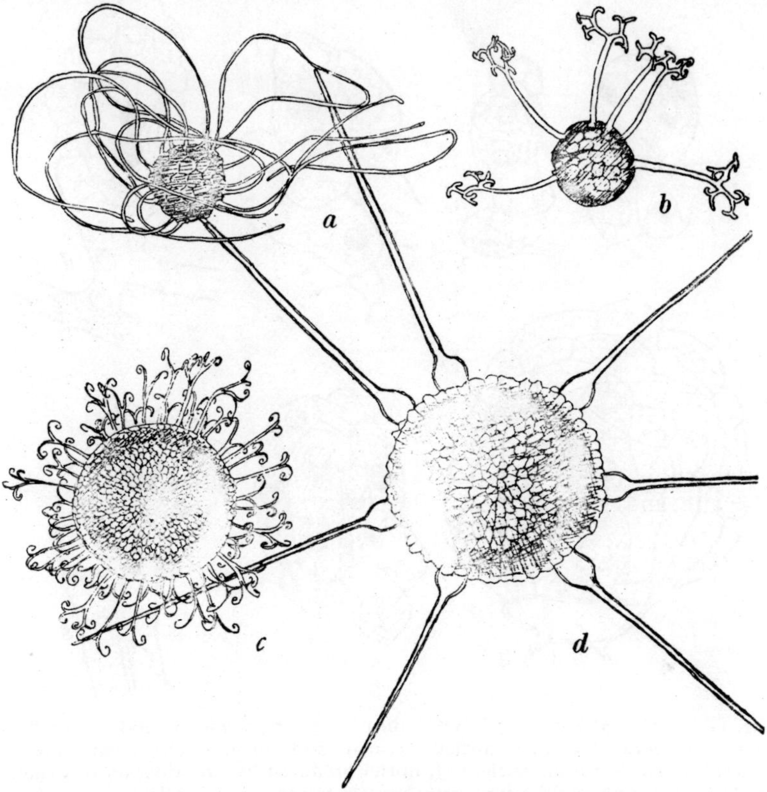

Fig. 116. Perithecia of *a*, *Erysiphe tortilis* (Wallr.) Fr.; *b*, *Microsphaera* sp.; *c*, *Uncinula Aceris* (DC.) Sacc.; *d*, *Phyllactinia Corylea* (Pers.) Karst.; × 120.

of thin-walled cells rich in cytoplasm; it forms a packing round the developing asci and supplies them with food material. The function of the outer layer is protective; its cells have scanty contents, and their walls undergo a change analogous to lignification; from this layer the characteristic appendages are derived.

A single ascus is formed in the fructification of *Sphaerotheca* and *Podosphaera*; in other genera several asci are formed; the spores in each ascus number two to eight. The development of the spores begins in the summer or autumn,[1] but they are not ordinarily set free by the decay of the perithecial wall until the following spring when they produce the first infections of the season.

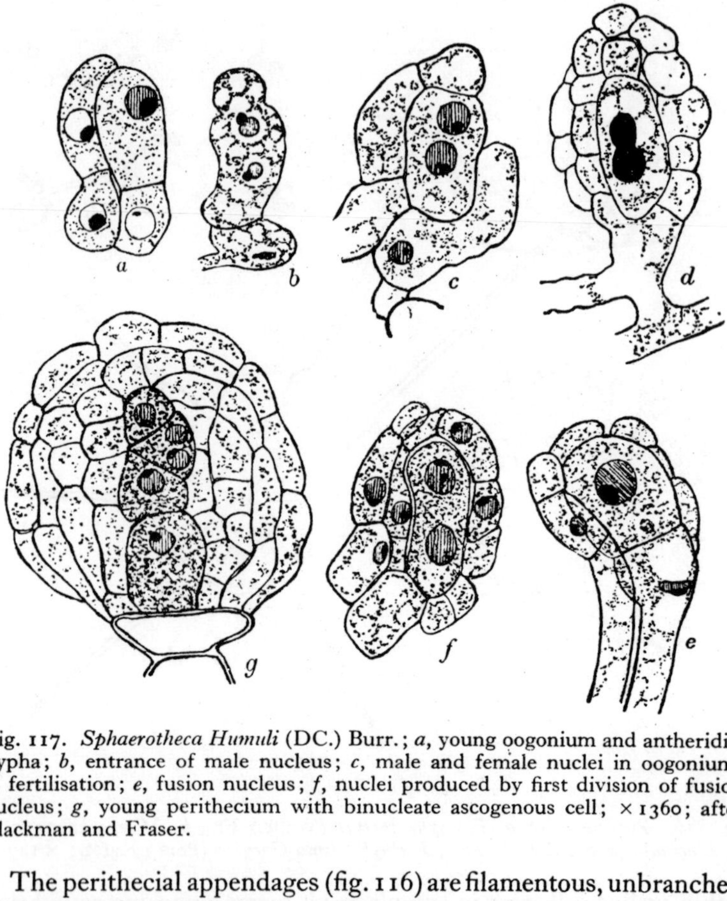

Fig. 117. *Sphaerotheca Humuli* (DC.) Burr.; *a*, young oogonium and antheridial hypha; *b*, entrance of male nucleus; *c*, male and female nuclei in oogonium; *d*, fertilisation; *e*, fusion nucleus; *f*, nuclei produced by first division of fusion nucleus; *g*, young perithecium with binucleate ascogenous cell; × 1360; after Blackman and Fraser.

The perithecial appendages (fig. 116) are filamentous, unbranched or branched irregularly in *Sphaerotheca* and *Erysiphe*; in *Podosphaera* and *Microsphaera* they are dichotomously branched; in *Uncinula* their apices are spirally coiled; and in *Phyllactinia* the perithecia bear stiff, pointed hairs with swollen bases. In the last-named genus the top of the young ascocarp is furnished with a ring

[1] Salmon, 1913 i, ii, iii.

of short, richly branched hyphae. At about the time of spore formation these break down, forming a sticky, gelatinous cap by which the perithecium, when first set free, adheres upside down to its original host or to other objects. In view of this peculiarity *Phyllactinia* must not be assumed to be parasitic on any plant on which its perithecia happen to be found. The usual function of the appendages is to anchor the perithecium to the host during development, but in *Uncinula* the tips become mucilaginous and serve the same purpose as the branched cells of *Phyllactinia*. In *Phyllactinia* the bulb at the base of the appendage executes hygroscopic movements, its underside being pushed in and therefore shortened as water is lost; in this way, when the perithecium is ripe and ceases to absorb water, it is loosened from the host leaf.

The first account of development in this family, as in so many others, is due to de Bary, who described *Sphaerotheca Humuli* in 1863; his work has since been confirmed and extended.[1] This species occurs on a number of common plants, and is the cause of hop mildew and of the mildew on strawberry. The sexual organs arise as lateral branches from the mycelium and project at right angles to the infected surface. The oogonium is an oval, uninucleate cell; it is cut off from the parent hypha and a stalk cell may be differentiated below it (fig. 117 *a*). The antheridium is much smaller and is borne on an elongated, narrow stalk (fig. 117 *c*); like the oogonium it contains a single nucleus. The wall between the oogonium and antheridium now breaks

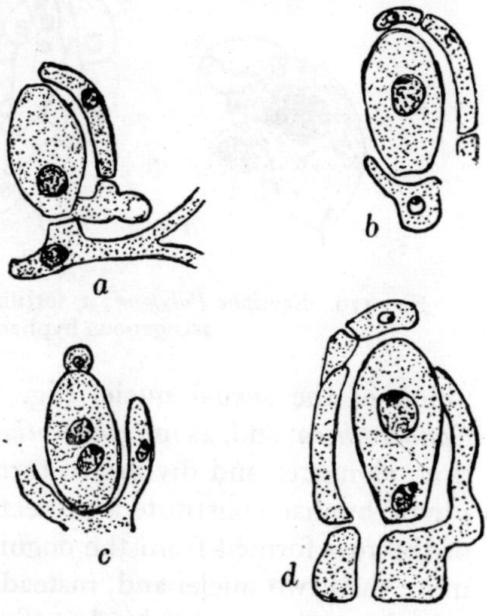

Fig. 118. *Sphaerotheca Humuli* (DC.) Burr.; *a–d*, development of sexual apparatus; in *c*, two nuclei, regarded as the product of division, are shown in the oogonium, while a cell at the top of the oogonium, regarded as the antheridium, still contains a nucleus; after Winge.

[1] de Bary, 1863 ii; Harper, 1895, 1897; Blackman and Fraser, 1905; Hein, 1927.

down, the male nucleus passes into the oogonium, and unites with the female nucleus (fig. 117 *d*); mitosis follows, and the oogonium is divided into a row of cells (fig. 117 *g*), the penultimate of which contains two nuclei and becomes the single ascus. Meanwhile a sheath of gametophytic hyphae has grown up around the sexual apparatus and forms the wall of the ascocarp. Specimens have been described in which the male nucleus fails to enter the oogonium (fig. 118), and it seems likely that apogamy is not uncommon.[1]

Erysiphe Polygoni[2] occurs on the leaves and stems of a considerable variety of hosts. The development of the sexual organs and the

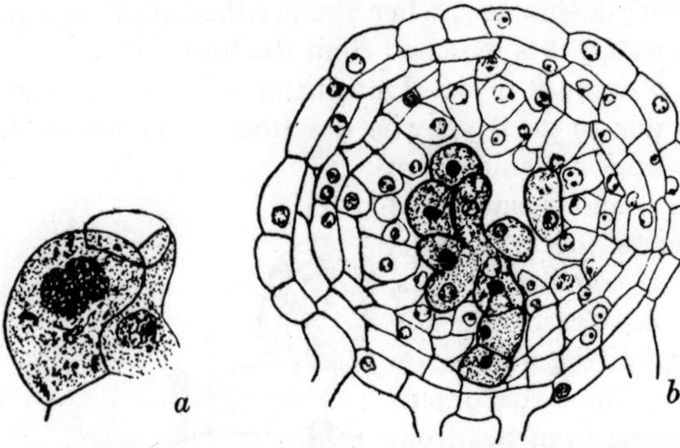

Fig. 119. *Erysiphe Polygoni*; *a*, fertilisation; *b*, young perithecium with ascogenous hyphae; after Harper.

union of the sexual nuclei (fig. 119 *a*) takes place much as in *Sphaerotheca*, and, as in *Sphaerotheca*, the oogonium, after fertilisation, elongates and divides to form a row of cells, while gametophytic hyphae constitute a protective sheath. The penultimate cell of the row formed from the oogonium, however, usually contains more than two nuclei and, instead of becoming the ascus directly, it buds out ascogenous hyphae (fig. 119 *b*), from the cells of which asci are formed.

Phyllactinia Corylea[3] infects the leaves of deciduous trees and shrubs, including ash, oak, beech, hazel and hornbeam. The sexual organs arise where two hyphae meet; they become closely applied,

[1] Dangeard, 1897, 1907; Winge, 1911; Bezssonof, 1914.
[2] Harper, 1896, 1905. [3] Harper, 1905; Raymond, 1934.

and the oogonium, as it grows, sometimes becomes twisted around the antheridium. In due course the male nucleus enters the female organ (fig. 121 *a*), fusion takes place, the fusion nucleus divides,

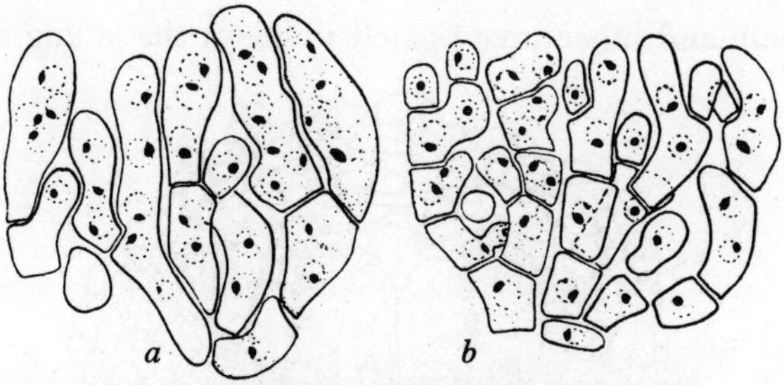

Fig. 120. *Phyllactinia Corylea* (Pers.) Karst.; groups of ascogenous hyphae; *a*, young; *b*, older; × 800; B. Colson.

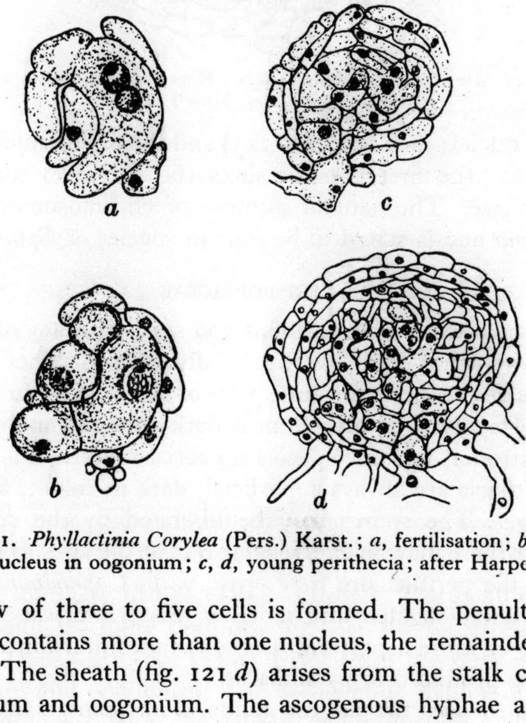

Fig. 121. *Phyllactinia Corylea* (Pers.) Karst.; *a*, fertilisation; *b*, fusion nucleus in oogonium; *c*, *d*, young perithecia; after Harper.

and a row of three to five cells is formed. The penultimate cell regularly contains more than one nucleus, the remainder are uninucleate. The sheath (fig. 121 *d*) arises from the stalk cells of the antheridium and oogonium. The ascogenous hyphae are formed as lateral branches from the septate oogonium; all or most being derived from the penultimate cell; they are made up of binucleate cells with end cells containing one nucleus (fig. 120). Eight nuclei are formed as usual in each ascus, but spores are organised around only two.

In these and other investigated mildews the young ascus con-

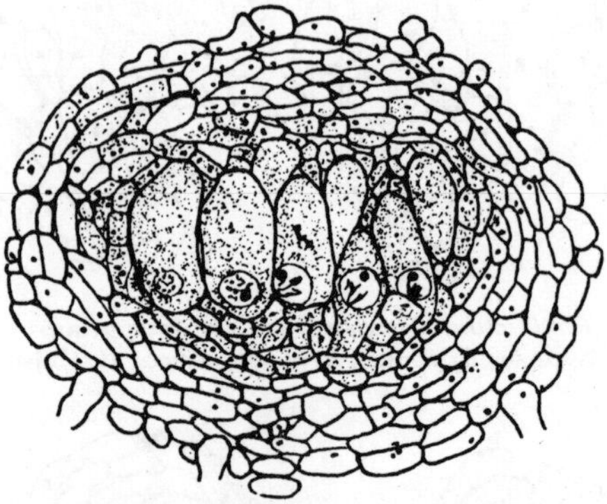

Fig. 122. *Phyllactinia Corylea* (Pers.) Karst.; perithecium containing uninucleate asci; after Harper.

tains two nuclei; they fuse (fig. 123) and the fusion nucleus divides three times, the first two divisions (fig. 124 *a–c*) constituting a meiotic phase. The haploid number of chromosomes is eight in *Phyllactinia* and is stated to be four in species of *Sphaerotheca*.[1]

PERISPORIACEAE

The Perisporiaceae include about 300 species, many of which are little known while few have been fully studied. They develop as epiphytes on the leaves or young parts of plants, or occur on decaying plant substances. The mycelium is dark-coloured and occasionally forms a stroma; in some species no aerial mycelium is developed. The perithecia are always superficial, dark in colour, and without appendages. The spores may be liberated by the decay of the sheath, or there may be an irregular rent at the apex as in *Antennularia*, or the perithecium may open, as in *Capnodium*, by valves. The asci are elongated and more or less cylindrical. The spores have one or more septa and are sometimes muriform; accessory spores are produced in pycnidia.

Dimerosporium, the largest genus, with some sixty species, is

[1] Winge, 1911; Bezssonof, 1914.

epiphytic on the leaves of Angiosperms; *D. Collinsii* forms witches' brooms on the service-berry.

The species of *Capnodium* and *Antennularia* are among the soot fungi. These, like the wild yeasts, are epiphytic saprophytes; they occur on leaves frequented by aphids and obtain their food from the

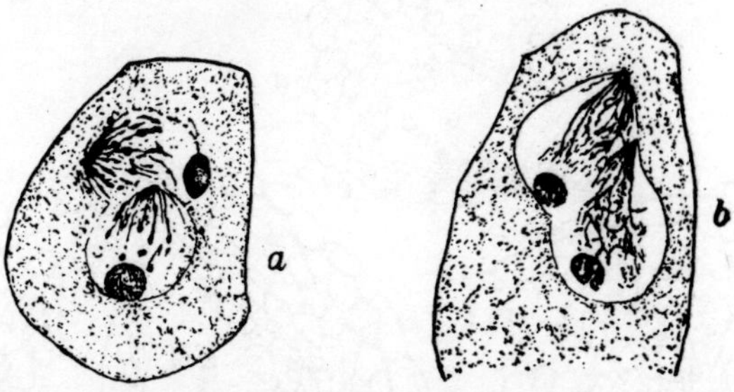

Fig. 123. *Phyllactinia Corylea* (Pers.) Karst.; *a*, *b*, fusion in ascus; after Harper.

"honey dew" secreted by these insects. The dark-coloured mycelium forms a sooty covering on the leaves of the host, but is not thick enough to injure them by excluding light.

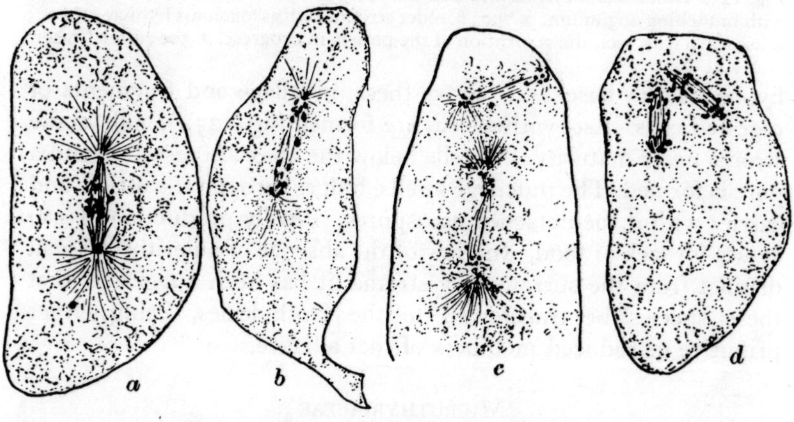

Fig. 124. *Phyllactinia Corylea* (Pers.) Karst.; *a*, metaphase of first division in ascus; *b*, anaphase of first division; *c*, anaphase of second division in ascus; *d*, anaphase of third division; after Harper.

In *Aithaloderma* and *Capnodium*[1] the formation of ascocarps is preceded by the development of a stroma (fig. 125 *a*) in which the

[1] Fraser, L., 1933, 1934, 1935 i, ii, iii.

oogonium appears as a multinucleate cell. Its nuclei divide and ascogenous hyphae grow out (fig. 125 *b*). Meanwhile digestion of the overlying cells takes place, accompanied by further growth of the stroma, till a hollow, globular body is produced with the ascogenous

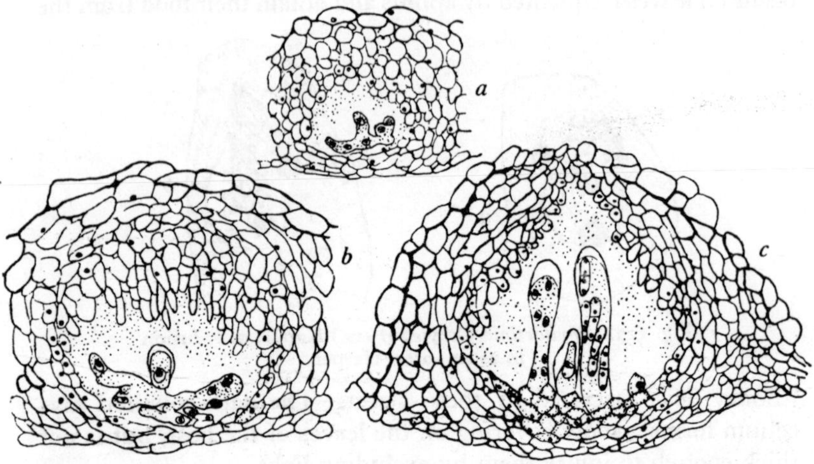

Fig. 125. *Aithaloderma ferruginea* Fraser; *a*, longitudinal section through stroma with branching oogonium, × 700; *b*, older stroma with ascogenous hyphae, × 700; *c*, ascocarp with asci, the resorption of the pore is in progress, × 500; after Fraser.

hyphae at its base and, above these, mucilage and fragments of digested cells. Asci with spores are formed (fig. 125 *c*), and, as the stroma nears maturity, the cells below the apex are resorbed and a pore is formed. The mucilage swells, bulges through the ostiole and carries with it the detached ascospores. Finally production of asci ceases for lack of food. In view of the absence of a perithecial wall distinct from the surrounding stroma, it has been suggested that these species be transferred to the Dothideales, being either primitive or reduced members of that alliance.

Microthyriaceae

The aerial mycelium of the Microthyriaceae is dark-coloured and superficial; the perithecia are flattened and shield-shaped, with only the upper part of the sheath fully developed. The asci are cylindrical or pyriform, the spores are frequently bicellular. The fructification may open by a definite ostiole, as in *Microthyrium*, or may become torn at the apex, as in *Asterina*. These two genera, with about forty and ninety species respectively, and *Asterella*, with sixty, are the largest in the family. The species are mainly tropical epiphytes on leaves, with few representatives in Europe or North America.[1]

EXOASCALES

The Exoascales include the single family Exoascaceae, characterised by the type and arrangement of the asci, which are cylindrical in form, parallel in arrangement, and unprotected by a sheath. No sexual organs are known. The cells of the mycelium on which the asci arise sometimes contain two nuclei, and fusion occurs in the young ascus. It is open to question whether the parallel arrangement and cylindrical form of the asci have any phylogenetic significance, and are not rather the result of their development as a dense palisade near the surface of the infected carpel or leaf. A derivation of the alliance from the Protomycetales through *Taphridium* has been proposed, the ascus being regarded as equivalent to the sporangial sac, and the thick-walled segment at its base in *Taphrina aurea* and other species being homologised with the chlamydospore. There does not appear at present to be sufficient evidence to support this ascription. The cylindrical form of the ascus does not suggest a primitive group, and might be held to indicate that the Exoascales are derived from one of the alliances with parallel, protected asci.

The Ascocorticiaceae, containing the saprophytic genus *Ascocorticium* with five species, are sometimes associated with the Exoascaceae; here also the asci are cylindrical, parallel, and unprotected.

Exoascaceae

The Exoascaceae[2] are obligate parasites upon vascular plants; they cause hypertrophy of the infected region, stimulating cells of the permanent tissue to renewed growth and division, and producing red, yellow or purple discolorations, blisters, curling of the leaf, malformation of the fruit, or the development of the bunches of twigs known as witches' brooms. They are responsible for several diseases of economic importance, including peach leaf curl, induced

[1] Theissen, 1913, 1914.
[2] Knowles, 1887; Juel, 1901–2, 1921; Ikeno, 1903 i; Martin, E. M., 1924, 1925.

by *Exoascus deformans*, a witch's broom on cherries caused by *E. Cerasi*, and the distortion known as pocket plums, due to the presence of *E. Pruni*, which infects the flesh of the fruit and inhibits the development of the stone.

The principal genera are *Exoascus* and *Taphrina*; in the former eight ascospores (fig. 126) are shed, in the latter (fig. 128) the spores bud while still in the ascus, giving rise to numerous small conidia,

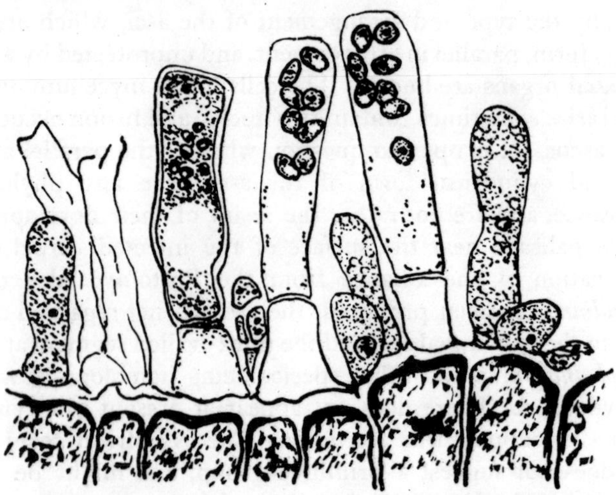

Fig. 126. *Exoascus deformans* (Berk.) Fuck., × 1000.

which caused early investigators to suggest a relationship between the Exoascaceae and the yeasts.

Infection is due to ascospores or to conidia produced from them; they fall on the twigs and bud scales of the host and may survive in that position for more than a year. In cold, damp weather, when the young leaves are in a state of lowered vitality, the fungus readily enters, its germ tubes pushing through the walls which separate the epidermal cells into the zone below. Once in the leaf, the hyphae, in most species, ramify between the cells of the host, but, in *Taphrinopsis Laurencia* on *Pteris biaurita*, they are intracellular. Haustoria are not formed. The mycelium may be annual or perennial.

In investigated species the germ tubes and cells of the hyphae are binucleate or have their nuclei in pairs. *Taphrina epiphylla* and

T. Klebahni[1] are heterothallic, with spores of different complement in the same ascus. Before producing infection the conidia fuse in pairs and the germ tube receives both nuclei. In *Exoascus deformans*[2] the binucleate condition of the germ tube is due to nuclear division.

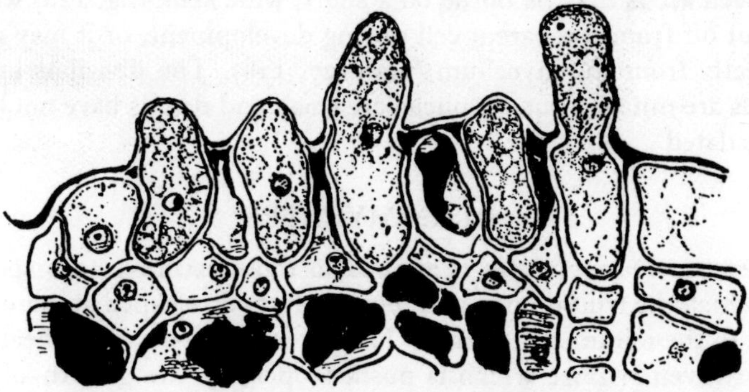

Fig. 127. *Taphrina aurea* (Pers.) Fr.; young asci, × 500.

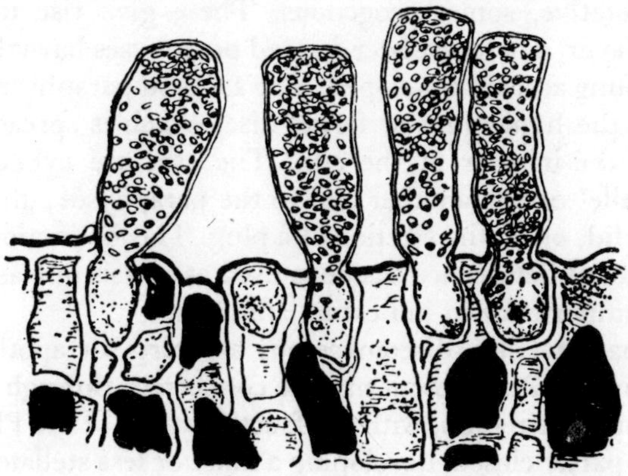

Fig. 128. *Taphrina aurea* (Pers.) Fr.; mature asci, × 500.

Fertile hyphae are found mainly in the leaves and carpels. Sometimes the mycelium permeates the whole tissue, and asci arise below the epidermal cells, and push up among them, as in *Taphrina aurea* on the poplar. Sometimes the asci originate between the

[1] Wieben, 1927; Mix, 1935.
[2] Fitzpatrick, R. E., 1934; Mix, 1935.

epidermal cells, as in *Magnusiella Potentillae*, but in most species, as in *Exoascus deformans*, they are developed above the epidermal cells and just below the cuticle. In *Taphrinopsis*, on *Pteris*, the asci are formed inside the epidermal cells.

Each ascus may be borne on a short, wide stalk (fig. 126) which is cut off from the parent cell during development, or it may arise directly from the mycelium (figs. 127, 128). The divisions in the ascus are mitotic, but the nuclei are small and details have not been elucidated.

DISCOMYCETES

The term Discomycetes is applied to those ascomycetous species in which the fruit is open and more or less cup-shaped at maturity, and to their immediate allies. The ascocarp possesses a sheath of interwoven hyphae which is pushed open by the growth of the paraphyses so that it forms the outer wall of the cup. The lower part of the cup is filled by the **hypothecium**, a weft of hyphae, some vegetative, some ascogenous. These give rise to the sub-hymenial layer, where the later formed paraphyses have their origin and the young asci are developed. The asci and paraphyses together constitute the hymenium or fertile disc, which is spread over the surface of the interior of the cup. The asci are cylindrical and stand parallel one to another and to the paraphyses; they usually open by a lid, or by the ejection of a plug. The distinction between these modes of dehiscence has been used as a basis of classification[1] and is of importance in indicating affinities.

This characteristic, discomycetous ascocarp, or apothecium, is well seen in the Pezizales; it may be connected, through the Patellariaceae and their allies, with the fructifications of the Phacidiales, which are partly closed, developing a more or less stellate aperture, and with the elongated fructifications of the Hysteriales, which open by a narrow slit. It may be related also to the reflexed ascocarp of *Rhizina* and *Sphaerosoma*, and, by invagination of the fertile surface, with the secondarily closed fruit of the truffles. In many species the ascocarp is stalked, and from such forms the stalked Helvellales may perhaps be derived.

The sexual apparatus is known in detail only among the Pezi-

[1] Boudier, 1907; Ramsbottom, 1913.

zales. Here the coenocytic antheridium is the terminal cell of the male branch and never becomes detached as a spermatium. In the simplest forms the female branch consists of a short stalk, a unicellular, coenocytic oogonium and a short, unicellular trichogyne. This arrangement is seen in *Ascodesmis*, where the oogonium becomes septate after fertilisation. A very similar oogonium is found in *Eurotium* and other members of the Plectascales, and there is reason to regard it as the primitive female organ of the Discomycetes. From it may be derived the globose oogonia (fig. 132) which, no doubt because of their shape, give rise to ascogenous hyphae without undergoing septation, and the narrower coils (fig. 157) and oogonial regions in some of which ascogenous hyphae grow out from several cells. The distal region of the female branch may serve as a trichogyne, or, as in *Pyronema*, the trichogyne may be a secondary outgrowth from a terminal oogonium. In the Discomycetes, as in other groups of fungi, apogamy is common.

The Discomycetes include over 4000 species and may be subdivided as follows:

Hymenium fully exposed at maturity	
Mature ascophore cup-shaped	PEZIZALES
Mature ascophore reflexed; or elongated with fertile region then spread over upper surface and often convoluted	HELVELLALES
Hymenium covered at maturity	TUBERALES
Hymenium incompletely exposed at maturity	
Ascophore round, aperture usually stellate	PHACIDIALES
Ascophore elongated, opening by a slit	HYSTERIALES

PEZIZALES

The Pezizales are characterised by their fleshy or sometimes leathery ascocarp, bounded, except in the Pyronemaceae, by a more or less definite peridium which is closed at first, and is later pushed open by the growth of a conical mass of paraphyses, giving the mature fruit its cup or saucer shape.

The asci contain usually eight, but sometimes four, or sixteen, or a larger number of spores, which, in a few species, give rise by budding to conidia. Accessory spores, including conidia of several kinds, chlamydospores, and oidia are also produced. The mycelium is well developed. Sclerotia are rare.

The following families are included in the alliance:

Peridium fleshy, continuous with hypothecium	
Peridium incomplete; ascocarps usually compound	PYRONEMACEAE
Peridium well-developed	
Asci not rising above the surface when ripe; ascospores usually uniseriate	PEZIZACEAE
Asci rising above the surface when ripe; ascospores often coloured and biseriate	ASCOBOLACEAE
Peridium fleshy, distinct from hypothecium	
Peridium of elongated hyphae (pseudoprosenchymatous)	HELOTIACEAE
Peridium pseudoparenchymatous	MOLLISIACEAE
Peridium absent or ill-defined; epithecium[1] formed	CELIDIACEAE
Peridium tough; epithecium formed	
Ascocarp free	PATELLARIACEAE
Ascocarp embedded when young	CENANGIACEAE
Apothecia numerous, sunk in a stroma	CYTTARIACEAE

PYRONEMACEAE

The Pyronemaceae are a small family distinguished from other Pezizales by the slight development of the peridium. The principal genera are *Ascodesmis* and *Pyronema*. In both the ascocarp is compound, arising from several pairs of sexual branches.

Ascodesmis nigricans[2] (fig. 129) is a small, self-fertile, coprophilous form which grows readily in artificial culture. About forty-eight hours after the germination of the spore, stout, multinucleate hyphae grow up from the mycelium (fig. 130 *a*) and dichotomise to form a group of six to eight oogonial branches. Near these, usually from the same filament, one or two antheridial hyphae arise, grow towards the female branches (fig. 130 *b*) and dichotomise (fig. 130 *c*), while around

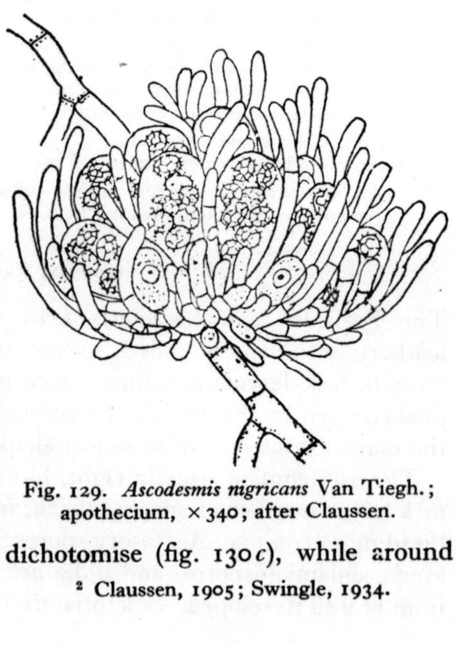

Fig. 129. *Ascodesmis nigricans* Van Tiegh.; apothecium, × 340; after Claussen.

[1] Cf. p. 228.

[2] Claussen, 1905; Swingle, 1934.

each of their terminal cells, or antheridia, a female branch becomes wrapped (fig. 130 *d*). Walls are laid down, so that each

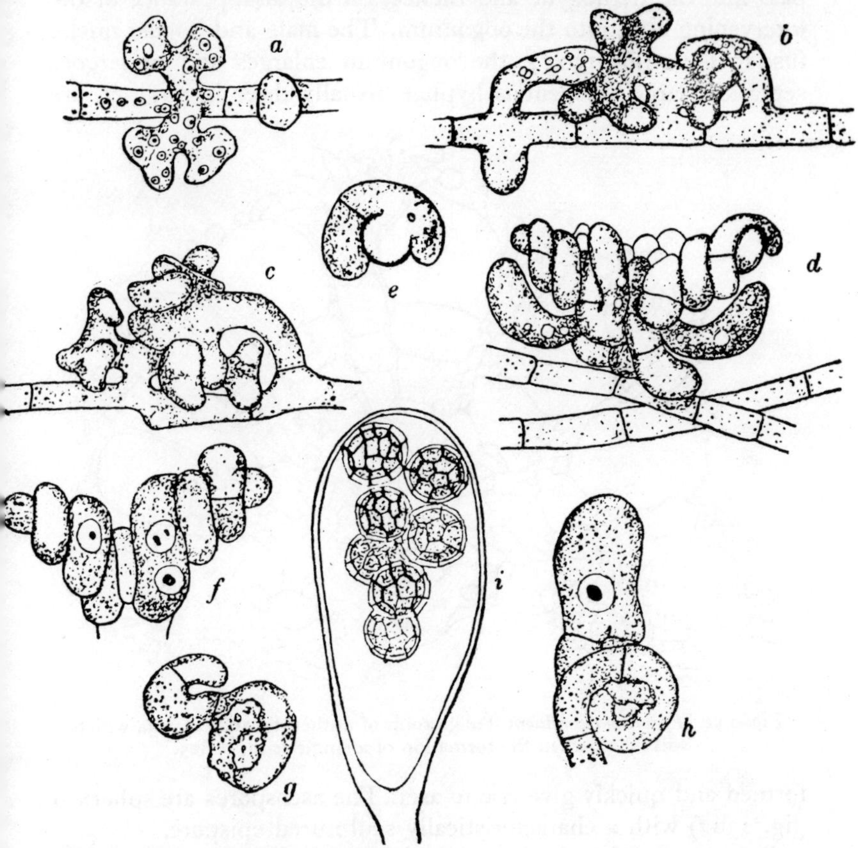

Fig. 130. *Ascodesmis nigricans* Van Tiegh.; *a*, *b*, *c*, *d*, development of the sexual apparatus, *a* and *b*, ×1000, *c*, ×1100, *d*, ×800; *e*, communication between antheridium and trichogyne, ×1300; *f*, fusion in oogonium, ×1600; *g*, septate oogonium and ascogenous hypha, antheridium and trichogyne shrivelled, ×1000; *h*, uninucleate ascus attached to coiled and septate oogonium, ×1100; *i*, sculptured spores in ascus, ×750; after Claussen.

antheridial and oogonial branch is cut off from its parent cell, and each female branch divides transversely to form a trichogyne, oogonium, and stalk. The trichogyne usually contains two nuclei, the oogonium five or six, and the antheridium about the same number. The nuclei of the trichogyne degenerate, the wall between this cell and the antheridium breaks down, the male nuclei pass into the trichogyne and thence, on the disappearance of the intervening wall, into the oogonium. The male and female nuclei fuse in pairs (fig. 130 *f*), the oogonium enlarges and undergoes septation, large ascogenous hyphae, usually three in number, are

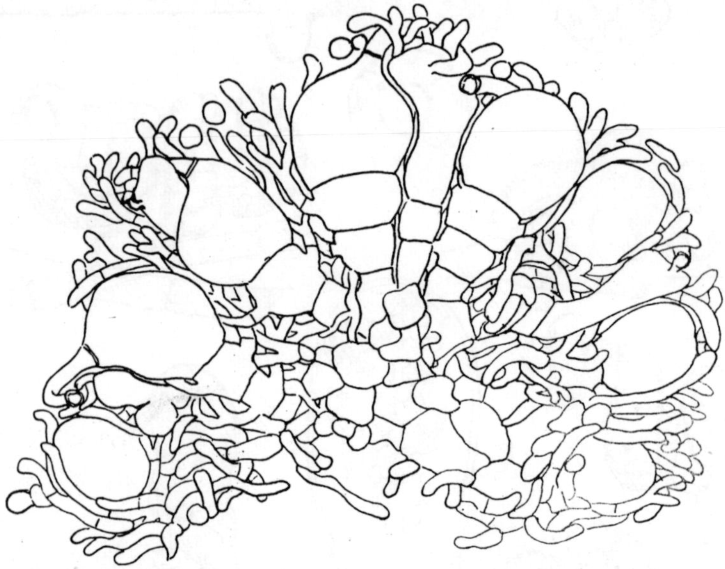

Fig. 131. *Pyronema confluens* Tul.; group of antheridia and oogonia which will take part in the formation of a single fruit, ×625.

formed and quickly give rise to asci. The ascospores are spherical (fig. 130 *i*) with a characteristically sculptured epispore.

Pyronema confluens[1] occurs on burnt ground or decaying leaves; the mycelium is colourless and superficial, bearing pink ascocarps. There are no accessory spores. Six chromosomes may be observed in the nuclei of the germinating spores and of the vegetative hyphae (fig. 133 *a*). The species is self-fertile.

The male and female organs are multinucleate from their first formation; several of each take part in the production of a single ascocarp. They arise from separate hyphae (fig. 132 *a*) which are

[1] de Bary, 1863 ii; Tulasne, 1866; Harper, 1900; Brown, W. H., 1909, 1915; Claussen, 1912; Gwynne-Vaughan and Williamson, 1931.

often dichotomously branched. Soon a slight elevation appears on the surface of the oogonium and elongates to form the multinucleate, unicellular trichogyne, which is cut off from the oogonium by a transverse wall. The trichogyne and antheridium make contact, the trichogyne meeting sometimes the apex but more often the flank of the male organ (fig. 132 *a*). The nuclei of the trichogyne begin to degenerate, the walls separating the antheridium and tricho-

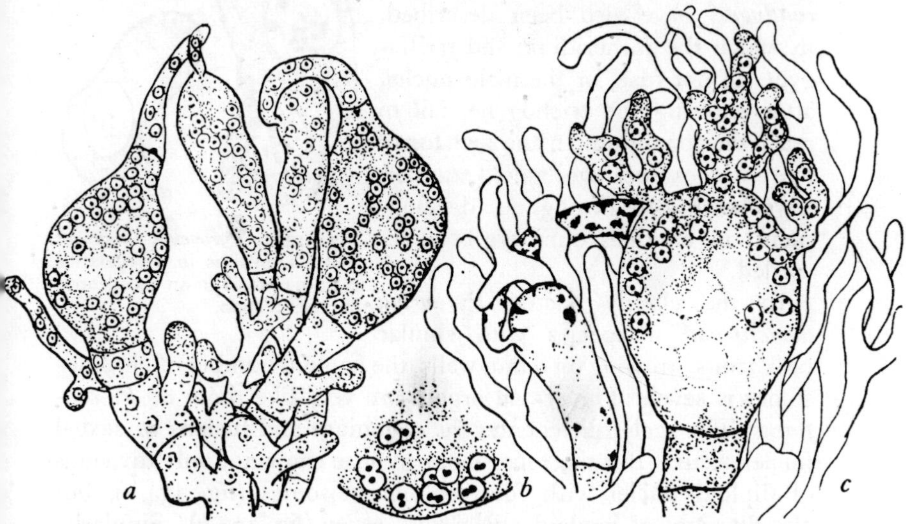

Fig. 132. *Pyronema confluens* Tul.; *a*, antheridia and oogonia before fertilisation, × 875; *b*, fusion of male and female nuclei, × 1400; *c*, antheridium and oogonium after fertilisation, the antheridium is nearly empty, the oogonium bears ascogenous hyphae, protective vegetative filaments are growing up, × 875.

gyne disappear, and a pore is formed through which the male nuclei stream into the trichogyne. The wall at the base of the trichogyne is next ruptured, the male nuclei enter the oogonium, a wall is again formed across the base of the trichogyne and the contents of the latter soon begin to degenerate.

Meanwhile the sexual nuclei fuse in pairs (fig. 132 *b*) and the fusion nuclei begin to pass into the ascogenous hyphae (fig. 132 *c*) where the next mitoses show twelve chromosomes, the diploid number. Asci are formed and a second nuclear fusion takes place in the young ascus, so that the ascus nucleus is tetraploid. Twelve gemini are present in the first mitotic division and twelve chromosomes in the second, while, after the third division, six chromosomes, the haploid number, may be counted in the telophase.

During the development of the sexual organs growth has also taken place in the vegetative hyphae, a weft of these has formed the hypothecium and paraphyses have appeared as narrow, parallel branches among which the young asci push up.

Apogamous varieties of *Pyronema confluens*[1] have also been described, in which the antheridium and trichogyne do not fuse, or the male nuclei, after entering the trichogyne, fail to reach the oogonium. In one such form[2] five chromosomes, the haploid number, have been seen throughout development, in another four have been recorded.[3]

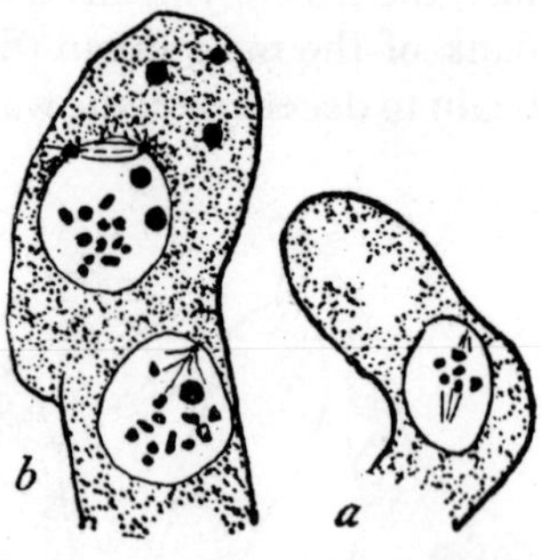

Fig. 133. *Pyronema confluens* Tul.; *a*, mitosis in a vegetative cell; *b*, mitosis in an ascogenous hypha; ×2600.

In the closely related *Pyronema domesticum*[4] a species with similar pink fruits, growing on damp walls, the haploid number of chromosomes is seven. The sexual organs are very like those of *P. confluens*. The male nuclei enter the oogonium and fusions of sexual nuclei occur. But the ascogenous hyphae show not only divisions of diploid nuclei with fourteen chromosomes (fig. 134 *a*), but also divisions of haploid nuclei with seven (fig. 134 *b*). Similarly, after the fusion in the ascus, the definitive nucleus may have fourteen (fig. 134 *c*) or seven (fig. 134 *d*) gemini. After the third division, however, seven chromosomes, the haploid number (fig. 134 *g*), are present in all nuclei. In this species some sexual nuclei fuse in the oogonium while others enter the ascogenous hyphae in the haploid condition. In the former circumstance the definitive nucleus, after the second fusion, is tetraploid and brachymeiosis takes place as usual, in the latter it is diploid and brachymeiosis is omitted.

[1] Dangeard, 1907; Moreau and Moreau, 1930, 1931; Raymond, 1934.
[2] Brown, W. H., 1915. [3] Raymond, 1934. [4] Tandy, 1927.

Pezizaceae

In the large family Pezizaceae the ascocarp is superficial, sessile or stalked, usually with a well-marked peridium of fleshy or waxy consistency, decaying soon after maturity. The spores, though

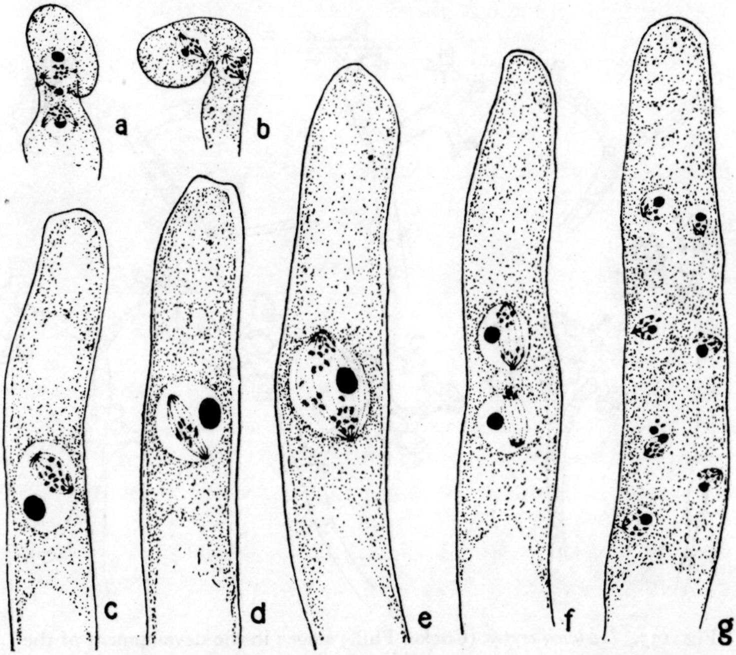

Fig. 134. *Pyronema domesticum* (Sow.) Sacc.; *a*, ascogenous hypha bending over to form ascus, nuclei in metaphase with 14 chromosomes, the diploid number; *b*, the same, nuclei in metaphase, showing 7 chromosomes, the haploid number; *c*, young ascus, metaphase of first meiotic division with 14 gemini (28 chromosomes); *d*, same, but with 7 gemini (14 chromosomes); *e*, anaphase of first meiotic division, with 14 chromosomes going to each pole after reduction is accomplished; *f*, anaphase of second meiotic division in nuclei which are the product of only one fusion, 7 chromosomes going to each pole; *g*, telophase of third division in the ascus, showing 7 chromosomes in the uncut nuclei; × 1600; after Tandy.

septate in some small species, are usually continuous and hyaline; their arrangement is uniseriate. The ripe asci do not project above the level of the disc as they do in the Ascobolaceae. Most of the species are saprophytic, on the ground, on wood, or on dung. Classification depends on the shape of the spores, the size and consistency of the ascocarp, the presence or absence of hairs, and the method of dehiscence of the ascus.

In most of the species the fruit is fleshy and without hairs; such species were formerly grouped together in the genus *Peziza*. That name is now retained for large species with a sessile or sub-sessile cup, regular in form and often two centimetres or more in diameter.

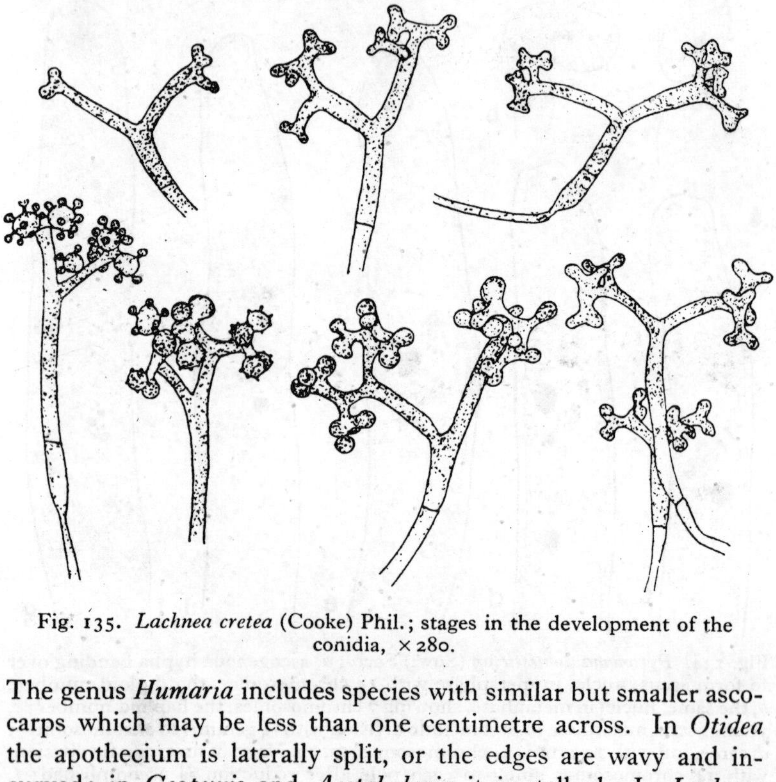

Fig. 135. *Lachnea cretea* (Cooke) Phil.; stages in the development of the conidia, ×280.

The genus *Humaria* includes species with similar but smaller ascocarps which may be less than one centimetre across. In *Otidea* the apothecium is laterally split, or the edges are wavy and incurved; in *Geopyxis* and *Acetabularia* it is stalked. In *Lachnea* and in some other genera the fruit is beset with hairs, while in *Sepultaria* it is hairy and more or less sunk in the soil. The peridium in *Lachnea*, though much better developed than in *Pyronema*, is never completely closed, as in *Humaria* or *Peziza*, across the top of the ascocarp.

Lachnea cretea[1] has a pale buff apothecium (fig. 136 *a*); like

[1] Fraser, H. C. I., 1913; Dodge, 1922; Gwynne-Vaughan and Williamson, 1927.

many other saprophytic fungi, it grows readily in culture. The mycelium produces conidia on branched conidiophores (fig. 135), the female branch has a multicellular stalk, an oogonial region of three or four coenocytic cells (fig. 136 *d*), and a long, branched,

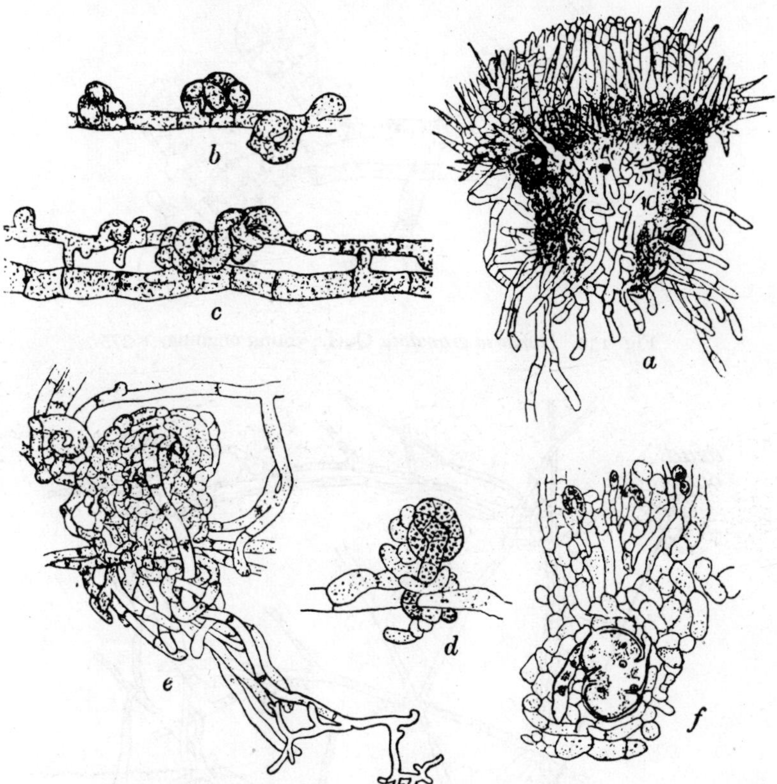

Fig. 136. *Lachnea cretea* (Cooke) Phil.; *a*, mature ascocarp, ×90; *b*, *c*, development of female branch, ×300; *d*, older female branch showing crowded nuclei, ×400; *e*, young ascocarp with elaborately branched trichogyne, ×400; *f*, oogonial region of three cells united by very large pores, ×400.

multicellular trichogyne (fig. 136 *e*). No antheridium has been observed. In the oogonial region the transverse septa are broken down (fig. 136 *f*), so that the cells are in free communication; ascogenous hyphae arise from them all.

Lachnea melaloma,[1] a species on burnt ground, is heterothallic;[2]

[1] Gwynne-Vaughan, 1937. [2] Cf. p. 40.

it lacks an antheridium, but possesses a long, coiled, female branch from the oogonial region of which ascogenous hyphae arise.

The orange red fruits of *Lachnea scutellata*[1] occur on old

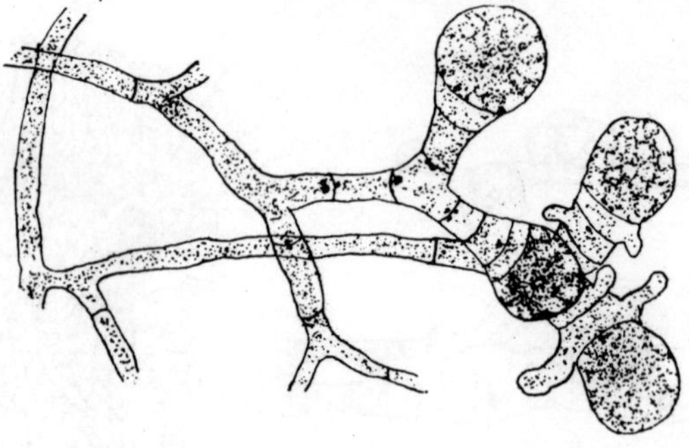

Fig. 137. *Humaria granulata* Quél.; young oogonia, × 375.

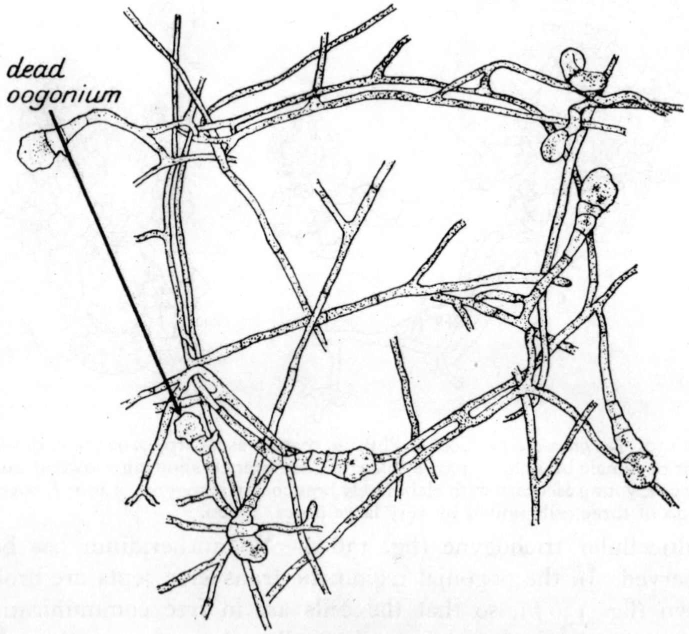

Fig. 138. *Humaria granulata* Quél.; region of intermingling of two *B* mycelia seven days after inoculation, × 125.

[1] Gwynne-Vaughan and Williamson, 1933 i.

wood; they can be grown on various dung media and probably depend in nature on the excreta of birds or mammals which contaminate their habitat. The gametophyte is self-fertile and bears

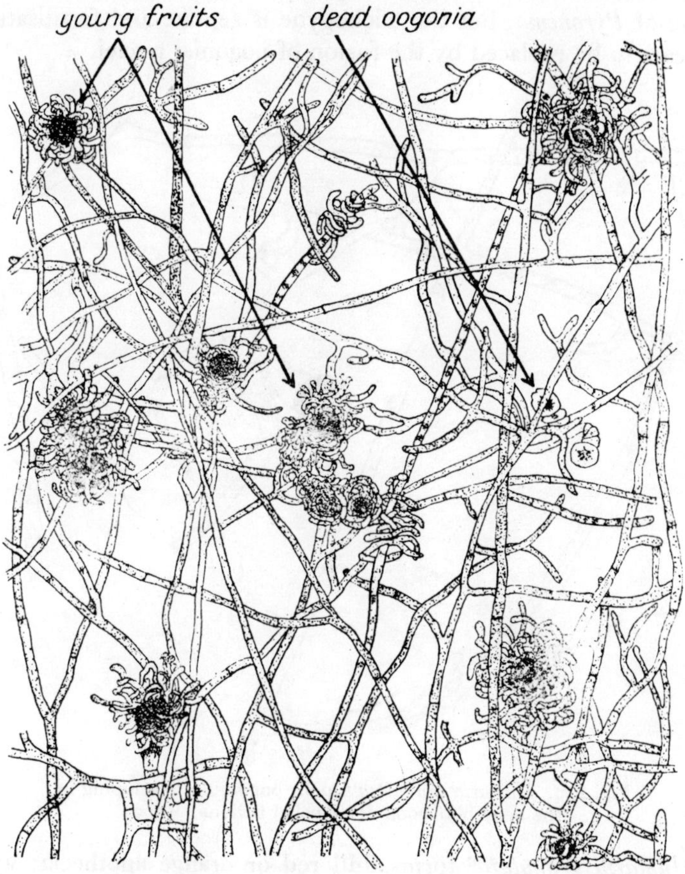

Fig. 139. *Humaria granulata* Quél.; for comparison with fig. 138; region of intermingling of *A* and *B* mycelia seven days after inoculation, × 125.

numerous chlamydospores. Female branches arise either from these or from ordinary cells; a male branch has not been identified. The oogonium is usually the second or third of a row of about nine cells; it gives rise to very wide ascogenous hyphae producing the usual, narrow branches on which the asci are borne. The nucleus of the ascus is tetraploid, two long and ten short gemini are visible in meiosis, and the haploid group of chromosomes shows one longer than the other five.

In *Lachnea stercorea*[1] the antheridium and oogonium are like those of *Pyronema*, but the trichogyne is septate and fertilisation appears to be replaced by the fusion of oogonial nuclei.

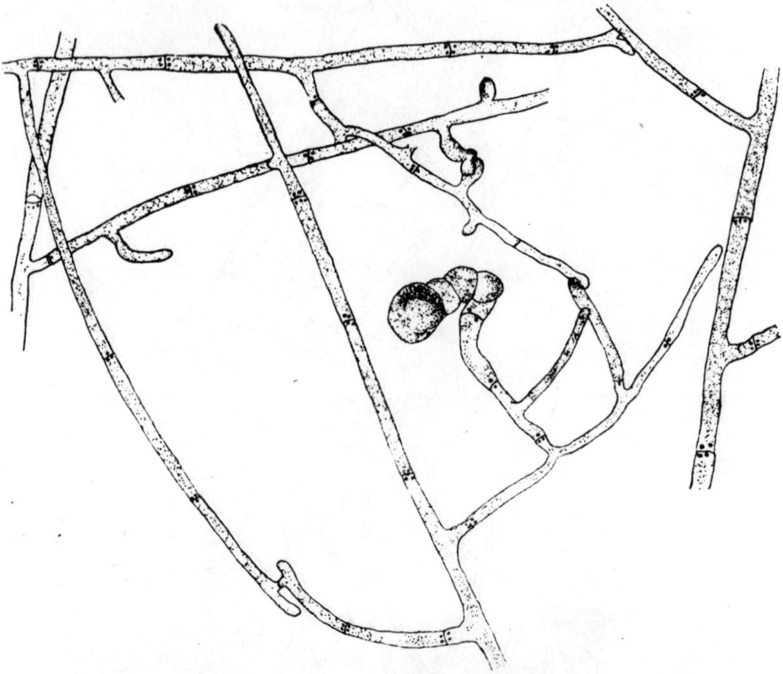

Fig. 140. *Humaria granulata* Quél.; oogonium developing in the neighbourhood of mycelial fusions, × 300.

Humaria granulata[2] forms dull red or orange apothecia; with *Lachnea stercorea* and *Ascobolus stercorarius* it is among the common coprophilous Discomycetes, occurring with great regularity on cow dung after the first crop of *Piloboli* has died down and before the substratum begins to dry and bear pyrenomycetous and hymenomycetous forms. The oogonium has a multicellular stalk and con-

[1] Fraser, H. C. I., 1907; Raymond, 1934.
[2] Blackman and Fraser, 1906 ii; Gwynne-Vaughan and Williamson, 1930.

tains, when fully grown, from six hundred to fourteen hundred nuclei. The species is self-sterile; there is no trichogyne and no antheridium. Young oogonia (fig. 137) are present in single spore culture, but die (fig. 138) without completing their development. When two complementary mycelia, *A* and *B*, are brought together,

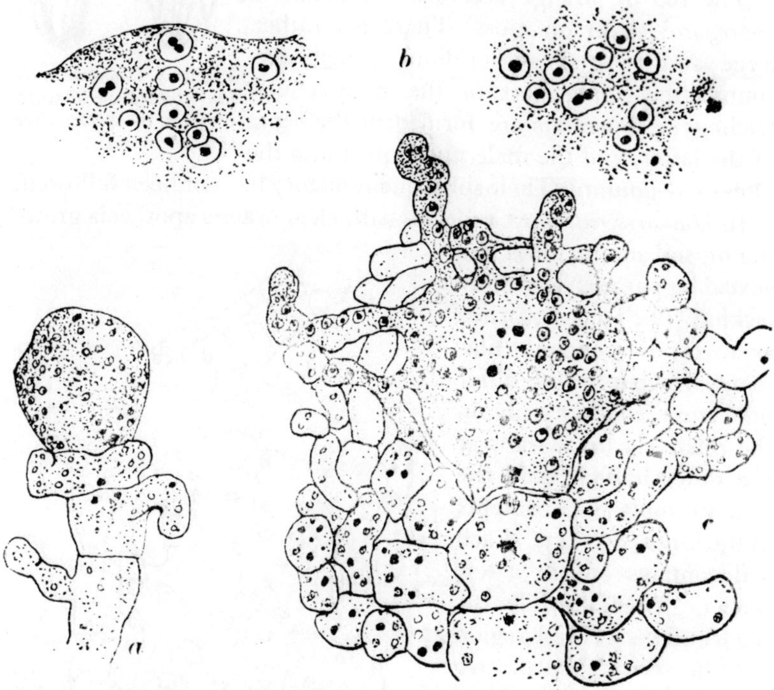

Fig. 141. *Humaria granulata* Quél.; *a*, young oogonium, the stalk branching to form a sheath, × 700; *b*, fusion of nuclei in oogonium, × 1600; *c*, oogonium giving rise to ascogenous hyphae × 700.

mycelial fusions occur (fig. 140) and thereafter a sheath surrounds the oogonium (figs. 139, 141 *c*) and the latter buds out ascogenous hyphae. In the oogonium the nuclei fuse, not simultaneously, as in species where male nuclei have entered the female organ, but pair by pair (fig. 141 *b*), and, as usual, two diploid nuclei fuse in the young ascus. Spores giving *A* and spores giving *B* mycelia are present in the same ascus, showing that the definitive nucleus contained both *A* and *B* factors, and therefore that both *A* and *B* nuclei had entered the oogonium and that fusion between them had taken place. The spores germinate under natural conditions when they are about seven months old. They can be induced to germinate earlier by the application of heat. The haploid number of chromosomes is four.

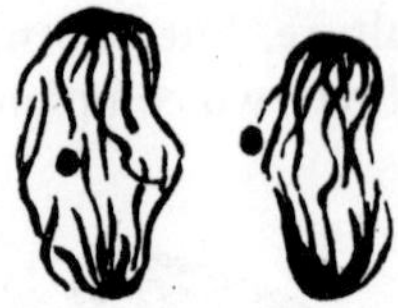

Fig. 142. *Humaria rutilans*; telophase of third division in ascus; after Guilliermond.

The red or orange ascocarps of *Humaria aggregata*[1] occur on moss. There is a rather large and twisted antheridium which surrounds the upper part of the multicellular trichogyne, openings are formed in the septa of the latter, and the male nuclei pass into the globose oogonium. Their subsequent history has not been followed.

In *Humaria rutilans*,[2] a species with clear orange apothecia growing on soil among moss, the sexual apparatus is wholly lacking. The ascocarp arises as a weft of branching filaments which for a time differ one from another only in the relatively thick walls of the outer hyphae, and the rich protoplasmic contents of the inner (fig. 143). Each cell contains one or a few nuclei. After a while the nuclei of the central region may be seen to be of two sizes, the smaller fusing in pairs (fig. 144 *a*) to give rise to the larger; sometimes, in connection with this process, a nucleus migrates through the cell wall (fig. 144 *b*), as in the prothalli of apogamous ferns. Thus normal fertilisation is replaced by the union of vegetative nuclei in pairs; the cells which

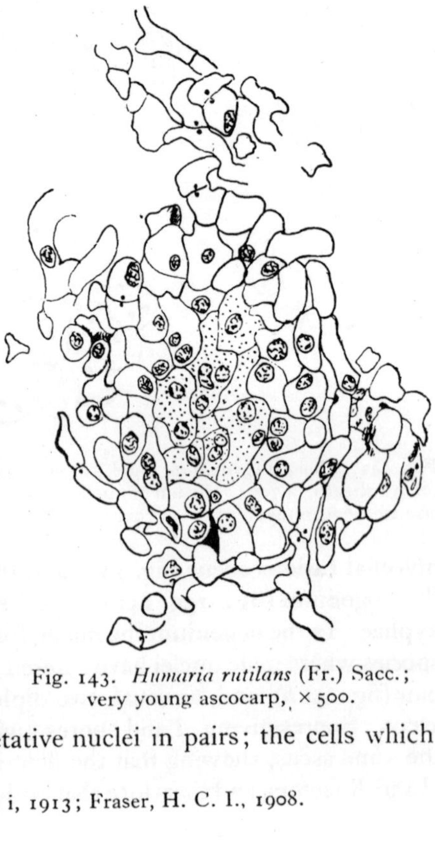

Fig. 143. *Humaria rutilans* (Fr.) Sacc.; very young ascocarp, × 500.

[1] Rosenberg, 1933.
[2] Guilliermond, 1905 i, 1913; Fraser, H. C. I., 1908.

contain fusion nuclei give rise to ascogenous hyphae, while, from the rest, the paraphyses and cells of the peridium arise.

Sixteen chromosomes are present in the mitoses in the ascogenous hyphae (fig. 147 *a*); there are sixteen gemini in the definitive

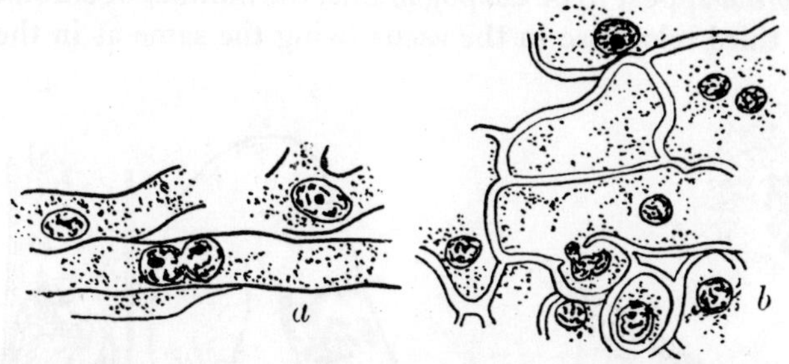

Fig. 144. *Humaria rutilans* (Fr.) Sacc.; *a*, fusion in a vegetative hypha; *b*, migration of nucleus from one vegetative cell to another; both ×1100.

nucleus of the ascus (fig. 147 *c*), sixteen chromosomes in the telophase of the second division (fig. 145) and eight (fig. 146) in the telophase of the third. Sixteen chromosomes have also been re-

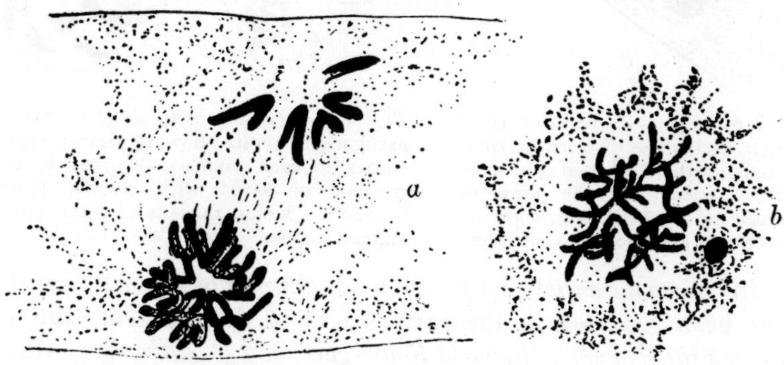

Fig. 145. *Humaria rutilans* (Fr.) Sacc.; *a*, telophase of second division in ascus, ×3370; *b*, prophase of third division in ascus, showing sixteen curved chromosomes, ×2808.

corded (fig. 142) in the third telophase; if this observation is confirmed it may indicate that, as in so many angiosperms, there exists in this species an apogamous variety with double the haploid number of chromosomes.

In several other members of the Pezizaceae, such as *Peziza tectoria*, *Peziza vesiculosa*,[1] *Pustularia bolarioides*[2] and *Peziza subumbrina*,[3] development takes place without the formation of sexual organs. *Peziza vesiculosa*, like *Humaria rutilans*, shows evidence of pseudapogamous fusions, but *Pustularia bolarioides* and *Peziza subumbrina* appear to be euapogamous, the number of chromosomes in the third telophase in the ascus being the same as in the first.

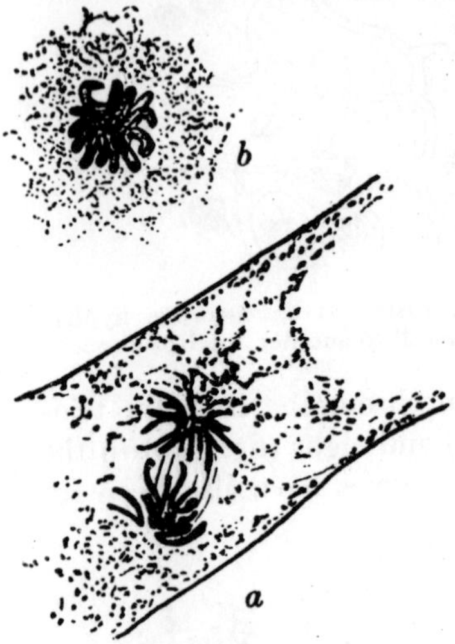

Fig. 146. *Humaria rutilans* (Fr.) Sacc.; *a*, metaphase of third division in ascus with V-shaped chromosomes, ×2080; *b*, polar view of telophase of third division, showing eight curved chromosomes, ×3100.

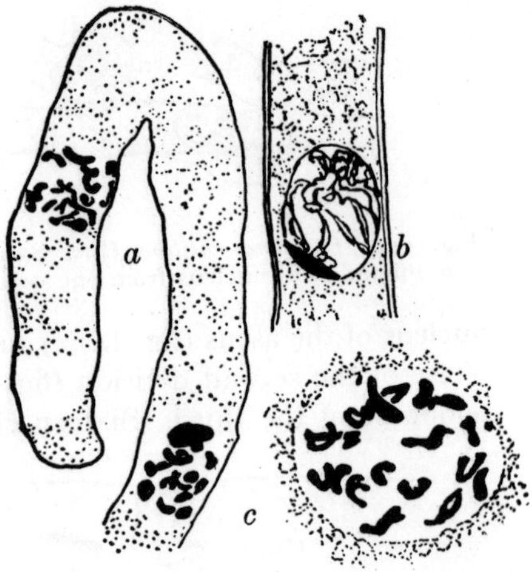

Fig. 147. *Humaria rutilans* (Fr.) Sacc.; *a*, ascogenous hypha showing sixteen chromosomes in each nucleus, ×1950; *b*, fusion nucleus of ascus with double spireme, ×1300; *c*, fusion nucleus of ascus showing sixteen gemini, ×1950.

In *Otidea aurantia*[1] a large cell, no doubt part of a female branch, has been recorded in the early stages of development, and in *Peziza theleboloides*, *Humaria Roumegueri* and *H. carbonigena* there is a well-marked oogonial region of one or more cells.

ASCOBOLACEAE

The ascocarp of the Ascobolaceae is soft and fleshy or somewhat gelatinous, and possesses a well-marked peridium by which the fruit is closed at first. The family is distinguished from the Peziza-

[1] Fraser, H. C. I., and Welsford, 1908. [2] Bagchee, 1925.
[3] Matsuura and Gondo, 1935.

ceae by the usually multiseriate arrangement of the spores and by the fact that the ripe asci stand well above the level of the hymenium before their spores are discharged. Often the asci are large and few in number; they have a peculiar, two-lipped method of dehiscence in *Ascozonus* (fig. 148), but in other genera they open by a lid. The spores are brown or violet in *Ascobolus*, *Saccobolus* and *Boudiera*, hyaline in most of the other genera; they are usually ellipsoid, but round in *Boudiera* and *Cubonia*; in *Saccobolus* they are enclosed within a special membrane and ejected together; in *Thecotheus*, *Rhyparobius*, *Ascozonus* and *Thelebolus* they are sixteen or more in number.

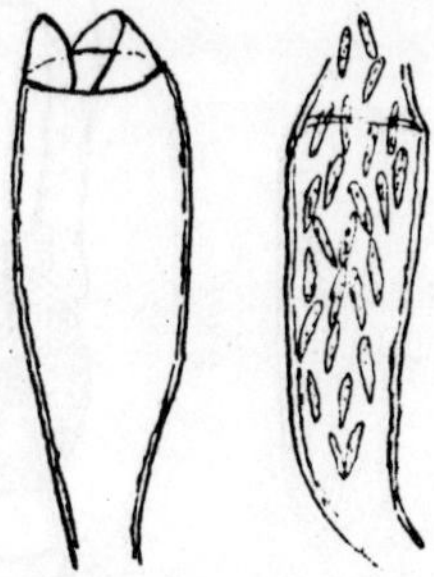

Fig. 148. *Ascozonus* sp.; ascus during and just after dehiscence, × 335; W. M. Page del.

As in the Pezizaceae, there is considerable variety in the form of the sexual apparatus. *Ascophanus Aurora*,[1] with orange-brown ascocarps and smooth, colourless spores, occurs not uncommonly on sheep pellets and grows readily in culture. Sexual organs appear six or seven days after the germination of the spores and are very like those of *Pyronema confluens*. As in that species several pairs are concerned in the development of each fructification (fig. 150), the oogonium is similar, but usually twisted or curved, and a similar, unicellular trichogyne is formed as a secondary outgrowth (fig. 149). The whole apparatus, however, is much smaller (fig. 82), the oogonium contains only eight to twenty nuclei, the antheridium is drumstick-shaped and contains four to ten. Normal fertilisation takes place in the oogonium (figs. 152 *b*, *c*), the antheridium, when its stalk withers (fig. 149 *b*), often remaining attached to the tip of the trichogyne (fig. 149 *c*). The haploid number of chromosomes is two. A balancing hypha (figs. 149 *a*, 150 *a*) often grows out from the stalk of the sexual branch, counteracting the weight of the antheridium or oogonium.

The sexual apparatus of *Ascophanus ochraceus*[2] appears to be very similar to that of *A. Aurora*.

Ascophanus carneus,[3] with pink or orange apothecia, is common on the dung of cows and rabbits, on old leather, rope and similar

[1] Gwynne-Vaughan and Williamson, 1934.
[2] Dangeard, 1907.
[3] Cutting, 1909; Ramlow, 1914.

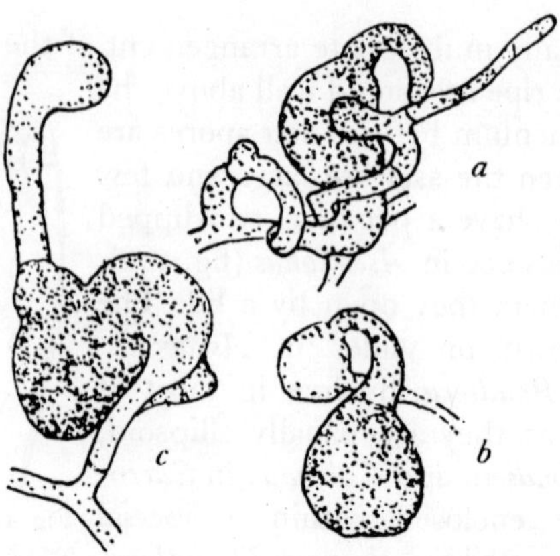

Fig. 149. *Ascophanus Aurora* (Crouan) Boud.; *a*, sexual branches before fertilisation, the male branch has thrown out a balancing hypha; *b*, trichogyne in communication with nearly empty antheridium; *c*, oogonium after fertilisation with empty antheridium detached and carried up at end of trichogyne; ×1600.

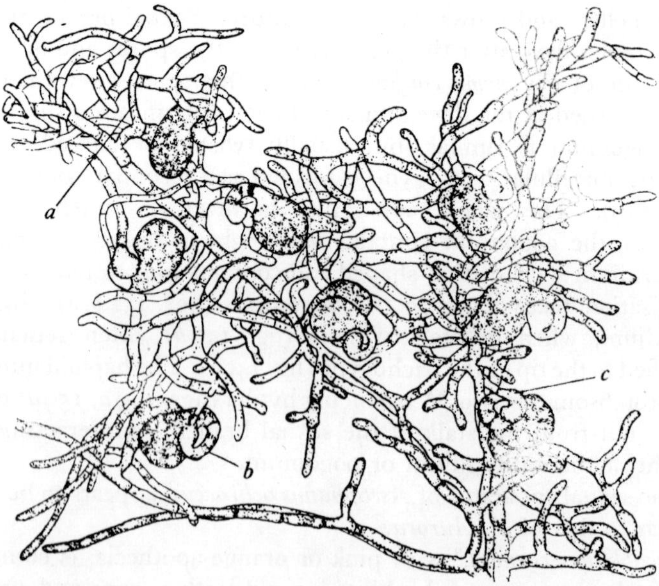

Fig. 150. *Ascophanus Aurora* (Crouan) Boud.; group of sexual branches, most of the antheridia are already empty, ×750. At *a* the oogonial stalk has thrown out a balancing hypha, at *b* the stalk of the oogonium has begun to wither, at *c* fertilisation has just occurred.

habitats. The male branch is lacking, the female branch is a coiled, multicellular filament with a central oogonial region of three to seven cells in which nuclear fusion takes place and from which ascogenous hyphae grow out.

Ascobolus strobilinus,[1] on sheep pellets, is not unlike *Ascophanus Aurora*, each fructification having several pairs of sexual organs somewhat similar in form. Male nuclei pass through the unicellular trichogyne to the oogonium and are stated to pair but not to fuse

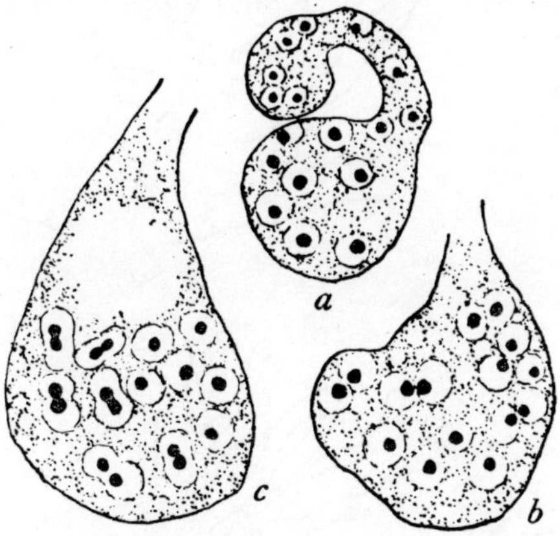

Fig. 151. *Ascophanus Aurora* (Crouan) Boud.; *a*, sexual organs in transverse section with male nuclei passing through trichogyne from small antheridium to larger oogonium; *b*, oogonium in longitudinal section with paired sexual nuclei; *c*, sexual fusion, unpaired female nuclei are present in *b* and *c*; $\times 2600$.

with the female nuclei, fusion being delayed till the formation of the ascus. The male nuclei are smaller than the female.

All these species are self-fertile.

Ascobolus magnificus[2] is a monoecious, self-sterile form in which normal fertilisation occurs. It has large ascocarps, found on horse dung in New York State and Porto Rico and shows two complementary strains, *A* and *B*. Each, when grown alone, produces accessory spores but neither is able to form sexual branches. When they meet, male organs arise on the younger and female organs on the older hyphae of both strains, but union only takes place between an *A* antheridium and a *B* oogonium or between a *B* antheridium

[1] Schweizer, G., 1931.
[2] Dodge, 1920; Gwynne-Vaughan and Williamson, 1932.

and an oogonium of *A*. The antheridium is cylindrical or clavate. The female branch cuts off a terminal trichogyne (fig. 152 *c*) which

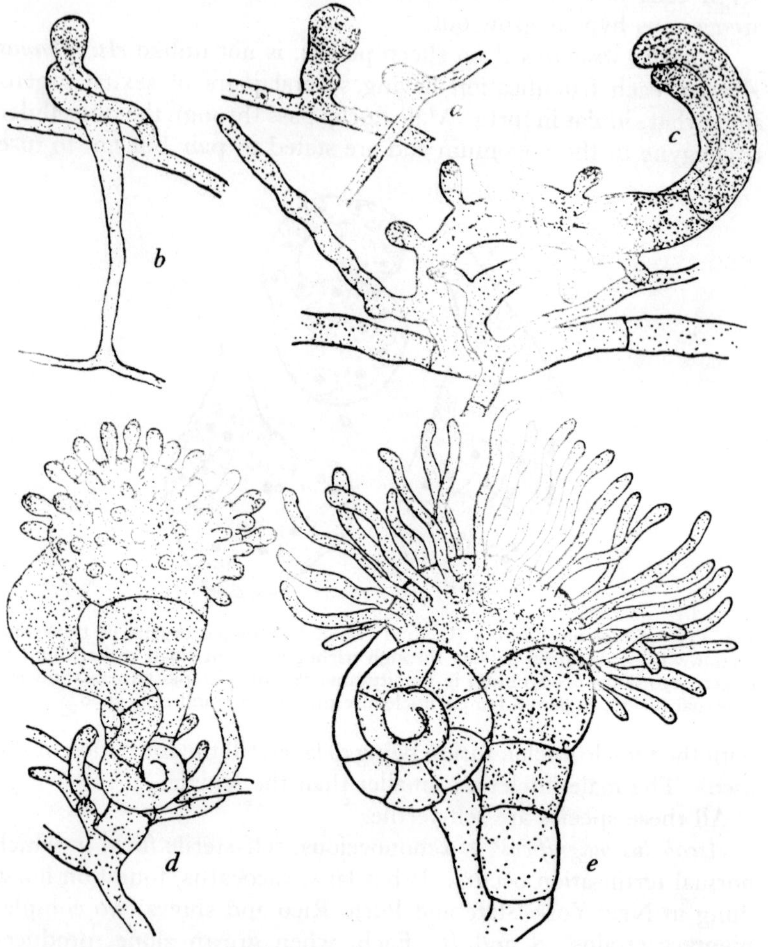

Fig. 152. *Ascobolus magnificus* Dodge; *a*, very young sexual branches; *b*, older sexual branches; *c*, male branch with antheridium, female branch with oogonium beginning to swell and terminal trichogyne not yet septate; *d*, *e*, sexual branches after fertilisation, the empty antheridium is still attached to the septate trichogyne, ascogenous hyphae are growing from the oogonium; × 440.

later divides into seven or more cells. Below the trichogyne the oogonium enlarges, becoming almost globose (fig. 152 *d*). The stalk of each sexual branch is multicellular and outgrowths of each share alike in the formation of the sheath. Since the apices of the sexual branches are attached, the elongation of the female branch causes the trichogyne to become twisted round and round the antheridium (fig. 152*e*). Continuity is established between the antheridium and the end cell of the trichogyne, and the movement of the male nuclei from cell to cell of the latter begins, the septa being perforated to allow their passage. The male nuclei, in this fungus, are characterised by

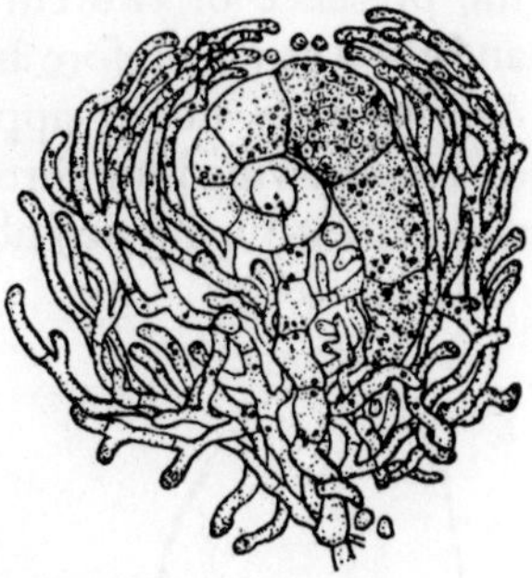

Fig. 153. *Ascobolus magnificus* Dodge; section through ascocarp before fertilisation, × 200.

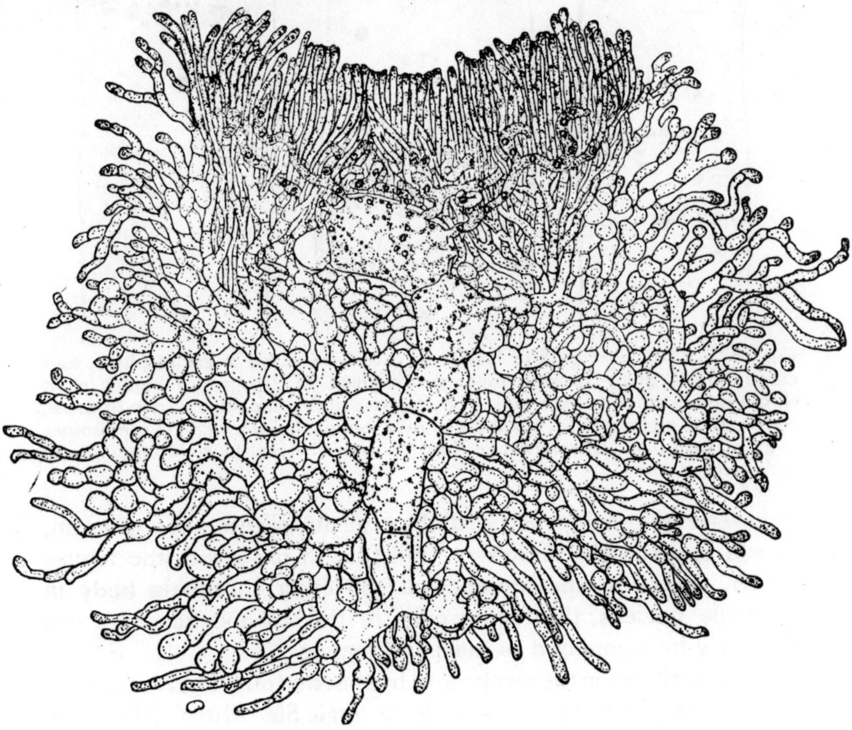

Fig. 154. *Ascobolus magnificus* Dodge; section through ascocarp after fertilisation, × 200.

the presence of two chromatin bodies, instead of the usual one, and they can therefore be identified at all stages of their journey. As the male nuclei approach the oogonium, the female nuclei divide, showing four chromosomes (fig. 155 *a*), the haploid number, and retreat to the neighbourhood of the oogonial wall. The last

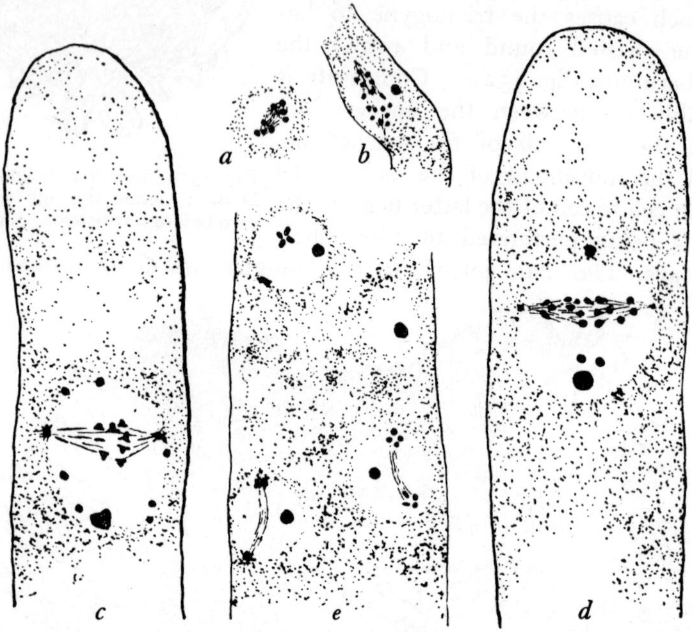

Fig. 155. *Ascobolus magnificus* Dodge; *a*, anaphase in cell of gametophyte, four chromosomes going to each pole; *b*, anaphase in ascogenous hypha, eight chromosomes going to each pole; *c*, metaphase of first meiotic division in ascus, eight gemini on spindle; *d*, anaphase of first meiotic division, eight chromosomes going to each pole; *e*, telophase of third division in ascus, four chromosomes at each pole; ×2600.

transverse septum is penetrated, the male nuclei enter the oogonium, spread towards the periphery, and fuse in pairs with the female nuclei. Owing to the presence of the extra chromatin body in the male nucleus, the fusing pairs in this coenocytic organ can definitely be recognised as male and female.

After fertilisation the ascogenous hyphae bud out. They are narrow from the first (fig. 152 *e*) with nuclei in single file. Mitosis (fig. 155 *b*) shows eight chromosomes, the diploid number. Binucleate cells and asci are formed in the usual way. In the ascus, after a second nuclear fusion, the definitive nucleus is tetraploid with eight gemini (fig. 155 *c*). Meiosis is followed by a brachymeiotic reduction and four chromosomes (fig. 155 *e*) are present in the third telophase. The sheath is often slow to develop, the ascogenous hyphae at first spreading freely into the air (fig. 152 *e*).

Ascobolus carbonarius[1] occurs on burnt ground among charcoal. The ascocarp is scurfy and green or brown. Numerous conidia are formed on the mycelium, and from one of these, while still

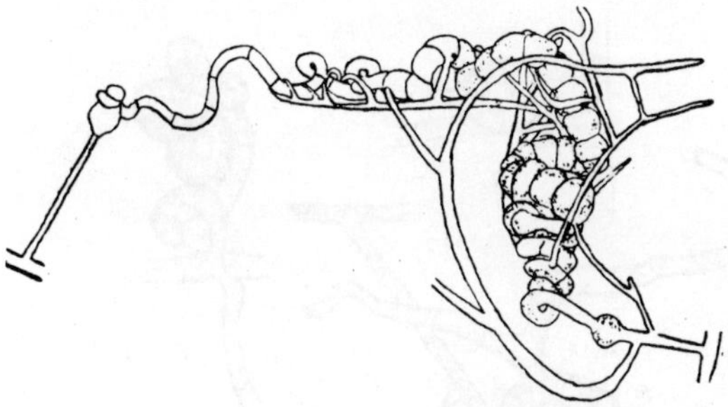

Fig. 156. *Ascobolus carbonarius* Karst.; fertile branch, ×280; after Dodge.

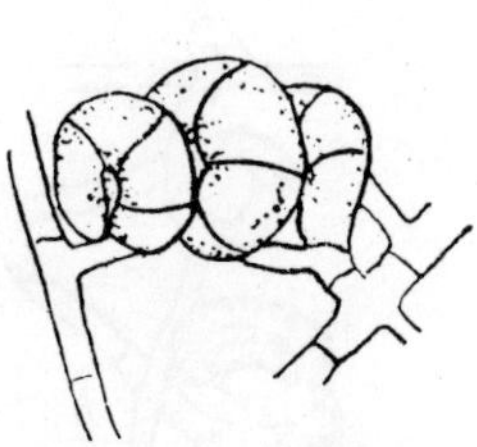

Fig. 157. *Ascobolus stercorarius* (Bull.) Schröt.; fertile branch, ×740; after Dodge.

attached to its stalk, the female branch may arise. It has an oogonial region (fig. 156) of some twenty to forty cells and a terminal trichogyne of ten to twenty cells. The apex of the trichogyne wraps itself round a second conidium or conidium-like body, recalling the antheridium of *Ascophanus Aurora*, and possibly functioning in the same way. Cytological details would be of great interest.

Ascobolus stercorarius[2] is one of the common, coprophilous species, occurring on horse or cow dung in spring and summer. The green or brown ascocarps have a characteristically scurfy margin. Mature spores germinate readily at ordinary temperatures and give rise to complementary mycelia, *A* and *B*. Both strains bear oidia, and their association, under natural conditions, may occur when these are transported by mites. When the complements are brought together, ascocarps appear in about fourteen days, but they may

[1] Dodge, 1912.

[2] Harper, 1896; Dangeard, 1907; Welsford, 1907; Fraser, H. C. I., and Brooks, 1909; Green, 1931; Dowding, 1931 ii.

also develop in single spore culture after a prolonged delay. The female branch (fig. 157) consists of seven to twenty cells; one near the middle enlarges to form the oogonium and communicates with its neighbours by means of large pores. Additional nuclei thus reach the oogonium and in it nuclear fusions take place, but ascogenous hyphae arise from this cell only. There is no antheridium, association of *A* and *B* nuclei being brought about, as in *Humaria granulata*, by mycelial fusion. The haploid number of chromosomes is four.

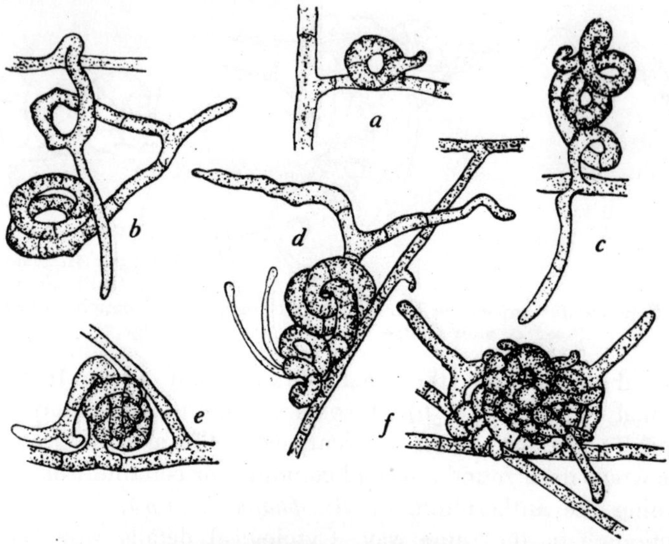

Fig. 158. *Ascobolus viridulus* Phil. & Plowr.; *a–f*, stages in development of fertile branch, ×420.

In *Ascobolus citrinus*,[1] as in *A. stercorarius*, only one cell gives rise to ascogenous hyphae. Similar fertile branches are formed in *A. glaber*,[2] *A. viridulus* (fig. 158), *A. Leveillei* and *A. equinus*,[3] and also in *Lasiobolus pulcherimus*,[4] but the full details of development have not yet been followed. None of them possesses an antheridium, *A. viridulus*, *A. Leveillei* and *A. equinus* are self-fertile.

[1] Schweizer, G., 1923.
[2] Gwynne-Vaughan and Williamson, 1933 ii.
[3] Green, 1931.
[4] Delitsch, 1926.

In *Saccobolus depauperatus*,[1] a self-fertile, coprophilous species, the male branch is absent, the female branch is small and forms an open coil. A somewhat similar structure has been reported[2] in *S. obscurus*.

The species of *Rhyparobius*[3] and *Thelebolus*,[4] two genera with many spored asci, are minute, coprophilous forms. They are distinguished by the fact that the apothecium of *Rhyparobius* produces several large asci, and that of *Thelebolus* only one (fig. 159).

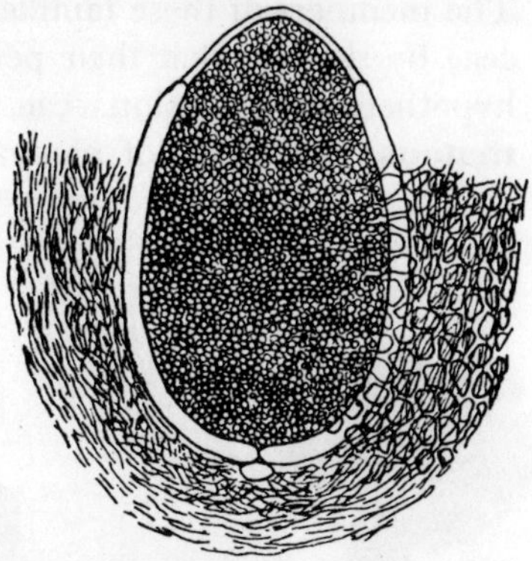

Fig. 159. *Thelebolus stercoreus* Tde.; ascocarp with single ascus, ×250; after Brefeld.

In *Thelebolus stercoreus* the primordium of the ascocarp is a thick branch containing a single nucleus. Later (fig. 160), two, four, and finally eight nuclei are seen and septation takes place so that a row of cells is formed. One of these contains two nuclei; it becomes the single ascus and the nuclei fuse. The definitive nucleus divides

Fig. 160. *Thelebolus stercoreus* Tde.; development of fertile branch, ×1750; after Ramlow.

karyokinetically, sometimes as many as ten times, so that 1024 nuclei are formed. The septate branch may perhaps be regarded as a single ascogenous hypha of vegetative origin.

[1] Gwynne-Vaughan and Williamson, 1933 ii. [2] Green, 1931.
[3] Barker, 1903–4; Overton, 1906.
[4] Brefeld, 1891; Ramlow, 1906.

Helotiaceae and Mollisiaceae

The members of these families are distinguished from the Pezizaceae by the fact that their peridium differs in structure from the hypothecium. In Helotiaceae the peridium is **pseudoprosenchymatous**, composed of elongated, parallel hyphae, usually light

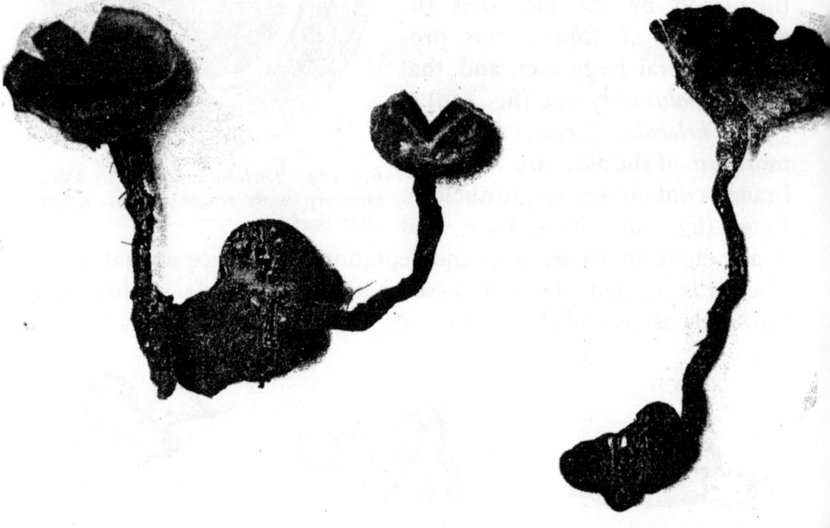

Fig. 161. *Sclerotinia tuberosa* (Hedw.) Fuck.; sclerotia and apothecia, nat. size.

in colour and thin walled. In Mollisiaceae it is **pseudoparenchymatous**, of round or polygonal cells, usually thick walled and dark coloured. In both families the ascus opens by the ejection of a plug, and not, as in most of the Discomycetes, by a lid.

The apothecia in both families are small, frequently stalked, sometimes growing out from a sclerotium; they are waxy in consistency and may be glabrous or hairy. Most species are saprophytes, often occurring on dead plants. In Helotiaceae the apothecium is sunk in the substratum, in Mollisiaceae it is frequently superficial. *Helotium ciliatosporum*[1] is a saprophyte, very sensitive to moisture, with attenuated ends to the spores.

Among the Mollisiaceae *Pseudopeziza Trifolii* is parasitic and causes the leaf spot disease of clover. The leaves become spotted and finally die, so that the crop is injured. There are many other species of *Pseudopeziza*, several on dead stems and leaves, a few on the living tissues of wild plants.

Another parasite of some importance is *Dasyscypha Willkommii*,[2] the well-known larch canker. The apothecia are externally yellow with an orange disc; the ascospores give rise to germ tubes which cannot penetrate the bark, but obtain entrance through wounds caused by hail, ice or snow, or by the attacks of insects. The mycelium ramifies chiefly in the soft bast, but may penetrate the wood as far as the pith. Where it spreads into the bark the tissues shrivel, producing depressed canker spots in which the white stromata develop. These give rise to minute, unicellular conidia, and to ascocarps later if conditions are moist.

In the genus *Sclerotinia* the stalked ascocarps arise from sclerotia (fig. 161). Many of the species are parasitic; *S. tuberosa* on *Anemone nodosa*, *S. trifoliorum*,[3] the fungus of clover rot, on clover, *S. sclerotiorum* on the potato, cabbage and other hosts, *S. fructigena* and *S. cinerea* on species of *Prunus* and *Pyrus* where they give rise to brown rot, blossom wilt and other pathological conditions. *S. cinerea*,[4] the cause of brown rot on plums, kills the shoots of the infected tree soon after their leaves unfold, and the mycelium extends from the dead shoots into the twigs that bear them, producing cankers above and below which the xylem elements are destroyed. Conidia occur on the leaves and may infect new trees; the ascigerous stage is found on mummified plums. In *S. sclerotiorum* and *S. trifoliorum* sclerotia appear in the stems; in the latter species the apothecia which develop from them expand only in the presence of light. Conidia are produced in bunches such as characterise the form genus[5] *Botrytis*, or in chains as in the form genus *Monilia*. *S. Gladioli*,[6] a common parasite on ornamental bulbs and corms, is self-incompatible with two com-

[1] Barnes, 1933. [2] Hiley, 1919. [3] Wadham, 1925.
[4] Wormald, 1919, 1921, 1922. [5] Cf. p. 359.
[6] Drayton, 1932, 1934.

plementary strains. Each of these produces microconidia (fig. 162) and also forms a stroma from which arise receptive bodies (fig. 163) of varying form, the primordia of the ascocarps. Intermingling of the complementary mycelia is without effect, but, when microconidia of either complement are applied to the receptive bodies of the other, fructifications are formed. It seems possible, therefore, that the microconidia act as male organs, though this cannot be determined without a knowledge of the cytological details of the process.

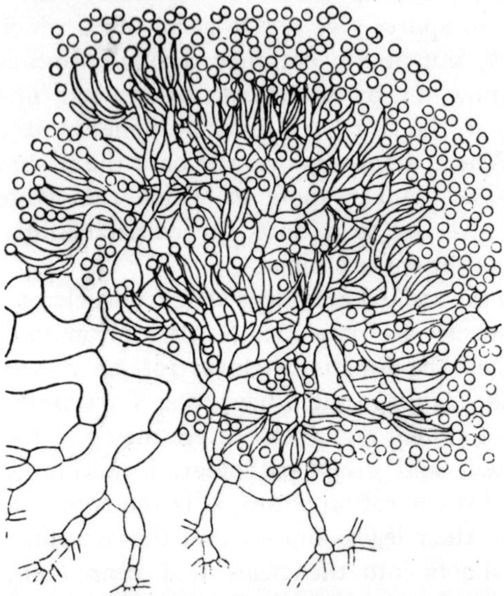

Fig. 162. *Sclerotinia Gladioli* (Massey) Drayton; part of a group of hyphae bearing conidia, such groups are surrounded by mucilage, ×850; after Drayton.

Fig. 163. *Sclerotinia Gladioli* (Massey) Drayton; receptive body, a column of hyphae, some deeply stained, ×70; after Drayton.

Celidiaceae, Patellariaceae and Cenangiaceae

In the families already described the ascocarp is either fleshy or waxy. In Celidiaceae, Patellariaceae and Cenangiaceae it is leathery, horny, or cartilaginous, and the ends of the paraphyses are interwoven above the asci to form a layer known as the **epi-**

thecium. The ascospores are sometimes more or less than eight in each ascus, and may be one to many celled; in some species pycnidia are present. The three families are sometimes grouped together as Dermateaceae.

Bulgaria polymorpha,[1] one of the best known of the Cenangiaceae, occurs on dead trunks of trees, especially beech. The ascocarps burst through the bark as small, rusty brown, scurfy knobs, which gradually expand, exposing the black and shiny hymenium; the substance is soft and tough, resembling india-rubber in consistency and appearance. The species is readily distinguished by its four, slightly curved, brown ascospores associated with four abortive spores.

The genus *Coryne* is possibly related to *Bulgaria*. *C. sarcoides* and *C. urnalis*[2] are common species on rotten trunks and stumps. The apothecia are crowded together and are dull red or purple in colour. The conidial fructifications occur among them or alone; their stromata are rather paler than the apothecia and bear fertile hyphae from the ends of which minute conidia are abstricted.

Cyttariaceae

This very curious family[3] contains the single genus *Cyttaria* (fig. 164). Six species are known, occurring in New Zealand, Tasmania and South America; all are parasitic on species of *Nothofagus*.

Cyttaria Darwinii was collected in Tierra del Fuego by Darwin in 1833.

"In the beech forests," he says, "the trees are much diseased; on the rough excrescences grow vast numbers of yellow balls. They are of the colour of the yolk of an egg, and vary in size from that of a bullet to that of a small apple; in shape they are globular, but a little produced towards the point of attachment. They grow both on the branches and stems in groups. When young they contain much fluid and are quite tasteless, but in their older and altered state they form a very essential article of food for the

[1] Biffen, 1901.
[2] Howarth and Chippindale, 1930–1.
[3] Berkeley, 1842, 1847 ii, 1848; Fischer, E., 1886; Buchanan, 1885.

Fuegian." He observed that the whole surface is "honeycombed by regular cells". These are the separate apothecia, considerable

Fig. 164. *Cyttaria Gunnii* Berk.; *a*, twig of *Nothofagus Cunninghami* bearing the fungus, $\times \frac{3}{5}$; *b*, group of stromata; *c*, single stroma cut across; all after Berkeley.

numbers of which occur on the same stroma. Unfortunately the details of development are still unknown.

HELVELLALES

The members of this alliance are saprophytes, growing chiefly on the ground, sometimes on decayed wood. Most have large, fleshy and stipitate ascophores with the hymenium spread over the upper surface and covered at first by a veil or membrane. Such a fructification may perhaps be homologised with a stalked *Peziza*, in which the cup is turned inside out during the early stages of development; the structure is often complicated by extensive convolutions of the fertile surface.

There are three families:

Ascophore flattened, not stalked	RHIZINACEAE
Ascophore stalked	
Fertile region of head distinct from stalk, ascus opening by a lid	HELVELLACEAE
Fertile region not always distinct from stalk, ascus opening by a plug	GEOGLOSSACEAE

RHIZINACEAE

The Rhizinaceae include two genera, *Rhizina* and *Sphaerosoma*. *Rhizina* has a flattened, crust-like, superficial ascocarp, more or less concave below and attached to the soil by strands of mycelium, while in *Sphaerosoma* the ascocarp is partly sunk in the substratum and the rooting hyphae are sometimes grouped on a short pedicel. It is concave when young but later becomes strongly reflexed (fig. 165).

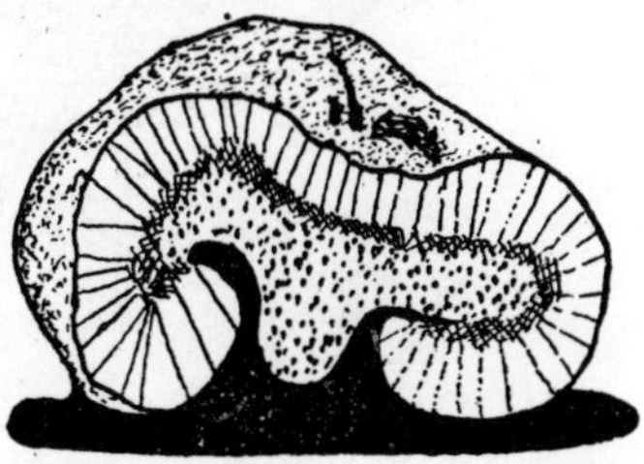

Fig. 165. *Sphaerosoma fuscescens* (Klotz.) Roup.; apothecium, ×6; after Rouppert.

Rhizina inflata is known to occur in Great Britain only as a saprophyte growing on soil, but, both in France and Germany, it has been found to attack conifers. The mycelium ramifies in the intercellular spaces and forms masses of pseudoparenchyma in dead and diseased tissues of the host. The trees lose their needles and die.

In *Rhizina undulata*[1] the development of the apothecium is pre-

[1] Fitzpatrick, H. M., 1918.

ceded by the appearance of a long, multicellular female branch recalling that of the Ascobolaceae.

In *Sphaerosoma Janczewskianum*[1] (fig. 166) a large oogonial cell has been described from which the ascogenous hyphae originate.

Helvellaceae

The Helvellaceae include six genera, *Helvella* (fig. 167 *a*), *Morchella* (fig. 167 *b*, *c*), *Verpa*, *Gyromitra*, *Cidaris* and *Cudoniopsis*;[2] of these the first four are British. In all a definite fertile head is

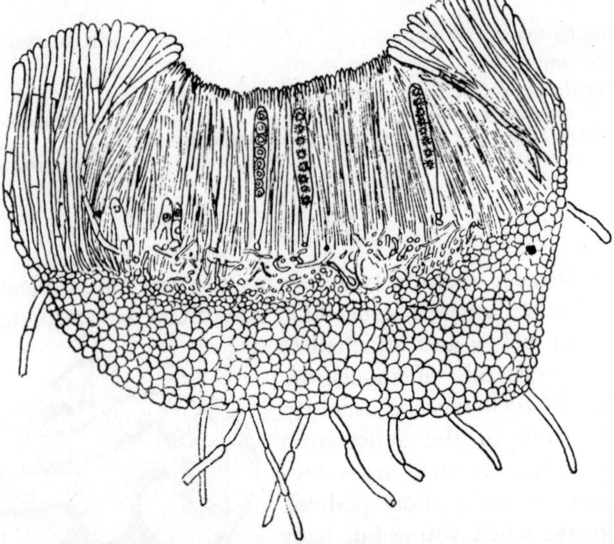

Fig. 166. *Sphaerosoma Janczewskianum* Roup.; apothecium showing asci and paraphyses, × 70; after Rouppert.

distinguished from the stalk, and the hymenium extends over its more or less convoluted surface. Development has been studied only in *Helvella*,[3] where the fruit arises as a tuft of branching, septate hyphae, the outer of which form a protective membrane, broken as growth proceeds, to expose a palisade of club-shaped filaments. An oogonium has not been observed.

In *Helvella crispa*[4] apogamous fusions, similar to those in *Humaria rutilans*, have been recorded in the vegetative cells of the hymenium, and both in this species and in *Morchella esculenta*[5]

[1] Rouppert, 1909. [2] Spegazzini, 1925. [3] McCubbin, 1910. [4] Carruthers, 1911. [5] Maire, 1905.

the third division in the ascus appears to bring about a second reduction in the number of chromosomes.

GEOGLOSSACEAE

The Geoglossaceae grow usually in damp or moist situations such as low, wet woods or shady slopes. They occur on soil, or on dead branches or leaves, and two species of *Mitrula* are parasitic on moss. There are eight genera, five of which are British.

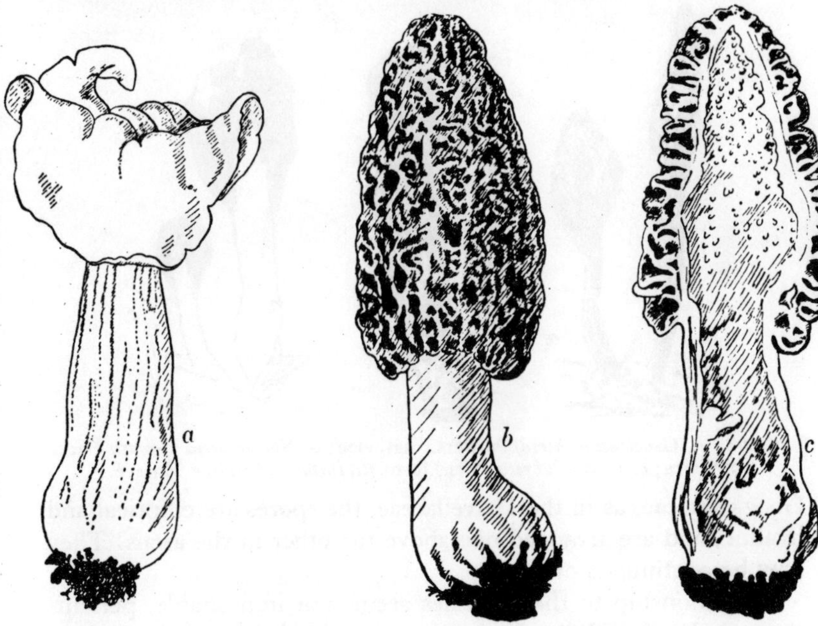

Fig. 167. *a*, *Helvella crispa* (Scop.) Fr.; *b* and *c*, *Morchella vulgaris* Pers.; after Boudier.

The ascophore is erect and stipitate, the terminal, fertile region being club-shaped (fig. 168 *a*), laterally compressed, or forming a cup, or a pileus (fig. 168 *c*). In some of the simpler forms, as in *Geoglossum hirsutum*, there is no clear line of demarcation between the fertile and sterile regions. The ascus contains eight spores and opens by the ejection of a plug.

The young ascocarp consists, as in *Helvella*, of a tangle of vegetative hyphae and is commonly protected, as in that genus, by an outer membrane or veil. In *Leotia lubrica*[1] a large, branching

[1] Massee, G., 1897; Durand, 1908; Brown, W. H., 1910; Duff, 1922.

cell occurs at the base of the young ascocarp; in *Spathularia velutipes* and *Cudonia lutea* the ascogenous hyphae arise from cells with dense contents just below the hymenium.[1]

In the species of *Spathularia* and *Vibrissea*, as in *Geoglossum*, the spores are very long, narrow and septate, lying side by side in the ascus. *Geoglossum* (figs. 168 *a*, 169 *a*) is distinguished by its coloured spores, the other two genera, in which the spores are hyaline (fig. 169 *b*), by the form of the fructification. In the rest of the

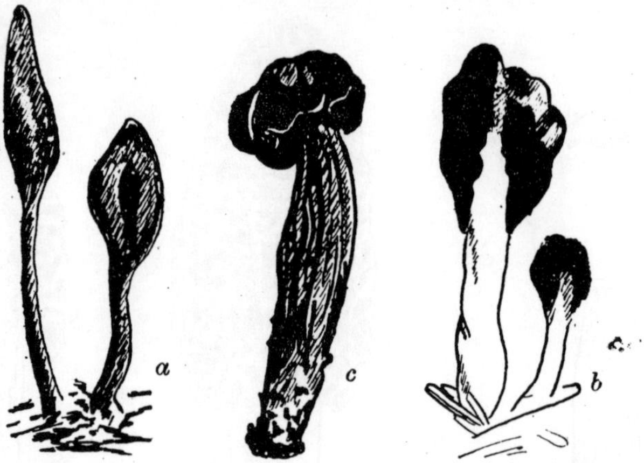

Fig. 168. *a*, *Geoglossum hirsutum* Pers., nat. size; *b*, *Spathularia clavata* Sacc., nat. size; *c*, *Leotia lubrica* Pers., form *stipitata*, $\times \frac{2}{3}$; after Massee.

Geoglossaceae, as in the Helvellaceae, the spores are elliptical and hyaline, and are arranged one above the other in the ascus. They may be continuous or septate.

A relationship to the Pezizales seems not improbable, perhaps through *Leotia*, where the fertile region is pileate, to the Helotiaceae or Mollisiaceae, which resemble the Geoglossaceae in the means of dehiscence of the ascus.

TUBERALES

The Tuberales are characteristically subterranean, though some species are imperfectly buried and others grow among decaying leaves. When mature the fruits emit a powerful odour by which rodents are apprised of their whereabouts. The ascocarp is eaten and the spores dispersed after passing through the alimentary canal of the animal.

[1] Duff, 1922.

The ascocarp is more or less globose, sometimes completely closed, sometimes with a small opening. The hymenium may form a smooth lining to the fruit, or may be thrown into elaborate folds, so that the fertile region is divided into chambers. The asci contain one to eight spores; the epispore is often elaborately ornamented at maturity.

Early investigators classed the Tuberaceae with the hypogeal

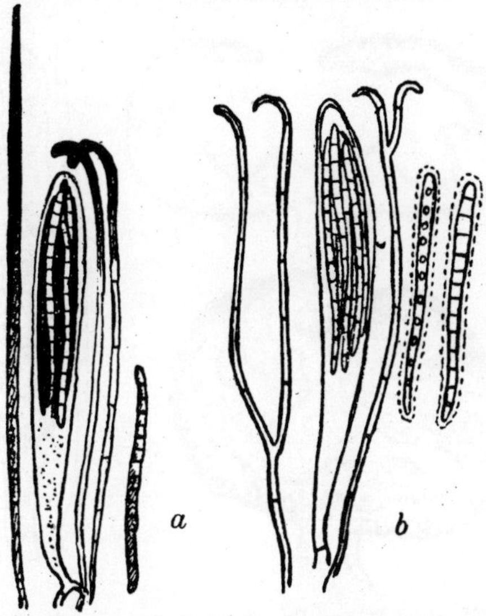

Fig. 169. *a*, *Geoglossum hirsutum* Pers., hair, ascus, paraphyses and spore, ×230; *b*, *Spathularia clavata* Sacc., paraphyses, ascus and spores, ×400; after Massee.

Gasteromycetales; a consequence of this survives in the use of the term **gleba** to describe the contents of the fructification, including both the vegetative hyphae and the fertile cells.

The Tuberales include a single family, the Tuberaceae; their relationship is probably to the Pezizaceae and Rhizinaceae. One or more series can be traced from these families to the Tuberales, the principal modifications being in the direction of adaptation to subterranean conditions by the increased protection of the hymenium. This has been achieved either by retaining at maturity the closed form of the young pezizaceous ascocarp, as in *Genea* and *Pachyphloeus*, or by the invagination of a prematurely exposed fertile layer, as in *Tuber*, and its protection by a secondary sheath.

TUBERACEAE

The Tuberaceae[1] include eight genera; in *Genea* the ascocarp is of the pezizaceous type, but with a somewhat convoluted hymenium (fig. 170 *b*, *c*); the asci are cylindrical and the spores uniseriate. In *Stephensia* and *Pachyphloeus* the convolutions of the hymenium are more marked; the asci of *Pachyphloeus* are stouter and the spores irregularly biseriate.

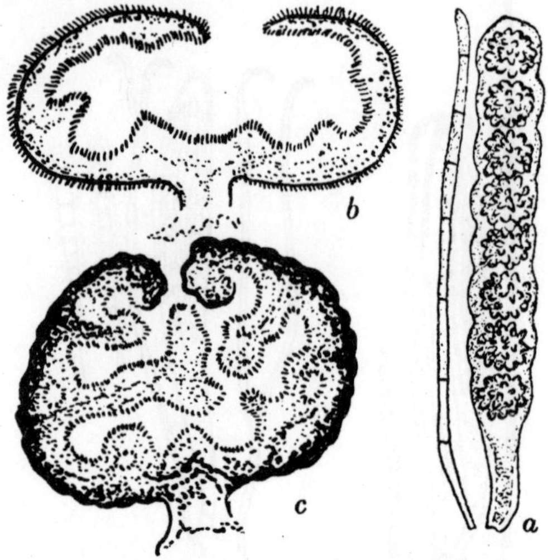

Fig. 170. *Genea Klotzchii* B. and Br., *a*, ascus and paraphysis; *Genea hispidula* Vitt., *b*, apothecium; *Genea sphaerica* Vitt., *c*, apothecium; after Massee.

In *Balsamia* the asci are broadly oblong or subglobose (fig. 171); the convolutions of the hymenium give the appearance in section of separate chambers (fig. 172). The mature ascocarp is enclosed by a special sheath. During development the hymenial chambers communicate at several points with the exterior, indicating that they are formed by the invagination of a reflexed fertile disc similar to that of *Rhizina*.

In *Tuber* this type of structure is clearly marked. The ascocarp is irregularly globose, fleshy or in some species almost woody; internally the walls which divide the gleba are extensively branched, and the free space between them is diminished, so that opposite

[1] Vittadini, 1831; Boulanger, 1904, 1906 i, ii; Bucholtz, 1908, 1910; Massee, G., 1909.

layers of the hymenium are brought close together and constitute the so-called fertile "veins". Other "veins", white and sterile, run between the hymenial layers (fig. 173). The asci are often globose (fig. 174); the spores are usually four in number but are sometimes reduced to two or one.

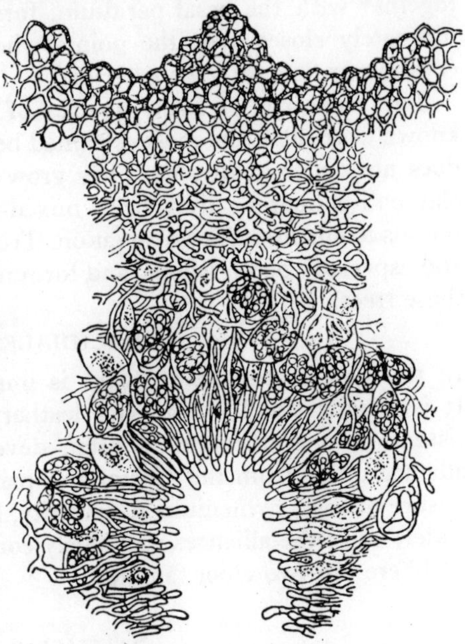

Fig. 171. *Balsamia vulgaris* Vitt.; section through hymenium; after Tulasne.

In *Tuber puberulum*[1] (fig. 175), the very young apothecium consists of a mass of hyphae, the outer rather more loosely interwoven than the inner. Around the lower part a dense basal sheath is differentiated, corresponding to the peridium of a cup-shaped ascocarp. Soon the first signs of the fertile veins appear as invaginations of the upper surface, which, though as yet showing no trace of asci, corresponds to the fertile disc. Owing to the

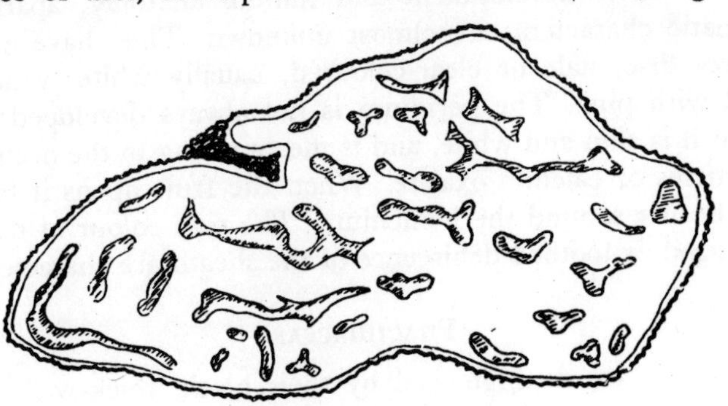

Fig. 172. *Balsamia vulgaris* Vitt.; after Tulasne.

[1] Bucholtz, 1903.

rapid growth of the upper portion of the fruit, the basal sheath is bent backwards, while along the fertile veins the young asci begin to appear. Later the peripheral tissues become thickened, and, together with the basal peridium, form the special sheath. This ultimately closes over the points where the fertile veins are in communication with the exterior.

The ascocarps of many species of *Tuber* are edible, and are known as truffles, the most esteemed being *T. melanosporum* which does not occur in Britain. They grow chiefly in soils mixed with clay and containing iron, or in mixed alluvium; the soil must be porous to ensure sufficient aeration. Truffles occur in chestnut, oak, and especially beech woods, and form mycorrhiza with the roots of these trees.

PHACIDIALES

In the Phacidiales the ascocarp is immersed in the substratum. It is usually small in size and leathery, waxy, or corky in consistency; an epithecium is often developed. Members of this alliance differ from the Hysteriales principally in the greater exposure of their hymenium at maturity, but the cup never opens as widely as in the alliances previously considered.

There are two chief families.

STICTACEAE

The Stictaceae form a considerable assemblage of species with small ascocarps, occurring saprophytically on wood or other plant remains. Their development and minute anatomy, apart from systematic characters, are almost unknown. They have a fleshy or waxy disc, pale or clear-coloured, usually white, yellow, or tinged with pink. The peridium is not always developed: when present it is thin and white, and is mealy owing to the occurrence of particles of calcium oxalate. When the fruit opens it forms a white border around the hymenium. The pale colour of the fruit and ragged or toothed dehiscence of the sheath are characteristic.

PHACIDIACEAE

These fungi are distinguished by their black, thick-walled apothecia, usually scattered, sometimes, as in *Rhytisma*, grouped on a black stroma. Where the hymenium is circular the sheath splits

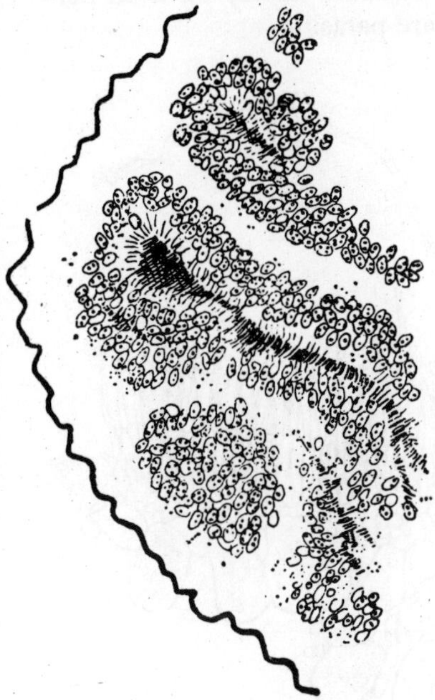

Fig. 173. *Tuber rufum* Pico; general view of fertile region; after Tulasne.

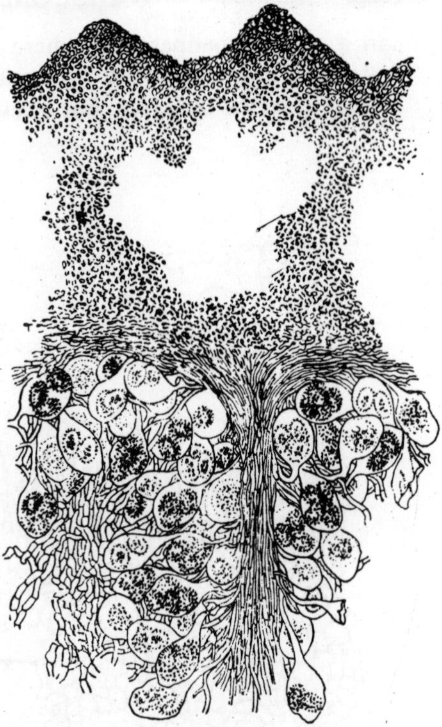

Fig. 174. *Tuber rufum* Pico; section through hymenium; after Tulasne.

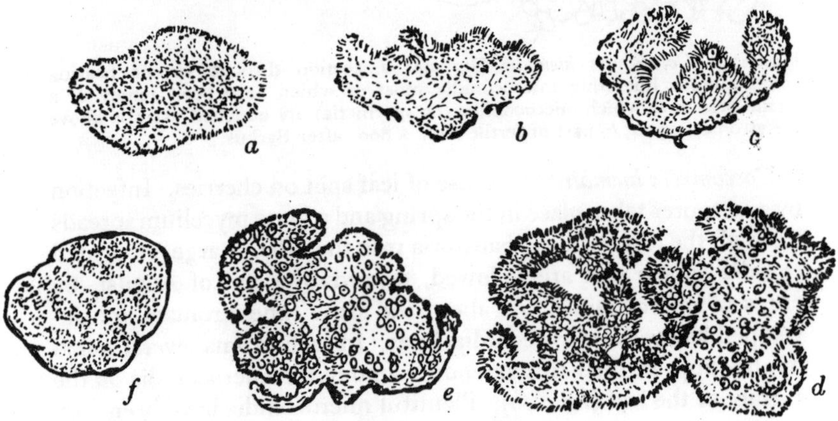

Fig. 175. *Tuber puberulum* (B. and Br.) Ed. Fisch.; *a–e*, development of ascocarp, *a*, ×52, *b* and *c*, × 28, *d* and *e*, ×21; *f*, section through mature ascocarp, ×6; all after Bucholtz.

in a stellate manner, but where it is elongated dehiscence takes place by means of a slit. The species occur mainly on dead herbaceous stems or leaves, but a few are parasitic.

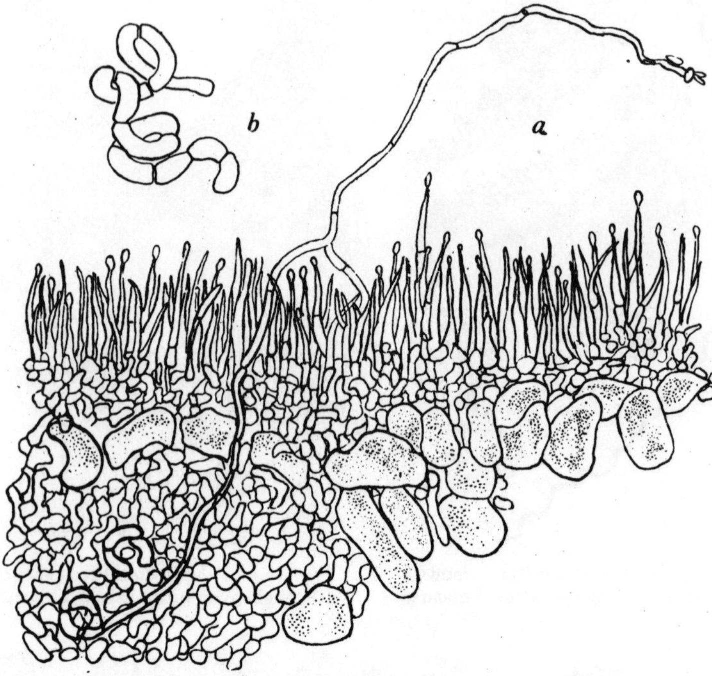

Fig. 176. *Coccomyces hiemalis* Higgins; *a*, section through stroma bearing conidiophores liberating microconidia most of which have been removed, a fertile branch to which microconidia (or spermatia) are attached extends above the stroma, × 500; *b*, part of fertile coil, × 800; after Backus.

Coccomyces hiemalis[1] is a cause of leaf spot on cherries. Infection by ascospores takes place in the spring and a loose mycelium spreads through the leaf, sending haustoria into the cells. Large, elongated conidia appear and are followed, towards the end of August, by crops of microconidia. At about the same time stromata are developed below the microconidial beds. In each stroma several coiled hyphae appear, their tips extending among the microconidia on the surface of the leaf (fig. 176). Plentiful microconidia have been seen

[1] Backus, 1933, 1934.

adhering to them. It is suggested that the coiled hyphae are female branches and the microconidia functional spermatia. As in many other fungi further cytological evidence is required. After the leaves are shed the fungus continues its development as a saprophyte, giving rise next spring to asci with elongated, septate spores. Four chromosomes are described in all the divisions in the ascus.

Rhytisma Acerinum[1] (fig. 177) infects the leaves of species of

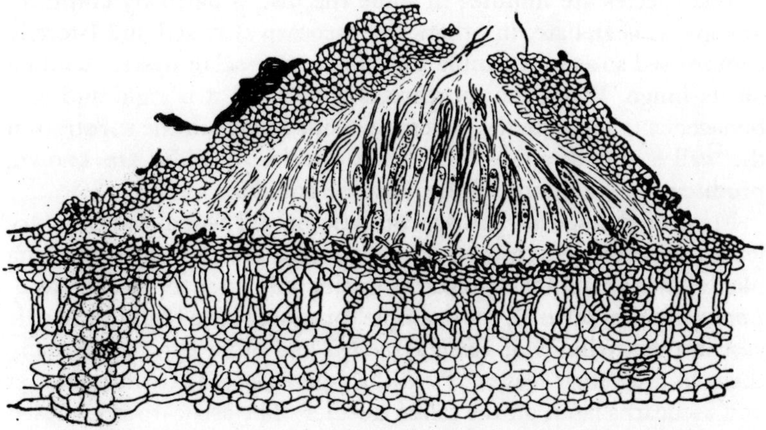

Fig. 177. *Rhytisma Acerinum* (Pers.) Fr.; apothecium in longitudinal section, × 160.

Acer. The mycelium ramifies in the tissues of the host and causes yellow spots about three weeks after infection. Some five weeks later pycnidia develop under the cuticle in these areas and produce small, unicellular conidia. The epidermis and underlying tissues of the host become filled with hyphae and a dense, black sclerotium is completed. In this state the leaf falls off and next spring the sclerotia thicken, become wrinkled, and finally burst by elongated fissures to expose the discs of the apothecia. The ascospores are filiform and aseptate, they are ejected with some force and are probably conveyed to the young leaves by the wind; their germ tubes enter through the stomata.

Both in this species and in *Cryptomyces Pteridis*[2] swollen cells

[1] Bracher, 1924; Jones, S. G., 1925. [2] Killian, 1918.

have been described in the sub-hymenium, but in view of their unusual position further evidence is required before their sexual significance can be accepted.

HYSTERIALES

The Hysteriales are characterised by a black, elongated ascocarp, which opens by a longitudinal slit so narrow that the greater part of the disc remains concealed.

The species are minute; in some the disc is narrowly elliptical, in some it is stellate, in others the ascocarp is raised and laterally compressed so as to resemble a miniature mussel or oyster standing on its hinge. When the ascocarp is superficial it is rigid and carbonaceous in consistency, when developed beneath the substratum the wall is membranous. In a few species pycnidia are known, producing oblong, unicellular, hyaline conidia.

Most of the species are saprophytic on old wood, bark, or dry leaves. The mycelium is sometimes found in the tissues of living plants, though apothecia reach maturity only after the host is dead. *Lophodermium pinastri*, for example, produces pine blight or needle cast in the seedlings of *Pinus sylvestris* and other conifers, causing them to drop their leaves. Pycnidia (spermogonia) are formed first and ascocarps later, on the fallen needles. The ascocarp is enclosed by a thick, dark-coloured peridium interrupted in the middle region by a group of loosely locked periphyses with swollen tips. With the absorption of water these swell, become mucilaginous and exert pressure by which dehiscence is ultimately caused. Structures regarded as oogonia have been described.[1] Mature trees are attacked and, though not themselves seriously injured, act as centres of infection in the neighbourhood of seed-beds and nurseries.

The subdivision of the Hysteriales depends on the consistency of the sheath, the form of the ascocarp, and its superficial or immersed position. The Hysteriaceae and Hypodermataceae are the principal families.

In the form of their fructification the members of the Hysteriales are intermediate between the Discomycetes and Pyrenomycetes. Species differ from certain of the Phacidiales chiefly in the narrow opening of the ascocarp, and from the simpler Sphaeriales in the

[1] Jones, S. G., 1935.

fact that this opening is elongated instead of round. Such ascocarps might be said to resemble not a cup or a flask but a laterally compressed jug. Further research is required to determine their true relationship, but, in view of their small size and hard consistency, detailed investigation will not be easy.

PYRENOMYCETES

The Pyrenomycetes[1] include over 10,000 species; they possess a flask-shaped perithecium opening by a small pore, the ostiole, and containing an hymenium with parallel asci spread over the floor and lower parts of the sides. There are four **alliances**, the Hypocreales, Dothideales, Sphaeriales and Laboulbeniales. The Laboulbeniales are true pyrenomycetous fungi in that they produce regularly arranged asci and a perithecium opening by an ostiole, but they are distinguished by a number of special characters, some of which may be related to their peculiar habitat as external parasites on insects.

In the Dothideales the perithecia are always immersed in a stroma or cushion of fungal hyphae; in the Hypocreales and Sphaeriales they are sometimes isolated and free, sometimes sunk in the tissues of the host, sometimes embedded in a stroma. The perithecium is lined by delicate filaments, some of which, the **periphyses**, grow along and partly close the neck, and may protrude through the ostiole, while others, the paraphyses, are mingled with the asci in the venter of the fruit.

Accessory multiplication in these alliances is by chlamydospores, and by conidia which may be borne on free conidiophores or grouped in pycnidia. In certain species there is evidence that the so-called pycnidia are spermogonia and the cells they set free spermatia. In some this function has been lost, others would repay more detailed investigation.

In the Hypocreales, Dothideales and Sphaeriales the female branch is of the coiled type found in *Eurotium* and in many of the Discomycetes. In most of the forms in which the perithecia are free it is comparatively short and simple, in species with immersed perithecia it may be elaborately coiled, furnished with a many-celled trichogyne and associated with the development of spermatia

[1] Seaver, 1909; Arnaud, 1920.

in spermogonia; in many such forms it is known that the trichogyne does not function and in several that the ascogenous hyphae arise from vegetative cells. Finally, in *Xylaria* and others of the higher members of the group, the female branch is reduced to a short row of cells, sometimes known after its discoverer as Woronin's hypha, or, as in *Cordyceps*, no trace of sexual apparatus is found.

The Pyrenomycetes do not seem to have given rise to any higher forms; some of their number have a greater vegetative development than any other ascomycetous fungi.

They may be subdivided as follows:

Wall of perithecium differentiated from stroma;
- Perithecium wall and stroma, if present, soft in texture, colourless or light coloured — HYPOCREALES
- Perithecium wall and stroma, if present, firm, leathery or brittle, dark in colour — SPHAERIALES

Perithecium always sunk in a stroma from the tissue of which its wall is not differentiated; colour of stroma black or dark brown — DOTHIDEALES

Minute, external parasites on insects, perithecium borne on a receptacle which also bears appendages, spores two-celled — LABOULBENIALES

HYPOCREALES

The Hypocreales are readily distinguished by the clear colour and fleshy consistency of the perithecium or stroma. In most of the Pyrenomycetes the colour is black or dark brown, here bright red, yellow, purple and various pale shades are found; it is only occasionally that so dull a tint as brown or dirty violet appears.

The asci contain usually eight, in some species four, and in some many spores. The spores are usually hyaline, but are dark coloured in *Melanospora* and its allies; they are elliptical or filiform and may be one or more celled.

There are some sixty genera of Hypocreales. In the simplest forms a stroma is absent, and the separate perithecia may or may not be partly sunk in the substratum; in others a fleshy stroma appears and the perithecia are more or less embedded in it. In the highest members the perithecia originate deep in the stroma and remain immersed throughout their development. Upon these characters the subdivision of the alliance is based:

Stroma absent, or when present with perithecia entirely superficial	NECTRIACEAE
Stroma forming a conspicuous matrix in which the perithecia are partially or entirely immersed	HYPOCREACEAE

NECTRIACEAE

The family includes both parasites and saprophytes. *Nectria* grows on living and dead wood, *Hypomyces* and *Melanospora* attack hymenomycetous fungi, *Neurospora* is a pest in bakeries. In *Hypo-*

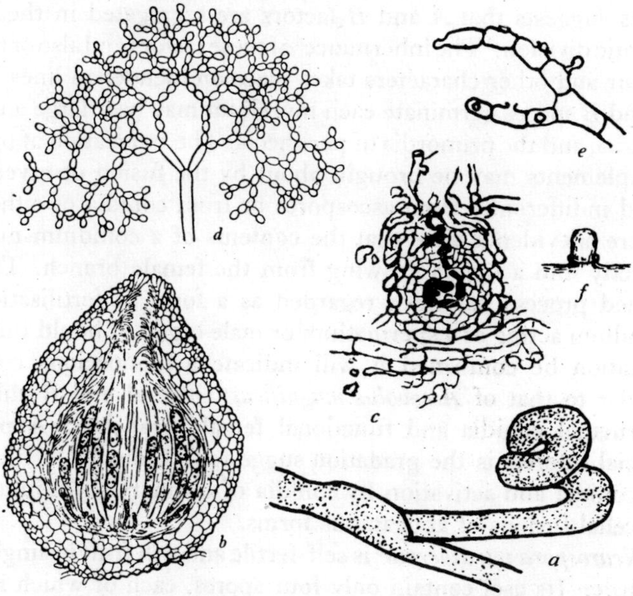

Fig. 178. *Neurospora tetrasperma* Dodge, *a*, young fertile branch, × 1600; after Colson; *b*, section through ascocarp containing asci, × 100; *Neurospora sitophila* Dodge, *c*, coiled fertile branch surrounded by sheath; after Dodge; *d*, moniliform conidia; after Dodge; *e*, *f*, microconidia; after Dodge.

myces[1] the perithecia are free, in *Melanospora*[2] there is a filamentous stroma.

The species of *Neurospora*,[3] the red bread mould, bear large conidia, which are moniliform in arrangement (fig. 178 *d*), and microconidia

[1] Moreau, 1914.
[2] Kihlmann, 1885; Nichols, 1896; Cookson, 1928.
[3] Shear and Dodge, 1927; Dodge, 1927, 1928 i, ii.

(fig. 178 *e*, *f*) growing singly or in short chains. *Neurospora sitophila* is self-incompatible, perithecia reaching maturity only when two complementary mycelia, *A* and *B*, have become associated. The young perithecium[1] contains a coiled female branch (fig. 178 *c*) from which hyphae may grow to the exterior, meeting hyphae or conidia of the complementary strain. When the perithecium ripens *A* and *B* spores are present in the same ascus, showing that the definitive nucleus contained both *A* and *B* factors, and, therefore, that fusion of *A* and *B* nuclei has occurred. The arrangement of spores in the ascus suggests that *A* and *B* factors are segregated in the second meiotic division.[2] The inheritance[3] of these factors and also of conidial colour and other characters takes place on mendelian lines. When *A* and *B* spores germinate each mycelium may bear large and small conidia, and the primordia of perithecia. The association of opposite complements may be brought about by the fusion of mycelia derived indifferently from ascospores or from conidia of either size. There is evidence also[4] that the contents of a conidium may pass directly into a hypha growing from the female branch. The last-named process has been regarded as a form of fertilisation, the conidium acting as a spermatium or male organ. Should this interpretation be confirmed it will indicate a monoecious condition similar to that of *Ascobolus magnificus*, since *A* and *B* alike bear spermatial conidia and functional female branches. A point of special interest is the gradation suggested, through "fertilisation" by conidia and activation by conidia of ordinary mycelium, to the mycelial fusions of apogamous forms.

Neurospora tetrasperma[5] is self-fertile and will fruit in single spore culture. Its asci contain only four spores, each of which is binucleate at its first formation, though later the nuclei divide. Occasionally, however, two or more small spores are formed around single nuclei, and these produce self-sterile mycelia which can fruit only when appropriately mated. Two such small spores occurring in the same ascus are respectively *A* and *B*. These facts make it evident that mycelia from the normal, large spores produce fruits because they contain both *A* and *B* nuclei. In the self-incompatible form of this species the cytology has been fully studied. The cells are

[1] Dodge, 1935 ii.
[2] Dodge, 1927; Wilcox, 1928.
[3] Dodge, 1928 i, 1935 i; Lindegren, 1932–6.
[4] Dodge, 1932, 1935 ii.
[5] Dodge, 1927; Colson, 1934.

multinucleate, mycelial fusions between *A* and *B* take place freely and both thalli bear coiled, multicellular female branches (fig. 178 *a*), round which the vegetative cells grow up to form a sheath. There is no sign of an antheridium. In single spore culture the ascocarp rudiment forms a small, resting sclerotium from which, under suitable conditions, vegetative branches may grow out. When mycelial fusions between *A* and *B* have occurred the perithecia reach maturity and give rise to asci in which spores are formed around two nuclei each, or occasionally around one. The haploid number of chromosomes is six, two being longer than the others. In the prophase of meiosis twelve chromosomes are recognised in six pairs showing that the definitive nucleus is diploid and that

Fig. 179. *Nectria cinnabarina* (Tde.) Fr. on a fallen twig; *a*, conidial stroma; *b*, young perithecia; ×6; E. J. Welsford del.

only one nuclear fusion, that in the ascus, and only one reduction, the meiotic phase, occur in this life history.

In *Nectria* (fig. 179) the usually red or yellow perithecia are produced on a fleshy stroma of the same colour. The genus is large, with some 250 species, of which *Nectria cinnabarina*, the commonest in Great Britain, occurs on the branches of deciduous trees. The mycelium from the germinating spores is unable to penetrate the bark of the host, and either enters through open wounds or spreads along the xylem elements from a dead region to the living tissue; as the mycelium grows, blocking of the wood causes wilting and death.[1] Meanwhile stromata appear; in the conidial stage they are bright pink and occur at all seasons on dead or living branches; perithecia are produced only in the autumn and winter and only

[1] Mayr, 1882; Line, 1922.

after the tissues have been killed. They are deep red in colour, and are partly immersed in deep red stromata. When a perithecium is about to be formed a coil of hyphae larger than the other elements of the stroma appears a little below the surface, and probably represents the remains of whatever sexual apparatus originally gave rise to ascogenous hyphae.

Nectria galligena[1] is the cause of apple canker, the perithecia developing on withered apples and serving to tide the fungus over the winter; oogonial cells have been observed but they degenerate early and the ascogenous hyphae arise from vegetative filaments.

Both species are examples of the rather numerous fungi which produce conidia during their parasitic phase, and ascospores only when the death of the host has rendered them saprophytic. Several species of *Nectria* are parasites on scale insects.[2]

HYPOCREACEAE

Among the Hypocreaceae *Polystigma*[3] is a small genus, the best known member of which, *P. rubrum*, develops on the leaves of *Prunus spinosa*, of *Prunus insititia*, and of the cultivated plum, where it produces conspicuous orange, yellow, or scarlet stromata. Each stroma is the result of a separate infection and spreads over only a small part of the leaf, so that, in autumn, when the leaves are shed, the host is freed from the disease. The fungus hibernates in the fallen leaves, and next spring, unless these are destroyed, the ascospores mature, reach the young foliage, and cause fresh infections.

Within the leaf the hyphae ramify among the cells of the host, becoming massed in the intercellular spaces below the stomata and often pushing their way to the exterior between the guard-cells. Finally the stroma may extend from the lower to the upper epidermis and only a few isolated cells of the mesophyll be found in the infected region. During the summer, flask-shaped spermogonia appear and open on the under side of the leaf, usually in the position of a stoma. The wall of the spermogonium consists of densely interwoven filaments and is lined by thin, uninucleate, spermatial hyphae (fig. 180), from the ends of which spermatia are abstricted.

[1] Cayley, 1921; Dillon Weston, 1925. [2] Petch, 1921.

[3] Fisch, 1882; Frank, 1883; Blackman and Welsford, 1912; Nienberg, 1914; Higgins, 1920.

The mature spermatium is a curved, filiform cell, containing a single, elongated nucleus. All attempts to bring about the germination of the spermatia have failed, and no relation has been demonstrated between them and the female organs; consequently their original use can be inferred only from their structure. Their scanty contents and large nucleus suggest that they are better constituted to act as agents of fertilisation than as a means of vegetative increase.

The female branches originate as multinucleate hyphae which

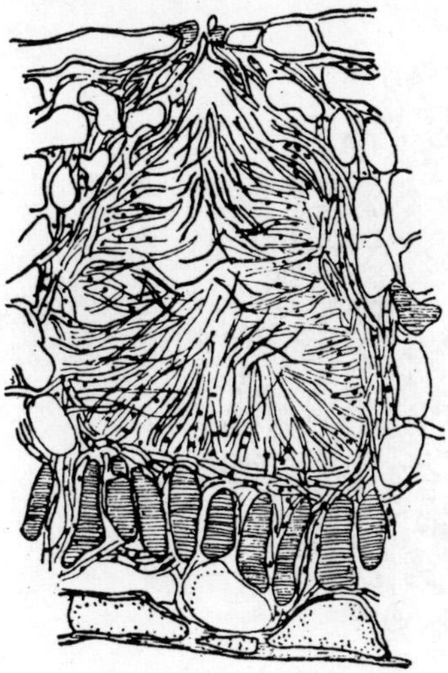

Fig. 180. *Polystigma rubrum* DC.; spermogonium, × 250; after Blackman and Welsford.

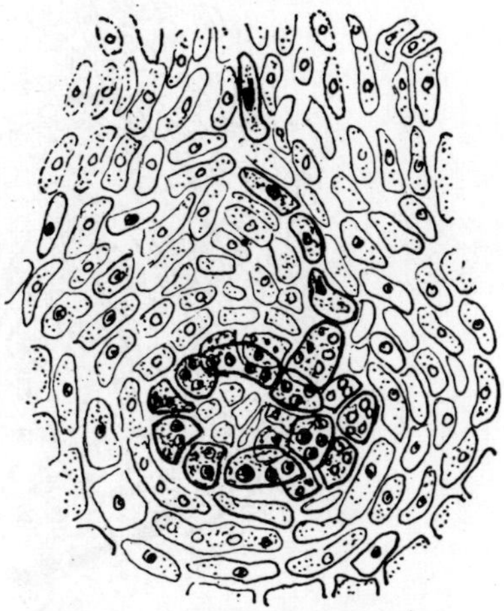

Fig. 181. *Polystigma rubrum* DC.; mature fertile branch, × 800; after Blackman and Welsford.

become septate and elaborately coiled (fig. 181), but degenerate without giving rise to a sporophyte. The ascogenous hyphae are derived from the surrounding vegetative cells, among which the remains of the female branch can sometimes be seen (fig. 182). There is some evidence that pseudapogamous fusions take place in the cells of the ascogenous hyphae before the appearance of asci. A perithecium (fig. 183) arises in the position of each female branch.

In the genus *Podocrea* the stroma is erect and sometimes branched; in *Hypocrea* it is usually hemispherical or bolster-shaped; in both genera and their immediate allies the ascospores are two or more celled. Most of the species occur saprophytically on wood or as parasites on the larger fungi. The systematic position of *Podocrea* has undergone some strange vicissitudes. In consideration of its form *P. alutacea*[1] was first placed in the Clavariaceae; later, when the perithecium-bearing head of the stroma had been recognised as pyrenomycetous, it was regarded as parasitic on its own stalk, which was classified as a *Clavaria* or as one of the Geoglossaceae. Finally the upright stroma was grown in pure culture from the ascospores and thus shown to be a single fungus.

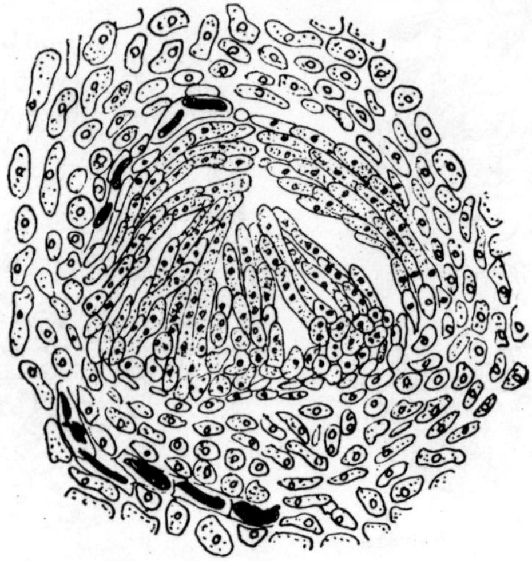

Fig. 182. *Polystigma rubrum* DC.; young perithecium; the ascogenous hyphae are not yet clearly distinguished, the darkly stained remains of the fertile branch are visible near the periphery, ×680; after Blackman and Welsford.

The species of *Epichloë* occur parasitically on the stems of grasses which become coated by the white or yellow stroma. During development oval conidia are produced, later the perithecia, which are completely immersed in the stroma, reach maturity. As in the remaining genera of the Hypocreaceae, the ascospores are filiform.

The genus *Cordyceps*[2] (fig. 184) includes about sixty species; these are mainly tropical parasites on insects, the bodies of which they transform into sclerotia. The peculiar appearance of these

[1] Atkinson, 1905.
[2] Berkeley, 1843; Tulasne, 1861–5; Massee, G., 1895.

structures and of the stromata derived from them has given rise to curious views as to their significance and medicinal value. *Cordyceps sinensis*, for example, was a celebrated drug in the Chinese pharmacopoeia. The quaint belief that it was "a herb in summer and a worm in winter" may sufficiently account for the esteem in which it was held.

The ascospores are filiform and multicellular and, when shed, break up into their separate segments. The germ tubes from these

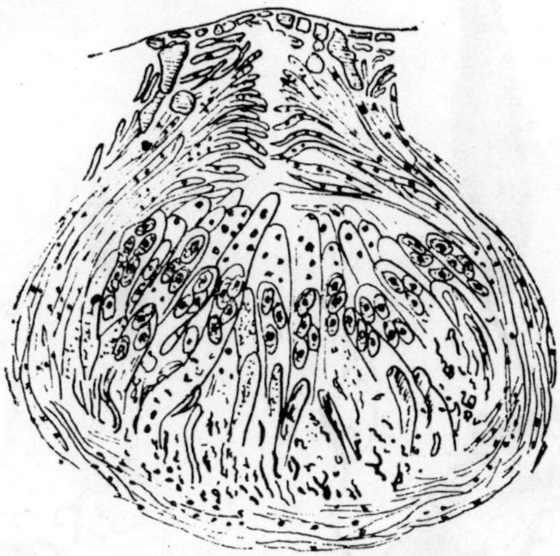

Fig. 183. *Polystigma rubrum* DC.; mature perithecium, ×270; after Blackman and Welsford.

or from the conidia penetrate the insect and cut off cylindrical cells which enter the blood stream and increase by yeast-like budding until the creature dies. A mycelium then appears and the formation of the sclerotium begins, while chains of conidia may be produced on subaerial conidiophores arranged in a fascicle. The mature sclerotium is a compact mass of interwoven hyphae whose cells are rich in glycogen and oil; its development destroys the internal organs of the host, the skin alone remaining intact, so that the appearance of the insect is maintained. From this mummified structure the stromata emerge; they are pale or bright coloured, red in the best known British species, *Cordyceps militaris*, in others purple, flesh-coloured, lemon-yellow or brown. The perithecia (fig. 185) are embedded in the upper portion.

Two species, *Cordyceps ophioglossoides*[1] (fig. 184 *b*) and *C. capitata*, are parasitic on subterranean fungi of the genus *Elaphomyces*; in contrast to the forms on insects, they do not produce true sclerotia.

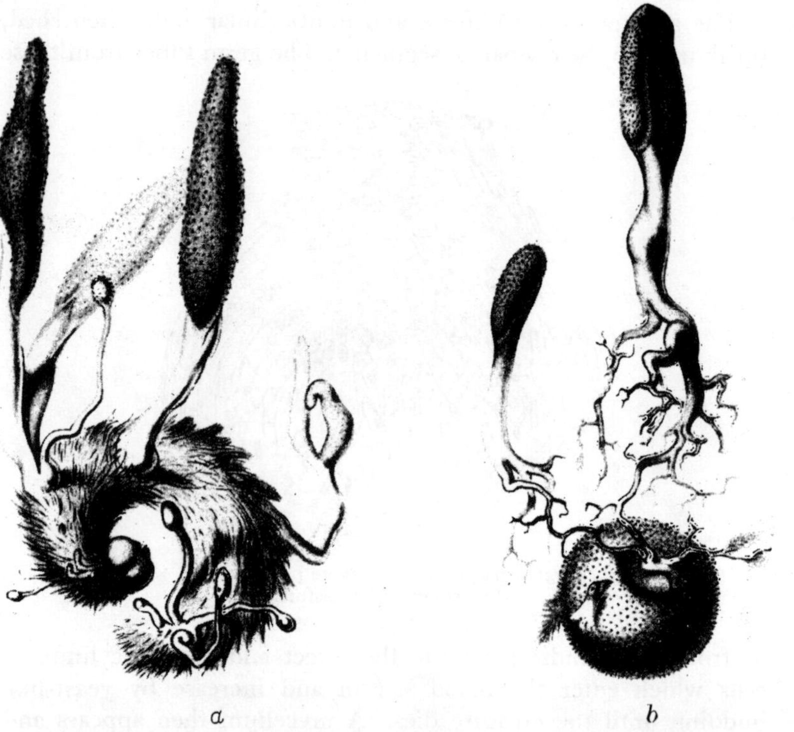

Fig. 184. *a*, *Cordyceps militaris* (L.) Link; *b*, *Cordyceps ophioglossoides* (Ehrh.) Link; after Tulasne.[2]

The species of *Claviceps*, like those of *Cordyceps*, possess filiform ascospores; they also form sclerotia from which stromata arise. The genus includes six species parasitic on members of the Gramineae. The best known is the almost cosmopolitan *Claviceps purpurea*,[3] the ergot, on rye and other grasses.

The ascospores germinate on the flowers of the host, and the

[1] Lewton-Brain, 1901. [2] Tulasne, 1861–5. [3] McCrea, 1931.

mycelium ramifies in the ovary, ultimately forming a sclerotium; meanwhile hyphae reach the exterior, bud off conidia to which the generic name *Sphacelia* was formerly applied, and excrete a sweet fluid. This attracts insects which carry the conidia with them to other flowers where they may germinate and produce infection. On the completion of the conidial stage the sclerotia become dark purple or bluish-black. If they fall to the ground or are sown with the seed they give rise next spring to numerous stromata with violet stalks and rose-pink heads in which the perithecia are immersed.

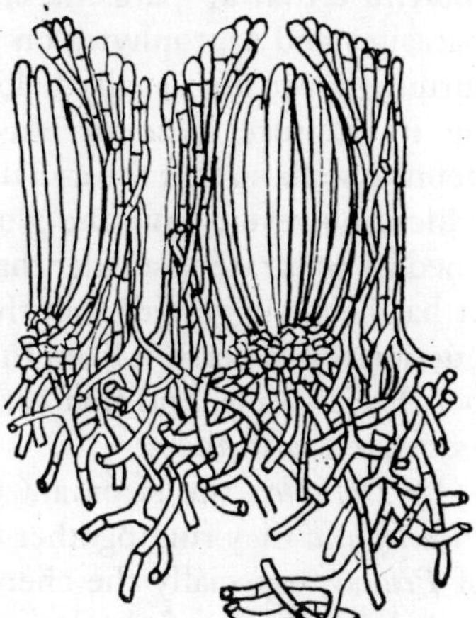

Fig. 185. *Cordyceps Barnesii* Thwaites; young perithecia, × 170; after Massee.

During the formation of the perithecium a coenocytic antheridium and oogonium are stated to develop[1] from a common branch and undergo fusion; they appear to degenerate without giving rise to ascogenous hyphae; these, as in *Cordyceps*, are derived from the vegetative cells.

The sclerotia[2] are well supplied with reserve materials, including fats and proteins, and contain a number of alkaloids one of which, ergometrine, causes rapid contraction of the uterus. This fact explains the former use of the sclerotia to hasten childbirth and their present employment to control haemorrhage after delivery. Two of the other alkaloids, ergotoxine and ergotamine, have an action similar to that of ergometrine, but produce a slower, though more prolonged response. When eaten by cows or sheep the sclerotia are liable to cause abortion, and many local traditions as to the prevelance of abortion in certain fields or byres are probably due to this cause.

Owing to the presence of ergotoxine the sclerotia, if included in the grain used for bread, give rise to serious disease. When the grain was less carefully purified than at present, the inhabitants of whole districts sometimes became afflicted with gangrene, especially

[1] Killian, 1919. [2] Barger, 1932.

if the presence of sclerotia in the food was associated with a deficiency of the vitamins available in milk and eggs.

DOTHIDEALES

The Dothideales are a small alliance of some four hundred species included in a single family, the Dothideaceae. Except for *Phyllochorella oceanica*,[1] parasitic on the gulf weed, *Sargassum*, they are parasites and saprophytes on the leaves or stems of higher plants, forming stromata usually below the epidermis and finally exposed by its rupture. The perithecia are immersed and are without definite walls so that the asci develop in mere cavities in the stroma, which, however, have the globose form of a perithecium and are lined by cells rather smaller than those of the surrounding mycelium. It has been suggested[2] that the members of this family form a reduction series derived from the stromatal Sphaeriales by loss of the perithecial walls, and that certain of the Perisporiaceae should be associated with them.

In *Dothidea* the stromata form black, projecting cushions; in *Plowrightia* they run together in masses. *P. morbosa* attacks species of *Prunus*, especially the cherry and plum, and produces swelling and deformation of the branches. *Phyllachora* is a cause of leaf spot on various angiosperms and, in one or two species, on ferns and fungi. *P. Lathyri* in North Africa[3] infects leaves of *Hedysarum* in the spring and disappears with the onset of hot weather. After the conidial stage, to which the name *Placosphaeria* was given, stromata are formed in which coiled, septate hyphae can be recognised. Fusion may take place between pairs of these, followed by the disappearance of one and an increase in the number of nuclei in the other which ultimately gives rise to asci.

SPHAERIALES

The Sphaeriales are distinguished by the dark colour and membraneous, corky, or carbonaceous texture of their perithecia, and of their stromata if present. They include considerably over 6000 species and new species are constantly brought to light, so that

[1] Ferdinandsen and Winge, 1920.
[2] Orton, 1924.
[3] Killian, 1931.

further search, especially for tropical forms, will no doubt increase their number. Not only the number of species, but also the number of individuals, is very great; the majority are saprophytes and serve a useful purpose in bringing about the first stages of decay in such resistant materials as wood or straw. They reatly outnumber the Hypocreales and Dothideales, and it is from their black or brown colour and often charred appearance that the name Pyrenomycetes is derived.

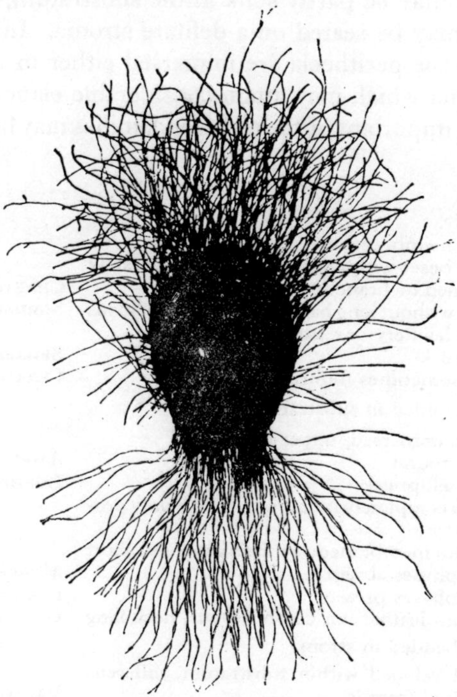

Fig. 186. *Chaetomium pannosum* Wallr.; mature ascocarp, × 50; W. M. Page del.

The perithecia of the simplest forms are borne singly, free or partly embedded in the substratum; from these may be traced an intermediate series culminating in the elaborate stromata and sunken perithecia of the highest species. There is, indeed, a marked parallelism between the Hypocreales and Sphaeriales and it is by no means clear that the colour and texture of the stromata and perithecial walls are sufficiently important as criteria of relationship to justify the separation of the two alliances. Considerably more knowledge will, however, be required before a natural system of classification can be elaborated. In the meantime the subdivisions of the Sphaeriales rest on the form of the ostiole, the colour and septation of the spores, and the structure and development of the stroma. As in the Hypocreales accessory fructification of several kinds may be present.

In the first eight families the perithecia are more or less free, though they may be partly sunk in the substratum, or in a weft of hyphae, or may be seated on a definite stroma. In the remaining ten families the perithecia are immersed either in the substratum or in a stroma which may attain considerable elaboration.

The most important of the eighteen families may be distinguished as follows:

Perithecia free	
Peridium membraneous	
Ostiole beset with long hairs often elaborately coiled or branched	CHAETOMIACEAE
Ostiole without long hairs, mainly coprophilous	SORDARIACEAE
Peridium leathery or carbonaceous	
Short neck	SPHAERIACEAE
Long, sometimes filiform neck	CERATOSTOMATACEAE
Perithecia embedded in substratum	
Perithecia immersed, upper part free	
Ostiole round	AMPHISPHAERIACEAE
Ostiole elliptical	LOPHIOSTOMATACEAE
Perithecia completely immersed, ostiole only projecting	
Peridium membraneous or leathery, neck short	
Paraphyses absent	MYCOSPHAERELLACEAE
Paraphyses present	PLEOSPORACEAE
Peridium leathery or carbonaceous, neck long	GNOMONIACEAE
Perithecia embedded in stroma	
Stroma developed within substratum, differentiated from it	VALSACEAE
Stroma free, ascospores dark brown	XYLARIACEAE

CHAETOMIACEAE

The members of the Chaetomiaceae occur on straw, paper, dung, and other waste materials; they possess free, thin-walled perithecia beset with characteristic long hairs (fig. 186), which are often branched or coiled. On these hairs, or on the hyphae of the mycelium, conidia may be produced. An ostiole is lacking in

Chaetomium fimeti; in the remaining species it is present and the perithecium is of the usual flask shape (fig. 188).

In all species studied the perithecium is initiated[1] by the forma-

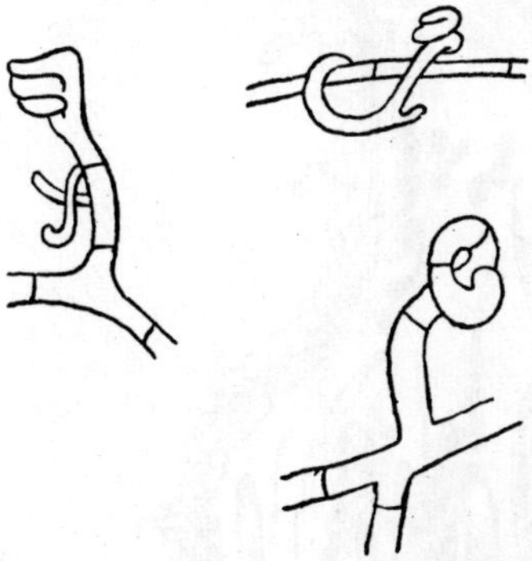

Fig. 187. *Chaetomium Kunzeanum* Zopf; primordia of perithecia; after Oltmanns.

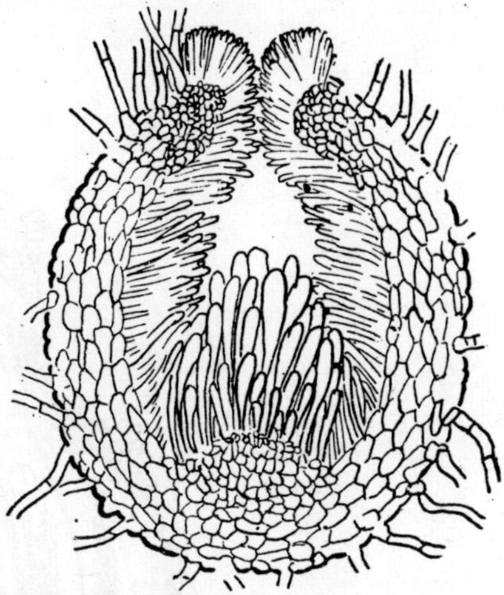

Fig. 188. *Chaetomium globosum* Kunze; perithecium in longitudinal section, × 200; after Zopf.

tion of a coiled branch (fig. 187), which divides into several cells; there is no sign of an antheridium. When the asci reach maturity their walls deliquesce and the lumen of the perithecium is filled with spores embedded in mucilage. The mucilage absorbs moisture and swells,[2] causing the spores to exude through the ostiole like paste squeezed from a tube.

Sordariaceae

The members of the Sordariaceae[3] are mainly coprophilous; their perithecia are for the most part free and superficial, but are sometimes so deeply sunk in the substratum that little more than the neck is to be seen. The genus *Hypocopra* is exceptional in possessing a small stroma in which the perithecium is immersed, it resembles *Sordaria* in other characters. The Sordariaceae differ from the Chaetomiaceae in not bearing long hairs round the ostiole

[1] Zopf, 1881; Oltmanns, 1887; Vallory, 1911; Page, 1925.
[2] Ingold, 1933.
[3] Woronin, 1886; Lewis, I. M., 1911.

and from the Sphaeriaceae in their habitat and type of spore as well as in the character of the peridium.

In *Sordaria* the ascus contains eight uninucleate spores, but the common species *Sordaria fimicola* has a variety[1] with four binucleate

Fig. 189. *Sordaria* sp.; two ascocarps in longitudinal section, left, very young, right, mature, showing asci, paraphyses and periphyses, × 400.

spores in most of its asci, though larger spores with four to eight nuclei are sometimes found and also small, uninucleate spores. The binucleate and larger spores give rise to self-fertile mycelia, but mycelia from the dwarf spores are self-incompatible and fruit only with the appropriate complementary strain. A closely similar form with self-incompatible dwarf spores has been described by Dowding[2] under the name of *Gelasinospora tetrasperma*. As in *Neurospora*

[1] Page, 1933. [2] Dowding, 1933.

tetrasperma, dwarf spores occurring in the same ascus are of opposite complement. In *Podospora anserina*[1] also and in *P. minuta*,[2] where the normal content of the ascus is four binucleate spores, dwarf and giant spores are not uncommon. Here, too, the normal and giant spores give rise to self-fertile, the uninucleate dwarfs to self-incompatible mycelia which can also be obtained by isolating uninucleate hyphal tips.[3] In the genus *Podospora* microconidia[3, 4] are to be found on the self-fertile mycelia and also on the two self-incompatible strains.

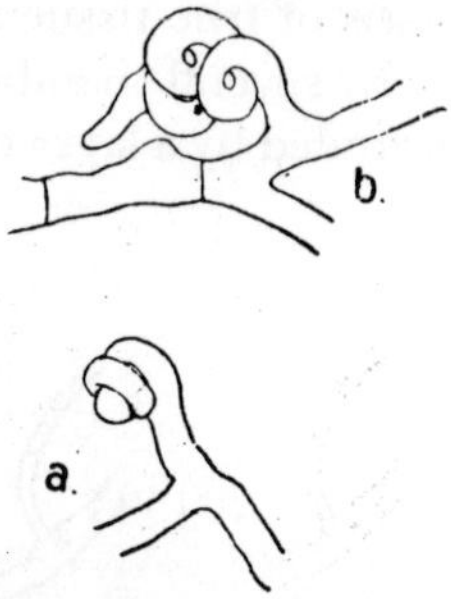

Fig. 190. *Podospora anserina* (Rabh.) Wint.; *a*, very young female branch; *b*, older female branch, ×400; after Page.

The development of the perithecium is initiated by the appearance of a coiled, septate hypha (fig. 190) which becomes surrounded by a sheath of vegetative filaments with projecting branches (fig. 191). In the self-incompatible form of *Podospora anserina*[4] microconidia have been seen attached to these outgrowths, and, if the microconidium and perithecium are of opposite complement

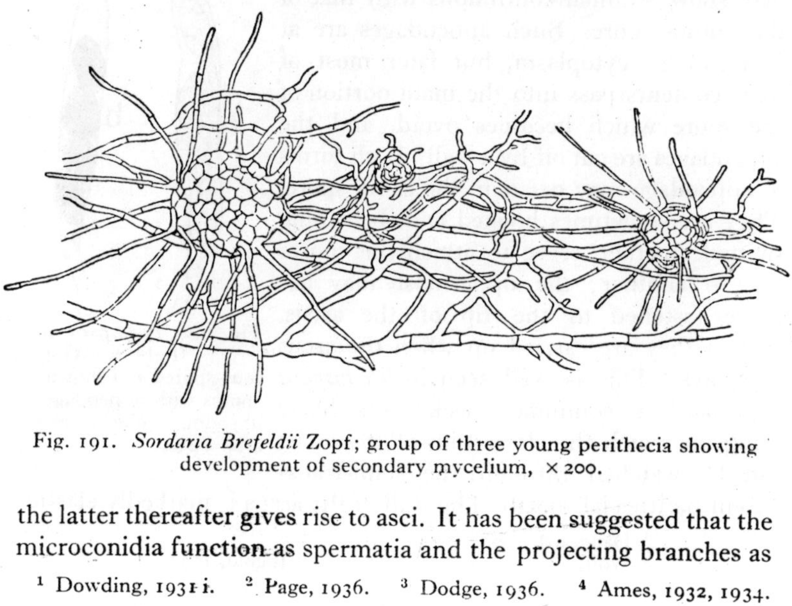

Fig. 191. *Sordaria Brefeldii* Zopf; group of three young perithecia showing development of secondary mycelium, ×200.

the latter thereafter gives rise to asci. It has been suggested that the microconidia function as spermatia and the projecting branches as

[1] Dowding, 1931 i. [2] Page, 1936. [3] Dodge, 1936. [4] Ames, 1932, 1934.

trichogynes, but evidence from the behaviour of the nuclei is not as yet available.

In *Sporormia*[1] longitudinal as well as transverse septa appear in one or more of the cells of the perithecium initial, so that a small mass of true tissue is formed (fig. 192).

In several members of the Sordariaceae each ascospore is surrounded by a layer of mucilage; in others one or more appendages[2]

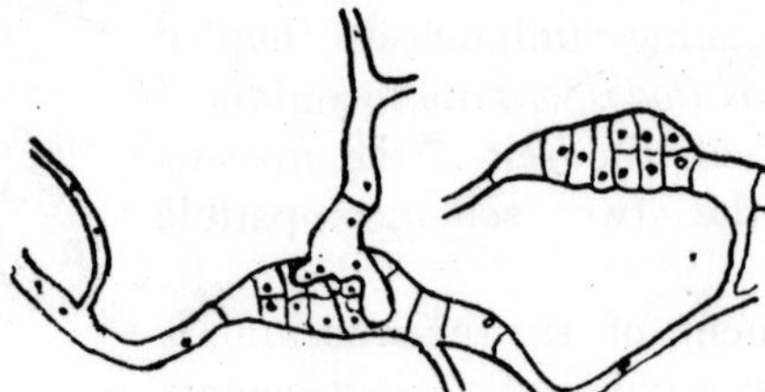

Fig. 192. *Sporormia intermedia* Auersw.; initial cells of perithecium; after Dangeard.

(fig. 193) are produced. These may be gelatinous and derived partly or wholly from the epiplasm, apparently much as the ordinary thickening of the spore coat is derived, or they may show a lumen continuous with that of the young spore. Such appendages are at first rich in cytoplasm, but later most of their contents pass into the main portion of the spore which becomes ovoid, and the appendages are cut off by a wall. Both forms of appendage may occur in the same spore. They are sometimes hooked and, becoming twisted together, serve to attach the spores one to another; the uppermost may become fastened to the tip of the ascus, so that they are carried up when the ascus elongates. This is well seen in *Podospora curvicola*,[3] a common species on horse dung, in which the elongation of the ascus can be watched through the semi-transparent perithecial sheath. The wall of the ascus is markedly elastic

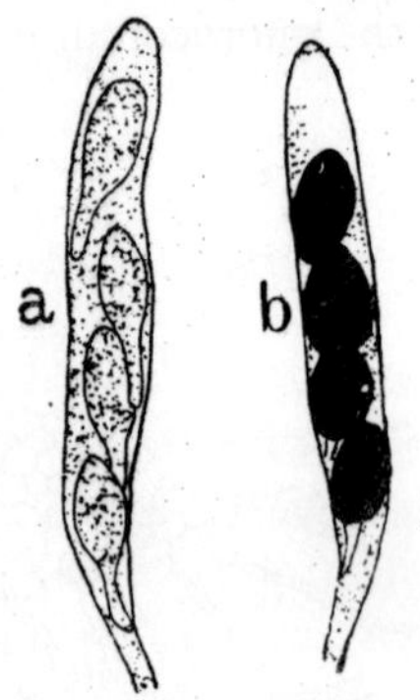

Fig. 193. *Podospora anserina* (Rabh.) Wint.; four-spored asci showing spores with appendages, *a*, young, *b*, older, × 280; after Page.

[1] Dangeard, 1907.
[2] Wolf, 1912.
[3] Ingold, 1933.

and, by the time that its tip protrudes through the ostiole for the discharge of the spores, it is seven or eight times its original length. In *Sporormia intermedia*,[1] a coprophilous form with four-celled spores, and in other species of the genus, the ascus wall is double, the outer layer being relatively rigid, the inner elastic and very thin. The outer wall is ruptured, its upper part separating as a thimble-shaped lid (fig. 194), the inner elongates, at first almost explosively, pushes through the neck and extends beyond the ostiole before liberating its spores.

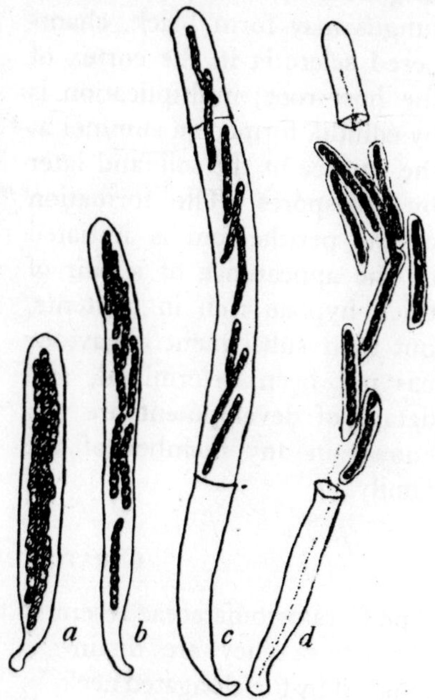

Fig. 194. *Sporormia bipartis* (Cane); asci showing dehiscence by means of a cap; *a, b, c, d,* stages of development; ×335; W. M. Page del.

In the germination[2] of the ascospore part of the contents emerge through a small aperture in the thick coat and form a swelling (fig. 195) from which the first hyphae of the mycelium grow out.

SPHAERIACEAE

The perithecia of the Sphaeriaceae are superficial, and borne singly or in groups; the neck is short, the peridium is leathery or carbonaceous, it may be smooth or beset with hairs. *Bombardia lunata*[3] is heterothallic.

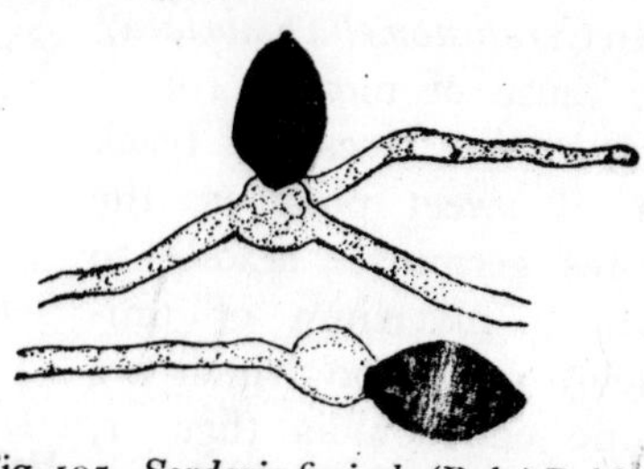

Fig. 195. *Sordaria fimicola* (Rob.) Rabh.; germinating spores, ×400; after Page.

Most of the species are saprophytes on plant remains, often on wood, but *Coleroa* (fig. 196) is parasitic on the leaves of *Potentilla*, *Rubus* and other angiosperms, and *Rosellina quercina*[4] attacks the roots of oak seedlings, entering the cells of the cortex and pith and

[1] Ingold, 1933. [2] Page, 1925. [3] Zickler, 1934. [4] Hartig, 1880.

forming interwoven strands of hyphae which grow out to infect neighbouring oak plants. This fungus may form black, chambered sclerotia in the cortex of the host root; multiplication is by conidia formed in summer at the surface of the soil, and later by ascospores. The formation of the perithecium is initiated by the appearance of a pair of thick hyphae rich in contents, but their subsequent behaviour has not been determined, and details of development are unknown in any member of the family.

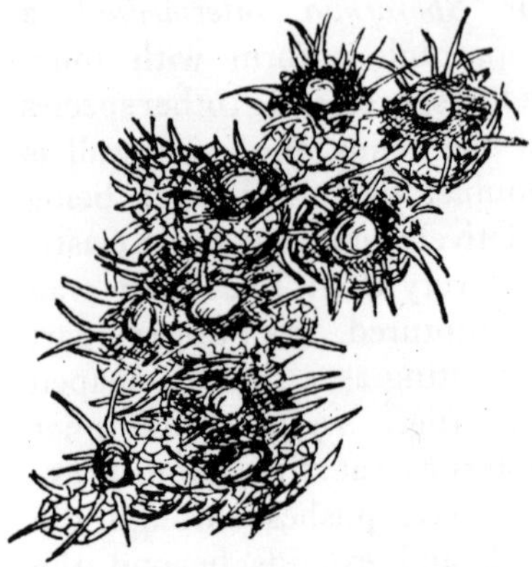

Fig. 196. *Coleroa Potentillae* (Fr.) Wint.; perithecia, × 190.

Ceratostomataceae

The Ceratostomataceae resemble the Sphaeriaceae in most of their characters; they are distinguished by the elongated neck of the perithecium, which is often drawn out to form a delicate, hair-like process.

In *Ceratostomella fimbriata*,[1] the cause of mouldy rot of Para rubber trees and black rot of sweet potatoes, the spores germinate readily to form a mycelium of uninucleate cells on which are borne endoconidia (figs. 1, 7), chlamydospores and hat-shaped ascospores in perithecia. The perithecium initial is a curved branch (fig. 197) one cell of which becomes multi-

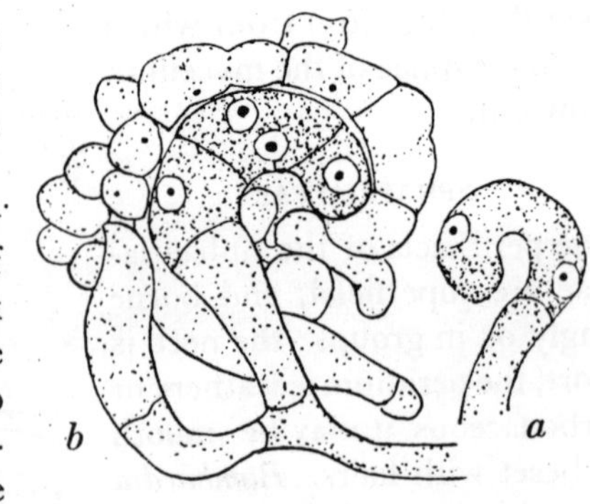

Fig. 197. *Ceratostomella fimbriata* (Ell. & Hals.) Elliott; *a*, young fertile branch; *b*, older fertile branch surrounded by sheath; × 1600.

[1] Elliott, 1925; Mittmann, 1932; Andrus and Harter, 1933; Gwynne-Vaughan and Broadhead, 1936.

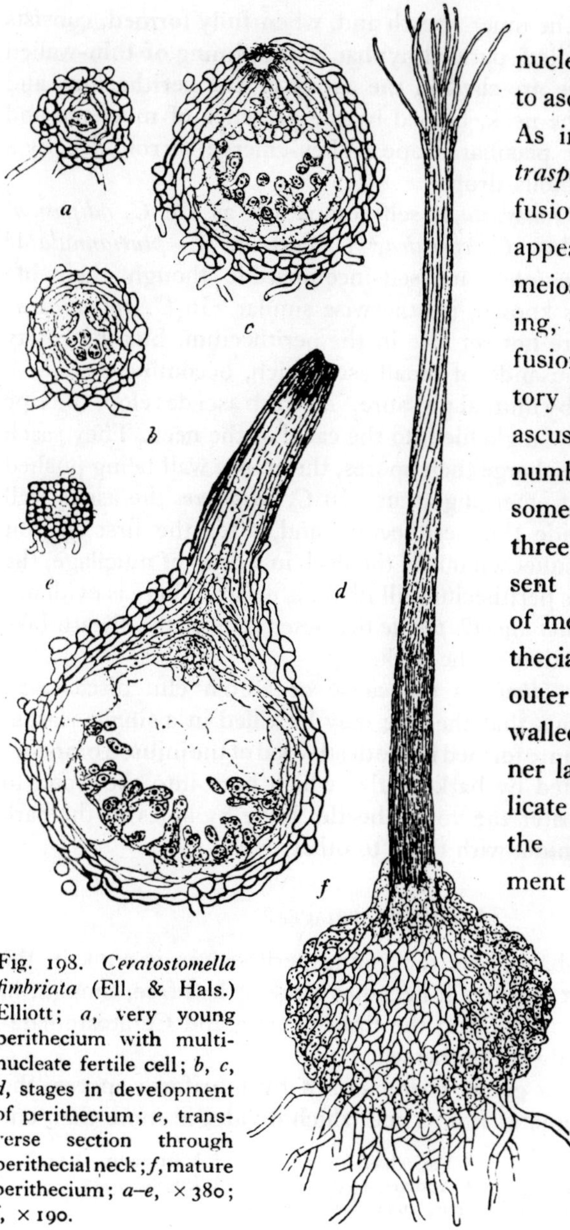

Fig. 198. *Ceratostomella fimbriata* (Ell. & Hals.) Elliott; *a*, very young perithecium with multinucleate fertile cell; *b*, *c*, *d*, stages in development of perithecium; *e*, transverse section through perithecial neck; *f*, mature perithecium; *a–e*, × 380; *f*, × 190.

nucleate and gives rise to ascogenous hyphae. As in *Neurospora tetrasperma*, the sexual fusion has wholly disappeared and brachymeiosis is also lacking, the only nuclear fusion in the life history being that in the ascus. The haploid number of chromosomes is three and three gemini are present in the prophase of meiosis. The perithecial sheath has an outer zone of thick-walled cells and an inner layer of more delicate hyphae. During the early development of the asci it is lined by turgid cushion cells (fig. 198 *c*, *d*) which later collapse. The long neck originates from a group of meristematic cells (fig. 198 *b, c*) in the inner sheath and, when fully formed, consists of long, thick-walled, parallel hyphae with a lining of thin-walled filaments. Spores are shed in the cavity of the perithecium and pass singly up the neck, helped by the presence of mucilage and possibly by their peculiar shape. They emerge surrounded by a shining mucilaginous drop.

Ceratostomella fimbriata is self-compatible as are *C. adiposum*[1] and *C. piceae*,[2] but *C. coerulea*,[3] *C. quercus*,[3] *C. pluriannulata*[4] and *C. multiannulata*[5] are self-incompatible though their life history, so far as known, is otherwise similar. In *C. ampullasca*[6] the ascospores are not set free in the perithecium, but the cavity is filled with thousands of small asci which, becoming detached, are held upright by mutual pressure. As fresh asci develop, the ripe ones are forced in single file into the canal of the neck. They reach the ostiole and discharge their spores, the empty wall being pushed aside by the next emerging ascus. In *C. adiposum* the ascus wall disintegrates inside the perithecium and, after the first crop of spores has been squeezed out of the neck in a drop of mucilage, the rest remain in the perithecium till its walls decay. There is evidence that in *C. adiposum* and *C. piceae* the outer cells of the sheath take part in the formation of the neck.

Ceratostomella Ulmi[7] is the cause of Dutch elm disease, an affection so serious that the tree may be killed in a single season. Masses of conidia are formed in the dead wood of the injured branches and are distributed by bark beetles which bore into the wood to lay their eggs. Later the young beetles leave by holes in the bark and carry the conidia with them to other trees.

AMPHISPHAERIACEAE

In the Amphisphaeriaceae the young perithecium is sunk in the substratum; as it matures it becomes more or less free, though, in contrast to the condition in the Sphaeriaceae and Ceratostomataceae, its base is always immersed.

In *Strickeria*,[8] a genus characterised by muriform spores, the perithecium originates from a cell which divides in more than one

[1] Sartoris, 1927. [2] Varitchak, 1931. [3] Mittmann, 1932.
[4] Gregor, 1932. [5] Andrus, 1936. [6] Ingold, 1933.
[7] Welch, Herrick and Curtis, 1934; Ledeboer, 1934.
[8] Nichols, 1896.

plane (fig. 199), as in *Sporormia*. The resultant parenchymatous mass appears to give rise to the perithecial walls as well as to ascogenous hyphae.

Lophiostomataceae

The Lophiostomataceae are distinguished from neighbouring families by the form of the ostiole, which is very large and laterally compressed, so that in external appearance they approach certain

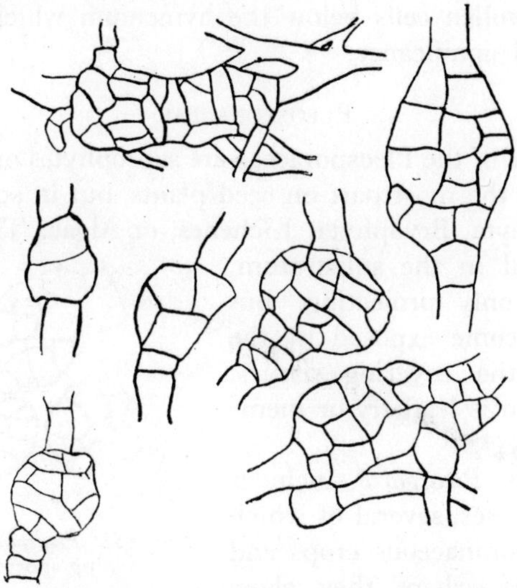

Fig. 199. *Strickeria* sp.; initial cells of ascocarps; after Nichols.

of the Hysteriales which they also resemble in their habitat on vegetable remains such as wood and bark. The perithecia are borne singly; during development they are embedded in the substratum, they may remain so or become partly free at maturity. There is no stroma; the peridium is black and brittle. None of the species has been investigated in detail.

Mycosphaerellaceae

The members of this family are parasitic, some on algae, some giving rise to various forms of leaf spot. The perithecia develop either under the cuticle or beneath the epidermis, breaking through at maturity; except in the genus *Stigmatea*, paraphyses are not developed. In several species the formation of the perithecia is preceded by a conidial stage.

Mycosphaerella nigerristigma[1] forms pycnidia on the living leaves of *Prunus pennsylvanicae* and perithecia after the leaves have fallen. A trichogyne like that of *Polystigma* has been recorded; it degenerates, leaving a basal cell, but whether this functions is not known. An oogonium, a septate trichogyne and spermatia have also been described in *M. Bolleana*[2] on the leaves of *Ficus carica*; and in *Stigmatea Robertiani*,[3] on *Geranium*, attention has been called to certain swollen cells below the hymenium which are said to have a sexual significance.

Pleosporaceae

The members of the Pleosporaceae are saprophytes or occasionally parasites, for the most part on seed plants, but in some instances on Pteridophyta, Bryophyta, Lichenes, or Algae. The perithecia are immersed in the substratum, the ostiole only projecting, but they may become exposed by the rupture of the covering tissues. The peridium is leathery or membraneous.

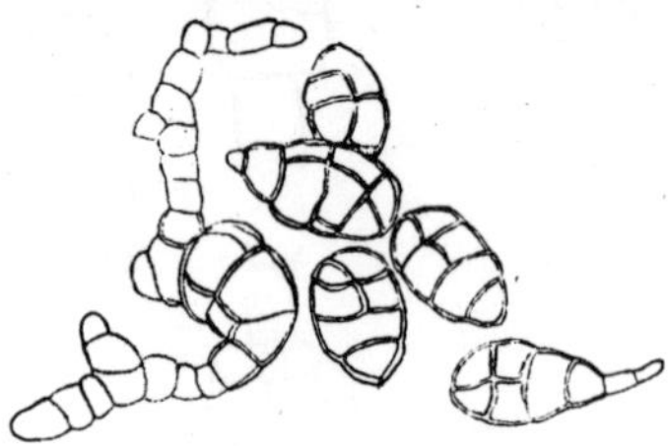

Fig. 200. *Pleospora* sp.; germinating spores, × 500.

The genus *Pleospora*[4] includes some 225 species, several of which occur on graminaceous crops and other grasses where they show biologic specialisation. *P. herbarum* is a facultative parasite upon the leaves of angiosperms; the perithecium is developed from a mass of parenchyma, the asci originating from the same cells as the paraphyses. The ascospores (fig. 200) are muriform. Multicellular conidia on branched conidiophores are also produced. In some of the species of this genus and of the closely related *Pyrenophora* the ascus possesses a double wall; when dehiscence is about to occur the outer wall is ruptured and slips down to form a ring about the middle of the ascus,[5] the inner wall opens just above this ring, and the spores are shot out with considerable force.

[1] Higgins, 1914. [2] Higgins, 1920. [3] Killian, 1922.
[4] Diedicke, 1902. [5] Atanasoff, 1919.

There are some fifty species of *Venturia*, several of which are parasitic on leaves; the perithecium is immersed and the large ostiole beset with stiff hairs. An antheridium and a coiled oogonial branch have been recorded for *V. inaequalis*[1] and the passage of male nuclei through the trichogyne reported. The conidial stage of this species, known as *Fusicladium dendriticum*, attacks the fruits and vegetative organs of the apple in summer, while the stromata are found in autumn on the fallen leaves and give rise to perithecia in the following spring.

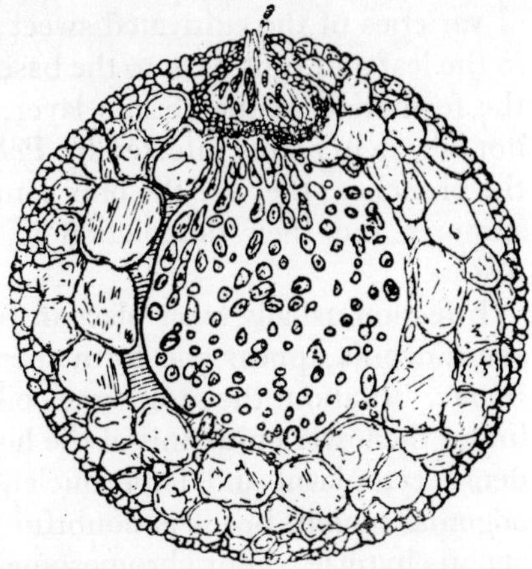

Fig. 201. *Leptosphaeria Lemaneae* (Cohn) Brierley; transverse section through thallus of *Lemanea*, showing perithecium, ×125; after Brierley.

Leptosphaeria includes some 500 species characterised by the papillate or conical ostiole, usually free from hairs. The majority are saprophytes on plant remains, some are parasites on land plants, and *L. Lemaneae*[2] (fig. 201) infects the thallus of species of the red alga, *Lemanea*.

Gnomoniaceae

The Gnomoniaceae are for the most part saprophytic on the leaves or other parts of land plants. The perithecia are embedded in the substratum, from which their long necks project. The asci are characterised by their thick apices through which a canal allows the exit of the spores. The spores are hyaline; paraphyses are not usually developed. The family differs from the Pleosporaceae in the long neck of the perithecium and the thickened tip of the ascus. There is no stroma, and this fact, as well as the dark colour, distinguishes *Gnomonia* from the similar genus *Polystigma* among the Hypocreales.

[1] Killian, 1917; Frey, 1924.
[2] Woronin, 1886; Brierley, 1913.

Gnomonia erythrostoma[1] is the cause of an epidemic disease known as cherry leaf-scorch, which attacks the foliage of *Prunus avium* and of varieties of the cultivated sweet cherry. The mycelium ramifies in the leaf and runs back to the base of the petiole where it prevents the formation of the absciss layer. In consequence the leaves do not fall, as in plants infected by *Polystigma*, but remain hanging on the branches; they are the only source of infection in the following summer and their destruction is a sure means of checking the disease.

Infection usually takes place in June; towards the end of August spermogonia appear and form spermatia similar to those of *Polystigma*. At about the same time coiled archicarps are differentiated under the lower epidermis of the host leaf; their lower cells contain dense cytoplasm and large nuclei, and presumably represent an oogonial region, but it is doubtful whether they give rise to ascogenous hyphae. Four chromosomes are present in the ascus nuclei and there is evidence of only one reduction, so that the fungus is in all likelihood completely apogamous.

VALSACEAE

The perithecia of the Valsaceae are frequently produced in compact groups on black stromata from which their long necks alone project. The stroma is developed within the substratum from which it is not always completely differentiated; it is very variable in form, sometimes indicated only by a black stain on wood or bark or by a black margin, sometimes extended as a thin black layer over a considerable area and ending irregularly, sometimes, as in species of *Valsa*, forming black cushions which break through the bark of the substratum. Conidia are often present, borne on free conidiophores or produced within pycnidia.

The genus *Valsa* includes some 400 species, for the most part saprophytic on wood and other resistant materials, and *Diaporthe* a rather larger number, the majority of which infect living plants. In one of the latter, *Diaporthe perniciosa*,[2] a coiled, multicellular fertile branch has been recorded, but its relation to the ascogenous hyphae is not known.

[1] Frank, 1886; Brooks, 1910; Dowson, 1925; Likhité, 1925.
[2] Cayley, 1923 i.

XYLARIACEAE

The Xylariaceae occur chiefly on wood; they constitute the highest development of the Sphaeriales and are characterised by the free, superficial stroma, though a few, including certain species of *Hypoxylon*, have stromata partly sunk in the substratum. The stromata

Fig. 202. *Hypoxylon coccineum* Bull.; the smallest stroma bears conidia, the others perithecia; after Tulasne.[1]

show every variety of form, from a spreading crust, as in *Nummularia*, to the almost spherical cushions of *Daldinia* or *Hypoxylon* (fig. 202), and the erect, simple or branched expansions of *Xylaria* (fig. 203) and its allies. The perithecia are arranged just below and at right angles to the surface of the stroma; their development may be preceded by the formation of conidia, which often cover young stromata with a white or brown powder.

Poronia punctata[2] occurs on old horse-dung. The stromata are

[1] Tulasne, 1861–5.
[2] Tulasne, 1861–5; Dawson, 1900.

Fig. 203. *Xylaria Hypoxylon* Grev.; after Tulasne.[1]

[1] Tulasne, 1861–5.

stalked, and expended above into a cup or disc (fig. 204), which, in the earlier stages of development, is covered by a greyish-white film of conidia; later the ostioles of the numerous perithecia (fig. 205) appear as black dots scattered over the surface of the disc. The asci, when ripe, protrude through the ostiole, so that the

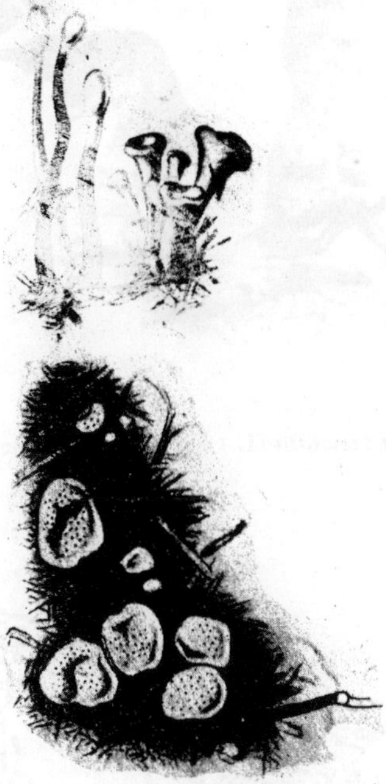

Fig. 204. *Poronia punctata* (L.) Fr.; *a*, superficial; *b*, lateral view; after Tulasne.

spores are shed outside the perithecium. Before the conidia have disappeared female branches are developed among the vegetative filaments as coils of deeply staining cells continued towards the surface in the form of a slender trichogyne (fig. 206 *a*). It is clear that the trichogyne does not function, as degeneration proceeds from its base and not from its apex, as might be expected if a male nucleus were travelling down; the relationship of the ascogenous hyphae to the oogonial region of the archicarp has not been demonstrated.

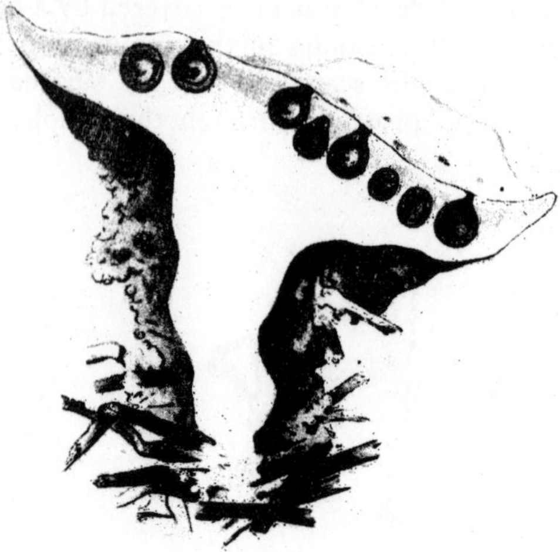

Fig. 205. *Poronia punctata* (L.) Fr.; stroma cut across; after Tulasne.

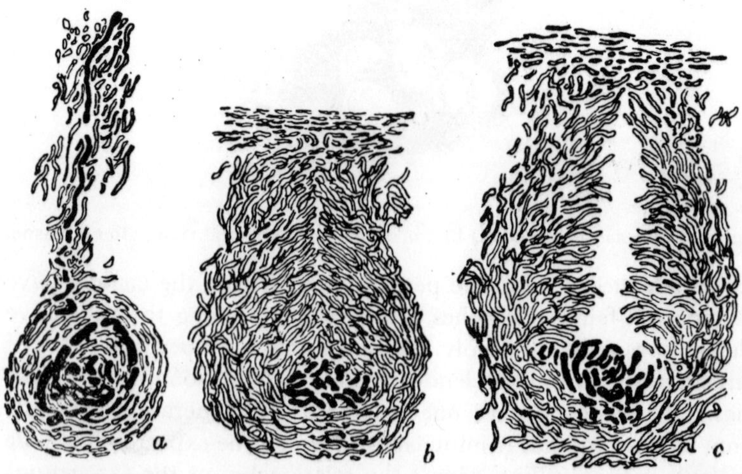

Fig. 206. *Poronia punctata* (L.) Fr.; *a*, archicarp, ×275; *b* and *c*, young perithecia, ×205; after Dawson.

In both *Xylaria*[1] and *Hypoxylon* the young stroma is covered by a turf of conidiophores from which the small, oval conidia are abstricted. In *Xylaria* these form a white coating in marked contrast to the exposed portions of the black stroma, and justify the

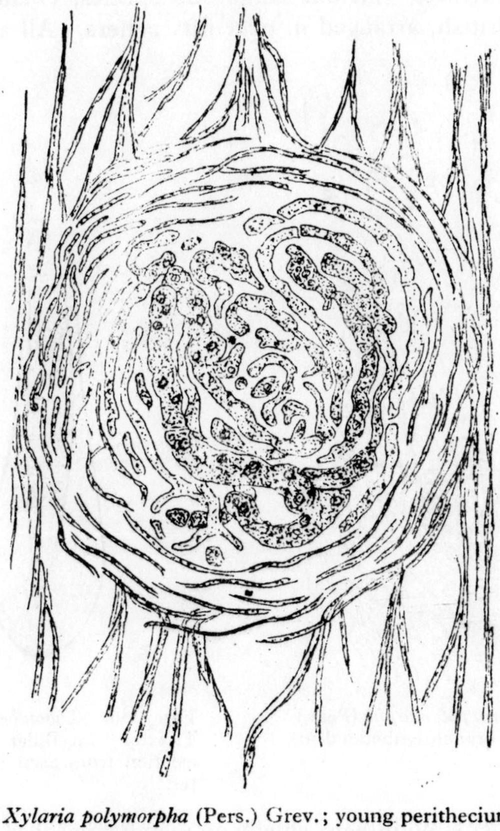

Fig. 207. *Xylaria polymorpha* (Pers.) Grev.; young perithecium, × 1000.

name candle-snuff fungus applied to the commoner species. If the stroma of *Xylaria* or *Hypoxylon* is sectioned during the conidial stage, nests of small hyphae are found, and form the first indications of perithecia (fig. 207). Still earlier a stout hypha with large nuclei, presumably a female branch, is recognisable (fig. 208). In *Xylaria* it has not been shown to function, but in *Hypoxylon*

[1] Freeman, D. L., 1910; Brown, H. B., 1913.

coccineum[1] the development of ascogenous hyphae from its cells has been described.

LABOULBENIALES

The Laboulbeniales[2] include some 600 species, twenty-eight of which are British, arranged in over fifty genera. All are minute,

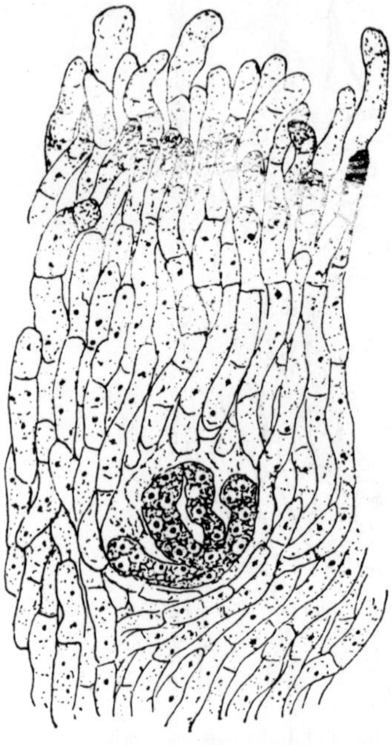

Fig. 208. *Xylaria polymorpha* (Pers.) Grev.; female branch embedded in stroma, × 1000.

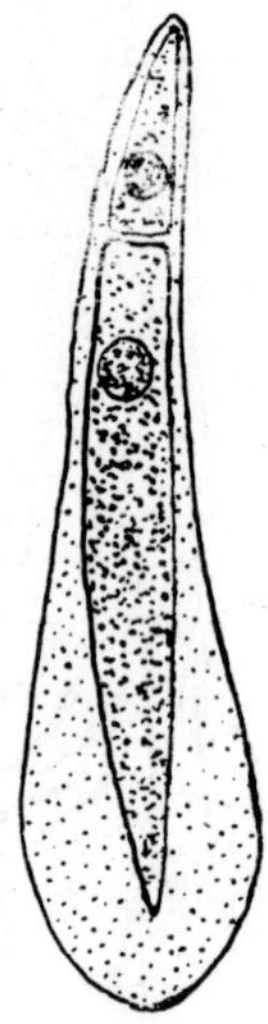

Fig. 209. *Laboulbenia elongata* Thaxter; bicellular spore after ejection from ascus; after Thaxter.

external parasites on insects, chiefly on members of the Coleoptera. They appear to do little injury to their host, inducing at most a slight irritation but never causing death; indeed, their own existence depends on the survival of the insect, since their life ends with that of their host. It is to the work of Thaxter that most of the existing knowledge of this group is due.

[1] Lupo, 1922.
[2] Thaxter, 1896 i, 1908, 1924, 1926; Biffen, 1908; Faull, 1911, 1912; Maire, 1912, 1916; Hake, 1923.

The Laboulbeniales are of simple structure and show an underlying similarity of type. The vegetative part consists of a **receptacle** (fig. 215), usually two celled, attached to the integument of the host by a blackened base or foot. From the receptacle grow out filamentous **appendages** on or among which the male organs are produced, and, with a few exceptions, the receptacle of the same individual also gives rise to a female organ from which a perithecium liberating ascospores is eventually developed. The fungus

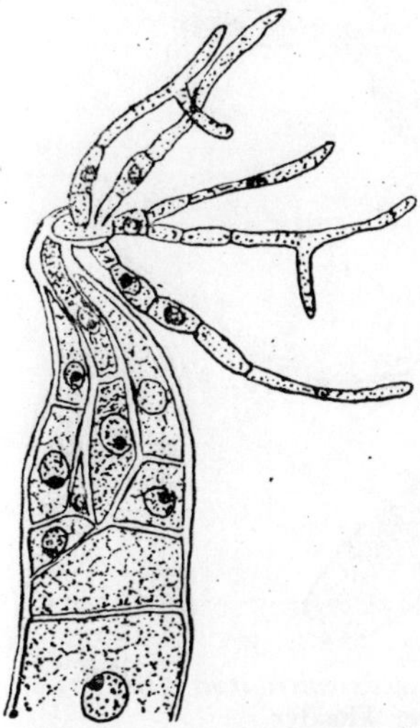

Fig. 210. *Laboulbenia chaetophora*; young perithecium and trichogyne, ×360; after Faull.

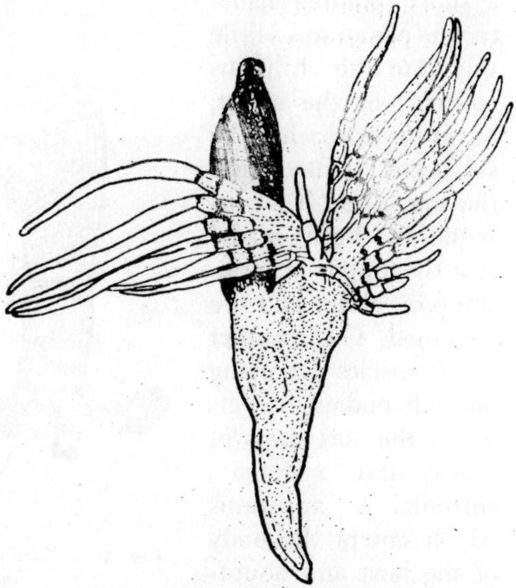

Fig. 211. *Laboulbenia triordinata* Thaxter, ×135; after Thaxter.

is covered by a thin, homogeneous membrane, which is exceedingly tough and impervious and is derived from the gelatinous coat of the spore. Within this coat the cells are uninucleate (fig. 210) with thick, laminated walls, sister cells being in communication by means of broad pits.

The spores are remarkably uniform throughout the alliance, being hyaline, fusiform or acicular, and almost always two celled (fig. 209). There is a gelatinous sheath, especially well developed around the larger cell, which is towards the tip of the ascus, and is therefore destined, when the spore is discharged, to come in contact with the integument of the host. Here the gelatinous mass enables the spore to take up the oblique position in which germination occurs, and later gives its attachment a certain elasticity, so that it lies back along the body of a rapidly swimming host. In *Stigmatomyces* a short, pointed haustorium penetrates a little way into the chitinous covering of the insect, but it never reaches the subjacent tissues, and the suggestion has consequently been made[1] that the fungus may decompose chitin and use it as food. On the other hand, species occurring on soft-bodied insects or on the soft parts of others have a definite rhizoidal apparatus which enters the body of the host and doubtless absorbs food from the fluid materials in which it is bathed. Closely related species which do not penetrate the covering of the host differ not at all in their cell contents or mode of life from forms with rhizoids; this suggests that all alike are supplied by the circulatory system which, by diffusion, or otherwise, nourishes in the living insect the structures on which they occur. This inference is borne out[2] by the greater luxuriance of specimens growing near the circulatory centres or along the circulatory channels, even when attached to a chitinous structure.

Fig. 212. *Zodiomyces vorticellarius* Thaxter; after Thaxter.

The receptacle develops from the larger cell of the spore. It

[1] Boedijn, 1923. [2] Thaxter, 1914 i.

consists, in the simplest examples, of two superposed cells and, in monoecious forms, bears appendages terminally and the perithecium laterally. More rarely it consists of a greater number of cells, attaining considerable complexity in *Zodiomyces vorticellarius* (fig. 212).

The appendages are filamentous and often elaborately branched. They bear the male organs, serve for the protection of the delicate

Fig. 213. *Ceratomyces rostratus* Thaxter; exogenous spermatia; after Thaxter.

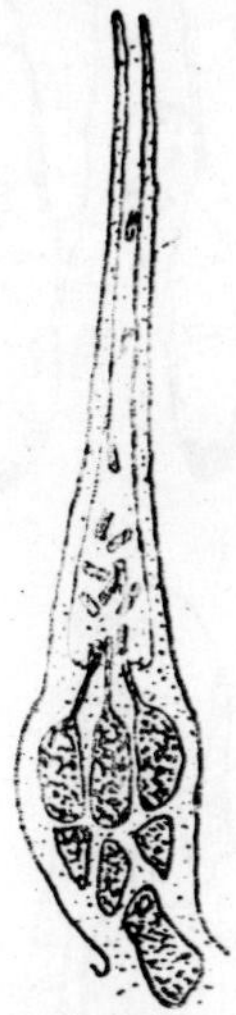

Fig. 214. *Dimeromyces Africanus* Thaxter; compound spermatial organ; after Thaxter.

trichogyne and may also facilitate fertilisation by holding water around the structures concerned. The first appendage is derived from the smaller segment of the spore, the later appendages arise from the receptacle.

The male elements are non-motile cells. In the simplest examples these are produced externally on the tips of more or less specialised branches (fig. 213); they are walled and seem to correspond exactly to the spermatia of other fungi, that is to say, they are antheridia each reduced to a single, uninucleate cell, and carried by external agencies to the female organ. In *Coreomyces*, instead of the segment of a branch becoming detached, its contents are extruded as a naked, uninucleate mass. This arrangement leads to more specialised endogenous organs which may be borne singly (fig. 215 *a*, *b*, *c*, *e*), or in groups (fig. 214). The naked mass of protoplasm may be re-

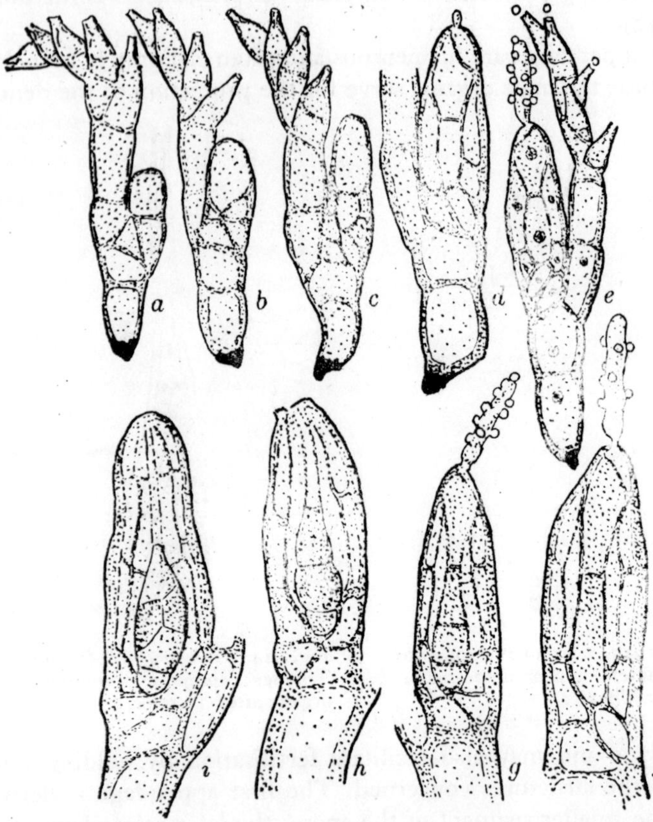

Fig. 215. *Stigmatomyces Baeri* Peyritsch; development of the perithecium; *a* shows the two-celled receptacle, a single appendage bearing five simple, endogenous spermatial organs, and the beginning of the perithecium; *b*–*i* indicate successive stages in the development of the perithecium; the trichogyne first appears in *d*; in *e*, spermatia are being shot out and some are attached to the trichogyne; in *i*, two of the four ascogenous cells are shown, with the superior sterile cell above them, and the primary and secondary inferior sterile cells below; after Thaxter.

garded as homologous with a spermatium, or the organ in which it is produced may be looked upon as antheridial and the naked mass itself as a non-motile male cell. In view of the close resemblance in other particulars between exogenous and endogenous forms, the former hypothesis seems the more probable and the term spermatium may conveniently be used.

The female organs are formed from the lower cell of the germinating spore and are thus necessarily lateral, though this condition is obscured in the mature plant, where the ripening perithecium may push the appendages aside and take up an apparently terminal position. Development is uniform for the greater number of species and has been fully described in *Stigmatomyces Baeri*.[1] Here the lower cell of the receptacle divides into two, and the upper of these grows out (fig. 215 *a*) to form the female branch. It divides transversely, the upper daughter cell gives rise to the oogonium, the lower or stalk cell divides several times (fig. 215 *b*), and ultimately forms the double wall of the perithecium, a function characteristic of the stalks of the sexual branches in Ascomycetes of richer development.

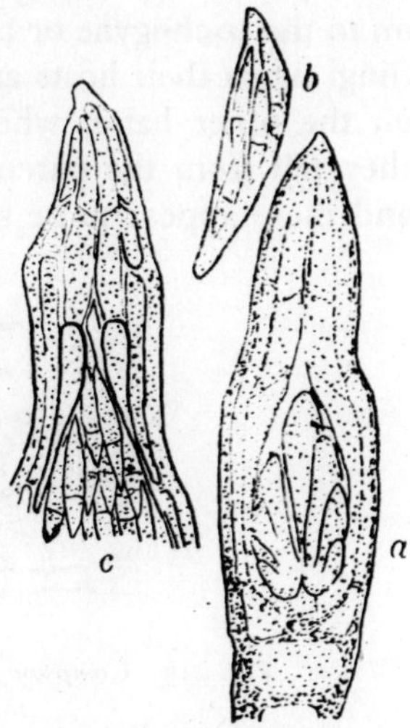

Fig. 216. *Stigmatomyces Baeri* Peyritsch; *a*, young asci; *b*, ascus containing four spores; *c*, mass of spores in perithecium; after Thaxter.

The upper cell divides, separating the oogonium below from a cell above, which divides again transversely to form the two-celled trichogyne (fig. 215 *d*, *e*).

After the fertilisation stage the oogonium divides into three (fig. 215 *g*) and then four superposed cells; the penultimate of these, next but one to the trichogyne, divides longitudinally into four ascogenous cells, two of which are shown in fig. 215 *i*, and from these the asci bud out (fig. 216 *a*). The young ascus is binucleate; the nuclei fuse, and three nuclear divisions take place in the usual way. As a rule, only four of the eight resulting nuclei function, and four spores are produced. The lack of ascogenous hyphae is doubtless due to restrictions of space; in other particulars the development of the female organ is closely comparable to the corresponding stages in the Erysiphaceae.

In *Stigmatomyces Baeri* the trichogyne is simple, but in many

[1] Thaxter, 1896 i.

other Laboulbeniales it undergoes frequent septation and branches freely (fig. 217). The apices of the branches alone are receptive and may be straight or spirally coiled.

In species with endogenous spermatia the latter are shot directly on to the trichogyne or carried to it by water which surrounds the fungi when their hosts are hiding in moist places. In *Zodiomyces*, on the other hand, where the spermatia are formed externally, they fall from the parent branches to the cup-shaped receptacle and there appear to be sought by the trichogyne which is at first

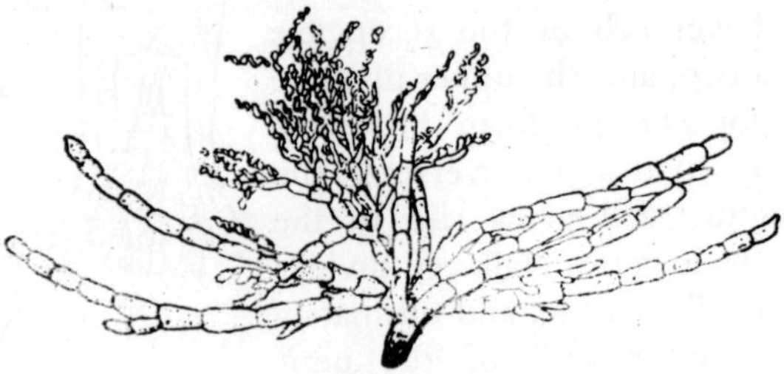

Fig. 217. *Compsomyces verticillatus* Thaxter; after Thaxter.

bent over (fig. 218 *a*), and later lifts itself when a spermatium has become attached (fig. 218 *b*).

Up to the present the cytology has been studied[1] only in two species of *Laboulbenia* and these, unfortunately, are both parthenogenetic, no spermatia being formed. Association takes place between the oogonial nucleus and a nucleus from the lower cell of the trichogyne, but only one fusion, that in the ascus, has been seen. The haploid number of chromosomes is four.

There is some variation in different species in the number and arrangement of the ascogenous cells and asci. In *Polyascomyces* (fig. 219) as many as thirty ascogenous cells are formed, covering a basal area from which the numerous asci bud upwards.

The ascospores are usually disposed more or less definitely in pairs and members of a pair are discharged together and germinate side by side. In monoecious species one member of a pair may produce a smaller individual than the other, while in *Laboulbenia inflata* the atrophy of one at an early stage of development is a regular phenomenon. In *Stigmatomyces Sarcophagae* (fig. 220) the

[1] Faull, 1912.

smaller individual is unisexual, producing only male cells, while the other is hermaphrodite.

In dioecious species, such as *Amorphomyces Falagriae* (figs. 222, 223), the paired spores are of different sizes (fig. 221); the smaller spore gives rise to a male individual, the larger to a female, so that their association insures the condition necessary for the perpetuation of the species. There is an obvious suggestion in these phenomena of a transition between the monoecious and dioecious

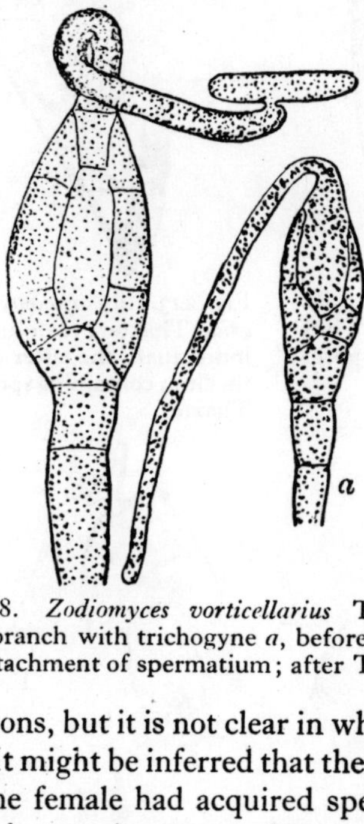

Fig. 218. *Zodiomyces vorticellarius* Thaxter; female branch with trichogyne *a*, before, and *b*, after, attachment of spermatium; after Thaxter.

Fig. 219. *Polyascomyces Trichophyae* Thaxter; after Thaxter.

conditions, but it is not clear in which direction the series should be read. It might be inferred that the male plant had become atrophied after the female had acquired spermatial organs, or, on the other hand, that, as in many other groups of plants, a hermaphrodite condition was primitive and segregation a later development. Any light on the cytology of dioecious species should be of special interest.

The systematic relations of the Laboulbeniales are not easy to determine. They are pretty evidently monophyletic and are highly specialised along lines appropriate to their peculiar habitat. The ascocarp is pyrenomycetous in character, but the production of asci from the subterminal cell only of the row formed after fertilisa-

Fig. 220. *Stigmatomyces Sarcophagae* Thaxter; male and hermaphrodite individuals, × 260; after Thaxter.

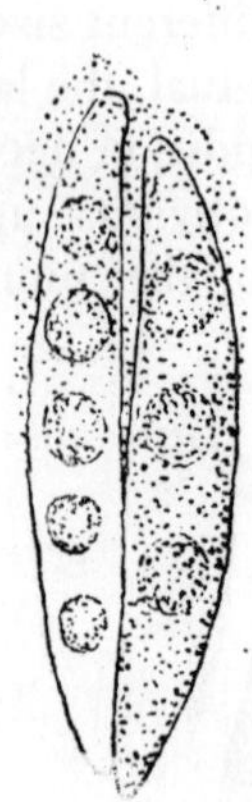

Fig. 221. *Amorphomyces Falagriae* Thaxter; paired spores; after Thaxter.

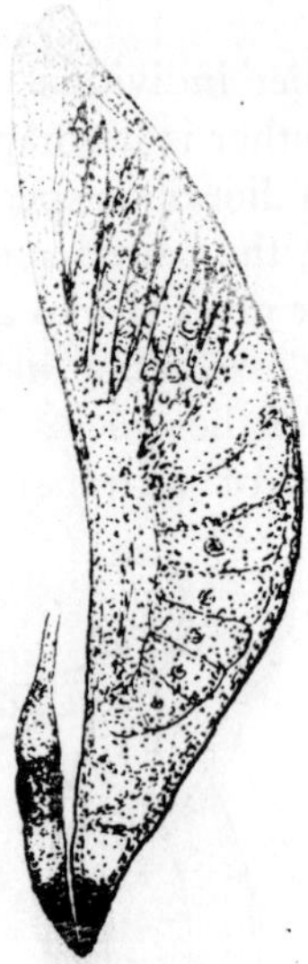

Fig. 223. *Amorphomyces Falagriae* Thaxter; male and female individuals, the latter with perithecium containing spores; after Thaxter.

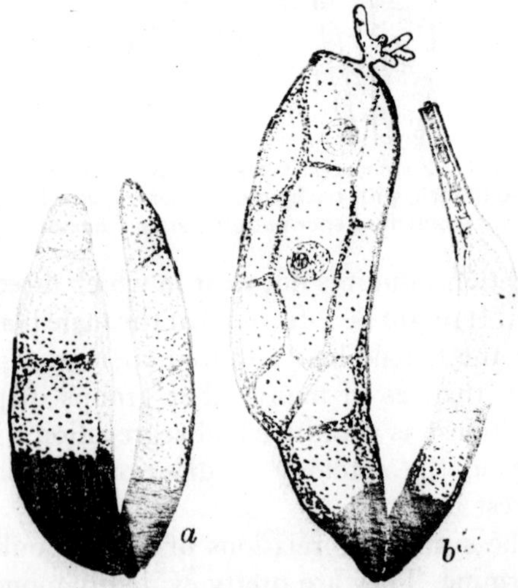

Fig. 222. *Amorphomyces Falagriae* Thaxter; male and female individuals; *a*, young; *b*, mature; after Thaxter.

tion recalls the Erysiphales, while the endogenous male element differs from anything found elsewhere among the fungi. The general inference is that the Laboulbeniales were derived from an ancestor already definitely ascomycetous but not otherwise highly specialised and that they have undergone considerable modification since branching off from the main line. It might be hazarded that the separation took place somewhere between the level of the Erysiphales and that of the lower Pyrenomycetes, while the ancestors of the latter still showed normal sexuality.

The Laboulbeniales are divided into three families, which, however, need not be separately considered here:

Spermatia endogenous	
Borne in compound organs	PEYRITSCHIELLACEAE
Borne in simple organs	LABOULBENIACEAE
Spermatia exogenous	CERATOMYCETACEAE

BASIDIOMYCETES

The Basidiomycetes include over 14,000 species possessing a well-developed mycelium which, among higher forms, builds up an elaborate fruit body, or **sporophore**, such as may be observed in the toadstools, bracket-fungi and puffballs.

The Basidium. These fungi are characterised by the fact that their principal spores, **the basidiospores**, are borne externally on the mother cell, or basidium. The young basidium contains two nuclei, these fuse, the fusion nucleus divides, providing the nuclei of the spores; the spore is formed at the end of a stalk, the sterigma, through which the nucleus passes from the basidium to enter the developing spore. Two successive divisions in the basidium constitute a meiotic phase. In the Autobasidiomycetes the basidia are without septa, and four spores arise from, or near, the apex of the basidium (fig. 224). In the Protobasidiomycetes the basidium is divided into four cells, each of which gives rise to a single spore; the walls are transverse in the Uredinales and Auriculariales, longitudinal or oblique in the Tremellales. In the Hemibasidiomycetes septa may or may not be present in the basidium, but more than two divisions follow nuclear fusion, and more than four spores are produced.

In several of the Uredinales and in *Sirobasidium* among the Tremellales the basidia are developed in chains, elsewhere they are borne singly. In the Ustilaginales and in most of the Uredinales the contents of the basidium are at first enclosed in a thick wall, forming the **brand spore** or **teleutospore cell**, which becomes detached, serving as an additional means for the distribution of the plant; later the contents are extruded as a thin-walled **promycelium** on which the basidiospores are produced. In other Basidiomycetes the basidia are thin-walled throughout their development and give rise to spores while still attached to the mycelium.

The Basidiospores. The basidiospores are unicellular, rounded or oval, asymmetrically attached to their sterigmata (fig. 224 *a*), usually with a smooth, rather thin wall. Echinulate spores are found in a few species, and, in many families, especially among gill-bearing fungi, dark or bright-coloured spores are common. In the

Protobasidiomycetes and Autobasidiomycetes the first indication[1] that a spore is about to be liberated is the appearance of a drop of

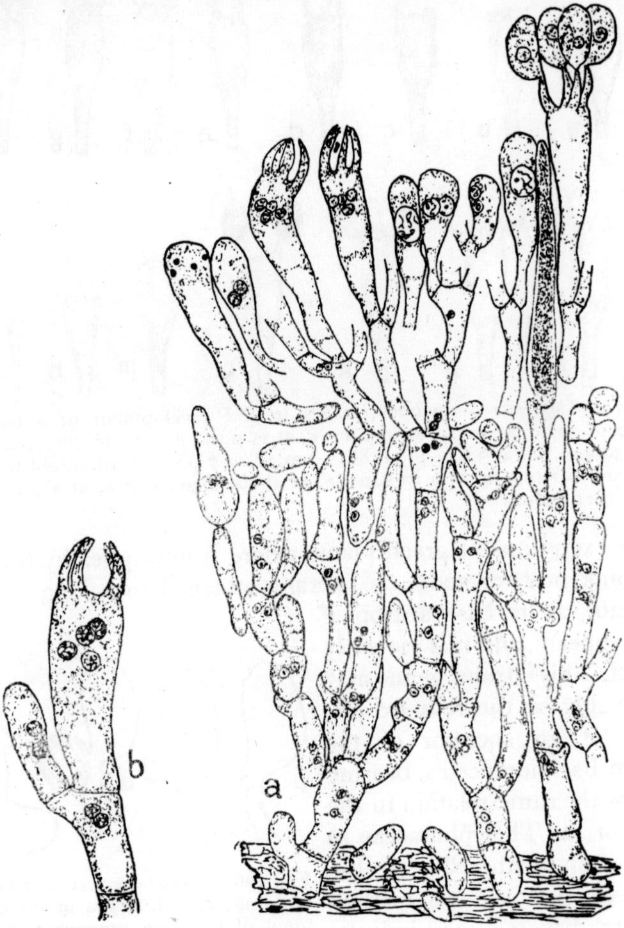

Fig. 224. *Hypochnus subtilis* Schroet.; *a*, hymenium of binucleate cells bearing basidia; *b*, young basidia; after Harper.[2]

fluid (fig. 225 *l*); the drop grows till its diameter is about one-quarter that of the spore, and is carried with the latter when it is shot off. The whole process takes a few seconds. In the Ustilaginales[3] conidia, produced on the basidiospores or on the mycelia to which they give

[1] Buller, 1922 ii. [2] Harper, 1902. [3] Cf. p. 294.

rise, are discharged in the same way, and a similar method has been observed in *Sporobolomyces*,[1] a unicellular fungus found on the

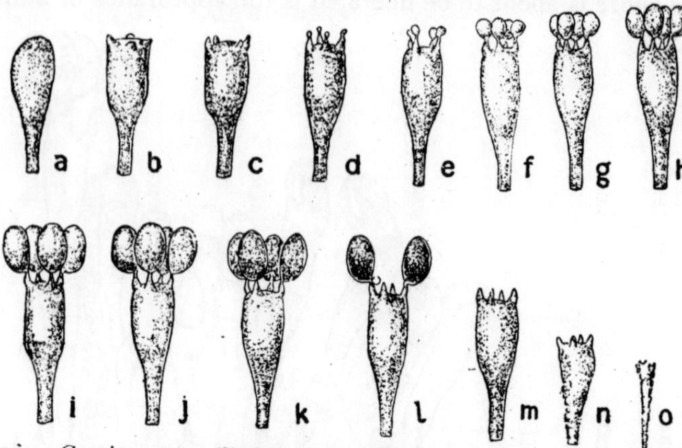

Fig. 225. *Coprinus sterquilinus*; stages in the development of a basidium; *a*, at 8 p.m.; *b*, at 9 p.m.; *c*, at 10 p.m.; *d*, at 11 p.m.; *e*, at midnight; *f*, at 2 a.m.; *g*, at 2.30 a.m.; *h*, at 3 a.m.; *i*, at 8–11 a.m.; *j*, at 7 p.m.; *k*, midnight to 6 a.m.; *l*, at about 6 a.m.; *m*, at 6.10 a.m.; *n*, at about 6.40 a.m.; *o*, at about 7 a.m.; ×272; after Buller.[2]

sugary excretions of green plants where it multiplies by budding. The pink bud, or spore, is laterally attached to a long sterigma (fig. 226), at its base a drop of water appears, and it is violently shot off. This curious genus may conceivably be the reduced offspring of an ancestor of the modern Basidiomycetes, bearing perhaps the same relation to the smuts or the Thelephoraceae as do the yeasts to other Plectascales, or its basidiomycetous method of spore dispersal may indicate a response to the same need rather than a relationship.

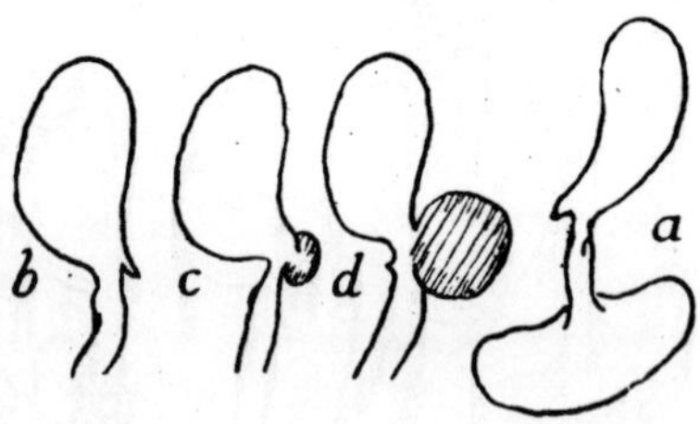

Fig. 226. *Sporobolomyces*; *a*, single cell budding; *b*, *c*, *d*, stages in the detachment of the bud, somewhat diagrammatic, the drop of water is shaded; after Kluyver and van Niel.

Reproduction. The rusts are the only basidiomycetous group in which sexual organs are found. Even there the nuclei do not

[1] Kluyver and van Niel, 1924; Lohwag, 1926; Guilliermond, 1928 ii; Derx, 1930; Buller, 1933, p. 171.
[2] Buller, 1924.

fuse when they become associated, but continue their development side by side. As commonly happens with nuclei in the same cell, and to that extent under the same conditions, they divide simultaneously, and, since their spindles are parallel, and the septa form across them, one daughter of each is located in each new cell, so that a mycelium of binucleate cells is developed, to the associated nuclei of which the term **synkaryon** is applied. The associated condition comes to an end in the basidium where the nuclei fuse in readiness for the inception of meiosis. The union of the nuclei thus takes place in two stages, nuclear association being separated by a longer or shorter series of vegetative divisions from nuclear fusion, and the sporophyte, or diplophase, showing paired nuclei instead of nuclei with the double number of chromosomes. Such a postponement of nuclear fusion, though usually of less extended duration, is not unknown in other plants and animals.

In the remainder of the Protobasidiomycetes, in the Hemibasidiomycetes and in the Autobasidiomycetes sexual organs are not found, but, as in the rusts, the young basidium and the cells of some part of the mycelium are constantly binucleate, and nuclear fusion occurs during the development of the basidium. The binucleate condition in the Hemibasidiomycetes arises at an early stage, when the basidiospores or their products become associated in pairs, and the nucleus of one passes into the other. In the Autobasidiomycetes, on the contrary, a considerable mycelium of uninucleate or multinucleate cells may be formed on the germination of the basidiospore; between the cells of such mycelia anastomoses are common, and in this way nuclei from different cells are brought together and a synkaryon formed. In most of the species examined the basidium is the terminal segment of a filament of binucleate cells.

Such cells make up the diplophase, dikaryophase or **secondary mycelium**,[1] in contrast to the **primary mycelium**, or haplophase, composed of cells containing one or several nuclei according to the species. In many of the Autobasidiomycetes the lower parts of the sporophore are formed by the primary mycelium, the synkaryon appearing first in the neighbourhood of the hymenium, in others the secondary mycelium gives rise to the whole sporophore.

The latter condition is well exemplified in the genus *Coprinus*,[2]

[1] Cf. p. 154.
[2] Bensaude, 1918; Kniep, 1920; Mounce, 1921–3; Vandendries, 1922–5.

some of the species of which are self-compatible, the diplophase arising from the growth of a single spore, while others, such as *C. fimetarius* and *C. radians*, are self-incompatible, and there is ordinarily no sign of a fructification, nor is a diplophase formed, except where two complementary strains of primary mycelium, *A* and *B*, have come into contact and anastomoses have taken place between them, initiating the synkaryon.

Often more than two strains exist, each of which is capable of forming secondary mycelium only with a limited number of other strains. Thus, *Ustilago longissima*, one of the smuts, has three strains,[1] and any arrangement may be fertile which brings two of these strains together. In *Coprinus lagopus*, *Aleurodiscus polygonus*, *Hypholoma Candolleanum* and many other species four strains[2] *AB*, *Ab*, *aB* and *ab*, are present, secondary mycelium being developed only when the combination *AaBb* is obtained. Such species are probably tetraploid. It is possible, by dexterous manipulation,[3] to germinate separately the four spores of a basidium and to discover that the same basidium may give rise to all the four above-mentioned strains, one from each basidiospore. Moreover, while each normal fruit body has its own four strains, *AB* giving secondary mycelium only with *ab*, and *Ab* only with *aB*, each of the four may be perfectly fertile[4] with several strains

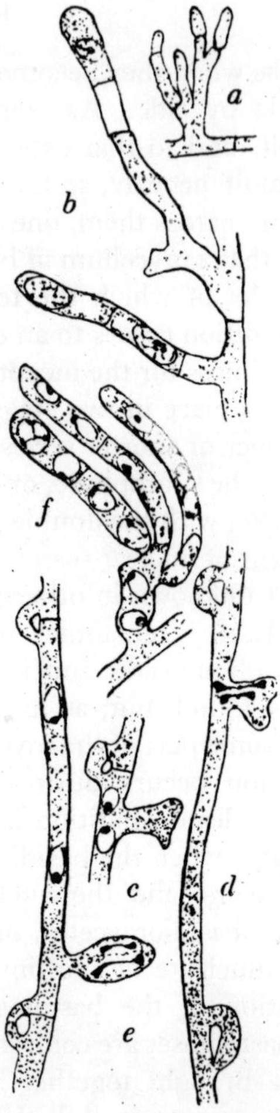

Fig. 227. *Pholiota aurivella* Batsch; *a*, branched oidiophore formed on primary mycelium, ×660; *b*, the same, showing uninucleate oidia, ×1930; *c*, *d*, *e*, stages in the development of an oidiophore on the secondary mycelium, ×1930; *f*, branched oidiophore on secondary mycelium with binucleate oidia, ×1930; after Martens and Vandendries.

[1] Bauch, 1923.

[2] Kniep, 1920–8; Hanna, 1925; Vandendries, 1932; Vandendries and Brodie, 1933.

[3] Hanna, 1925.

[4] Brunswik, 1924; Newton, D. E., 1926 i; Kniep, 1928; Quintanilha, 1933.

from another sporophore of the same species. In *Coprinus lagopus*[1] twenty-four strains have been recognised in this way, and larger numbers have been found in other forms.

As in *Glomerella cingulata* and *Ascobolus stercorarius* among the Ascomycetes, the primary mycelia of heterothallic forms are not wholly incapable of forming sporophores[2] or their equivalent, but, when this happens, the spores are all of one strain. Further, a monosporous culture may spontaneously become diploid,[3] even after remaining for weeks in the primary state.

Both in the **bipolar** species, with two strains, and in the **tetrapolar**, with four, the diplophase may arise, not only between two primary strains of appropriate complement, but when contact is made between a primary and a secondary mycelium.[4]

In the Hemibasidiomycetes and Protobasidiomycetes, where the cells of the haplophase are ordinarily uninucleate, the primary and secondary mycelia are readily distinguished; in the Autobasidiomycetes some structural differences exist in addition. The cells of the primary, and more rarely of the secondary mycelium may break up into short segments, or oidia[5] (fig. 227) which are readily distributed by insects, and doubtless often serve, in heterothallic species, to bring together the complementary strains. Even in homothallic forms, since mycelial fusions are common, they may facilitate the association of hyphae of different origin.

To the binucleate condition[6] is related the presence of clamp connections (figs. 224, 228) between adjacent cells. The formation of the clamp connection takes place in relation to the division of a synkaryon and the correlated deposition of transverse septa. When division is about to begin a beak or projection is put out from the cell and, curving downwards, fuses with it again. Meanwhile the upper of the two nuclei approaches the beak, while the other remains in the parent hypha. Both divide (figs. 228 *a*, *d*) and a daughter nucleus, passing through the beak, lies near the daughter of the other member of the synkaryon (fig. 228 *b*). Septa are now formed, both across the beak and across the original cell, each running at right angles to one of the spindles of the synkaryon. The

[1] Hanna, 1925. [2] Vandendries, 1925 i, ii, 1927; Newton, D. E., 1926 i, ii.
[3] Quintanilha, 1933, 1935; Chow, 1934.
[4] Buller, 1931; Brown, A. M., 1932. [5] Martens and Vandendries, 1932.
[6] Bensaude, 1918; Buller, 1922 i; Brodie, 1931, 1932; Vandendries and Brodie, 1933.

result of this somewhat complicated process is that each of the new cells is binucleate, receiving a descendant of either member of the synkaryon.

In some of the Autobasidiomycetes, primary and secondary

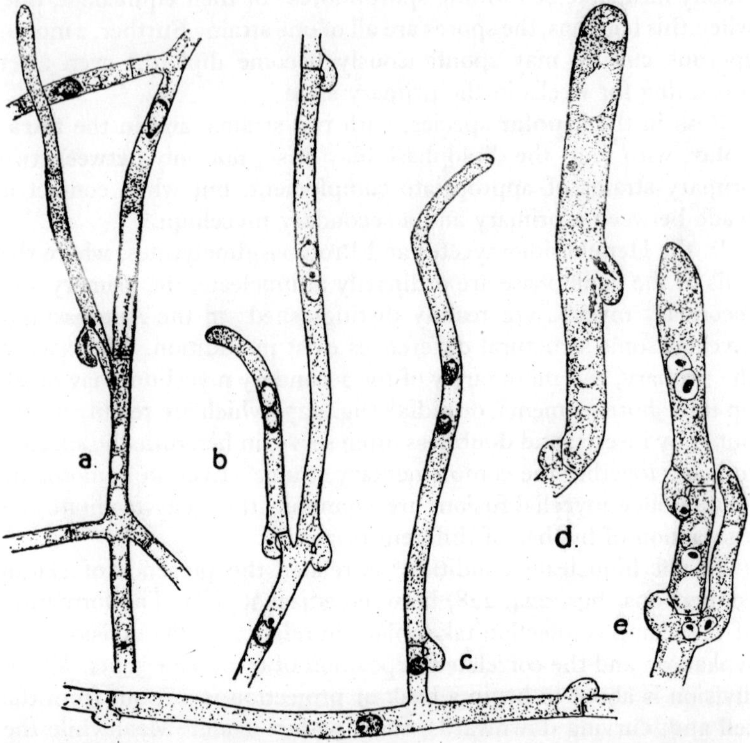

Fig. 228. *Collybia conigena* (Pers.) Quél., *a*, *b*, secondary mycelium with clamp connections, ×600; *Corticium serum* (Pers.) Fr., *c*, the same, ×600; *Armillaria mucida* (Schrad.) Quél., *d*, paired nuclei dividing before the basidium is cut off, a clamp connection in process of formation; *e*, young, binucleate basidium borne on a hypha showing clamp connections; all after Kniep.

mycelia can be recognised[1] by the fact that the lateral branches of the primary mycelium grow almost at right angles to the parent hypha, whereas, in the secondary mycelium, the angle is acute.

The Sporophore. The Hemibasidiomycetes and Uredinales are without exception obligate parasites, with a delicate, endo-

[1] Buller, 1931; Brodie, 1935 i.

phytic mycelium, and do not build a sporophore. In these respects they differ markedly from the Autobasidiomycetes, and even their general characters must be separately considered. In most of the Autobasidiomycetes the sporophore is large, often with a central stalk, the **stipe**, distinct from the **pileus**, or cap, and with the fertile region limited to the surface of teeth, or of **lamellae** or gills, or to the interior of deep pores or of closed chambers.

Phylogeny. Though showing less uniformity of structure than the ascus, the basidium possesses characters sufficiently well defined to suggest a common origin for the Basidiomycetes. Such an origin has often been proposed through the conidial forms of the Phycomycetes or Ascomycetes, but the regular fusion of two nuclei in the basidium followed by meiosis, and the regular production in the Protobasidiomycetes and Autobasidiomycetes of four basidiospores, point to the basidium as the tetraspore mother cell, or final cell of the sporophyte, and to the basidiospore as the first cell of the haploid generation. The association of nuclear fusion and reduction with the development of accessory spores would involve important changes, of which we have no indication.

The ancestor of the Basidiomycetes must have been a simple, filamentous fungus, probably with functional sexual organs not unlike those of the rusts, but perhaps with an antheridium still remaining attached. There may have been a well-marked alternation of independent, filamentous generations, the gametophyte originating from a basidiospore, and the sporophyte from a zygote, which may possibly have been a thick-walled resting cell such as occurs among Archimycetes to-day. All speculations of this kind, however, in view of our lack of knowledge of intermediate forms, must be of the most tentative character.

The Basidiomycetes may be divided as follows:

Number of basidiospores indefinite	Hemibasidiomycetes
Number of basidiospores definite, usually four	
Basidia septate	Protobasidiomycetes
Basidia continuous	Autobasidiomycetes

HEMIBASIDIOMYCETES

The Hemibasidiomycetes include a single alliance, the Ustilaginales; they are characterised by the fact that more than four spores are produced on the basidium, which may be septate or continuous.

The Ustilaginales, brand fungi, smuts or bunts[1] are an assemblage of some 400 species, obligate parasites on the higher plants, giving rise in the tissues of the host to thick-walled, usually dark-coloured resting spores, the **brand spores**, or teleutospores. These are developed in considerable quantities, either singly, in pairs, or in clusters known as **spore balls**; when ripe they break through the host tissue, forming a pustule or sorus. No distortion of the host is caused during the period of vegetative growth, but in preparation for the formation of spores very marked hypertrophy may be induced.

Fig. 229. *Ustilago Treubii* Solms; stem of *Polygonum* with "fruit gall", nat. size; after Solms-Laubach.[2]

Ustilago Treubii, on the stem of *Polygonum chinense* in Java, is responsible for the appearance of elaborate galls (fig. 229) provided with vascular tissue and growing by means of a cambium; *Ustilago Maydis* produces whitish swellings and blisters, often as large as a fist, on the cobs of *Zea Mays*; and *Urocystis Violae* deforms the stems and leaves of species of *Viola*. Several smuts form their spores in the ovary of the host, or infect the stamens, filling the anthers with brand spores and benefiting by the means of distribution provided for the pollen. *Ustilago antherarum* even induces development of the staminal rudiments in the pistillate flowers of *Lychnis dioica*; the stamens so formed undergo dehiscence as usual and differ from those of the male

[1] Prevost, 1807; Berkeley, 1847 i; Tulasne, 1847, 1854; de Bary, 1853; Fischer von Waldheim, 1867; Brefeld, 1883; Ward, 1888 i; Plowright, 1889; Dangeard, 1894 i; Lutman, 1911; Killian, 1924; Wang, 1934.

[2] Zu Solms-Laubach, 1887.

flowers in the presence of fungal spores instead of pollen in their anthers.

In all these species, and in most members of the alliance, spore formation is strictly localised, but in *Entyloma* and its allies spores may be formed in almost any part of the host.

The mature brand spore is uninucleate; it is surrounded by a delicate endospore and by an epispore which may be smooth or sculptured, and which usually contains pigment, giving the spore a black, brown or violet colour. Several may be grouped together to form a spore ball (fig. 230), with or without a coat of sterile cells.

Fig. 230. *Urocystis Fischeri*; spore ball, one spore producing basidiospores on a promycelium, × 500; after Plowright.

On renewal of growth the contents of the brand spore pass into a short, thin-walled tube, the promycelium (fig. 231), which thus becomes the effective basidium. Two or more nuclear divisions follow, after which the promycelium produces a number of basidiospores, sometimes termed **sporidia**. The promycelium may be unicellular at first, bearing a bunch of spores at its apex (figs. 230, 232), or multicellular, usually four-celled, forming one or more basidiospores from each cell (fig. 231 *e*). The nucleus of the parent cell does

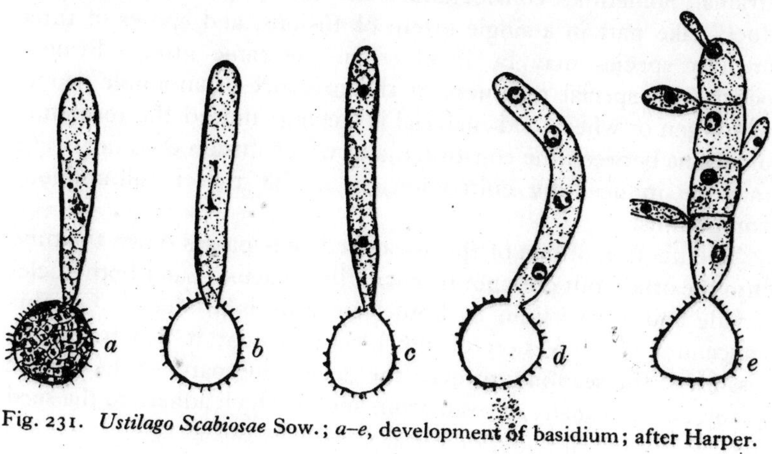

Fig. 231. *Ustilago Scabiosae* Sow.; *a–e*, development of basidium; after Harper.

not travel into the basidiospore, but divides, sending one daughter nucleus into the spore, while the other, remaining in the basidium, may divide again and thus furnish nuclei for successive spores.

Under suitable conditions the basidiospores are cut off in considerable numbers. They may multiply further by budding, giving rise to conidia, or a delicate mycelium may be formed from which conidia are abstricted as in *Tilletia*. These changes occur in nature in the damp, manured soil of the fields.

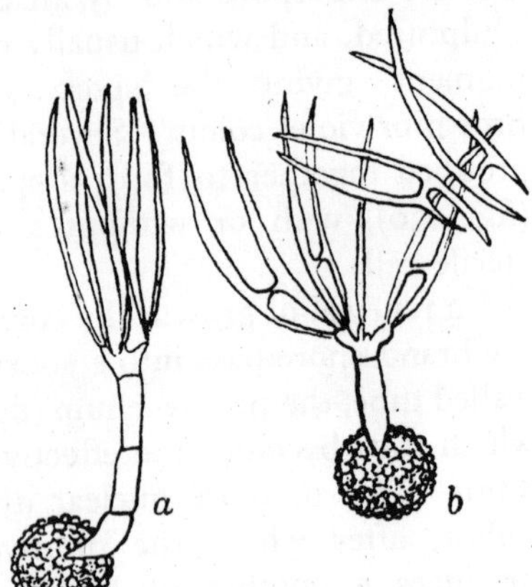

Fig. 232. *Tilletia caries* (DC) Tul.; *a*, basidium thirty hours after rupture of brand spore; *b*, after conjugation of basidiospores; × 300; after Plowright.

During their development the cells of the basidium, the basidiospores or the conidia may become united in pairs (figs. 232 *b*, 233) by means of a tube put out by one or both participants; the growth of these tubes is accurately directed and appears to depend on a chemical stimulus. In *Ustilago antherarum* and other species two or more strains are present,[1] and fusion is not indiscriminate but between cells of compatible strains. Sometimes considerable numbers of spores or their products take part in a single group of fusions, and spores of three or four species may be involved in the same group. Fusions occur with special readiness in the presence of an ample supply of oxygen or when food material is lacking; indeed the formation of strains between the constituents of which fusion does not occur may be induced by cultivation on media rich in albuminous compounds.

The nucleus of one of the associated cells passes down the tube into the other, but does not fuse with its nucleus. Later both nuclei divide and a mycelium of binucleate cells is produced. On this mycelium the infection of the host depends; it penetrates the tissues of the seedling, or even the developing parts of the mature plant, and is usually derived from spores which adhere to the seed

[1] Kniep, 1919, 1926; Bauch, 1923; Dickinson, 1927.

coat. These may be destroyed by dipping the seeds into hot water or formalin solution before sowing.

Once in the tissues the mycelium of binucleate cells penetrates in all directions between the cells of the host and sends haustoria into them. The internodes of the stem are traversed by long, unbranched hyphae, but, in the nodes, branching is frequent, and here also most of the haustoria are found. When the host is perennial the mycelium perennates with it, and, if the subaerial parts of the host die down in winter, remains quiescent in the stock

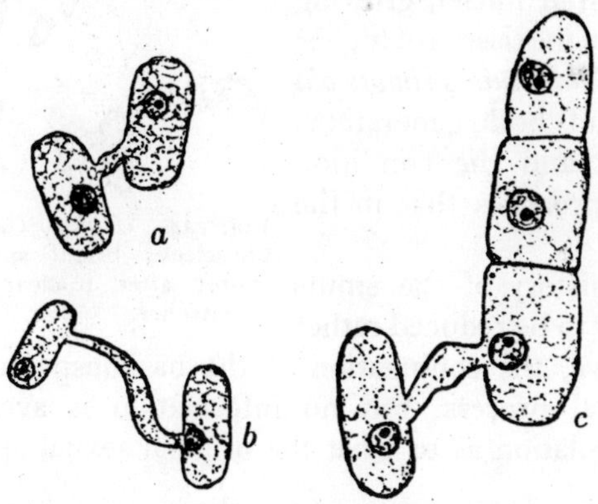

Fig. 233. *Ustilago antherarum*, Fr.; *a* and *b*, conjugating basidiospores; *c*, conjugation between a cell of the basidium and a basidiospore; after Harper.

till the growth of new shoots in spring gives a fresh opportunity for development.

Conidia on the parasitic mycelium are found in *Tuburcinia* and *Entyloma*, but are not of common occurrence in the alliance.

When the formation of brand spores is about to take place the mycelium becomes richly branched and often swollen and gelatinous; it may break into a number of short segments, the contents of each of which form a spore, or spores may be budded out laterally. Where a spore ball is produced it may be furnished with an outer investment of sterile cells.

The young brand spore, like the cells of the mycelium from which it is derived, contains two nuclei (fig. 234*a*). These fuse, so that the mature brand spore is uninucleate. The pairing of the nuclei, which begins with the association of the basidiospores, or of their conidia, is thus completed in the brand spore, which is homologous with the basidium of Autobasidiomycetes, the contents being extruded from the thick wall as a promycelium before the basidiospores are formed. There is here a definite alternation of a brief saprophytic gametophyte of uninucleate cells, and an extensive sporophytic mycelium of cells with paired nuclei, growing parasitically in the tissues of higher plants. In *Tuburcinia primulicola* the mycelia of both generations are parasitic and the condition therefore approaches that in the rusts.

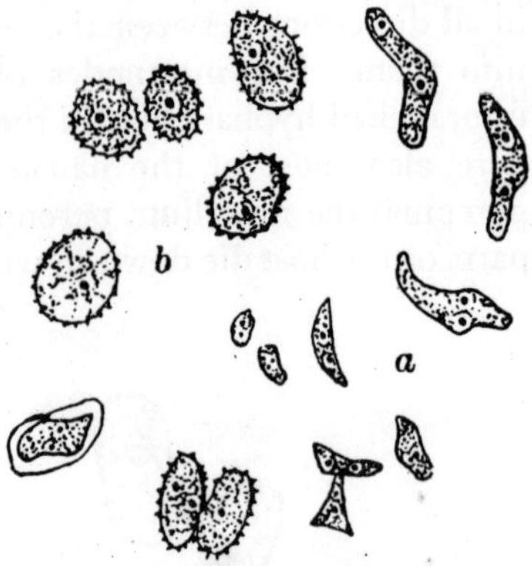

Fig. 234. *Ustilago Carbo*; *a*, young, binucleate brand spores; *b*, older spores after nuclear fusion; after Rawitscher.

The life history of the smuts would appear to be reduced rather than primitive, the conjugation of the basidiospores replacing a normal sexual process, but no information is available which justifies speculation as to what the original sexual apparatus may have been.

The Ustilaginales are divided into two families of about equal size which may be distinguished as follows:

Basidia septate	Ustilaginaceae.
Basidia continuous	Tilletiaceae

Ustilaginaceae

Ustilago, with nearly 200 species, is the most important genus of the Ustilaginaceae. It is cosmopolitan, occurring on all sorts of host plants, and is characterised by the fact that its brand spores are borne singly, not in pairs or groups.

Ustilago Carbo[1] infects species of *Avena*, *Triticum* and *Hordeum*, the form on each host being biologically distinct. The brand spore develops readily in dilute nutritive solutions, forming a three- or four-celled basidium from which basidiospores may be abstricted in the usual way. More commonly, however, the basidia give rise,

[1] Rawitscher, 1912.

without spore formation, to branched mycelia, between the cells of which conjugation takes place. Conjugation may occur between adjacent cells by means of a short branch like a clamp connection (figs. 235 *a*, 236), or between unrelated elements by means of a tube (fig. 235 *b*). In either case the nucleus of one of the associated cells passes into the other, and the nuclei lie close together without fusion.

Ustilago Maydis,[1] the smut of *Zea Mays*, produces considerable deformations containing the mass of gelatinous mycelium from

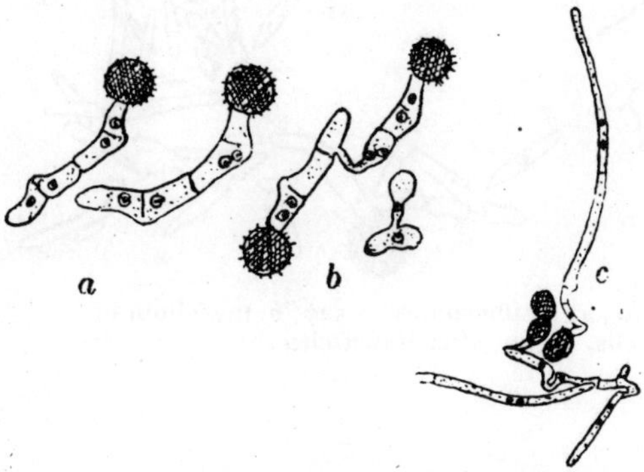

Fig. 235. *Ustilago Carbo*; *a*, *b*, conjugation; *c*, mycelium of binucleate cells; after Rawitscher.

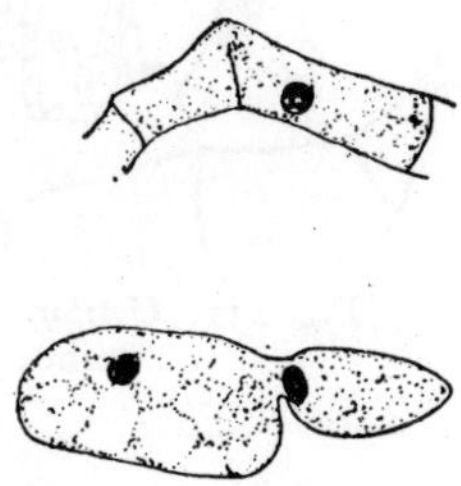

Fig. 236. *Ustilago Hordei*; conjugation; after Lutman.

which the brand spores are produced. When mature the spore mass swells, causing the rupture of the tissues, and the spores escape. The basidia abstrict uninucleate basidiospores; these multiply by budding, but do not conjugate, so that, when the host plant is infected, a mycelium of uninucleate cells is formed (fig. 237). The uninucleate condition is maintained until the hyphae begin to break up in preparation for spore formation; at this stage the walls between adjacent cells disintegrate (fig. 238), the nuclei become associated, and soon fuse. Thus the young brand spore in *U. Maydis*, as in other species, contains two nuclei, and the mature spore a single nucleus, although the parasitic mycelium consists of uninucleate cells. A similar condition has been observed in *U. Vaillantii*.[2]

[1] Rawitscher, 1912.
[2] Massee, I., 1914.

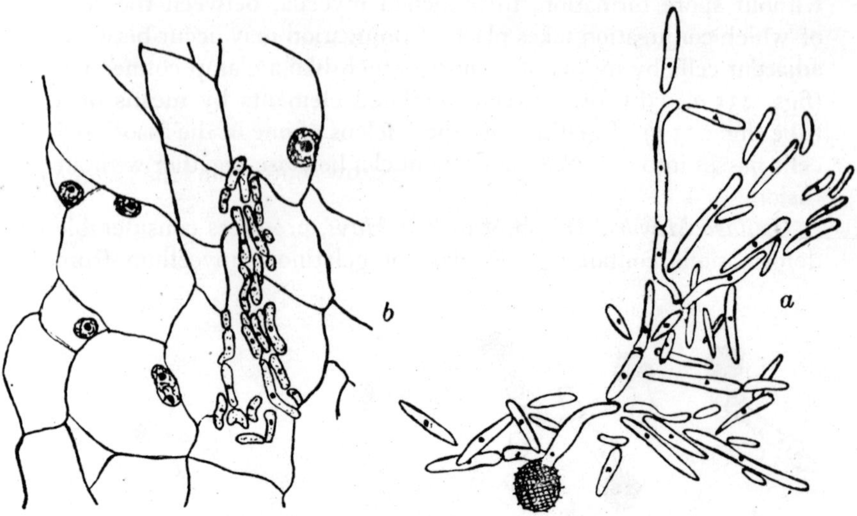

Fig. 237. *Ustilago Maydis*; *a*, basidiospores, ×540; *b*, mycelium of uninucleate cells, ×420; after Rawitscher.

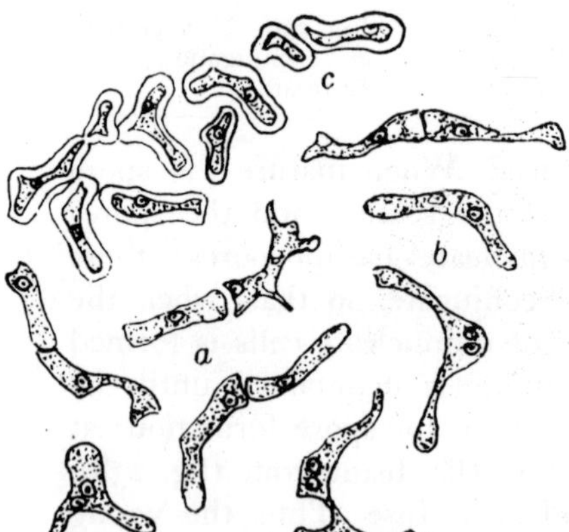

Fig. 238. *Ustilago Maydis*; *a*, uninucleate cells before spore formation; *b*, conjugation; *c*, young, uninucleate brand spores; after Rawitscher.

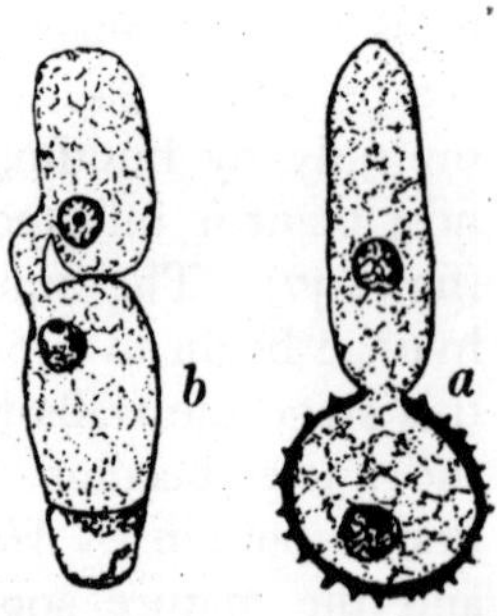

Fig. 239. *Ustilago antherarum* Fr.; *a*, development of brand spore; *b*, conjugation; after Harper.

There is evidence[1] that, while some forms of *Ustilago antherarum* (fig. 239) have a life history similar to that of *U. Carbo*, others are euapogamous, without any nuclear association or fusion. In

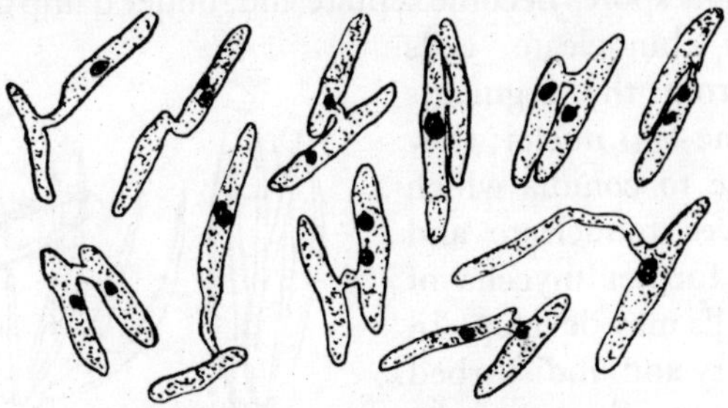

Fig. 240. *Ustilago Tragopogi-pratensis* (Pers.) Wint.; conjugation and nuclear fusion; after Federley.

U. Tragopogi-pratensis, on the other hand, fusion takes place between the basidiospores and the associated nuclei may unite at once[2] (fig. 240), so that the cells of the sporophytic mycelium have single, diploid nuclei.

TILLETIACEAE

The principal genera of the Tilletiaceae are *Tilletia*, *Entyloma*, *Tuburcinia*, *Urocystis* and *Doassansia*. They all show a basidium which is at first continuous, and a terminal group of spores.

Tilletia caries and *T. laevis* are the stink brands of wheat, so-called by reason of the strong smell of trimethylamine given out by the brand spores. The species differ in the character of the epispore, which is smooth in *T. laevis*, reticulate in *T. caries*. Both produce brand spores in the ovaries of the host, all tissues of which, except the outer coat, are destroyed. The spore masses, if garnered with the crop, damage the grain with which they are threshed or ground. The infected flour and chaff or straw are a cause of disease in man and animals.

On the germination of the brand spore of *Tilletia caries* the nucleus passes into the promycelium and three or more divisions follow. Eight basidiospores, or occasionally larger numbers, are budded off in a bunch at the tip of the basidium (fig. 241 *a*), and

[1] Harper, 1899 i; Werth and Ludwig, 1912.
[2] Federley, 1904.

each receives a single nucleus. Short conjugation tubes connect the neighbouring spores, usually before the latter are shed. The nucleus of one of the spores passes into the other, which thus becomes binucleate. The spores become septate and, under damp conditions, filaments of binucleate cells grow out from the segments containing the two nuclei; they may give rise to conidia which are themselves binucleate and from which further mycelia of binucleate cells may develop. In a relatively dry and undisturbed atmosphere conidia may be formed directly on the basidiospores while these are still attached. The conidia are shot off violently, reaching a maximum distance of about one millimetre.[1] Their ejection is associated with the appearance of a drop of water, as in the ejection of basidiospores;[2] for this reason it has been suggested[3] that the conidia are themselves basidiospores, and the structures ordinarily recognised as basidiospores, sterigmata.

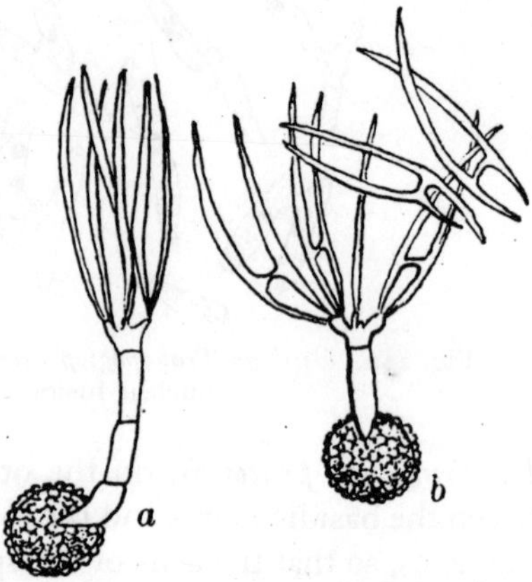

Fig. 241. *Tilletia caries* (DC) Tul.; *a*, basidium thirty hours after rupture of brand spore; *b*, conjugation of basidiospores; × 300; after Plowright.

The mycelia of binucleate cells bring about infection, pushing between the cells of the host seedling, and branching in the tissue. Nuclear fusion takes place in the young brand spores.[4] This is the usual life history, but there is evidence that in some specimens of *T. caries*, as of *Ustilago Tragopogi-pratensis*, nuclear fusion takes place soon after nuclear association,[5] and the parasitic mycelium consists of cells with single, diploid nuclei.

In species of *Doassansia* (fig. 242 *a*), *Entyloma* (fig. 242 *b*) and *Urocystis* (fig. 243), a mycelium of binucleate cells has been observed in the tissues of the host,[6] and nuclear fusion occurs in the

[1] Buller and Vanterpool, 1925; Buller, 1933, p. 217 *et seq.*
[2] Cf. p. 285.
[3] Buller, 1933, p. 234.
[4] Rawitscher, 1914.
[5] Dastur, 1921.
[6] Dangeard, 1894 i.

young brand spore; but *Doassansia Sagittaria*[1] has a life history similar to that of *Ustilago Maydis*.

Tuburcinia primulicola[2] forms a mycelium of uninucleate cells in the tissues of species of *Primula* and gives rise to uninucleate conidia on the young flower. When the flower opens the conidia conjugate

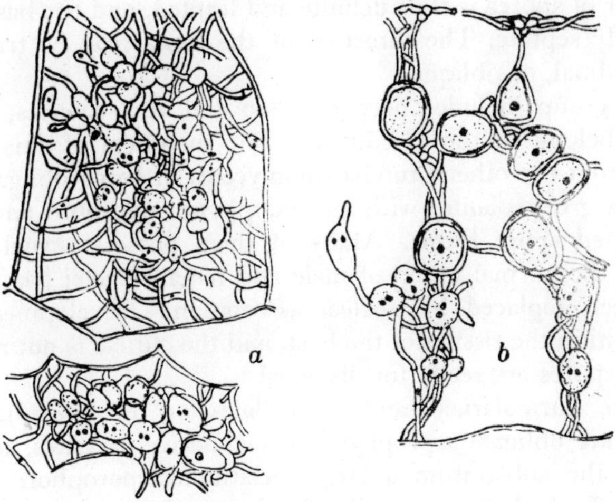

Fig. 242. Formation of brand spores; *a*, *Doassansia Alismatis* (Nees) Corn.; *b*, *Entyloma Glaucii* Dang.; after Dangeard.

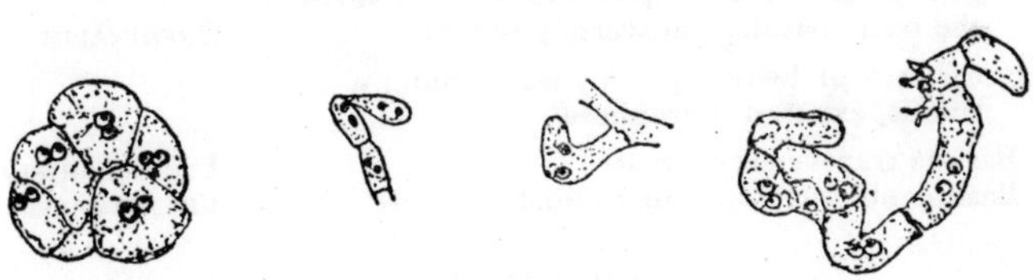

Fig. 243. *Urocystis Anemones* (Pers.) Wint.; young spore ball and mycelium; after Lutman.

in pairs, and those containing two nuclei germinate to form the sporophytic mycelium on which the brand spores are borne. The arrangement here differs from the ordinary life history of the smuts in the longer continuance and parasitic character of the gametophyte and in that respect suggests a comparison with the rusts.

[1] Rawitscher, 1922. [2] Wilson, 1915.

PROTOBASIDIOMYCETES

In the Protobasidiomycetes the nucleus of the basidium undergoes two successive divisions and the daughter nuclei are separated by walls cutting the basidium into four uninucleate cells. Each of these gives rise to a single spore into which the nucleus passes. The number of spores is thus definite and limited, and the basidium is regularly septate. The direction of the septa may be transverse, longitudinal, or oblique.

The group includes between 1800 and 1900 species, most of which belong to the Uredinales. The members of this alliance differ from the other Protobasidiomycetes in being obligate parasites on green plants, with an endophytic mycelium and highly specialised spore forms. Many of them possess sexual organs, though the normal union of male and female nuclei has in every case been replaced by nuclear association. Development takes place within the tissues of the host, and the cuticle is not ruptured till the spores are ready for dispersal.

In the Auriculariales and Tremellales, on the other hand, the species are obligate saprophytes or facultative parasites, and form outside the substratum a large, gelatinous sporophore, on the surface of which the spores develop.

The distinction between the three alliances may be summarised as follows:

Obligate parasites, developed within the tissues of the host; basidia transversely septate	UREDINALES
Saprophytes or hemi-saprophytes, forming a gelatinous, exposed sporophore	
Basidia transversely septate	AURICULARIALES
Basidia obliquely or longitudinally septate	TREMELLALES

UREDINALES

The rust fungi, members of the Uredinales,[1] or Aecidiomycetes, including some 1700 species, are without exception obligate parasites on the stems, the sporophylls, and especially the leaves of vascular plants, usually on angiosperms or gymnosperms, but in a few examples on ferns.

[1] Tulasne, 1854; de Bary, 1887; Plowright, 1889; Eriksson, 1894–1905; Sappin-Trouffy, 1896; Tranzschel, 1904; Maire, 1911; Ramsbottom, 1912; Grove, 1913 ii; Bailey, 19-

The mycelium ramifies between the cells of the host, sends haustoria[1] into the cells, and may induce hypertrophy with consequent curling and malformation of the infected part. The storage of starch may be stimulated; it is so abundant in the hypertrophies caused by the aecidial mycelium of *Puccinia Caricis* on *Urtica parvifolia* that these are used as food by the Himalayans. The products of one or two other species are similarly employed. Where the mycelium grows in the perennial tissues of the host, it is itself perennial.

The nuclear divisions of the rusts[2] have undergone considerable simplification; in the vegetative cells the chromatin appears to be

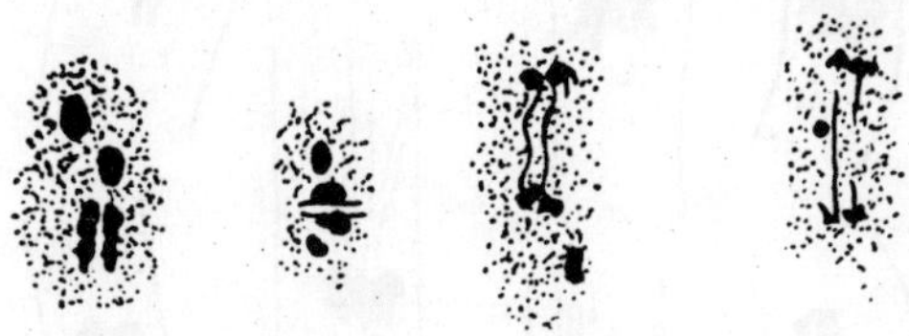

Fig. 244. *Uromyces Poae* Raben.; conjugate divisions in aecidium, × 1330; after Blackman and Fraser.

drawn apart as undifferentiated masses (fig. 244) between which a kinoplasmic thread represents the spindle; when conjugate division occurs the spindles of the members of the synkaryon lie parallel one to another. The divisions of the fusion nucleus of the teleutospore, however, still show some of the characters of a meiotic phase. The chromatin displays a synizetic contraction, and the spireme breaks into segments (fig. 245) which, in favourable instances, appear to be double throughout their length. The spindle is extra-nuclear, and lies free in the cytoplasm before coming into contact with the chromatin. In many species the chromosomes apparently lose their individuality after passing on to the spindle, and travel in irregular masses to the poles.

On the mycelium several kinds of spore are produced: minute spermatia in spermogonia, aecidiospores in aecidia, uredospores and teleutospores, sometimes mixed, sometimes separate, in more or less definite sori. One or more spore forms may be lacking, but

[1] Němec, 1911 ii.

[2] Poirault and Raciborski, 1895; Holden and Harper, 1903; Blackman, 1904; Blackman and Fraser, 1906 i; Arnaud, 1913; Moreau, Mme, 1914.

teleutospores are almost always present, and on them the classification of the alliance is based.

It was some time before spores of so many kinds were recognised as belonging to the same fungus, and the old generic names of stages other than the teleutospore, such as *Aecidium*, *Caeoma* and *Uredo*, still survive as descriptive terms.

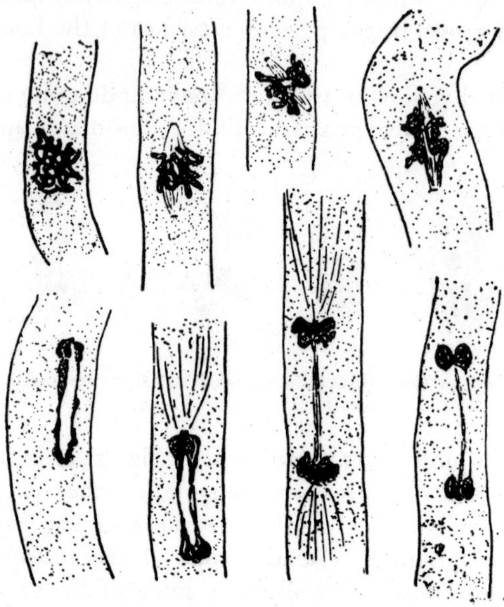

Fig. 245. *Gymnosporangium clavariaeforme* Rees; first meiotic division in basidium, × 1460; after Blackman.

The **teleutospores** (figs. 246, 247) may be unicellular or made up of two or more cells, forming a compound structure, each cell of which develops independently. One-celled teleutospores occurring exceptionally in species where two-celled teleutospores are normal are known as **mesospores**.

The teleutospores may be massed together and encrusted in the tissues of the host, undergoing further development *in situ* (fig. 248), or they may be readily detached and carried away by the wind or other agencies. Development may continue as soon as conditions are favourable, or only after a resting period, usually in the spring following the formation of the teleutospores. The nucleus of the

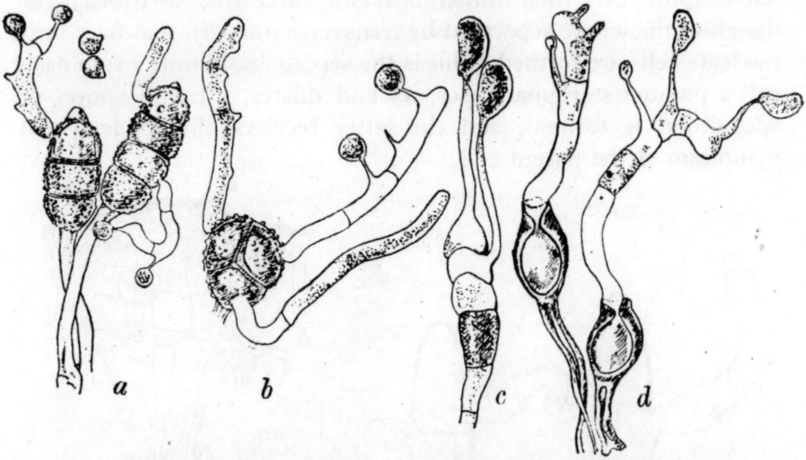

Fig. 246. Developing teleutospores; *a*, *Phragmidium bulbosum* Schm.; *b*, *Triphragmium Ulmariae* Lk.; *c*, *Coleosporium Sonchi* Lév., thin-walled teleutospore with internal septa; *d*, *Uromyces appendiculatus* (*Fabae*) Lév.; after Tulasne.

g. 247. *Cronartium lepiadeum* Fr.; teleu-pore mass with pro-celia and spores; er Tulasne.

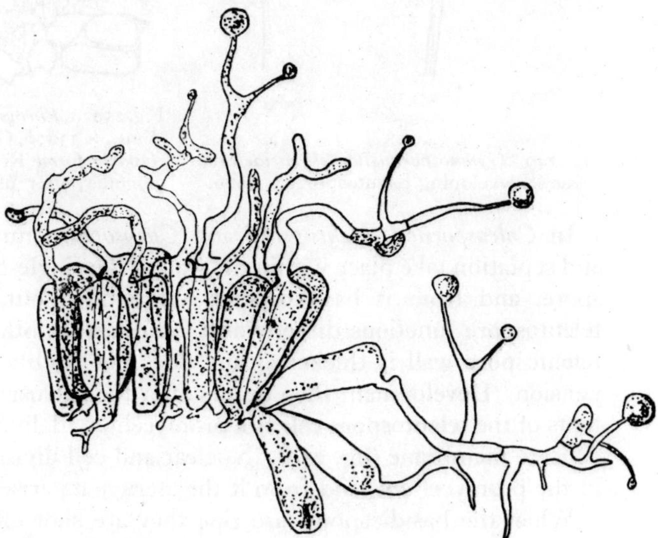

Fig. 248. *Melampsora betulina* Desmaz.; developing teleutospores; after Tulasne.

teleutospore cell then undergoes two successive divisions; the daughter nuclei are separated by transverse walls, so that four uninucleate cells are formed. This is the sep ate basidium. From each cell a pointed sterigma arises, its end dilates, a basidiospore, or sporidium, is formed, and the latter receives the nucleus and cytoplasm of the parent cell.

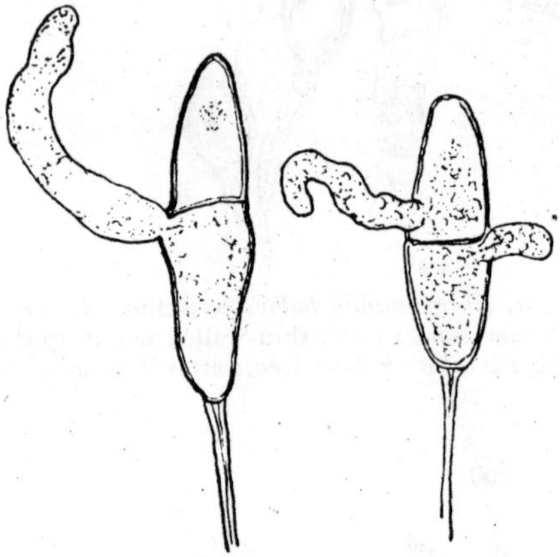

Fig. 249. *Gymnosporangium clavariaeforme* Rees; developing teleutospores; ×666.

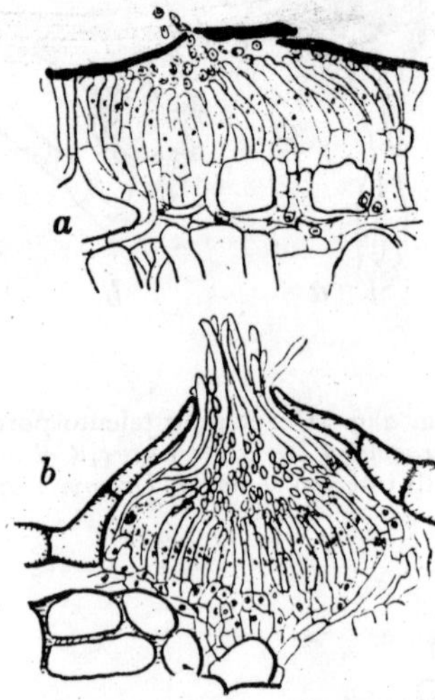

Fig. 250. *a*, *Phragmidium violaceum* Wint., ×330; *b*, *Gymnosporangium clavariaeforme* Rees, ×260; spermogonia; after Blackman.

In *Coleosporium*, *Ochropsora* and *Chrysopsora* nuclear division and septation take place within the wall of the single-celled teleutospore, and from it basidiospores are budded out, so that the teleutospore functions directly as a basidium; in other genera the teleutospore wall is thickened and does not readily admit of expansion. Development then occurs after the extrusion of the contents of the teleutospore cell as a promycelium, delimited only by a delicate membrane (fig. 249). Nuclear and cell division take place in the promycelium, and from it the sterigmata arise.

When the basidiospores are ripe they are shot off successively with some violence, reaching a distance from the sterigma of 0·2 to 0·9 mm.[1] before they begin to fall downwards under the influence

[1] Buller, 1924.

of gravity; just before ejection a drop of fluid appears at the base of the spore and is later carried away with it. Where the development of the teleutospore cell occurs *in situ*, the promycelium takes up a position which allows the basidiospores to be shot away from the sorus.

Should the basidiospore reach a suitable habitat, it germinates. Its germ tube penetrates the cuticle of the host and forms a mycelium of uninucleate, or in some cases, of multinucleate[1] cells bearing spermogonia and aecidia.

The **spermogonium** is usually found on the adaxial side of the leaf; it consists of a group of parallel, simple or branched,[2] outwardly directed spermatial hyphae, arising from a small-celled weft below the cuticle or epidermis of the host. In most species the outer hyphae of the group form paraphyses, so that the spermogonium is restricted in extent and assumes a flask-shaped outline; the paraphyses push between the epidermal cells of the host and project from a narrow ostiole (fig. 250 *b*). In simpler forms, such as *Phragmidium*, the spermogonium is indefinite in extent (fig. 250 *a*).

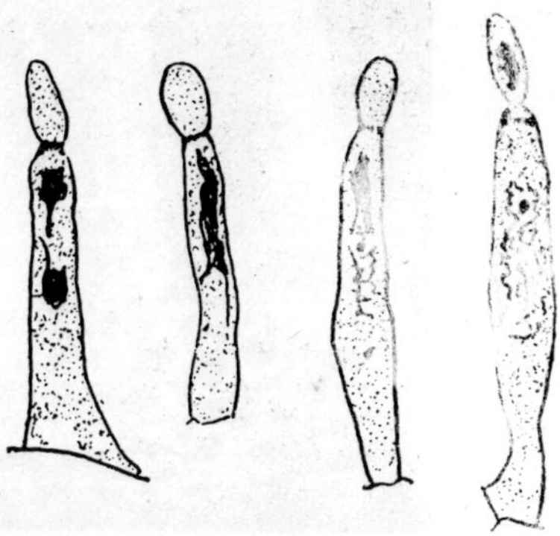

Fig. 251. *Gymnosporangium clavariaeforme* Rees; development of spermatia, × 1185; after Blackman.

The spermatial hypha is long and narrow, with an elongated nucleus; from its tip successive spermatia are cut off (fig. 251). The mature spermatium is a small, more or less ovoid cell enclosed in a thin wall. The cytoplasm is finely granular, without reserve material, and the nucleus is often of relatively large size. When cultivated in solution of cane sugar or honey, and also under natural conditions,[3] spermatia show yeast-like budding; in the presence of the sugary secretion from a spermogonium of opposite complement they form aseptate, uninucleate hyphae.[4] It was long ago suggested that spermatia might be male organs;[5] this is con-

[1] Lindfors, 1924.
[2] Fromme, 1912.
[3] Robinson, W., unpublished.
[4] Allen, 1933.
[5] Meyen, 1841.

sistent with their large nuclei and lack of reserve material. They may be distributed by insects[1] the latter being attracted by the sticky masses which often accumulate outside spermogonia. The scent of *Puccinia suaveolens* and the yellow or orange patches about the aecidia and spermogonia of several species may have a similar function.

The **aecidia** are cup-shaped structures (fig. 252) occurring in groups, usually on the abaxial side of the leaf, and known as

Fig. 252. *Uromyces Poae* Rabh.; groups of aecidia, × 10.

cluster cups; in them the **aecidiospores** are produced in basipetal rows (figs. 254, 270), alternating with small, intercalary cells which sooner or later disintegrate. The mature aecidiospore is usually subglobose or polyhedral; it is enclosed in a thick wall provided with several germ pores, and contains red, yellow or orange pigment, and always two nuclei. Under moist conditions it becomes turgid and rounded away from the neighbouring cells, its attachment to the intercalary cell immediately below it is weakened and at last overcome, and the spore is discharged with a force sufficient to carry it to a distance of from one to fifteen millimetres[2] away from the leaf. It may be noted in this connection that the aecidiospore is much larger and heavier than the basidiospore. Sometimes clumps

[1] Rathay, 1883; Craigie, 1927 ii.

[2] Buller, 1924.

of from sixty to a hundred and fifty spores are shot away together. In germination a hypha is put out which enters the host plant through a stoma, and penetrates into the intercellular spaces.

At the periphery of the aecidium the aecidiospore mother cells, instead of each dividing to form an aecidiospore and an intercalary

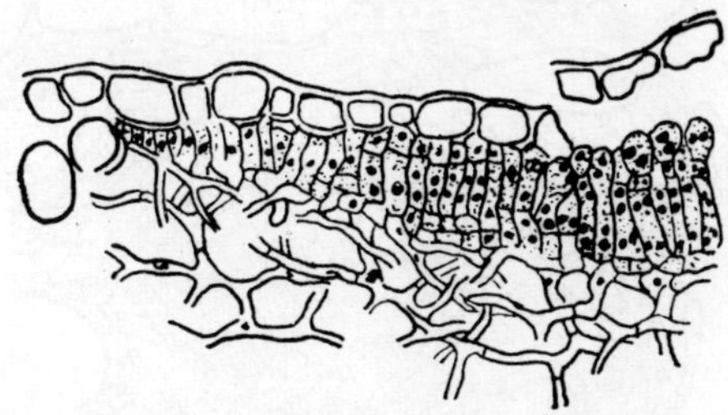

Fig. 253. *Phragmidium violaceum* Wint.; caeoma, ×240; after Blackman.

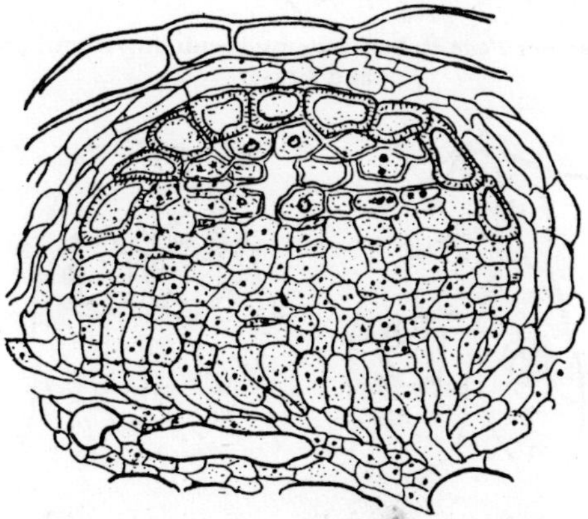

Fig. 254. *Uromyces Poae* Raben.; aecidium just before the epidermis is broken through, ×310; after Blackman and Fraser.

cell, acquire thick, striated walls, lose their contents, and form the sides[1] of a sheath, the **pseudoperidium**, about the sporogenous part. Centrally, where the pseudoperidium arches over the aecidial contents, it is derived from the cells first cut off in each row. When the uppermost functional spores are ripe the pseudoperidium pushes through the epidermis of the host, is itself ruptured, and

[1] Kursanov, 1914.

exposes the spores. It divides into teeth and becomes recurved, so that the characteristic cluster cup is produced. In some species the pseudoperidium is elongated and either cylindrical or inflated, giving the forms known as **roestelia** and **peridermium**, so-called from

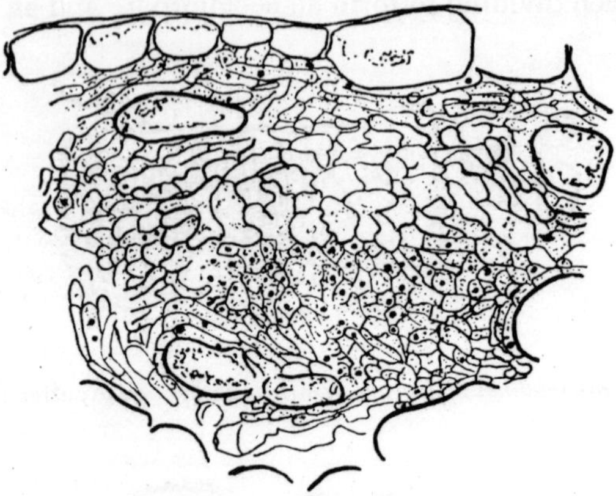

Fig. 255. *Uromyces Poae* Raben.; young aecidium, ×370; after Blackman and Fraser.

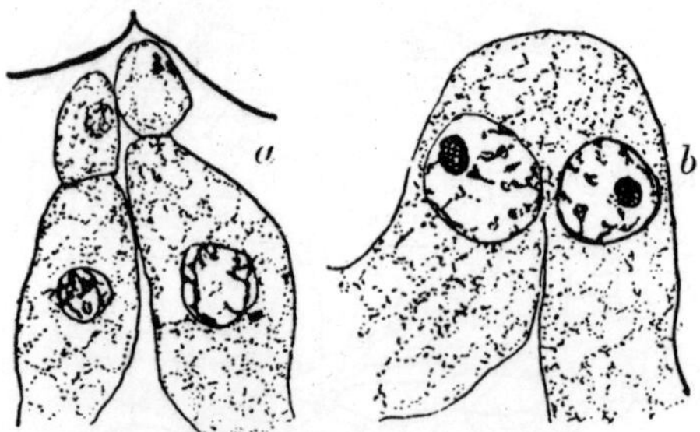

Fig. 256. *Phragmidium speciosum* Fr.; *a*, fertile and sterile cells; *b*, fusion of two fertile cells; after Christman.

their old generic names. In others, such as the species of *Phragmidium* and *Melampsora*, the developing spores are surrounded only by a few sterile filaments. Such aecidia (fig. 253), to which the old generic name **caeoma** is applied, are probably primitive, and occur in the same species as spermogonia of indefinite extent.

The development of the aecidium[1] begins either directly below

[1] Blackman, 1904; Christman, 1905; Blackman and Fraser, 1906 i; Olive, 1908; Grove, 1913 ii; Fromme, 1912, 1914; Welsford, 1915; Thurston, 1923.

the epidermis or deeper in the tissues of the host by the massing of hyphae (fig. 255), which give rise to a more or less regular plate of cells. Each of these usually cuts off a terminal **sterile** cell

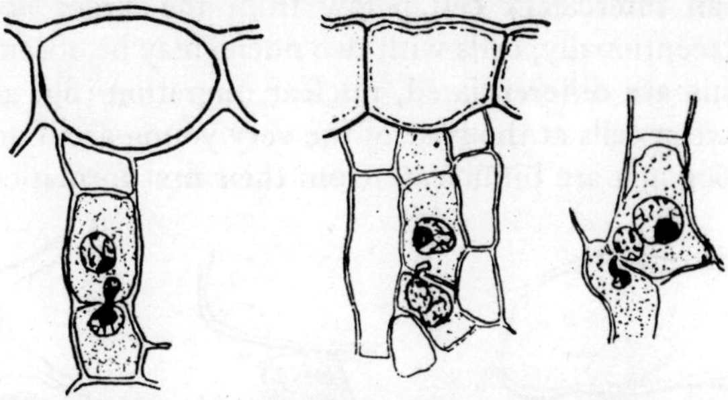

Fig. 257. *Phragmidium violaceum* Wint.; migration of second nucleus into fertile cell of caeoma, ×950; after Blackman.

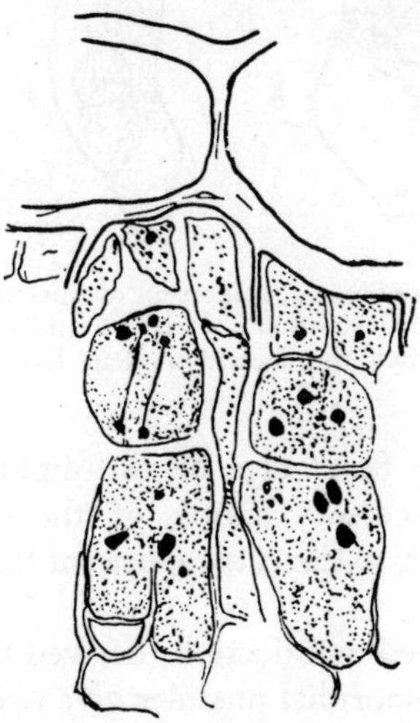

Fig. 258. *Melampsora Rostrupi* Wagn.; paired fertile cells, ×1200; after Blackman and Fraser.

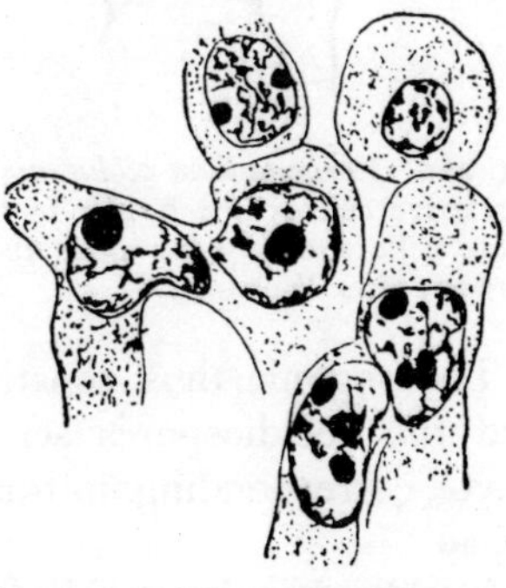

Fig. 259. *Triphragmium Ulmariae* (Schum.) Link; primary uredosorus; condition intermediate between migration and conjugation of fertile cells; after Olive.

(figs. 256, 257, 258) which ultimately degenerates. Its larger sister cell is the oogonium, sometimes known as the **fertile** or **basal** cell, and contains cytoplasm and a single nucleus (fig. 256 *a*). The oogonia may unite laterally in pairs (fig. 256 *b*), so that compound, binucleate cells are formed; they may similarly pair with cells below them (fig. 259), or may each receive a second nucleus by migration (figs. 257, 260 *a*). They proceed at once to cut off binucleate aecidiospore mother cells, each of which divides to separate an intercalary cell below from the larger aecidiospore above. Exceptionally, cells with two nuclei may be observed before the oogonia are differentiated, nuclear migration (fig. 261) taking place between cells at the base of the very young aecidium. In this case the oogonia are binucleate from their first formation.

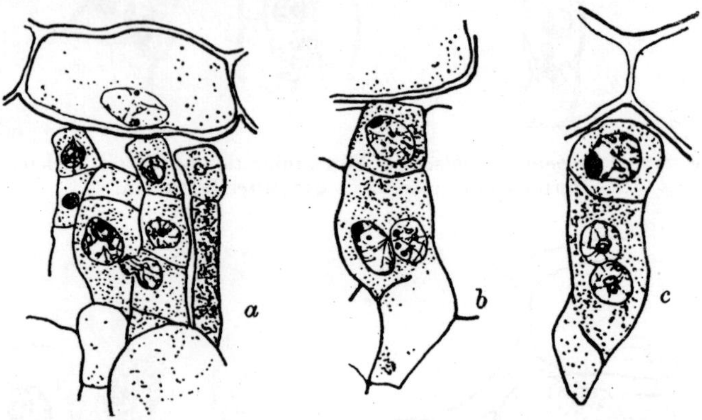

Fig. 260. *Phragmidium violaceum* Wint.; caeoma; *a*, migration of nucleus from vegetative cell of one hypha to fertile cell of another, × 1040; *b* and *c*, binucleate cells showing the pore through which the second nucleus has passed, × 1010; after Welsford.

The oogonia thus constitute the first stage of the diplophase and the aecidiospores serve as accessory spores of the sporophyte, corresponding, in this respect, to the carpospores of the Red Algae.

In many species aecidia are formed on infections derived from a single basidiospore,[1] in others monosporidial pustules give rise only to spermogonia. When such pustules overlap about half their number produce aecidia.[2] It has been shown that these species are heterothallic, possessing two self-incompatible strains, *A* and *B*, and forming aecidia when they come together. Their vegetative hyphae mingle, fusions take place between them, and presumably the two fertile cells which fuse, or the cell which receives a nucleus and the cell from which it migrates, originate on mycelia of opposite

[1] Ashworth, D., 1931. [2] Craigie, 1927 i, 1928; Brown, A. M., 1935.

complement. The formation of aecidia on a monosporidial pustule may also be induced by the presence of spermatia[1] from a complementary strain. There is evidence that the spermatium may become attached to and pass its contents into a hypha projecting from the pustule[2] and that the nucleus may travel to a fertile cell. In *Puccinia*

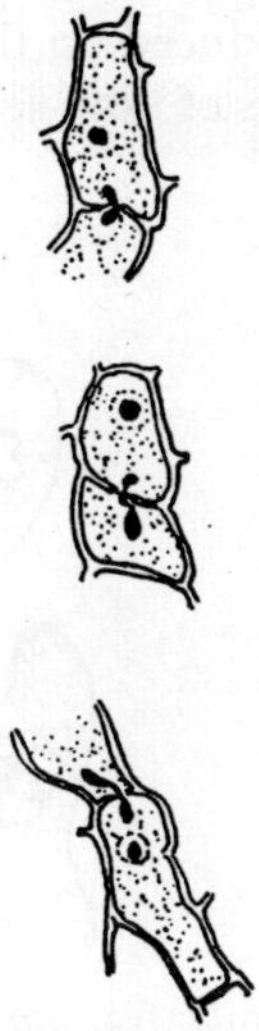

Fig. 261. *Uromyces Poae* Raben.; nuclear migrations in young aecidium, ×950; after Blackman and Fraser.

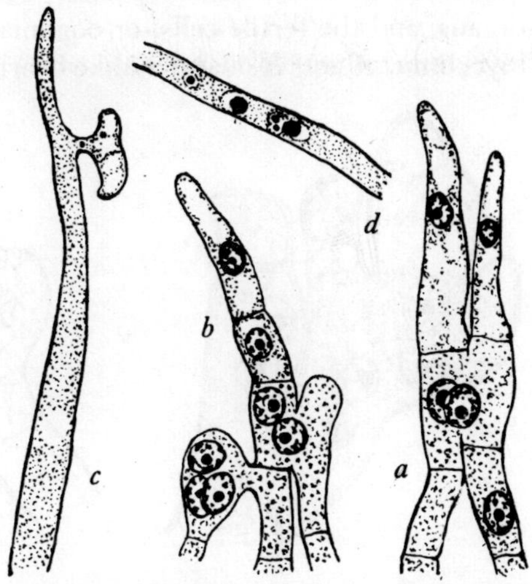

Fig. 262. *Puccinia Sonchi* Rob., *a*, *b*, cell fusions in developing uredosorus; *Puccinia Phragmites* (Schum.) Körn., *c*, paraphysis with attached spermatium; *d*, vegetative hypha with two large nuclei and three small (spermatial) nuclei; *a*, *b*, ×1300, *c*, *d*, ×750; after Lamb.

Phragmites,[3] where the spermatial nuclei are small, spermatia have been found attached in this way (fig. 262*c*) and small nuclei have been observed in proximity to the larger nuclei of the fertile cells (fig. 262*d*). Fusions of cells have been seen in *P. Graminis*[4] within forty-eight hours of the transfer of spermatia to the pustule. The spermatia of *P. Sorghi*[5] may germinate on the host leaf in the presence of exudate from a spermogonium of the complementary strain, and it is possible that they may thus provide hyphae which

[1] Craigie, 1927 ii.

[2] Allen, 1933, 1934 i, ii, 1935; Andrus, 1931; Craigie, 1933; Ashworth, D., 1935.

[3] Lamb, 1935. [4] Hanna, 1929. [5] Allen, 1933.

fuse with the appropriate cells. In the former circumstance the function of the spermatium is presumably antheridial and the association of nuclei in the fertile cell may be regarded as a syngamy, in the latter its behaviour would be that of conidia in other self-incompatible forms. It is to be noted that, in heterothallic as in homothallic species, both the spermatia, or male organs, and the fertile cells, or oogonia, are produced on the same mycelium, *A* and *B* pustules alike bearing organs of both sexes. As

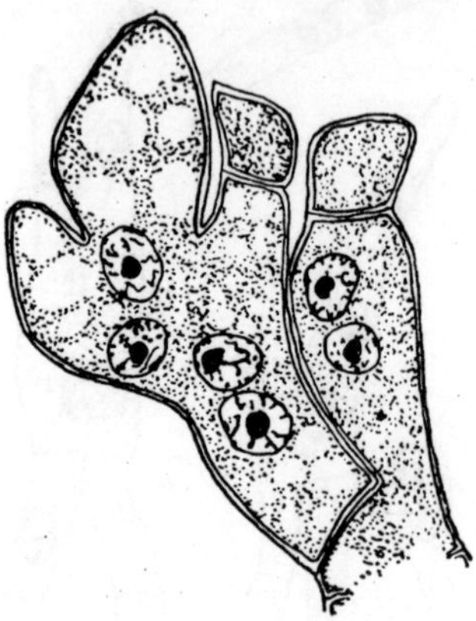

Fig. 263. *Puccinia Falcariae*; branched fertile cell of aecidium or primary uredosorus, × 1200; after Dittschlag.

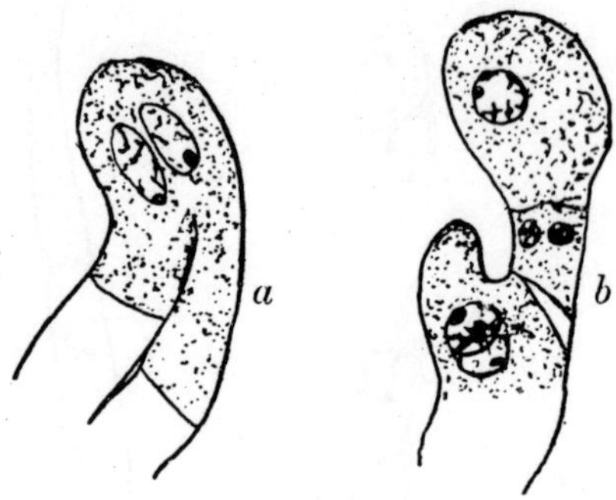

Fig. 264. *Phragmidium Potentillae-Canadensis* Diet.; *a*, conjugation; *b*, branched fertile cell; after Christman.

in many of the Ascomycetes, the difference between *A* and *B* mycelia is thus not one of sex, but of self-incompatibility in a monoecious form. This is borne out by the fact that, after a period of delay, functional aecidia may appear,[1] in heterothallic species, on isolated, monosporidial pustules.

In most species, each oogonium, after the fertilisation stage, gives rise to a single chain of spores; but occasionally the oogonia branch,[2] and thus produce two or more spore rows. In certain species branching is the normal condition, the oogonia regularly forming a number of buds (fig. 263), from each of which a single spore is produced. The spore mother cell divides in the usual way, separating the aecidiospore from its sister cell below, but the latter

[1] Craigie, 1927 i; Brown, A. M., 1935.
[2] Christman, 1907 i; Dittschlag, 1910.

here forms an elongated stalk instead of a short intercalary cell. The method of formation is closely similar to that of a uredo-

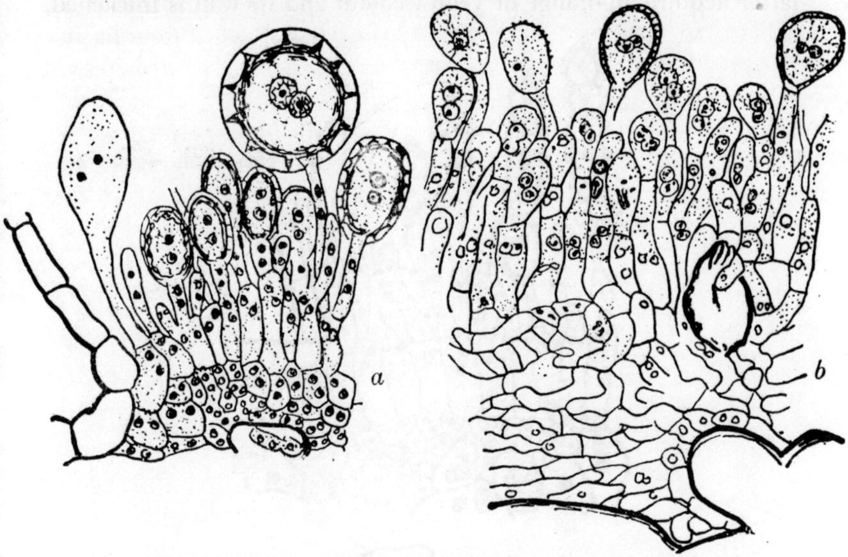

Fig. 265. *Phragmidium Rubi* Pers., *a*, uredosorus, ×600; after Sappin-Trouffy; *Phragmidium violaceum* Wint., *b*, uredosorus, ×480; after Blackman.

spore, and the fructification is still known as a **primary uredosorus** in reference to its appearance relatively early in the year.

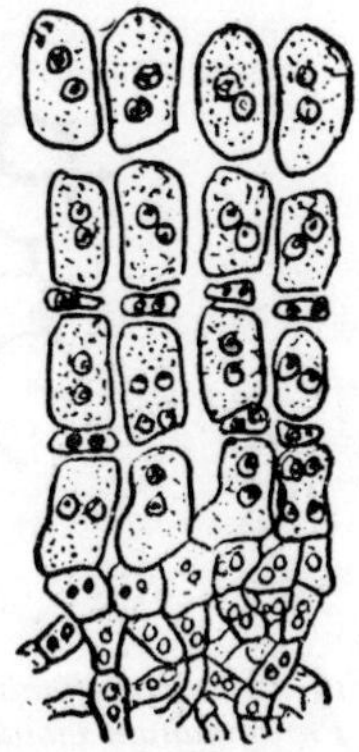

Fig. 266. *Coleosporium Sonchi*; uredosorus, ×545; after Holden and Harper.

The mycelium formed by the aecidiospores consists of binucleate cells; it may bear uredospores or teleutospores. The **uredospores** are produced in groups, or **uredosori** (fig. 265), which may be surrounded by paraphyses or by a pseudoperidium[1] similar to that found in aecidia. In the young sorus a regular layer of fertile or basal cells is formed, and from these the uredospore mother cells arise. In a few species they appear in vertical rows like aecidiospore mother cells, and divide to form uredospores and small intercalary cells (fig. 266), but in most they grow out as buds from the basal

[1] Moss, 1926.

cell, and give rise to a uredospore and stalk (fig. 265). The stalk remains narrow while the uredospore enlarges, the contents of the latter acquire an orange or yellow colour and its wall is thickened,

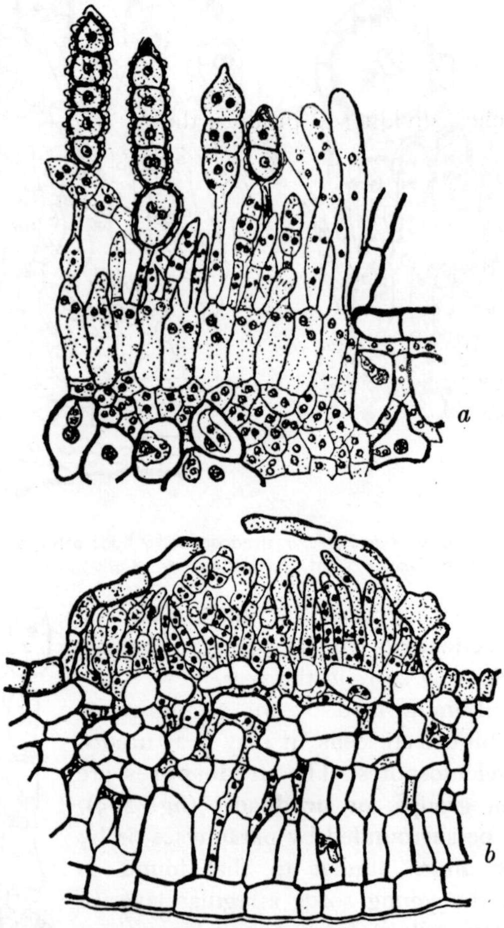

Fig. 267. *Phragmidium Rubi* Pers., *a*, teleutosorus, × 240; after Sappin-Trouffy; *Phragmidium violaceum* Wint., *b*, young teleutosorus, × 240; after Blackman.

and roughened by minute projections. Two or more germ pores are usually present, and the uredospore, like the cells from which it is developed, is invariably binucleate. It gives rise on germination to a mycelium of binucleate cells, on which teleutospores or further crops of uredospores are formed, and thus serves, like the aecidiospore, as an accessory spore of the sporophyte.

Rusts occurring under very dry conditions may produce a type of uredospore with specially thickened walls. These are known as **amphispores**.

Sooner or later the mycelium of binucleate cells gives rise to teleutospores, which are commonly grouped in **teleutosori** (fig. 267). Like the uredospores they may be associated with

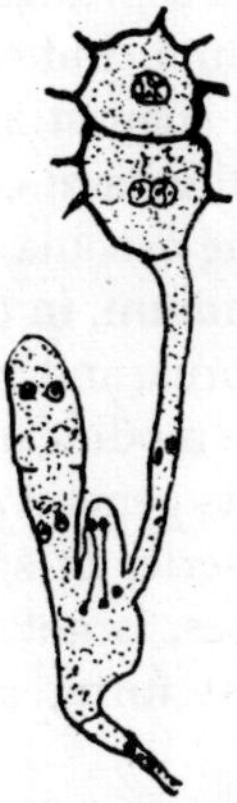

Fig. 268. *Puccinia Podophylli* S.; fertile cell of teleutosorus giving rise to teleutospores; after Christman.

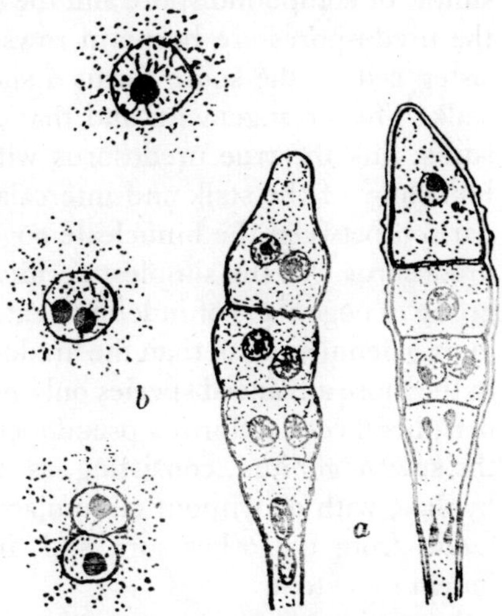

Fig. 269. *Phragmidium violaceum* Wint.; *a*, teleutospores, × 1080; *b*, fusion of nuclei in teleutospore cell, × 1520; after Blackman.

paraphyses, and, like them, they arise as branches from basal cells. The branch usually divides to cut off a stalk cell below, and the simple or compound teleutospore (fig. 268) may be derived from the terminal cell, or the latter may function as a buffer[1] and the teleutospore be formed from the penultimate cell. Like the segments of the mycelium on which it is borne, the teleutospore cell is binucleate (fig. 269). When the wall is fully thickened the two nuclei fuse, and the cell passes into a resting state. On the renewal of its development two nuclear divisions occur, the basidiospores are formed and the haploid phase initiated. In *Puccinia Arenariae*,[2] however,

[1] Dodge, 1918 i, ii.

[2] Lindfors, 1924.

the basidium, after the first mitosis, divides into two uninucleate cells; each nucleus divides again and each cell gives rise to one binucleate spore, so that the synkaryon begins in the basidiospore.

A striking similarity exists in the arrangement of the three kinds of sporogenous cells. The uredo- and teleutosori are clearly comparable; both are of indefinite extent, with or without a border of paraphyses, and both consist of plates of basal cells from which the spore mother cells arise as branches, dividing to produce the simple or compound spore and the stalk cell. Sometimes, however, the uredospores are borne in rows, one below the other, and the sister cell of the spore forms a short intercalary cell instead of a stalk. This arrangement, and that of the so-called primary uredosorus, link the true uredosorus with the aecidium, indicating the homology of the stalk and intercalary cells and suggesting a comparison between the binucleate oogonium and the basal cell of the uredosorus. In the simplest aecidia, those of the caeoma type, the group of oogonia is of indefinite extent. The aecidium, in fact, is no more a definite organ than the uredo- or teleutosorus, and appears so in the more advanced species only because of the modification of its peripheral cells to form a pseudoperidium. In its general structure the spermogonium, consisting, as it does, of a series of spermatial hyphae, with or without circumjacent paraphyses, is not very different from the other sori, and, in its simplest form, is also of indefinite extent.

In many rusts one or more spore forms are omitted; the substitution of a primary uredosorus for the ordinary aecidium has already been described. Such species, and those in which the aecidium alone is lacking, are distinguished by the prefix **brachy-**. Thus, if it were desired to indicate that a species of *Puccinia* lacked an aecidium, the term *Brachypuccinia* would be employed.

Hemi- indicates the presence of uredo- and teleutospores without aecidia or spermogonia.

The suffix **-opsis** is used for species with aecidia and teleutosori; they lack uredosori, but a few uredospores are sometimes found in the teleutosorus.

Micro- and **lepto-** species have teleutospores with a few uredospores and sometimes spermogonia. The teleutospores germinate in the micro- species only after a period of rest; in the lepto- species as soon as they reach maturity.

Species with the full complement of spores are distinguished by the prefix **eu-**.

In -opsis species and in brachy- species with primary uredosori the alternation of generations is the same as in eu- species, for the peculiarity of the second is that the aecidiospores are produced on stalks instead of in chains, while in the first only the uredospores, the accessory spores of the sporophyte, are lacking. In micro- and lepto- species the basidiospore germinates to produce, as in eu- forms, a mycelium of uninucleate cells on which spermogonia are sometimes borne. This mycelium also bears the teleutosori. At the base of the young teleutosorus, or below the uredosorus in hemi- species (fig. 262), cell fusions and nuclear migrations take place[1] like those below the aecidium, and, in self-sterile strains, this is similarly preceded by the association of mycelia of opposite complement or by the addition of compatible spermatia. Binucleate cells having been formed, the teleutospores, in which the nuclei fuse, arise in the usual way.

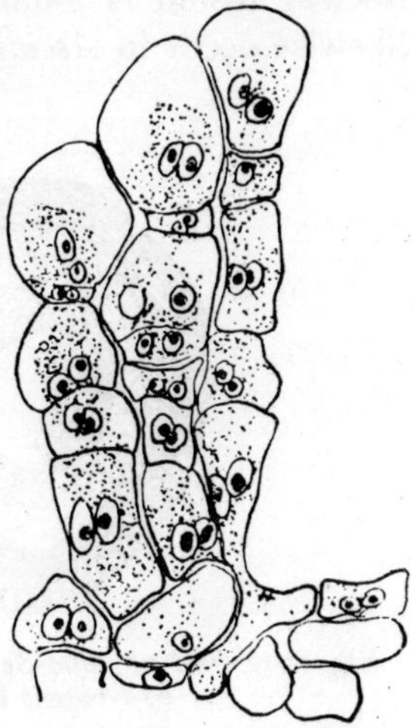

Fig. 270. *Endophyllum Sempervivi* Lév.; fertile cells and aecidiospores; after Hoffmann.

A diplophase of still briefer duration[2] is found in *Endophyllum* and *Kunkelia*. In both genera binucleate aecidiospores (fig. 270) are formed in the usual way, but their nuclei fuse (fig. 271) as soon as maturity is reached, and the aecidiospore develops like a teleutospore cell, with the protrusion of its contents as a promycelium (fig. 272), and the formation of four basidiospores. A clue to the origin of this condition may possibly be found in *Kunkelia nitens*,[3] the caeomata of which are morphologically indistinguishable from those of *Gymnoconia interstitialis*, a form with a normal life cycle. The two species differ markedly in their distribution, in America *Gymnoconia* occurring to the north and east of *Kunkelia*, with only slight overlap. It has been suggested that the warmer climate in which *Kunkelia* flourishes renders the thick-walled teleutospore

[1] Christman, 1907 ii; Werth and Ludwigs, 1912; Lamb, 1934 i, ii.
[2] Barclay, 1891; Hoffmann, A. W. H., 1912. [3] Kunkel, 1914; Moreau, 1920.

unnecessary.[1] Basidiospore formation without fusion or meiosis has been recorded for *K. nitens*.[2]

In *Endophyllum Euphorbiae*[3] on *Euphorbia sylvatica* the diplophase is wholly omitted; oogonia and aecidiospores are uninucleate throughout their development, the aecidiospore forms a promycelium of three or four cells, and neither nuclear association nor nuclear fusion is found at any stage. Uninucleate aecidiospores likewise occur in *Aecidium leucospermum*.[4]

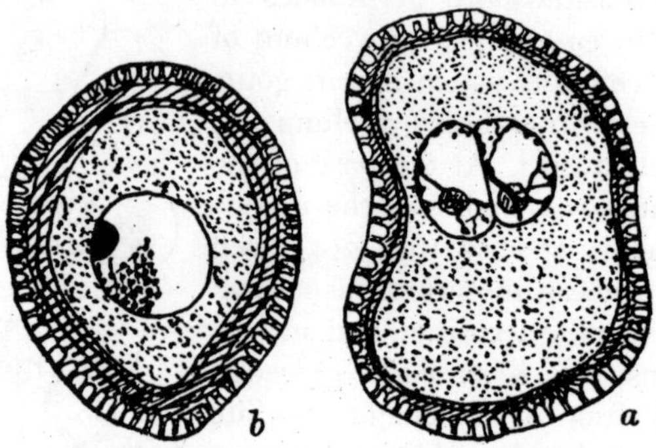

Fig. 271. *Endophyllum Sempervivi* Lév.; *a*, nuclear fusion in aecidiospore; *b*, synapsis in fusion nucleus; after Hoffmann.

In many rusts the gametophytic and sporophytic mycelia develop on different hosts. It is not surprising that the spore forms of such **heteroecious** species were recognised and described some time before they were identified as stages in the life of a single fungus. The final proof of the relationship of the aecidia and spermogonia on the one hand, and the uredo- and teleutosori on the other, was given by de Bary in 1865 for the wheat rust, *Puccinia Graminis*. In this plant the haplophase occurs on the leaves of the barberry, *Berberis vulgaris*, and the diplophase on wheat, rye, oats, and other grasses.

Long before this relationship was demonstrated, and even before the fungal origin of the disease was known, farmers had begun to suspect some harmful connection between barberry bushes and their crop, and had observed that dark areas of blackened and injured

[1] Arthur, 1917; Olive and Whetzel, 1917; Dodge, 1923.
[2] Dodge and Gaiser, 1926. [3] Moreau, 1920.
[4] Kursanov, 1917.

wheat were apt to occur in the neighbourhood of such plants. In the State of Massachusetts an Act was passed requiring the extirpation of all barberries before a given date in 1760, and writers of the period called attention to the belief among husbandmen that the barberry had "a mysterious power" of blighting the wheat near which it grew. It was not till 1797 that the wheat disease was identified as a fungus, and the first artificial infections were made in 1816.

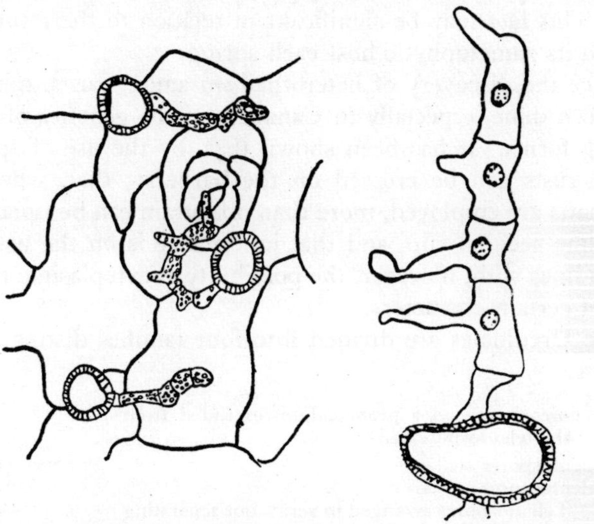

Fig. 272. *Endophyllum Sempervivi* Lév.; aecidiospores giving rise to promycelia; after Hoffmann.

Of the origin of the rusts very little is known; their ancestry has been postulated among the micro-forms,[1] among heterothallic, heteroecious eu- forms[2] and among eu- forms which are autoecious.[3] On the last-named hypothesis it has been suggested that heteroecism arose in relation to hosts with a short vegetative period. There would be then hardly time for the production of the full complement of spores, and the fungus might either shorten its life history, continuing in the autoecious condition as a micro- or other abbreviated form, or some of the spores might become adapted to life on a new host. Aecidiospores are produced relatively early in the

[1] Dietel, 1903; Christman, 1907 ii. [2] Orton, 1927; Jackson, 1931.
[3] Tranzschel, 1904; Olive, 1911; Grove, 1913 i.

season, they fall on hundreds of leaves besides those of their host, and the germ tubes in their case enter through stomata. If then an aecidiospore germinated and penetrated a satisfactory new host, a mycelium might develop, and further adaptations might fix the heteroecious habit. Gramineae and other hosts with refractory cuticles are easily infected by germ tubes from the aecidiospores or uredospores which enter through the stomata, but not by those of the basidiospores which usually penetrate the walls of the epidermal cells. This fact may be significant in relation to the return of the rust to its gametophytic host each spring.

Since the discovery of heterothallism among rusts, much work has been done, especially in Canada,[1] on the genetics of crop infecting forms. It has been shown that, by the use of spermatia, wheat rusts can be crossed on the barberry, that, when mixed spermatia are employed, more than one strain can be isolated from the same aecidial cup, and that inheritance is on the usual mendelian lines with, however, the possibility of cytoplasmic transmission of certain characters.

The Uredinales are divided into four families distinguished as follows:

Basidiospores borne on a promycelium extruded from the teleutospore cell
- Teleutospores stalked — PUCCINIACEAE
- Teleutospores sessile
  - Teleutospores arranged in series but separating later — CRONARTIACEAE
  - Teleutospores united in a flat layer under the epidermis — MELAMPSORACEAE

Basidiospores budded out from the teleutospore cell without a promycelium — COLEOSPORIACEAE

PUCCINIACEAE

The teleutospores of the Pucciniaceae are provided with a stalk which is often well developed, but is in some species short, or becomes detached at an early age. The teleutospores are one-celled in *Uromyces* (fig. 246 *d*) and *Hemileia*, two-celled in *Puccinia* and *Gymnosporangium* (fig. 249); they are made up of three cells in *Triphragmium* (fig. 246 *b*), and in *Phragmidium* (fig. 246 *a*) of three or more cells. *Gymnosporangium* is further characterised by the

[1] Johnson, T., and Newton, 1933; Johnson, T., Newton and Brown, 1934; Newton, M., and Brown, 1934; Newton, M., and Johnson, 1936.

long pedicels of its teleutospores and by the fact that they are embedded in a gelatinous mass; the uredospores are solitary. The aecidiospores of *Phragmidium* and *Triphragmium* are produced in caeomata, in other genera they are formed in aecidia, with pseudoperidia, which, in *Gymnosporangium*, are commonly elongated to form flask-shaped or cylindrical roestelia.

The family includes some of the most highly developed of the Uredinales, but it includes also genera with caeomata. It may perhaps be linked with the Coleosporiaceae through *Zaghouania*, where the development of the teleutospore is of an intermediate type.

Cronartiaceae

In the Cronartiaceae the teleutospores are unicellular and sessile, but so closely packed that they simulate multicellular spores. In *Chrysomyxa* they form wavy crusts, and in *Cronartium*[1] (fig. 247) a cylindrical body. A pseudoperidium is present around the aecidiospores. The genus *Endophyllum* is sometimes placed here, sometimes a separate family is created for it. It differs from the rest of the Cronartiaceae and from most of the rusts in the fact that its basidia originate as aecidiospores. This character is probably reduced rather than primitive.

Melampsoraceae

The teleutospores of the Melampsoraceae are sessile, loose in the tissue of the host in *Uredinopsis*, a genus infecting ferns. In other members of the family they are grouped in a flat layer under the epidermis. In *Melampsora* and its immediate allies they are unicellular, in *Calyptospora* and *Pucciniastrum* they are divided by longitudinal walls into two or four cells.

The aecidiospores may be surrounded by a pseudoperidium or developed in a caeoma; in some cases a pseudoperidium is present around the uredosorus also.

Coleosporiaceae

The outstanding character of the Coleosporiaceae is the method of development of the unicellular teleutospore, the contents of which undergo septation directly (fig. 246 *c*), and, in *Coleosporium*

[1] Colley, 1918.

and *Ochropsora*, without protrusion as a promycelium; in *Zaghouania*[1] the contents of the teleutospore divide within the teleutospore wall to form four cells, but emerge before the basidiospores appear.

The aecidia are cup-shaped in *Ochropsora*, but in *Coleosporium* and *Zaghouania* they are of the peridermium type with a cylindrical, more or less inflated pseudoperidium. This elaborate form of sorus makes it impossible to regard the Coleosporiaceae as primitive, though they may perhaps have branched off early from the line leading to the commoner rusts.

AURICULARIALES

The Auriculariales are an assemblage of some fifty species occurring as saprophytes or hemi-saprophytes on wood, especially in the tropics, where a considerable number probably still await discovery. They resemble the Uredinales in their transversely septate basidia (fig. 273 *a*), but differ from them in their saprophytic habit and in the formation of a gelatinous fruit body on the surface of which the spores are borne. In many species conidia as well as basidiospores are produced, and in some the basidiospores germinate to give rise to conidia, either directly or from the cells of a septate germ tube. There is no trace of a sexual apparatus, or any evidence of a sexual process except the fusion of two nuclei in the young basidium. In *Hirneola*[3] the mycelium shows numerous clamp connections which are especially frequent towards the sterile surface of the sporophore, but the nuclei are small and the details of their behaviour have not so far been elucidated.

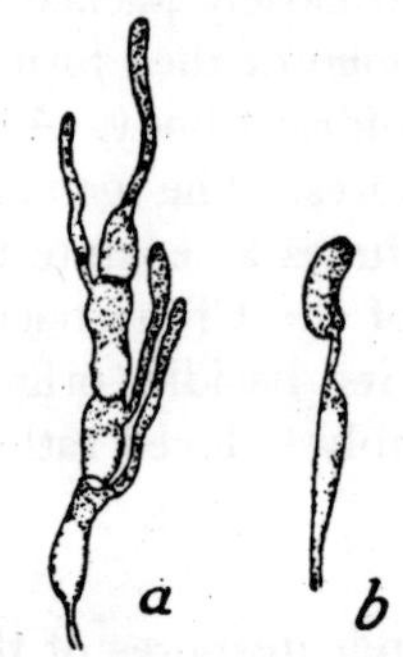

Fig. 273. *Auricularia sambucina* Mart.; *a*, transversely septate basidium with four sterigmata, ×300; *b*, a single, long sterigma with basidiospore attached, ×420; after Brefeld.[2]

In most of the Auriculariales the hymenium is fully exposed during development, but, in the genus *Pilacre*, a more or less definite peridium covers the young fructification and withers away

[1] Dumée and Maire, 1902.
[2] Brefeld, 1888.
[3] Green, 1925.

as maturity is reached. This genus and the closely related *Pilacrella* are accordingly placed in a separate family, the Pilacreaceae, while the remaining genera form the Auriculariaceae.

The best known member of the Auriculariaceae in Great Britain is *Hirneola Auricula-Judae*, the Jew's ear, which appears in winter on the branches of the elder, forming brown, gelatinous sporophores with some resemblance to a human ear.

Transversely septate basidia are also found in the genus *Septobasidium*[1] which is widely distributed in the tropics and the warmer regions of the temperate zone. It forms resupinate patches on the bark of trees and exists in a symbiotic relation with scale insects sheltering beneath the stroma. Some of the insects are parasitised by hyphae which grow in through or near one of the external apertures, mouth, anus or genital pore, and give rise to coiled haustoria; others survive and reproduce. Together the fungus and insects cause cracking of the bark and serious damage to the host. The mycelium is without clamp connections. Often the hymenium is raised above the stroma by upright pillars of hyphae. It bears thick-walled cells, known as **probasidia**, which correspond to the teleutospore cells of the rusts. Like them the probasidia either elongate, or empty their contents into cylindrical, thin-walled, promycelial outgrowths. Transverse septation takes place and from each of the four resultant cells a single basidiospore is cut off. The basidiospore may divide into two or more cells or may bud off conidia on which new infections depend.

TREMELLALES

The Tremellales[2] include about one hundred species; their distribution is world wide, but they are massed chiefly in the tropics. Like the Auriculariales they are mainly saprophytes with for the most part a gelatinous sporophore of indefinite outline such as occurs in *Tremella*; in *Tremellodon* a stipitate fructification is produced with basidia on downwardly directed teeth which resemble those of the Hydnaceae.

In the genus *Sirobasidium* the basidia are obliquely septate and arranged in chains; in other members of the alliance they are borne

[1] Couch, 1933, 1935 ii.

[2] Burt, 1901; Maire, 1902; Wager, 1914 ii.

singly and are divided by two longitudinal septa formed successively and at right angles one to another (fig. 274). In *Hyaloria*, as in *Pilacre* among the Auriculariales, the hymenium is protected by a sheath of sterile hyphae.

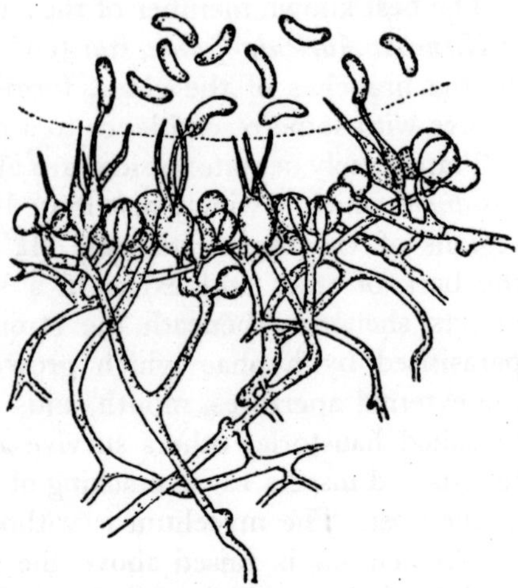

Fig. 274. *Exidia recisa* Fr.; group of longitudinally septate basidia; after Tulasne.[1]

As in the Auriculariales, no trace of a sexual apparatus has been found and the nuclei have so far proved too small for detailed study.

The South American genera, *Sirobasidium* and *Hyaloria*, are usually separated as distinct families, the Sirobasidiaceae and Hyaloriaceae; all other members of the alliance are included in the family Tremellaceae.

AUTOBASIDIOMYCETES

The Autobasidiomycetes include between 12,000 and 13,000 species; they are cosmopolitan in distribution, often with large and conspicuous fructifications; their basidia never become septate. Though the sporophores show an immense variety of form, the essential facts of the life history are similar throughout the group; there is no trace of a sexual apparatus, but the young basidium contains two nuclei (fig. 275 *a*, *b*), and is commonly borne on a mycelium of binucleate cells,[2] which may arise, as already described, by the association of two haploid mycelia. The nuclei of the basidium fuse; two successive divisions follow, constituting a meiotic phase (fig. 276 *a*), and the four haploid nuclei lie in a group at the top of the basidium. In the mushroom, *Psalliota campestris*,[3] their distance from the apex is soon increased, partly by the elongation of the basidium, partly by their own movement, and the sterig-

[1] Tulasne, 1853. [2] Cf. p. 287.
[3] Colson, 1935.

mata appear, first as bulges, then as horn-like projections at the tops of which the spores can soon be recognised (fig. 276 *b*). The spore

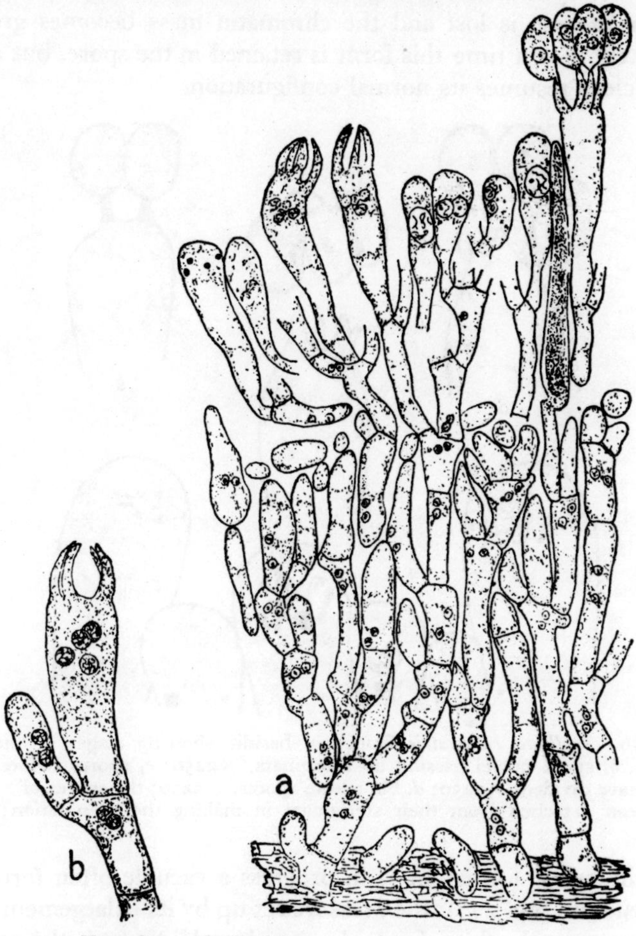

Fig. 275. *Hypochnus subtilis* Schroet.; *a*, hymenium of binucleate cells bearing basidia; *b*, young and older basidium; after Harper.[1]

is very small at first, its wall is delicate and it is densely filled with cytoplasm. It enlarges, the wall becomes thickened except at the point of junction with the sterigma. Meanwhile the nuclei move towards the sterigmata, the chromatin in each being arranged along

[1] Harper, 1902.

two sides of a triangle. One nucleus begins to pass into each sterigma, the apex of the triangle directed forwards (fig. 276 *b*), the nuclear vacuole is lost and the chromatin mass becomes greatly elongated. For a time this form is retained in the spore, but soon the nucleus resumes its normal configuration.

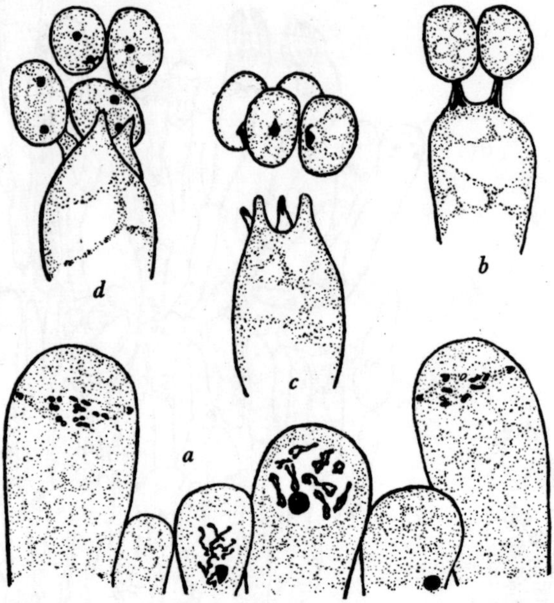

Fig. 276. *Psalliota campestris* Quél.; *a*, basidia showing stages of meiosis, × 3000; *b*, spore nuclei passing up sterigmata, × 2250; *c*, spores before their nuclei have divided, × 2250; *d*, binucleate spores, × 2250; the spores of *c* and *d* have been detached from their sterigmata in making the preparation; after Colson.

While spore formation is in progress a vacuole often forms in the basidium (fig. 277); the pressure set up by its enlargement may cause or assist the flow of cytoplasm and nuclei towards the spores.

The number of spores is almost always four, but there is one spore in *Pistillaria maculaecola*, two in varieties of the mushroom, in the genera *Calocera* and *Dacryomyces*, in *Coprinus bisporus*[1] and in several other species,[2] three in *Coprinus narcoticus*, five in *Cantharellus Friesii*, six in species of *Cantharellus*, *Hypochnus* and *Exobasidium*, and eight in several of the Thelephoraceae, where, also, basidia with seven spores sometimes occur. The ripe

[1] Lewis, C. E., 1906; Buller, 1922 ii. [2] Bauch, 1925, 1927; Buhr, 1932.

spore is aseptate, though septa may be formed, as in *Dacryomyces*, before a germ tube is produced. In many species the spores are coloured, and the epispore in some is sculptured. Conidia as well as basidiospores may be present, and chlamydospores are sometimes formed.

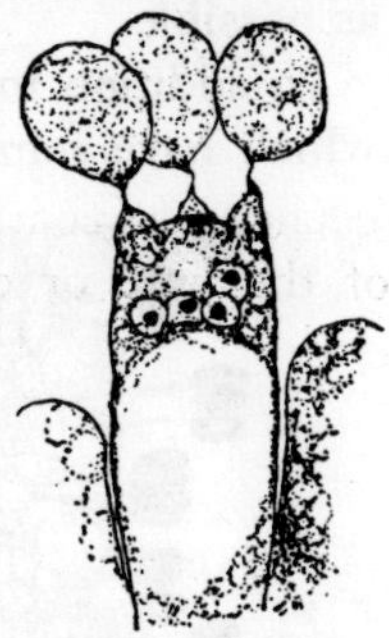

Fig. 277. *Stropharia semiglobata* Quél.; basidium during the formation of the spores, only three of the four spores are shown, × 500.

The hymenium, or fertile layer, includes basidia of all ages, from small, binucleate cells, the growth of which has only just begun, to mature, club-shaped specimens with sterigmata and spores projecting above the level of the hymenium, and older individuals from which the spores have been shed and which are in process of collapse. A young hymenium usually shows a preponderance of uninucleate basidia (fig. 278), indicating that this stage endures for some time. In *Coprinus*[1] the mature basidia may be of two (fig. 279), three, or even four lengths, thus allowing a very large number simultaneously to reach maturity in a limited space; this arrangement is doubtless

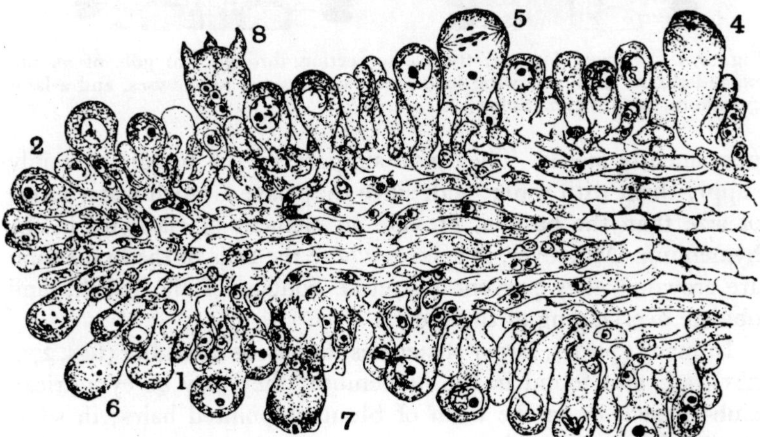

Fig. 278. *Panaeolus campanulatus* (L.) Quél.; longitudinal section through part of a young gill, showing young, uninucleate and binucleate basidia, basidia after the nuclei have fused, and basidia with sterigmata and four nuclei; the numbers indicate successive stages in development; somewhat diagrammatic, × 500.

[1] Buller, 1924.

correlated with the successive development in that genus of basidia in different regions of the gill. The spores on the basidia are always symmetrically arranged, and are thus as far distant one from another as possible.

In addition to the basidia, the hymenium includes paraphyses, which have sometimes the appearance of young basidia, but are

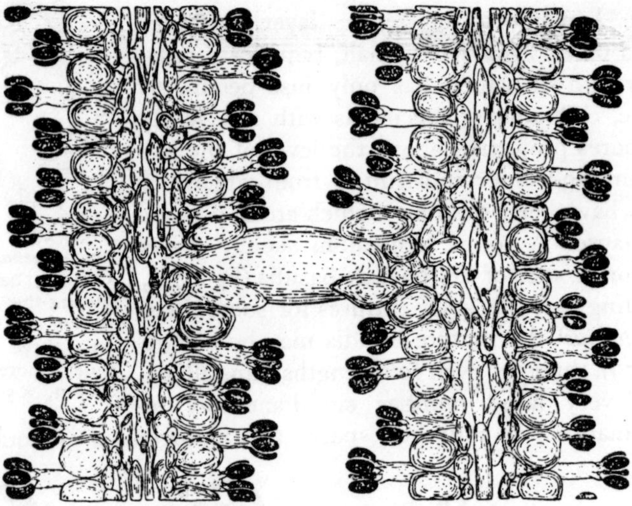

Fig. 279. *Coprinus lagopus*; transverse section through two gills of an unexpanded fruit body, showing long and short basidia, paraphyses, and a large cystidium, ×232; after Buller.[1]

permanently sterile. They give support to the basidia, possibly supply them with water, and, since they have a much wider range of size than the basidia, provide an element of elasticity in the hymenium. Moreover, in species where the cells of the hymenium are crowded, they prevent too close contact of the basidia and obviate overlapping of the spores.

In many species, larger structures, known as **cystidia**[2] (figs. 279, 281), are also found in the hymenium; they may be cylindrical, club-shaped, or in the form of blunt or pointed hairs; in some species they produce characteristic secretions; in others, where they project considerably from the hymenium, they serve to keep adjacent hymenia apart.

[1] Buller, 1924. [2] Knoll, 1912; Buller, 1924.

All these elements are the terminal cells of richly branched hyphae which ramify in and form part of the tissue of the sporophore; their arrangement becomes more compact as they approach the hymenium, and thus the subhymenial layer is formed. Where the hymenium is spread over definite teeth or gills, or lines definite pores or chambers, as in the Hydnaceae, the Agaricaceae, the Polyporaceae, and many of the Gasteromycetales, the central tissue of the gill or tooth, or limiting tissue of the pore or chamber, is known as the **trama**, and consists of parallel or interwoven hyphae, sending out lateral branches to form the subhymenial layer and the hymenium. In *Psalliota campestris*[1] and other species the pileus and stem are composed of multinucleate cells, binucleate cells appearing first in the lamella. In *Boletus granulatus*[2] the cells of the stipe are multinucleate, whereas those of the pileus and ring, as well as of the trama and its products, contain two nuclei. In most of the Autobasidiomycetes investigated, the binucleate condition supervenes early, before the primordium of the fructification is laid down. There is evidence[3] that in nature the sporophores of species of *Coprinus* and *Panaeolus* are commonly derived from the product of a single spore, or, in self-sterile forms, of two. In other species it is possible that several mycelia may contribute to the formation of a single sporophore, which is not then an individual in the usual sense. The sporophore of the Autobasidiomycetes differs from that of the Ascomycetes in that the fertile hyphae of the latter are sporophytic and the vegetative hyphae part of the gametophyte, whereas the whole tissue of the basidiomycetous fructification is normally of the same generation.

The efficiency of the apparatus may be gauged by the extraordinary quantity of spores set free. In round numbers a single fructification of *Lycoperdon Bovista*, the giant puffball, produces 7,000,000,000,000, that of *Fomes applanatus*, which continues to liberate 30,000,000,000 a day for six months, 5,500,000,000,000, that of the mushroom, with a productive period of six days, 16,000,000,000, and that of *Coprinus sterquilinus*, which sheds spores during only eight hours, as many as 100,000,000.[4] Nevertheless the number of fungi of any given species remains approximately constant, which means, not merely that only one spore

[1] Hirmer, 1920.
[2] Levine, 1913.
[3] Brefeld, 1877; Newton, D. E., 1926 i.
[4] Buller, 1922 ii.

from each sporophore is on the average successful in establishing itself and giving rise to a new specimen which reaches reproductive maturity, but that, where the mycelium gives rise to several fructifications, only one among all the spores from all the sporophores derived from the same plant attains success. The fact that many mycelia are perennial, and that those of fairy rings may endure even for hundreds of years, indicates a still smaller proportion of successful spores. On the other hand, a sporophore, if derived from several mycelia, may represent the effective development of more than one spore, and, in species where several strains exist, a considerable surplus of spores may be needed to ensure the association of appropriate complements. With such vast numbers of spores available, it is not surprising that, under favourable conditions, large quantities of saprophytes, or epidemics of parasites, sometimes appear.

The Autobasidiomycetes include two alliances which may be distinguished as follows:

Hymenium exposed at maturity, and often throughout development	HYMENOMYCETALES
Hymenium enclosed until after the spores are ripe	GASTEROMYCETALES

HYMENOMYCETALES

The Hymenomycetales constitute by far the largest subdivision of the Basidiomycetes, and include between 10,000 and 11,000 species, which may be parasites or saprophytes. Interest centres around the evolution of the sporophore, which, in the simpler species, is a thin, smooth layer, spread over the surface of twigs or branches, with the hymenium fully exposed; effective protection of the basidia is attained on the gills of toadstools and in the pores of bracket-fungi. Associated with this development is the efficient dispersal of the spores by wind, the orientation in response to light and gravity of the stalked or laterally attached sporophore, and the method of ejection of the basidiospore. The spore, if borne on a pendulous gill or similar outgrowth of the sporophore, is shot horizontally forward to a distance of about 0·1 mm., which brings it to the middle of the space between the gills or of the lumen of the pore; here, the momentum being exhausted, it executes a sharp turn through a right angle and falls vertically downwards under the influence of gravity. The fall is rapid at first, but, as soon as the spore leaves the shelter of the sporophore, it dries very quickly and, in dry air, the rate of fall is exceedingly slow, between 5 and 0·5 mm. a second,[1] according to the dimensions of the spore. The rate of fall

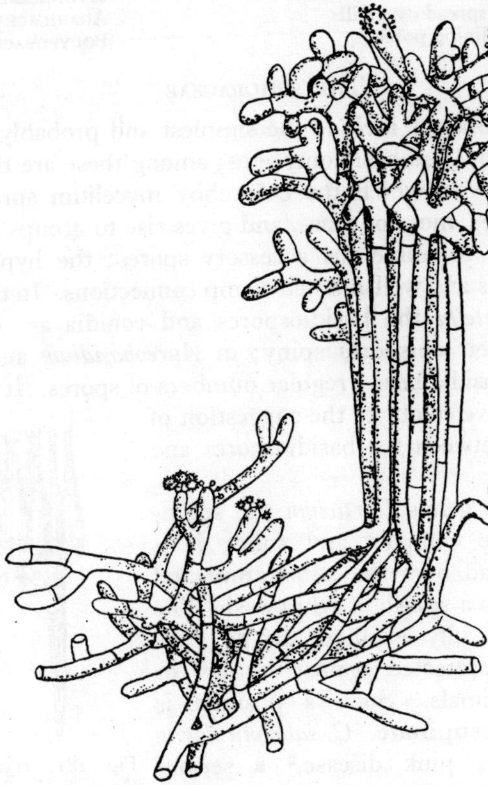

Fig. 280. *Tomentella flava* Bref.; group of small basidia and longer hyphae bearing conidia, × 202; after Brefeld.[2]

of thistledown is from thirty to three hundred times as rapid as that of a basidiospore.

A few of the Hymenomycetales, including members of the parasitic genus *Nyctalis*, multiply by means of chlamydospores as well as by basidiospores, and in *Dacryomyces*, *Tomentella* (fig. 280) and several others, conidia are produced

[1] Buller, 1922 ii. [2] Brefeld, 1888.

The alliance may be subdivided as follows:

Hymenium unilateral, spread over a smooth or corrugated surface	THELEPHORACEAE
Hymenium covering on all sides the smooth surface of branched or simple clubs	CLAVARIACEAE
Hymenium spread over spines	HYDNACEAE
Hymenium spread over gills	AGARICACEAE
Hymenium lining pores	POLYPORACEAE

THELEPHORACEAE

The Thelephoraceae include the simplest and probably the most primitive of the Autobasidiomycetes; among these are the species of *Hypochnus*,[1] in which the cobwebby mycelium spreads over rotten branches, moss or grass, and gives rise to groups of basidia (fig. 275) and to numerous accessory spores; the hyphae show binucleate cells and well-marked clamp connections. In the closely related *Tomentella* the basidiospores and conidia are brown or violet, and their epispores spiny; in *Aureobasidium* and *Pachysterigma* the basidia bear irregular numbers of spores. It was such facts which gave colour to the suggestion of a homology between the basidiospores and conidia.

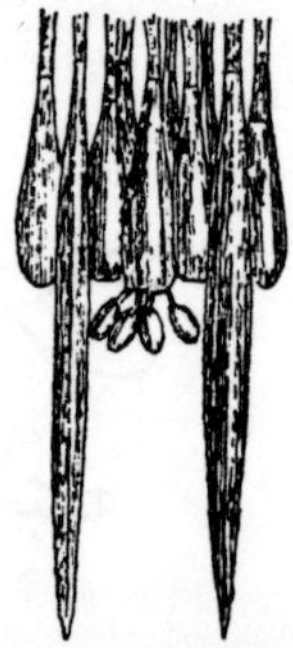

Fig. 281. *Hymenochaete Cacao* Berk.; basidia and cystidia; after Hennings.[3]

In the large genus *Corticium* the sporophore is more compact and solid; it is smoothly spread over the substratum and often resembles a splash of whitewash. The hymenium is fully exposed, facing outwards and unprotected against weather or marauding animals. Such a position is described as **resupinate**. *C. salmonicolor* is responsible for pink disease,[2] a serious affection on rubber, tea, and other crops. *C. anceps*[4] infects bracken and *C. fuciforme*[5] turf. The closely related genera *Peniophora* and *Hymenochaete* differ from *Corticium* mainly in the presence of cystidia, which are brown and thick walled in *Hymenochaete* (fig. 281), while in *Peniophora* the walls are thin and the tips encrusted with particles of lime; both are attractive objects under the microscope. In *Hymenochaete*, re-

[1] Harper, 1902; Kniep, 1915, 1917.
[2] Brooks and Sharples, 1923.
[3] Engler and Prantl, 1900.
[4] Gregor, 1935.
[5] Bennett, 1935.

supination is not always complete, part of the sporophore being free from the substratum and bent over (fig. 282 *b*, *c*), protecting the hymenium which is then downwardly directed. *H. rubiginosa*,[1] a common British species, is a cause of timber rot.

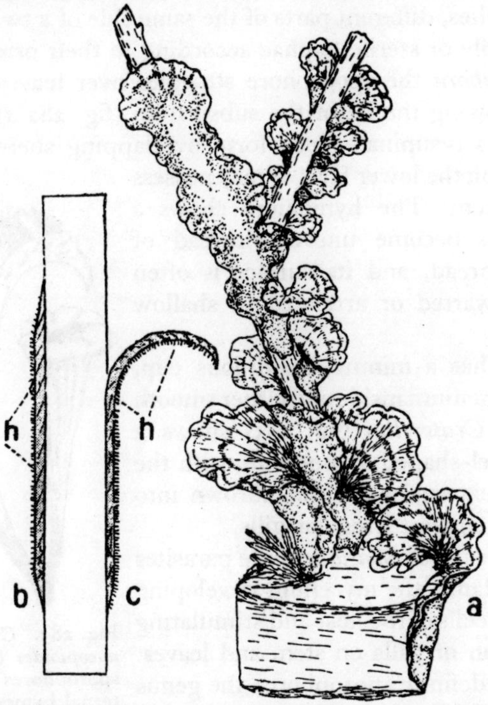

Fig. 282. *a*, *Thelephora tabacina* (Sow.) Lév.; partly resupinate sporophores; after Sowerby.[2] *b*, Diagrammatic representation of a longitudinal section through a resupinate sporophore. *c*, The same, part of the sporophore is free from the substratum and bent over, protecting the hymenium, *h*.

In *Stereum*,[3] another large genus, most of the species are laterally attached, with the sporophore partly or mainly free from the substratum, and the hymenium downwardly directed; the most elaborate forms possess a central stalk and a more or less funnel-shaped cap with the hymenium on its outer face. *S. purpureum* causes silver leaf disease[4] on apple and plum trees, the leaves

[1] Brown, H. P., 1915. [2] Sowerby, 1797.
[3] Ward, 1897; Mayo, 1925.
[4] Brooks, 1911–13; Brooks and Bailey, 1919; Brooks and Moore, 1926; Brooks and Storey, 1923; Brooks and Brenchley, 1929.

become silvered, the branches die, and on them the basidial fructifications develop. Even a non-living extract induces symptoms of disease. In *Sparassis*[1] the sporophore is richly branched, resembling a cauliflower or a bath sponge, and may reach nine inches in diameter; the hymenium is limited to the physiologically lower sides of the branches, different parts of the same side of a twisted branch bearing fertile or sterile hyphae according to their orientation.

In *Thelephora* the sporophore straggles over leaves and twigs, closely following the lie of the substratum (fig. 282 *a*); it may be more or less resupinate, may form overlapping shelves with the hymenium on the lower face, or may possess a central stem. The hymenium shows a tendency to become uneven, instead of smoothly spread, and its surface is often irregularly warted or arranged in shallow ridges.

Cyphella has a minute, pendulous cup, with the hymenium inside and either smooth or veined; *Craterellus* (fig. 283) shows a larger, funnel-shaped sporophore, with the hymenium external and often thrown into folds which suggest irregular gills.

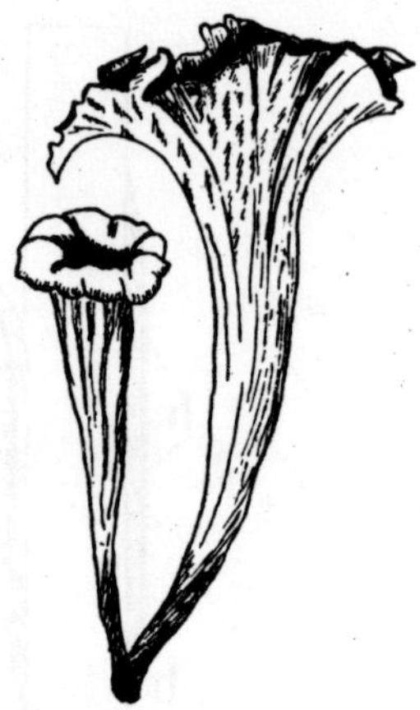

Fig. 283. *Craterellus cornucopioides* (L.) Fr.; two sporophores bearing external hymenium.

The species of *Exobasidium*[2] are parasites on higher plants, the mycelium developing between the cells of the host and stimulating the formation of galls on stem and leaves. There is no definite sporophore; the genus may be primitive or reduced, and is probably not closely related to the remaining Thelephoraceae. *E. Vaccinii*, on *Vaccinium Myrtillus* and other Ericaceae, is not uncommon, and may be recognised by the characteristic red or purple swellings induced on subaerial parts of the host.

Clavariaceae

The Clavariaceae are characterised by the erect sporophore, which may be simple or branched, and over all sides of which the hymenium is smoothly spread. In *Clavaria* (fig. 284), the so-called fairy club, the sporophore is club-shaped or shows coral-like

[1] Cotton, 1911. [2] Burt, 1915.

branching; it is usually white, buff or yellow and is from less than one inch to twelve inches high. Most species of *Pterula* have numerous, filiform branches, forming a brush-like tuft; *Typhula* consists of usually unbranched threads arising from sclerotia; and in *Pistillaria* the sporophore is a minute club, generally occurring on dead stems or leaves. *Typhula erythropus*[1] and *T. Trifolii*[2] are heterothallic.

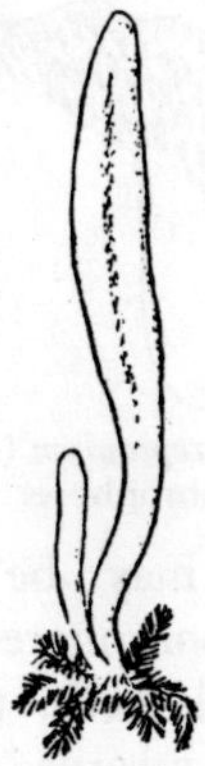

Fig. 284. *Clavaria cardinalis* Boud. & Pat.; after Boudier.[3]

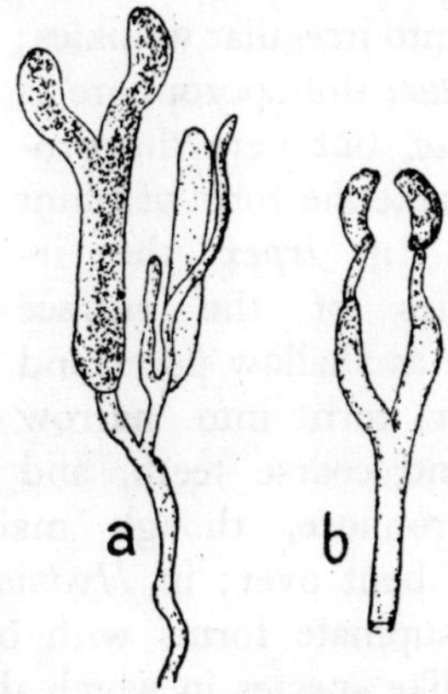

Fig. 285. *Dacryomyces chrysocomus* (Bull.) Tul., *a*, young basidium with sterigmata, ×400; *Dacryomyces ceribriformis* Bref., *b*, upper part of mature basidium with two spores, ×359; after Brefeld.[4]

Calocera resembles a small *Clavaria* in habit, but differs in the gelatinous consistency of its sporophore, becoming horny when dry, and in the specialised type of basidium (fig. 285), which has two very long sterigmata, about half as wide at the base as the basidium itself, and bears only two spores. *Dacryomyces*[5] forms small, subglobose, gelatinous masses; it is included here owing to the similarity of its basidium to that of *Calocera*; both genera occur on dead wood.

HYDNACEAE

The salient character of the Hydnaceae is the arrangement of the hymenium over warts or spines of sterile tissue; these excrescences may be very long and pointed, resembling stalactites,

[1] Lehfeldt, 1923. [2] Noble, 1937. [3] Boudier, 1905–10.
[4] Brefeld, 1888. [5] Wager, 1914 ii; Buller, 1922 i.

or may be rudimentary in character. There is a transition also from crust-like, resupinate forms to forms with a stem and pileus and the general appearance of a toadstool. In some cases the protuberances covered by the hymenium are united to form irregular plates.

Among the commoner genera, *Phlebia* has a soft and somewhat gelatinous sporophore, erect or lying more or less flat on the substratum, with the surface thrown into irregular wrinkles; in *Radulum* the sporophore is resupinate, but here the projections take the form of blunt fingers. In *Irpex* the irregularities of the surface originate as shallow pores and are later torn into narrow ridges and coarse teeth, and the sporophore, though mainly resupinate, may be partly free and bent over; in *Hydnum* the type of sporophore ranges from resupinate forms with outwardly directed teeth, through bracket-like species in which the teeth hang downwards, and are protected by the sterile tissue, to forms with a cap and central stalk (fig. 286); in all species of *Hydnum* the teeth are well-differentiated, awl-shaped, and distinct at the base.

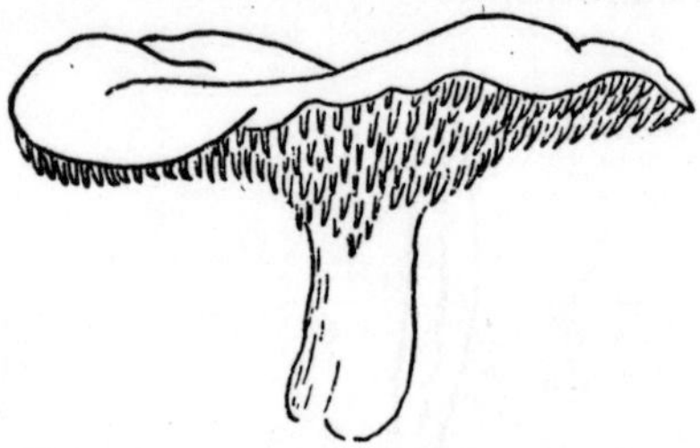

Fig. 286. *Hydnum repandum* (L.) Fr.; stipitate sporophore.

Agaricaceae

The Agaricaceae form a considerable assemblage of gill-bearing species most of which possess a cap and a central stem, though in some the stalk is lateral, and others are bracket-like in form; they are classified according to the colour of the spores (which may be white, pink, some shade of reddish brown, purple, or black), the shape of the gills, the shape of the cap, the method of protection of the young sporophore, and other details. From the evolutionary point of view they are all at much the same level of development; they may possibly have been derived from such forms as *Craterellus* or *Phlebia*, but do not appear to have given rise to any higher group, perhaps because they are well adapted to the environment in which they occur. It is a question whether some of the Polyporaceae or Gasteromycetales have reached a higher level of development, but they are not likely to have been derived from the Agaricaceae, though they doubtless had common ancestors.

The gills, or lamellae, are arranged on the under side of the cap, or pileus. In symmetrical species the stalk, or stipe, springs from the middle of the cap, and the gills radiate between the point of its insertion and the outer edge of the pileus; shorter gills, starting from the edge, reach only part of the way towards the centre, thus filling the gaps between the outer ends of those which form complete radii. A section of the gill (fig. 287) shows a central trama of

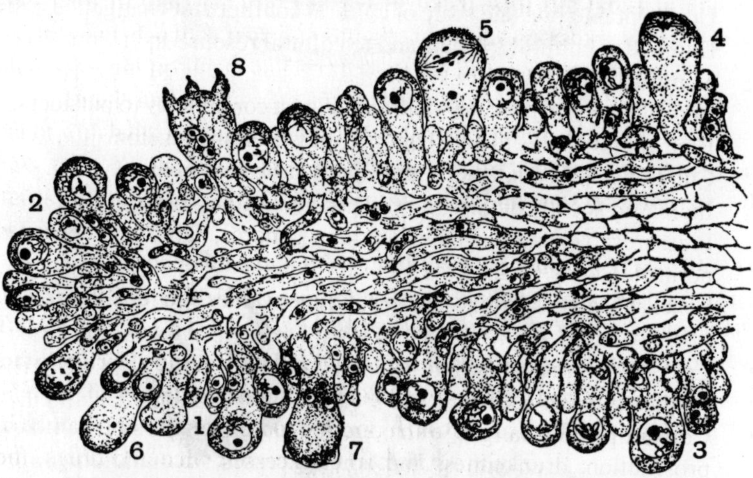

Fig. 287. *Panaeolus campanulatus* (L.) Quél.; longitudinal section through part of a young gill, showing young, uninucleate and binucleate basidia, basidia after the nuclei have fused, and basidia with sterigmata and four nuclei; the numbers indicate successive stages in development; somewhat diagrammatic, × 500.

parallel or interwoven hyphae, the ends of which bend outwards to form the subhymenium and ultimately to give rise to the basidia, paraphyses, and cystidia of the hymenial layer. In some species the developing gills are protected merely by the curvature of the edge of the pileus and its contact with the stipe; in others, as in the mushroom, a special weft of hyphae, the **velum partiale** or **partial veil**, connects the edge of the pileus with the stalk, and is torn apart when the pileus expands, leaving around the stipe a circular frill, the **annulus** or **ring** (fig. 288). Further protection is given in *Amanita* and a few other genera by the development of a special membrane, the **velum universale** or **volva**, which extends over the whole sporophore and within which development takes place. The volva is ruptured by the elongation of the stipe, torn fragments are carried up on the outer surface of the enlarging pileus, but the greater part remains about the base of the stem (fig. 288).

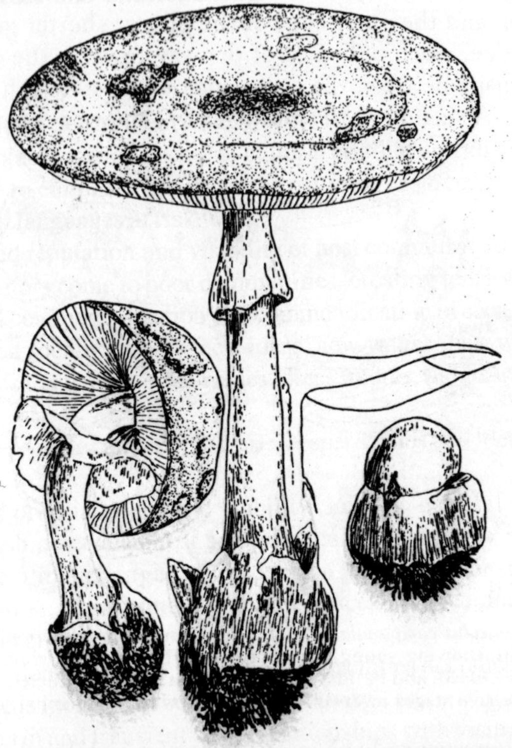

Fig. 288. *Amanita phalloides* Fr.; young and mature sporophores; after Cooke.[1]

The development[2] of the sporophore begins with the formation of a little knot of interwoven filaments; as growth proceeds the position of the fertile layer is early indicated by a ring of deeply staining hyphae; above this the tissue of the pileus grows rapidly, below, the growth of the region which will constitute the stipe ceases or is delayed; as a result an annular cavity is formed. This

[1] Cooke, 1881–3.

[2] Atkinson, 1906, 1914, 1916; Fischer, C. C. E., 1909; Sawyer, 1917; McDougall, 1919; Walker, 1919; Moss, 1923.

gill cavity is at first minute but rapidly enlarges as the pileus continues its growth, and into it push hyphae from the deeply staining layer, forming the gills; with the later expansion of the pileus these are horizontally extended. There is a good deal of variation in detail between different species, and comparatively few have as yet been fully studied; in *Amanita*, for example, the ring of gill-producing tissue is lacking and the gills arise from isolated primordia, while, in forms without a partial veil, the gill cavity may be open from the first to the exterior. *Psalliota*, *Armillaria*, and other genera which do not show a volva at maturity may be enclosed during development by a weft of hyphae corresponding to the primordium of the universal veil but disappearing at an early stage.

Among the many species in this family, *Psalliota campestris*, the mushroom, is probably the best known; the stem is central, the spores are purple, and there is a well-marked partial veil. This fungus is extensively cultivated for food either out of doors, or in caves, sheds or tunnels, the vast limestone caves beneath the suburbs of Paris being used for this purpose. There are many other edible forms among the Agaricaceae and their allies, but there are also some which are poisonous, and none should be eaten unless the species is definitely known. The most dangerous is *Amanita phalloides* (fig. 288), a white-spored, white or yellowish species with both a volva and a partial veil. It is probably responsible for ninety per cent of the deaths due to fungi; a very small quantity can be fatal, and, even when death does not occur, great pain and long illness are inevitable. The related *Amanita muscaria*, the fly agaric, characterised by the red pileus to which white fragments of the volva adhere, is also poisonous, and an extract of the dried cap produces intoxication.

Armillaria mellea, a white-spored form with a well-marked ring, and a conspicuous, brown, scurfy pileus and stalk, besides being a constituent of mycorrhiza,[1] is a harmful tree parasite. Spores germinate on dead stumps, and in addition to producing fresh sporophores, send out interwoven strands of hyphae, the rhizomorphs, which are easily recognisable from their resemblance to black bootlaces. If, under circumstances favourable to infection, they come in contact with the roots of living trees, they force their

[1] Cf. p. 25.

way in, spreading to form a white, felt-like mycelium just inside the bark; the cambium is infected and the xylem and phloem attacked by way of the medullary rays, thus bringing about the death of the tree and the subsequent decay of the wood. The young rhizomorphs are luminous, as are the hyphae of some other Basidio-

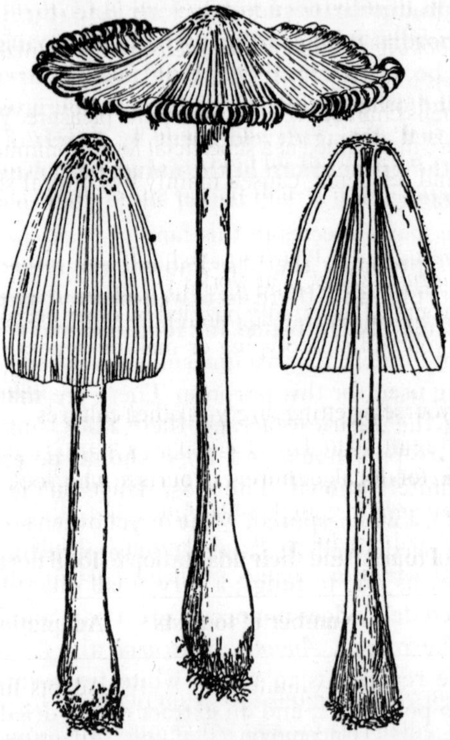

Fig. 289. *Coprinus aratus* Berk. & Br.; mature and nearly mature sporophores; after Cooke.[1]

mycetes, and pieces of wood impregnated with them were formerly used as a source of light in places, such as barns containing hay, into which a torch could not be carried.

The genus *Coprinus*, the ink cap (fig. 289), is one of the commonest forms with black spores; the spores are produced in succession,[2] first from basidia at the edge of the cap, later from those

[1] Cooke, 1881–3. [2] Buller, 1909 ii.

progressively nearer the stem. As the spores are shed the gills undergo deliquescence or autodigestion, disappearing in the form of inky drops. Since the ripe spores are always just above the region of deliquescence, and therefore at the edge of the gill, they fall free in this genus even if the gill is tilted. In the development of the basidiospores hyaline bodies have been described[1] connecting the nuclei in the basidium with the sterigmata.

The curious genus *Secotium*,[2] formerly referred to the Hymenogastraceae, should probably be placed in this family in the neighbourhood of *Psalliota*; the sporophore is surrounded by a thick volva or peridium which opens to disclose the much folded gills.

The species of *Nyctalis* are remarkable for being parasitic on other agarics, from the stalks and pilei of which they develop; they are disseminated not only by basidiospores, but by conidia produced on the surface of the cap.

Among the excentric species *Lentinus*[3] has a funnel-shaped sporophore; *L. lepideus*,[4] to the reactions of which reference has already been made, is a common cause of damage to railway sleepers and other woodwork. *Schizophyllum* has a lateral stem or none, and, in *S. commune*, shows a unique method of protection of the hymenium. Each gill is divided in the plane of its most extended surface; if desiccation sets in, the liberation of spores ceases, and the two halves curl one away from another till, when the sporophore is quite dry, segments of the larger gills completely cover the smaller ones. In this condition the fungus can retain its vitality for years. *Pleurotus* and *Lenzites* may be bracket-shaped or resupinate; in *Lenzites* the gills are sometimes united to form pores, and a connection with *Daedalea* may be surmised.

Polyporaceae

The Polyporaceae, as their name implies, are characterised by the presence of numerous pores, the cavities of which are lined by the hymenium. The pores may be merely shallow pits, as in the resupinate genera *Merulius* and *Poria*; they may be wavy and laterally elongated, sometimes almost resembling gills, as in species of *Daedalea*, or, as in *Boletus*, *Polyporus*, *Polystictus*, *Fistulina* and

[1] Vokes, 1931.
[2] Conard, 1915.
[3] Johnson, M. E. M., 1920.
[4] Buller, 1905, and cf. pp. 14, 35, 36.

Fomes, they may form long, narrow tubes, closely crowded together, so that the open ends, through which the spores escape, occupy the fertile surface of the sporophore.

In *Boletus* there is a central stem and a symmetrical pileus, usually large and very fleshy; the cap and pores are often bright coloured. In several species, the flesh turns blue when exposed to the air owing to the presence of a phenol known as boletol, with which is associated, as oxidising agent, its co-enzyme termed laccase, and suitable alkaline salts.[1] Laccase is a compound of manganese. In *Polyporus* a central stem is sometimes found, but most of the species are bracket-shaped, growing out of stumps or tree trunks, in the latter instance they are often destructive parasites, inducing heart rot; in several species the spores enter through wounds.[2] *Polystictus* differs from *Polyporus* in its thin flesh and velvety upper surface, and *Fomes* in the stratification of its tubes. This stratification is due to the fact that each spring a layer of sterile tissue closes the old tubes, and on it the tubes of the current year arise; *F. officinalis*, so called from its use as a purgative by the ancient Greeks and Romans, may show as many as forty-five strata, indicating an age of forty-five years. *F. applanatus*,[3] a wound parasite, is responsible for the decay of large quantities of wood, since it attacks all sorts of deciduous trees and several coniferous species, both in the living state and after death; delignification is brought about, and is followed by progressive solution of the remaining cellulose wall.[2] In *Fistulina*, a bracket form, the tubes containing the hymenium are separate one from another; *F. hepatica*, known as the beef steak fungus, is edible, it occurs mainly on old oak trunks.

Merulius lacrymans and *Poria vaporaria* are causes of dry rot, a most destructive affection of woodwork in badly ventilated houses; not only do the hyphae penetrate into and destroy the wood, but, under dry conditions, when this is not possible, they spread in strands or sheets over the surface, and even over brickwork and along bell wires till a fresh supply of moist wood is reached and attacked. Infection may often take place in the timber yard, before the wood is used. Spores of *Merulius* are stated to germinate readily on wood attacked by the thelephorous genus *Coniophora*,[4] or other fungi, and *Coniophora*, in turn, which is incapable of dealing with dry wood,

[1] Bertrand, 1902. [2] Buller, 1906; Brooks, 1909, 1925.
[3] White, 1919. [4] Falck, 1912.

can flourish in the presence of the moisture exuded by *Merulius lacrymans*.

Daedalea may be bracket-shaped or resupinate; in this genus, with its pits intermediate between rudimentary gills and pores, we have a suggestion of the origin of the higher members of the family, and possibly of a relationship between their ancestors and those of the Agaricaceae. *Daedalea*, in its turn, may have been derived from a form like *Poria* or *Merulius*, or like the young sporophore of *Irpex*, and these are not remote from the wrinkled hymenium of *Phlebia*.

GASTEROMYCETALES

The Gasteromycetales include between 600 and 700 species characterised by the retention of the fertile tissue within a closed peridium, corresponding to the volva of some of the Agaricaceae, until after the spores have reached maturity. The peridium may be of one, two or three layers; where there is more than one, the term **exoperidium** is applied to the outer, and **endoperidium** to the inner. The contents of the peridium are described as the **gleba**; they may form a continuous mass, or a series of closed chambers, or may be separated by wefts of thickened hyphae, forming accessory walls, into one or more **peridiola**. Usually the ripening of the spores is followed by the deliquescence of the basidia and other cells of the gleba, but in several species long, simple or branched hyphae remain unaltered; they are known as the **capillitium** and may help in the distribution of the spores.

In subterranean forms the peridium is never ruptured, but, when the spores are ripe, the sporophore emits a strong smell by which rodents are attracted; the sporophore is eaten and the spores distributed after passing through the alimentary canal. It is interesting to note the close parallelism between the distribution of these forms and of the subterranean Ascomycetes.

In the majority of subaerial species spore dispersal is by wind. After deliquescence the gleba dries to form a powdery mass, the peridium is ruptured and the spores escape. In *Lycoperdon*, the puff-ball, the emission of spores depends on contacts or disturbances of the air due to passing animals. In *Bovista* the mature sporophores break readily away from their support, and are blown about by the wind, distributing spores as they go. In *Scleroderma* the peridium opens and the spores are blown out in dry weather and washed out in wet.

In the Phallaceae and some other members of the alliance, the sporophore remains underground during development; when maturity is reached the peridium is ruptured and the rapid elongation of the stalk carries the hymenium into the air. In the stinkhorn, *Phallus*, and its allies, flies are attracted by the strong smell and sweet taste of the mucilage accompanying the spores, and the latter are carried away on the insects' legs or pass through their alimentary canal. In *Sphaerobolus* the whole contents of the peridium are shot out by a mechanical device.

In correlation with these methods of distribution the spores on a gasteromycetous basidium are often crowded together, the sterigmata are either short, as in species of *Scleroderma*, or long and very thin, as in *Lycoperdon*, and are sometimes variable in length. The basidium and sterigma, in fact, play no part in the distribution of the spores, as in Hymenomycetales, but only in their production.

All the Gasteromycetales are saprophytes, though the mycelium of some may form a constituent of mycorrhiza; in none of them are conidia produced; in most species the vegetative filaments are woven into rhizomorphs on which the sporophores are formed.

Little is known of the inter-relationships[1] of the members of this alliance, or of their probable origin, though, from the similarity of the basidium and general arrangement, it would appear that they and the Hymenomycetales arose from a common stock. Moreover, if the presence of clamp connections can be trusted as an indication of specialised nuclear conditions, the two groups must have diverged after this state of affairs was established. The Nidulariaceae and Phallaceae show a high degree of specialisation; there is reason to think that the Phallaceae may have been derived from the Hymenogastraceae.

The alliance may be subdivided as follows:

Developed underground
- The whole sporophore remaining underground — HYMENOGASTRACEAE
- The hymenium raised on a stalk or lattice at maturity — PHALLACEAE

Developed above ground
- Peridium opening at maturity to emit clouds of spores
  - Capillitium present — LYCOPERDACEAE
  - Capillitium absent — SCLERODERMACEAE
- Peridium opening to disclose one or more peridiola — NIDULARIACEAE

[1] Fischer, E., 1936.

Hymenogastraceae

The members of the Hymenogastraceae are subterranean, occurring usually under trees, and having possibly a symbiotic relation with the roots of the latter; the mycelium spreads through the neighbouring humus, sometimes forming rhizomorphs, from the tips of which the oblong or globose sporophores arise. In *Gautieria* (fig. 290) and *Hysterangium* the medulla of the rhizomorph is continuous with the columella of the sporophore; the fertile tissue originates from this strand (fig. 294), grows in the form of irregular fingers towards the periphery, and pushes outward the peridium. In *Hymenogaster* the fertile tissue originates at the periphery and grows inwards (figs. 291, 292); in all three genera the strands branch and anastomose and so form the chambers of the gleba. In *Rhizopogon*, on the other hand, the developing gleba is continuous and shows dense areas alternating with others where the hyphae are loosely interwoven; chambers are formed by the tearing apart of the latter.[2]

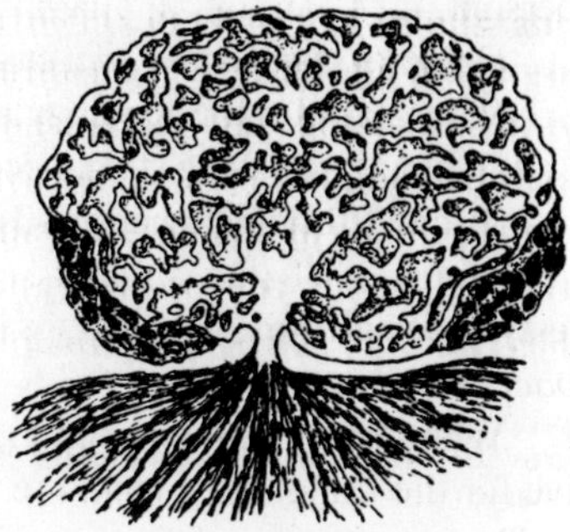

Fig. 290. *Gautieria morchellaeformis* Vitt.; sporophore cut longitudinally in half showing rooting hyphae and columella; after Vittadini.[1]

At maturity the gleba never becomes powdery; its general appearance is very like that of *Tuber*; some of the species are known as false truffles. As in the truffle, the peridium is thick, and, in *Rhizopogon*, three layers can be distinguished. In *Hysterangium*, *Melanogaster* and *Hymenogaster* the sporophore has a sterile base (fig. 293), in *Rhizopogon* and other genera the whole contents of the peridium are fertile.

Phallaceae

In the Phallaceae[3] the peridium is ruptured when the spores reach maturity and the gleba is carried up on a specially differentiated receptacle. In *Phallus impudicus*, known from its smell as stinkhorn or dead men's fingers, the exo- and endoperidium are

[1] Vittadini, 1831. [2] Rehsteiner, 1892.

[3] de Bary, 1887; Burt, 1894, 1896 ii, 1897; Petch, 1908; Fischer, E., 1910, 1923 i; Möller, 1895.

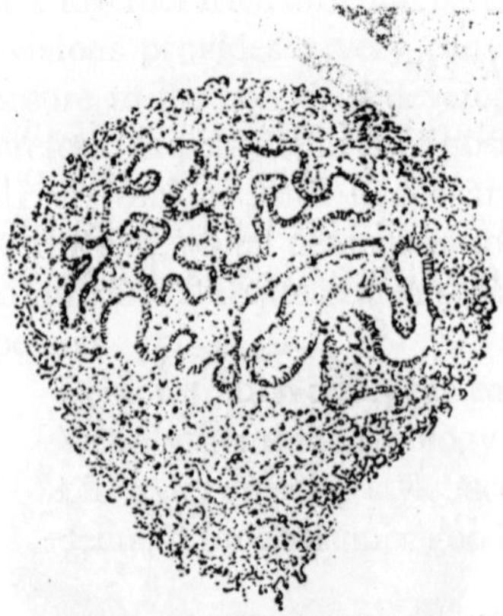

Fig. 291. *Hymenogaster decorus* Tul.; young sporophore with processes clothed by fertile tissue growing inwards from the periphery, ×28; after Rehsteiner.

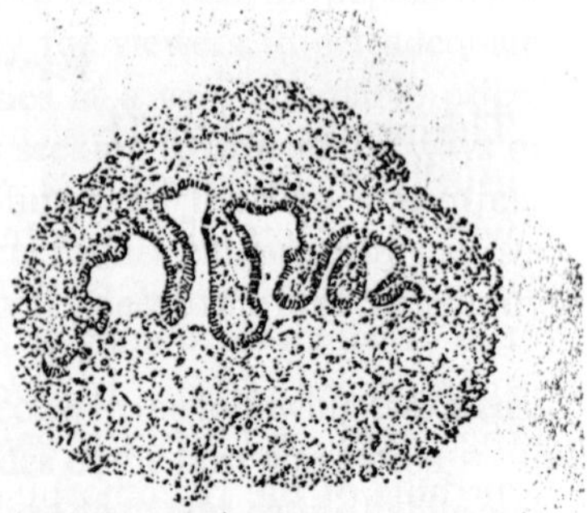

Fig. 292. *Hymenogaster decorus* Tul.; as fig. 291 but younger, ×28; after Rehsteiner.

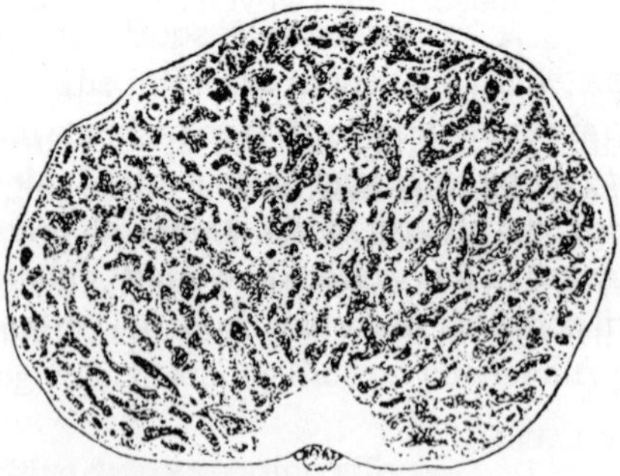

Fig. 293. *Hymenogaster tener* Berk.; mature sporophore in longitudinal section; after Tulasne.[1]

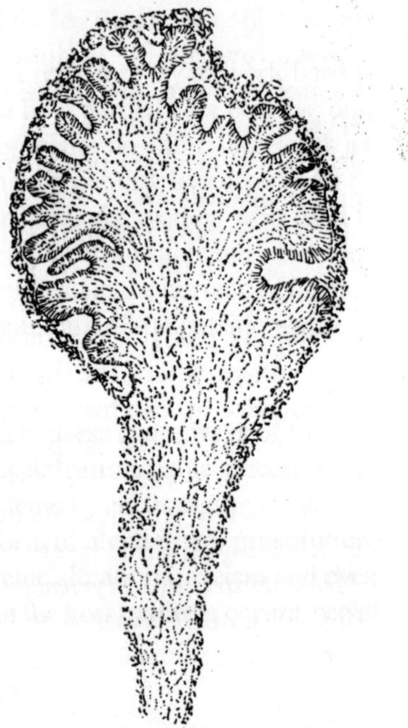

Fig. 294. *Hysterangium clathroides* Vitt.; young sporophore in longitudinal section, with processes clothed with fertile tissue pushing out from columella, ×28; after Rehsteiner.

[1] Tulasne, 1851.

tough, white membranes, between which is a gelatinous layer derived, at an early stage, from the disintegration of the hyphae; the exoperidium is continuous at the base with the receptacle, a sterile column which runs up the middle of the gleba, forming a hollow stipe filled with mucilage and developing elaborately chambered

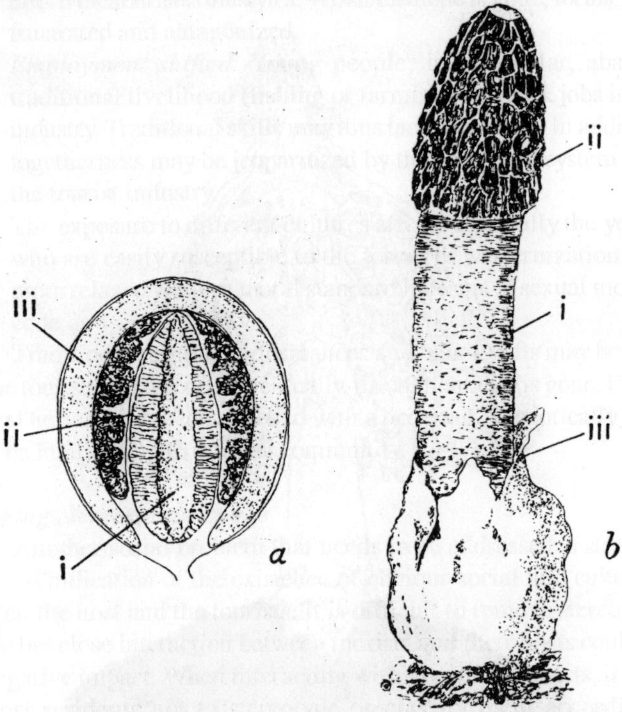

Fig. 295. *Phallus impudicus* L.; *a*, young sporophore in longitudinal section, $\times \frac{1}{2}$; after de Bary; *b*, older sporophore after the peridium has been ruptured and the stipe elongated; after Fischer;[1] i, stipe, ii, pileus bearing gleba, iii, gelatinous layer of peridium.

walls. The fertile tissue of the gleba arises from the peripheral zone just within the endoperidium. It is differentiated into chambers and supported on a tough membrane (fig. 295 ii) which later constitutes the pileus; the walls of the trama early become gelatinous. During development the sporophore is usually hidden under humus; it attains the size and shape of a hen's egg and is

[1] Engler and Prantl, 1900.

attached to a conspicuous, white rhizomorph, which may be traced for a considerable distance in loose soil. When the spores have ripened, the gleba breaks down, forming a viscid, olive-green slime, the elongation of the stipe begins (fig. 295 *b*), the peridium is torn, and the pileus, with the viscid spore mass, is carried up into the air. Given adequate water supply, the full size is reached

Fig. 296. *Dictyophora phalloidea* Desv.; mature sporophore; after Möller.

about half an hour after the rupture of the peridium. The characteristic smell and sweet taste of the disintegrated gleba attract bluebottles and other flies and by them the spores are distributed.

Mutinus[1] is very similar in structure to *Phallus*, but the sporophore is smaller, and the pileus is adherent to the stipe. In the genus *Dictyophora* a white network hangs below the pileus (fig. 296) and corresponds[2] to a rudimentary tissue between the pileus, and

[1] Burt, 1896 i; Fischer, E., 1923 i. [2] Atkinson, 1911.

stem of *Phallus impudicus*, which, in the latter, never reaches full development. The colours are brilliant, the pileus being snow-white, pink, orange, yellow or green, the mycelium sometimes violet, and the network white, green, yellow or lilac.[1] Doubtless insects are attracted by the display as well as by the foetid smell.

In *Lysurus* the pileus is replaced by five or six lobes, and in *Aseroë* by a lobed disc; in both the spore mass is situated on the inner, or upper face. This arrangement is also found in some of the tropical genera and in *Clathrus*, where the receptacle (fig. 297) has the form of a wide-meshed lattice, usually red in colour and contrasting with the white peridium.

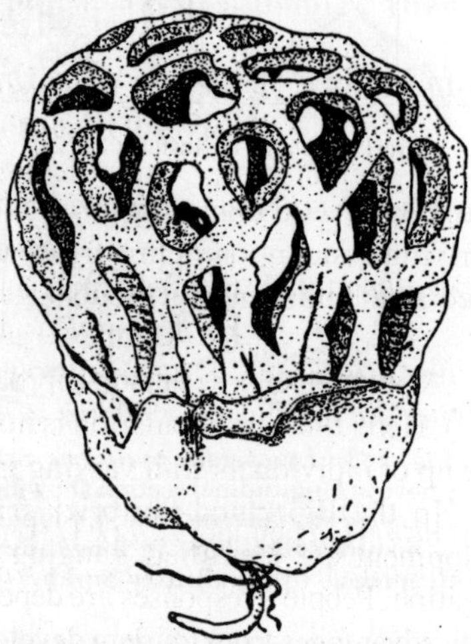

Fig. 297. *Clathrus cancellatus* Tour.; mature sporophore; after Rolland.[2]

In the developing sporophore of *Phallus* and its allies the basidia are peripheral in origin and face towards the axis; in *Clathrus*, and in other forms bearing the spore mass on the inner side of the receptacle, the basidia arise as outgrowths from the central tissue and are directed towards the periphery; these arrangements correspond to the conditions in *Hymenogaster* and *Hysterangium* respectively, and suggest that the Phallaceae fall naturally into two divisions with a separate origin among subterranean forms.

Lycoperdaceae

The members of the Lycoperdaceae are distinguished by the peridium of two layers, the outer of which may be further differentiated, and by the presence of a capillitium among the spores; the chambers of the gleba are produced, as in *Rhizopogon*, by the tearing apart of the tissue during growth.

Lycoperdon,[3] the puffball, has a well-marked sterile base, and a capillitium of irregularly branched threads which, in *L. pyriforme* (fig. 298 *a*) and others, forms a dense central mass, the columella,

[1] McLennan, 1932. [2] Rolland, 1910. [3] Lander, 1933.

remaining after the spores have blown away. The outer layer of the peridium is warted or spiny, and may disappear entirely, especially in wet weather, as the sporophore ripens; the endoperidium opens apically by a small, round hole when the spores are ready for dispersal. *Queletia* is somewhat similar, but the sterile base is elongated to form a stout, scaly stipe, and the endoperidium

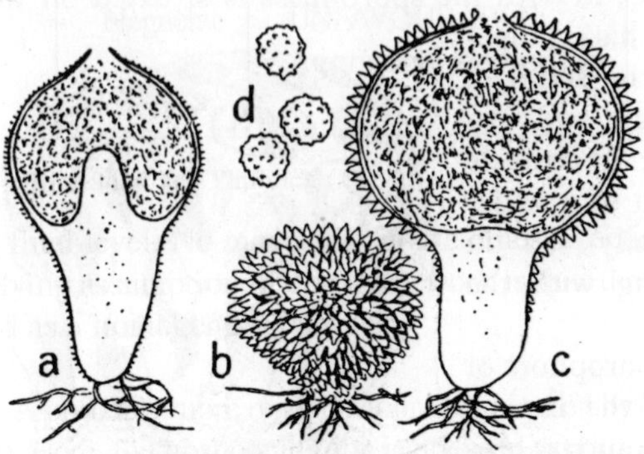

Fig. 298. *Lycoperdon pyriforme* Schaeff., *a*, sporophore in longitudinal section showing columella, $\times \frac{1}{2}$; *Lycoperdon echinatum* Pers., *b*, young sporophore, $\times \frac{1}{2}$; *c*, mature sporophore in longitudinal section, $\times \frac{1}{2}$; *d*, spores, $\times 750$; all after Smith.[1]

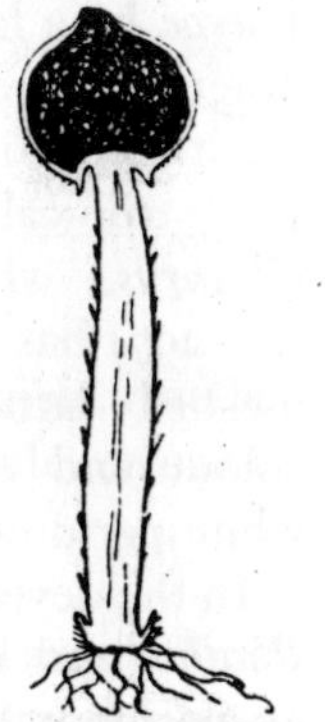

Fig. 299. *Tulostoma mammosum* Fr.; sporophore in longitudinal section, $\times \frac{2}{3}$; after Smith.[1]

opens irregularly. *Tulostoma* (fig. 299) also is stalked, and resembles a small puffball mounted on a stem. In *Bovista* the sterile base is absent, the exoperidium is papery and smooth, breaking away in irregular fragments, or remaining as a cup about the lower part of the fruit; the endoperidium opens by a definite pore.

In *Geaster* the young sporophore is more or less globose; the exoperidium is differentiated into a mycelial, a fibrous and a fleshy layer[2] and splits into pointed segments which spread outwards (fig. 300 *a*), leaving the gleba surrounded by the endoperidium which opens by a small apical pore, or, in the section of the genus sometimes separated as *Myriostoma*, by several apertures, giving a pepper-pot effect. The endoperidium may be raised on a stalk above the stellate segments of the outer layers. In some species the two layers of the exoperidium may separate (fig. 300 *b*), the outer remaining concave and the inner becoming convex.

[1] Smith, W. G., 1908. [2] Cunningham, G. H., 1927.

The sporophore of *Battarrea* develops underground; as in *Geaster* the exoperidium opens at maturity, but here the sterile base or stalk of the endoperidium is carried to a height of eight to fourteen inches, and is afterwards ruptured, exposing the spores.

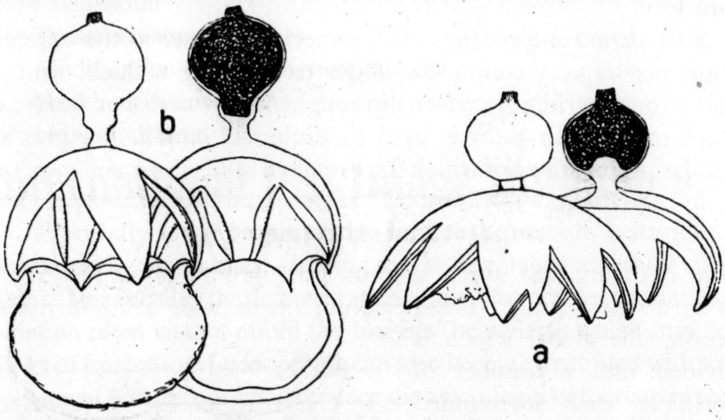

Fig. 300. *Geaster Berkeleyi* Mass., *a*, mature sporophore entire and in section; *Geaster fornicatus* Fr., *b*, mature sporophore entire and in section; $\times \frac{2}{5}$; after Smith.[1]

Sclerodermaceae

In the Sclerodermaceae the gleba is enclosed in a thick peridium, consisting of a single layer, and the sporophore is narrowed below

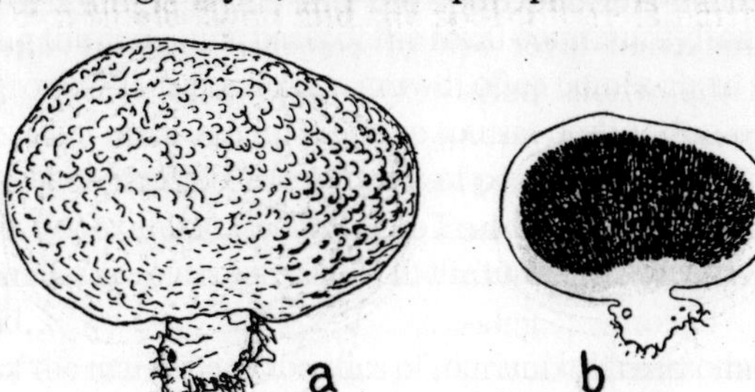

Fig. 301. *Scleroderma vulgare* Hornem.; *a*, mature sporophore; *b*, younger sporophore in section; $\times \frac{2}{5}$; after Smith.[1]

to form a sterile base. At maturity the peridium splits irregularly or partly disintegrates, liberating the spores. The gleba is divided by sterile tracts, and the capillitium is absent or rudimentary.

The species of *Scleroderma* are almost sessile, and are thus dis-

[1] Smith, W. G., 1908.

tinguished from the members of the genus *Polysaccum*, in which a definite stalk supports the fertile region. In *Polysaccum* the areas of the gleba, separated by sterile tracts, are ultimately rounded off to form peridiola, whereas, in *Scleroderma*, the whole gleba becomes powdery.

Scleroderma vulgare (fig. 301), sometimes known as the vegetable tripe, occurs very commonly under trees; it has a thick, white or pale brown peridium, externally roughened by scales or warts; the black or purplish gleba is used to adulterate *pâté de foie gras* and similar preparations of which the French truffle, *Tuber macrosporum*, is an ingredient.

NIDULARIACEAE

The Nidulariaceae are a curious little group, sometimes known as bird's nest fungi. The peridium opens wide at maturity, exposing the gleba, which is divided into a number of peridiola, each surrounded by a thick wall, and resembling a clutch of eggs in a nest. Each peridiolum is hollow, the basidia projecting inwards from the wall, which, developmentally, is a thickened layer of the trama, the whole peridiolum corresponding to a single cavity of the gleba. Between the peridiola the mycelial threads, as growth proceeds, become gelatinised and break down, leaving the peridiola exposed. In *Nidularia* the peridium is a single, thin membrane, opening at maturity to form a cup in which the peridiola lie free and enveloped in mucilage. In *Crucibulum*[3] (fig. 302) the peridium is differentiated into two layers, a loosely woven exoperidium, bearing

Fig. 302. *Crucibulum vulgare* Tul.; group of sporophores, nat. size; after Fischer.[1]

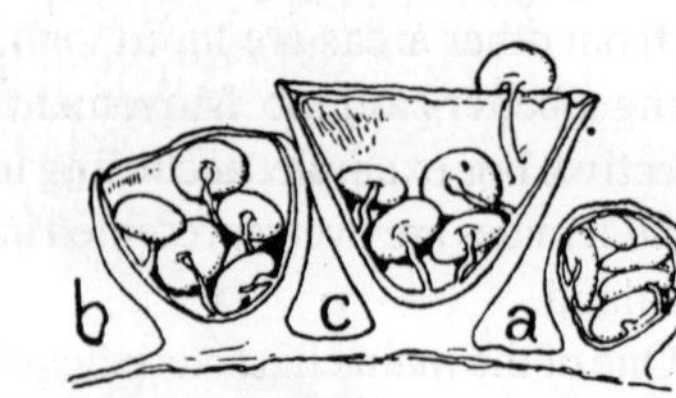

Fig. 303. *Crucibulum vulgare* Tul.; *a-c* young and mature sporophores in longitudinal section, × 3; after Smith.[2]

[1] Engler and Prantl, 1900. [2] Smith, W. G., 1908.
[3] Molliard, 1909; Walker, 1920.

long hairs, and an endoperidium of more or less gelatinised hyphae. Before maturity the peridium opens, exposing the remains of the undifferentiated tissue of the young sporophore, which conceals the peridiola (fig. 303 *b*) and is known as the **epiphragm**; it eventually becomes gelatinised and torn apart. On the flattened

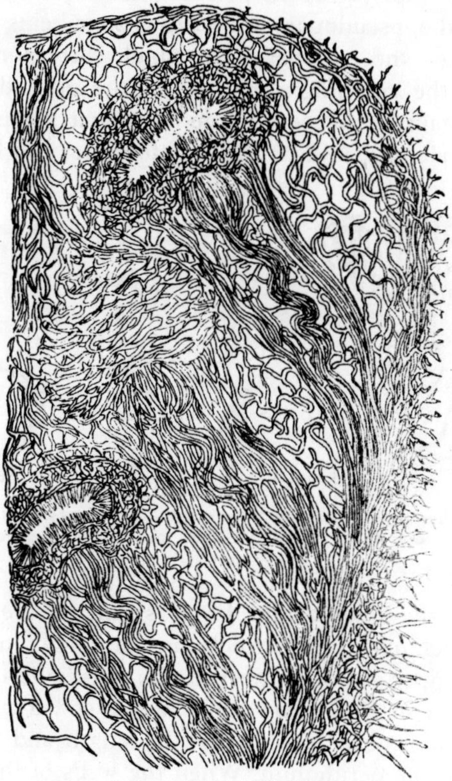

Fig. 304. *Crucibulum vulgare* Tul., section through part of a young sporophore, showing the developing peridiola; after Sachs.[1]

side of each peridiolum (fig. 304), next the peridium, is a median depression, the centre of which is occupied by a nipple-shaped mass of coiled and twisted hyphae, continued outwards as a sinuous thread, enclosed in a loose bag of partly gelatinised hyphae, and attaching the peridiolum to the peridium. When these hyphae and

[1] Sachs, 1855.

the surrounding mucilage are moistened they swell, and the sinuous thread becomes softened and can be pulled out to a length of three or four centimetres. It is difficult to indicate any function fulfilled by this curious structure, but it has been suggested that the mucilaginous threads may assist in dispersal by animals.

In *Cyathus*[1] (fig. 305 *a*) the peridium is differentiated into three layers, a middle, pseudoparenchymatous zone being developed between the exo- and endoperidium which correspond to those of *Crucibulum*; the epiphragm consists of the ground tissue of the gleba. The stalk of the peridiolum is at first 2 mm. long and is divided by a slender, middle region into upper and lower portions

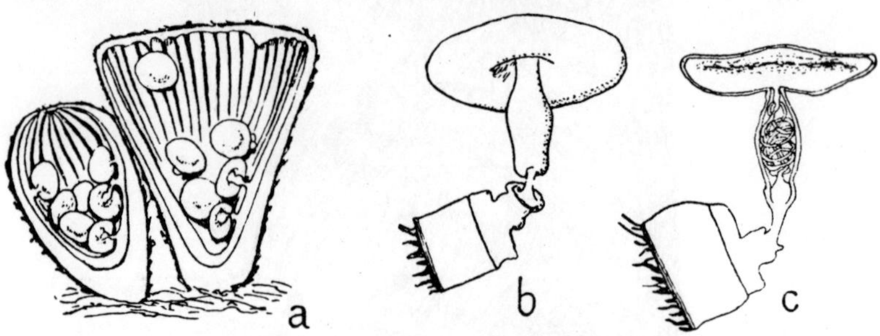

Fig. 305. *Cyathus striatus* Hoffm.; *a*, young and mature sporophores in longitudinal section, ×3; *b*, *c*, single peridiola entire and in section, ×10; after Smith.[2]

(fig. 305 *b*); the middle and lower segments are composed of narrow, branching hyphae, which become tough and elastic when moist; the upper segment, nearest the peridiolum, forms a bag, similar to that in *Crucibulum*, in which is coiled a slender thread (fig. 305 *c*) some three centimetres in length and attached at its upper end to the peridiolum. When the walls of the bag are ruptured, the whole stalk may be drawn out to a length of as much as eight centimetres; the hyphae of the extensible tissue are thick walled, their lumina having almost disappeared.

The genus *Sphaerobolus* differs from the other members of the family in having a single peridiolum (fig. 306 *a*, *b*) and in the function and elaboration of the peridium. The exoperidium is

[1] Walker, 1920.
[2] Smith, W. G., 1908.

white and floccose; the endoperidium consists of an external, fibrillar zone, and an internal zone of radially elongated cells known as the collenchymatous layer; the cells of the latter are rich in glycogen. The whole peridium, when mature, divides into six or seven stellate lobes, exposing the peridiolum. Growth continues in the collenchymatous zone and, as the other layers do not grow with it, a state of tension is produced and at last brings about the separation of the collenchymatous layer from the exoperidium

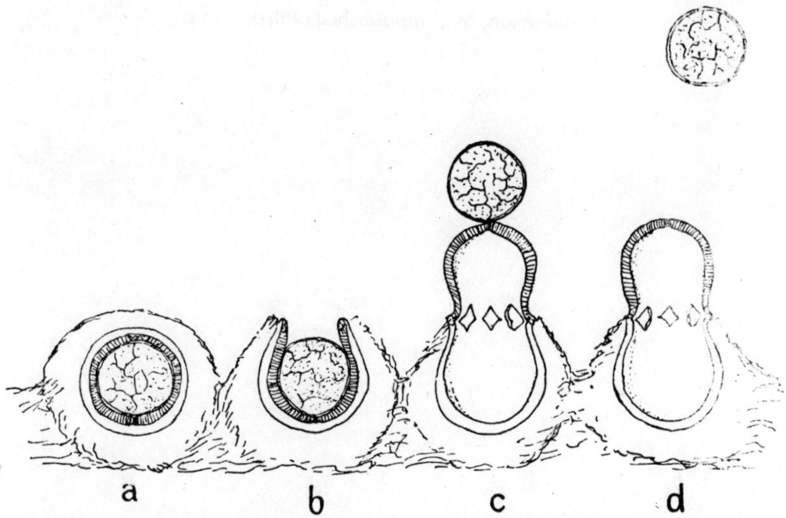

Fig. 306. *Sphaerobolus stellatus* Tode.; *a-d*, stages in the liberation of the peridiolum, ×12; after Smith.[1]

except at the tips of the lobes; it becomes arched outwards, carrying with it the fibrillar layer. If this takes place slowly, the peridiolum remains poised on the arch (fig. 306 *c*); if, as usually happens, the process is sudden, the endoperidium changing its position with a jerk, the peridiolum is shot away (fig. 306 *d*) to a distance of as much as fourteen feet,[2] and, being mucilaginous, adheres to an object on which it alights. Energy appears to be obtained by the transformation into sugars of the glycogen of the collenchymatous layer, with consequent increase of osmotic pressure; the change

[1] Smith, W. G., 1908.
[2] Walker and Andersen, 1925.

is gradual if the temperature is low and the light dim, but at a temperature of 32·5° C. and in bright light it takes place rapidly and violent ejection results. Internally the peridiolum shows a number of imperfectly divided chambers, and thus differs from the peridiola of *Cyathus* and its allies. The basidiospores[1] are uninucleate and give rise to short, aseptate germ tubes containing several nuclei. Very early in development binucleate cells are formed and, at the same time, clamp connections appear.

[1] Robinson, W., unpublished; Pillay, 1923.

FUNGI IMPERFECTI

Many thousands of fungi are known in which conidia are the characteristic unit of multiplication, though in some oidia and chlamydospores may be present. Neither sporangiospores, ascospores nor basidiospores are produced, nor are sexual organs found; accordingly such forms cannot be referred to any of the main subdivisions of the fungi from which they are probably derived, and of which they may be regarded as reduced or incompletely known representatives, showing only a gametophytic mycelium with accessory organs of multiplication. These species are described as *Fungi imperfecti*. They are an artificial assemblage, and many of the genera include members which are grouped together owing to a resemblance in their method of growth, but are not related; these are known as **form genera**. Thus the form genus *Oedocephalum* contained species which have now been shown to be stages in the life history of Phycomycetes, Ascomycetes and Basidiomycetes, as well as others which are known to produce only the *Oedocephalum* form; *Botrytis* is a conidial phase in the development of such Ascomycetes as *Sclerotinia* and such Basidiomycetes as *Hydnum omnivorum*,[1] while other species of *Sclerotinia* have conidia of the *Monilia* form, and are thus actually more closely related to species of *Botrytis* than species of *Botrytis* are one to another. Again the hyphomycetous genera *Penicillium* and *Coremium* indicate different methods of growth of the same fungus; *Penicillium* has the characteristic form of the conidial stage belonging to the plectomycetous fungus of that name, and most of its species are probably members of the Aspergillaceae which produce no sexual stage or ascus fructification under ordinary conditions; in *Coremium*, conidiophores, in no way different from those of *Penicillium*, are bunched together in response to the peculiarities of the environment.

The occurrence of similar systems of branching in the conidiophores of unrelated forms is doubtless due to the fact that each has reached in the course of evolution a similar solution of a common problem, the efficient exposure of spores to the air for dispersal. The conidiophores of *Botrytis*, the sporangiophores of *Thamnidium* and *Blakeslea* among Zygomycetes, and the corresponding structures in the Peronosporaceae may be compared from this point of view. Many of the *Fungi imperfecti*, under circum-

[1] Shear, 1925.

stances favourable to their growth, increase with great rapidity and some are the cause of serious plant disease.

In classifying the *Fungi imperfecti* it is convenient to include with them the conidial forms of fungi, the other phases in the life history of which are known, and to subdivide them according to the peculiarities of their conidial development and without reference to any sexual stage. Such a grouping is purely artificial; it is a key, not a natural system, and has no relation to origin or evolution, only to form. Since different spores of the same fungus are developed in response to different conditions of environment it must prove a valuable adjunct to attempts at a natural classification.

On this basis the Sphaeropsidales include forms with short, simple or branched conidiophores contained in flask-shaped receptacles known as pycnidia. The pycnidia may arise directly from the mycelium as in *Phoma*, or may be produced in or on a stroma. When mature the conidia escape through a definite pore, the ostiole, or are set free by a rent in the pycnidial wall; they are often held together by mucilage, forming long, twisted strings. Pycnidia are found in several members of the Pyrenomycetes, and it is possible that many of the Sphaeropsidales belong to that class.

In the Melanconiales, also, the conidiophores are short, but in this group there are no pycnidia. The species are parasitic, or are saprophytes on dead plants, and a more or less definite stroma is developed below the epidermis or periderm of the host, which is ruptured when the conidia are mature. The arrangement recalls the conidial stage of *Rhytisma* as well as of other Discomycetes.

The Hyphomycetales possess neither pycnidia nor stromata; they include most of the saprophytic *Fungi imperfecti*, as well as many parasites. The conidiophores are simple or branched, and may arise singly from the mycelium or be assembled in a brush-like **coremium**, or in a hemispherical **sporodochium**. The conidia are borne singly or in groups or chains, and the whole structure is usually exposed on the surface of the substratum. Similar growth forms are found in all the great groups of fungi.

These distinctions may be summarised as follows:

Conidiophores not grouped in pycnidia	
Stroma not present	HYPHOMYCETALES
Stroma present	MELANCONIALES
Conidiophores grouped in pycnidia	SPHAEROPSIDALES

MYCOLOGICAL TECHNIQUE

CULTIVATION

As the study of fungi necessitates cultivation on artificial media and special applications of micro-technique, a record of some of the methods employed by the authors and their colleagues may be useful; by no means all of these are original, but most of them have been tested by several years' work.

Fungi, like all other organisms, should be studied as far as possible in their natural habitat, but the mycelia of many of the smaller species are so intermingled in these circumstances that the stages of their development cannot with certainty be separated, and the use of pure cultures becomes essential. The common obligate saprophytes and such facultative saprophytes as *Pythium* and *Phytophthora* grow readily in culture. Many coprophilous forms develop freely on sterilised dung or on dung agar to which scraps of grass or filter paper have been added. Aquatic species may sometimes be obtained by leaving dead insects floating in pond water, and, less often, by placing pieces of twig, taken from ponds, and still bearing bark, in about five hundred times their volume of cool, clear water. A search through dying material of *Spirogyra* or *Oedogonium* may yield some of the smaller forms.

In the laboratory, fungi are most conveniently grown in Petri dishes; tall species may be grown in beakers covered with half a Petri dish, or in conical flasks plugged with cotton wool. Cultures may also be made in test tubes containing a suitable medium; these cultures are useful for retaining a stock of the fungus, but they are inconvenient for purposes of examination.

Isolation. Isolation of a tall species is usually easy, as a platinum wire can be thrust among the spores, or, as in *Mucor*, a sporangium can be pinched off with sterilised forceps; subsequent transfer to a medium usually gives a clean culture. When a mixture of low-growing species is to be sorted out, the material is examined and the spores distinguished; either a single spore is isolated, or a few spores taken up in sterilised water in a sterilised pipette, and spilled on to prepared media; instead of water a pipette of Barnes' medium without agar or of centinormal aqueous solution of sodium carbonate is often useful. Where bacteria are likely to be troublesome the medium is slightly acidified with citric acid. As soon as growth is visible to the naked eye, the colonies are cut out with a hot scalpel, and separately transferred to fresh media; it may be necessary to make several transfers before a clean culture is obtained. If bacteria are present, the hyphal tips alone should be transferred, as they are usually in advance of the film of bacteria.

Solid Media. For rough cultures of *Mucor* and *Rhizopus* it is only necessary to scatter a little dust upon a slab of moist bread. This is placed under a bell jar, and, at ordinary temperatures, sporangia will be present in about a week. If the bread is left for a longer period, the Mucors disappear and Aspergillaceae and *Fungi imperfecti* may be obtained. Dung treated in similar fashion often yields a number of interesting forms, including *Mucor*, *Piptocephalis*, *Pilobolus*, various Ascomycetes and later, Basidiomycetes.

For more critical work, fungi are isolated from rough cultures and grown in Petri dishes, using a medium based on agar-agar.

A combination of the liquor from boiled potatoes with agar will support the growth of some parasites and most of the saprophytes found on plant material, coprophilous species usually flourish best on an appropriate dung agar, prune agar is of use in obtaining the perithecia of *Eurotium herbariorum*. For a fungus being brought into culture for the first time, it is well to try several media, including some made up from the substratum on which the species is usually found, with or without mineral salts. The methods of preparation given on p. 378 can be applied to a number of other substances.

Among synthetic mixtures Barnes' medium has been found generally useful. The variant known as "M" is of value when it is desired to obtain ascocarps and to avoid accessory spores, as the latter appear sparsely on this substratum. The addition of scraps of filter paper encourages fruiting in a number of species. Many fungi, however, especially those of the dung flora, refuse to grow, or refuse to fruit on synthetic media, though they flourish on media derived from their natural habitat. If it is desired to cut sections, a dung agar is preferable to sterilised dung, but when a species has been growing long in culture it may be useful to return it occasionally to dung covered by a dung agar film. Fungi from fallen branches of trees or from debris of various kinds will often develop well on dung agar; doubtless their normal substratum is apt to be fouled by animals. The larger soil fungi are usually difficult to grow in culture and, for many of them, no satisfactory medium has as yet been found.

If it be desired merely to obtain a supply of the fungus or to study its life history, media can be made up with tap water and, for media other than synthetic, approximate quantities can be used. For physiological work pure chemicals are required and must be made up with distilled water, and with due regard to accuracy in weighing. Care is needed that only one of the substances in the medium, or one of the external conditions, should be altered at a time.

Liquid Media. When agar is omitted from the media just described, nutrient solutions are obtained; these are sterilised in flasks, and support the growth of a number of fungi.

Sterilisation. If Petri dishes are to be used, the medium is poured into flasks which are then plugged. These may be placed, with the required number of dishes, in a cold autoclave. Heat is applied, and, when steam issues freely from the valve, the latter is closed and a pressure of thirty pounds is maintained for about twenty minutes. The gas is then turned out, and, as soon as the gauge has fallen to zero, the valve of the outlet pipe is cautiously unscrewed. The Petri dishes when removed from the autoclave are placed flat on paper, and at once poured. This is done by lifting the lid of the dish sufficiently to allow the entrance of the neck of the flask containing the medium, pouring in enough to make a layer 5 mm. deep, and shutting down; it is sometimes desirable to pour in less, in order to get a shallow layer of medium which will allow of the examination of the culture through the bottom of the dish. It is well to avoid a draughty place for this operation, and the comfort of the operator will be increased by the use of a pair of thick woollen gloves. The plates are allowed to cool, and, when the medium has set, may be inoculated. If time permits it is better, before inoculation, to incubate the plates at 20° C. for three days or so. This dries off the water of condensation and allows time for contaminations introduced during pouring to become visible; contaminated dishes should be rejected.

If a dry oven is available, the dishes are sterilised in hot air at about 140° C. The medium, after sterilisation in an autoclave, is left there till required, and is poured as cool as possible, to avoid water of condensation and currents of air.

Inoculation. It is usually most convenient to inoculate by means of a piece of platinum wire set in a glass rod. The wire is heated in a flame, and, as soon as it is cool, pushed rapidly among the spores of the fungus, and, as quickly, through a narrow chink between the side and lifted lid of the prepared dish, into the centre of the medium. It is best to avoid opening a dish more than once for this purpose; when an inoculation has been made, the dish should be put aside and opened no more until the fungus is to be examined. When possible a number of cultures of the same fungus should be made at the same time, and, unless there be good reason to the contrary, once a dish has been opened for examination, its contents should be preserved in a suitable fluid, or thrown away.

Some fungi, such as *Phytophthora* and *Basidiobolus*, often do not form spores in large numbers. For their transfer, a piece of the medium, bearing the fungus, is cut out with a hot scalpel and rapidly carried to a fresh dish. The lid of the latter is opened only sufficiently to permit the introduction of the scalpel and its load, and, if the portion of inoculum be dropped on the way, or touch the outside of the dish, it is rejected, the scalpel reheated, and a fresh portion transferred. Scrupulous care is taken to avoid the introduction of anything except the fungus and the piece of substratum, and, if it is suspected that extraneous material has been introduced, the dish is marked and specially watched.

Tube and Flask Cultures. When tubes and flasks are to be prepared for cultures, the medium is made up as for dishes, but it is usual to sterilise in the flask or tube. Accordingly, the agar is melted, the whole well shaken, and a sufficient quantity of the medium poured into each vessel. The use of a funnel to which a short glass tube is attached by rubber tubing furnished with a clip makes it easy to carry out this operation without wetting the mouth or sides of the vessel, positions in which traces of the medium result only too readily in contamination. Before use the funnel is warmed sufficiently to keep the agar in a liquid state. The receptacles are loosely plugged with cotton wool, and sterilisation is carried out. When removed from the autoclave, tubes are placed in a slanting position until cool, so that the medium sets in a slope. It is well to incubate for several days, in order to dry the plugs and give intruders an opportunity to reveal their presence. Inoculation is carried out as with dishes, but, before the plug is removed, its upper part is passed through a flame; the plug is lifted between the fingers so that it stands out from the back of the hand, and is so held whilst inoculation is performed. The lower part of the plug should be passed through a flame before being replaced.

Single Spore Cultures. Two methods have been found simple and useful.

In one, a dummy objective fitted with a small sharp-edged tube in place of the lens, is attached to the microscope. The spores are spread thinly in a little water on stiff agar, and the surface examined with a two-thirds inch objective. A spore lying at some distance from its neighbours is centred under the objective, the tube of the microscope is raised, the dummy objective swung into position, and carefully racked down until a circular cut is made in the agar. A second examination is made with the two-thirds inch objective to ensure that the circle is around the spore, the piece of agar is taken out on a sterile platinum wire, and is quickly transferred to a Petri dish.

The second method depends on dilution. A quantity of spores is shaken up in sterile water, and small drops of the suspension are examined, being taken with a fine pipette which is heated between each drop. If the drop contains several spores, further dilution is needed; dilution is continued until about half the small drops contain one spore, the remainder, none. When the suspension is of the necessary strength, a drop is squirted on to the under side of a small piece of heated cover glass, suspended over a hole $\frac{3}{16}$ in. diameter in a piece of hot sheet brass. Rapid examination is made, and, if a spore is present, the glass with its drop is transferred with hot forceps to a culture dish. All such examinations are facilitated by the use of a binocular microscope.

MICROSCOPIC EXAMINATION

Examination. The preliminary examination of the larger fungi, with or without a hand lens, requires no special comment; a similar examination of living micro-fungi should whenever possible be undertaken under a one-inch, or, if a one-inch is not available, a two-thirds objective. For this purpose the material is mounted on a slide, its long axis lying by preference parallel to the glass, and the slide is placed on a piece of cardboard or paper which is twice folded (fig. 307) so that its upright sides reflect the light. A postcard cut longitudinally in half is just the right size for this purpose. The source of illumination, and, if necessary, a bull's eye condenser are so arranged as to throw light downwards on the specimen, which is then examined under the low powers of the microscope. When the material is pale in colour, black or tinted paper may with advantage be placed between the cardboard and the slide. Material growing on agar can effectively be studied in surface view by placing a block of agar with the undisturbed fungus on the slide; a lateral view is obtained by cutting a thick slice of agar and laying it on its side. With transparent media in thin layers, it is often possible to make out a good deal without opening the dish, by turning it upside down. If it is necessary to retain for future use a culture from which a specimen has been removed, care must be taken to cut out the latter with a sterilised scalpel; a dish which is wanted again should not be fully uncovered, or its lid placed upon the bench, or turned upside down.

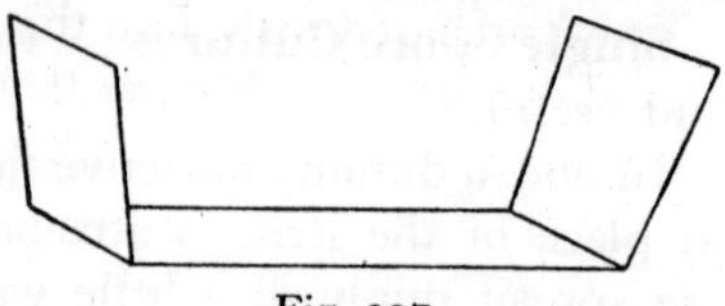

Fig. 307.

For examination in transmitted light, material may be dipped in 70 or 90 per cent. spirit to remove air, washed and mounted in water, or it may be examined in glacial acetic acid; this method has the advantage of ensuring that conidia remain attached. Moist specimens can be examined in water without preliminary treatment, and so can asci of all ages.

Material for permanent preparations, to show habit or general characters, should be fixed in 70 per cent. spirit, or, better, in 70 per cent. spirit to which 15 to 20 per cent. glacial acetic acid has been added; after about ten minutes in the latter it is well washed in 70 per cent. spirit. Alternatively one of the chromacetic mixtures[1] may be employed. Such material is usually best mounted in glycerine or glycerine jelly. An excellent method, especially for hyphae and young fructifications growing on agar, is to immerse the fixed material, after transferring to 50 per cent. alcohol or

[1] For formulae of fixatives and stains see pp. 378–81.

to Calberla's fluid, in a mixture of one part water, two parts glycerine and one part saturated aqueous solution of erythrosin. After about twenty minutes the material is well washed in 50 per cent. glycerine and left in 50 per cent. glycerine for one or two days. The exact times for any fungus or stage of development can be found only by experiment. The hyphae should be a brilliant pink, the agar nearly colourless. Washing out is helped by moderate warmth. As much as possible of the agar below the hyphae is now cut away, and the remaining thin slice, with the undisturbed fungus, is placed on a warm slide, mounted in glycerine jelly and covered with a warm cover slip. The agar melts into the warm jelly and a very clear preparation is obtained. If examination with a one-twelfth objective is likely to be needed, it is important to keep the film of jelly above the fungus extremely thin. It must be emphasised that on no account should material on agar be disturbed at any stage of preparation. It is seldom advisable to take loose hyphae or hand sections of fleshy fungi into balsam; glycerine or glycerine jelly are likely to cause much less contraction, but some of the parasitic forms, such as *Epichloë* or *Peronospora*, are sufficiently strengthened by the tissues of their host to make effective balsam mounts; they stain well with safranin and light green.

In objects so minute as fungal hyphae, or their spore mother cells or spores, cytological methods are required for detailed work, and even these may prove inadequate to elucidate the finer structure of the cells or the behaviour of the nuclei; in particular there is a danger of confusing small nuclei with metaplasmic granules which take up a similar stain.

Fixation. Flemming's strong fluid diluted with an equal quantity of water has proved the most satisfactory fixative for many fungi. The fluid known as 2BD is also good, and the transparency which it gives to the cytoplasm is especially useful in photography. Where there is much stored food material, as in the oögonia of *Pyronema*, Merkel's fluid gives readily stainable preparations, while particularly clear material of rusts and some other parasites can be obtained with acetic alcohol. Other fixatives are of value for particular fungi or for purposes of comparison; in research work two or more fixing fluids should be employed and the results compared.

There should be no delay between gathering and fixing specimens; fungi growing under natural conditions must be fixed on the spot. Even if samples are carried home on their substratum, normal growth is apt to be prejudiced. The best results are usually obtained during the warmer hours of the day, from noon onwards, or occasionally at midnight. Abundant nuclear divisions may be secured at midnight, for example, in various Hymenomycetes. For fungi growing in culture the time of day seems less important, doubtless as a result of their more uniform conditions.

When the life history is being studied, active material, as indicated by the presence of mitoses in vegetative or reproductive cells, is of special importance. It is essential to ascertain the conditions of food, temperature and illumination by which growth is encouraged and to select for fixation a period of active development. When the requirements of the fungus are known, specimens should be fixed at regular intervals, say once an hour, throughout the day and night. Equally important are the fixation of undisturbed material and the use of rapid methods. If a fungus is removed from the warmth and light in which it is growing to the laboratory bench while preparations for fixation are made, any fusions and divisions in progress are likely to be completed and others will certainly not begin. When possible the best procedure is to flood the Petri dish or other vessel in which the fungus is growing with fixing fluid while still in its accustomed position.

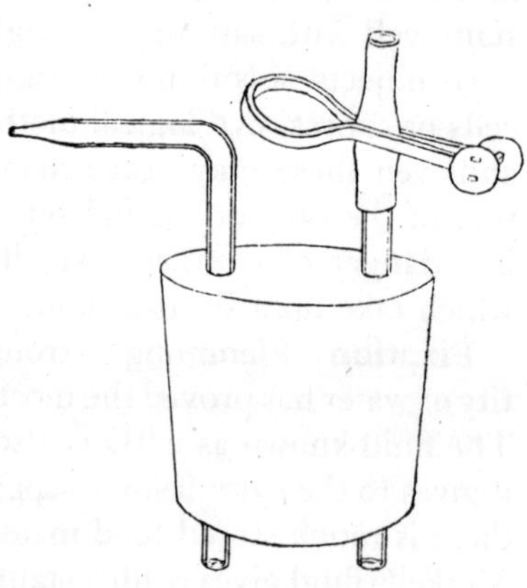

Fig. 308.

If a fixative is used which does not contain alcohol, and if the material is not itself saturated with water, air must be pumped out as soon as the material is immersed in the fixative. For laboratory purposes a foot pump, such as Hearson's patent air pump, is convenient; for field work a small exhaust syringe can be carried. The pump is attached to a tube passing through a rubber stopper (fig. 308), and an exit tube is provided with a clip, which can be quickly opened, allowing the sudden entrance of air, and causing the pieces of material to sink. During the entrance of air it is well to keep the material in motion.

For fixation in the field, a box or case fitted with half a dozen bottles or tubes containing fixatives is convenient; a rubber stopper should be carried which fits them all and is provided with the necessary connections for attachment to the syringe. If acetic alcohol is to be used, spirit for washing must be carried.

Everything should be ready for rapid work before the material is gathered. To expedite matters only a few pieces should be picked at a time, these are put in the fixative and pumped, and others added and pumped again. Small specimens may be fixed whole, larger ones should be cut into blocks about five millimetres square. To avoid delay, adherent particles are better fixed with the fungus and cleaned away under a lens when the material reaches spirit or Calberla's fluid.

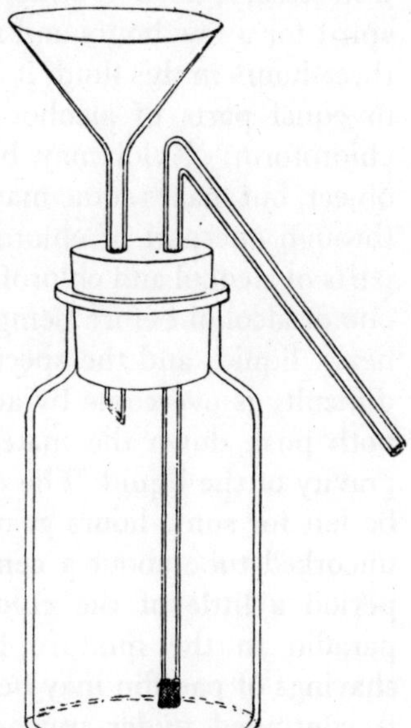

Fig. 309.

After Flemming's fluid, 2BD, Merkel's fluid, and the chromacetic mixtures, material is washed in running water for six to twelve hours. A convenient washing bottle (fig. 309) has a funnel through which water enters, and an escape tube closed by a piece of fine muslin; this can be used either for objects of ordinary size, or for minute specimens enclosed in muslin bags. When the fungus is lightly attached to the substratum, as often happens with ascocarps growing on agar, a large basin is used and the water siphoned away and replaced from time to time, or the material is gently lifted out in a Petri dish while the water is being changed. After washing, the material is placed in 20 per cent. spirit for half an hour or more, and successively in 40, 60 and 80 per cent. spirit for three or more hours each; where nuclear structures are to be studied the strengths of spirit should be increased by 10 per cent. instead of by 20 per cent. From 60 or 70 per cent. spirit the material may be transferred for storage to Calberla's fluid, consisting of equal parts of water, absolute alcohol and glycerine, or taken up through 80 and

90 per cent. spirit to absolute alcohol and embedded. Most fungi cut better if left in Calberla's fluid for a few days.

Specimens fixed in acetic alcohol are left in the fixing fluid for fifteen to thirty minutes and well washed in 90 or 95 per cent. spirit; from this they may be transferred to Calberla's fluid or embedded at once.

Preparation of Slides. Material for embedding is selected and, if necessary, cleaned under a lens; it is carried to 90 or 95 per cent. spirit for a few hours and thence to absolute alcohol. After about three hours in this fluid, it is transferred for a similar length of time to equal parts of alcohol and chloroform and thence to pure chloroform. Xylol may be similarly used when economy is an object, but makes some material brittle. Delicate material is passed through one part of chloroform with three parts of alcohol, equal parts of alcohol and chloroform, and three parts of chloroform with one of alcohol before being put into chloroform. Chloroform is a heavy liquid, and the specimens as a rule will not sink in it; this difficulty is overcome by adding a few shavings of paraffin, which both push down the material, and, as they dissolve, reduce the gravity of the liquid. The chloroform and paraffin mixture should be left for some hours at ordinary temperatures, preferably in an uncorked tube about 3 centimetres high by 2 across; during this period a little of the chloroform evaporates, the proportion of paraffin in the mixture being thereby increased, and further shavings of paraffin may be added from time to time. The process is continued under warmer conditions on the top of a water-jacketed oven, and finally for about twenty minutes inside. The contents of the tube are then emptied into a warm watch glass and left in the oven for another ten minutes; by this time, if the earlier stages have been sufficiently prolonged, all sign of chloroform should have disappeared. If any trace remains, the used paraffin is poured off and replaced by a fresh supply. Meanwhile a stoneware saucer, two inches in diameter, is warmed, smeared evenly with pure glycerine, placed in the oven and filled with fresh paraffin; the material is transferred to it with warm forceps, and the paraffin solidified by holding the saucer over a dish of cold water, blowing gently on the surface till a film is formed, and finally immersing the saucer and its contents in the water. The block of paraffin, when cold, should float out of the saucer; this may be ensured by the addition of up to 0·1 grm. of saponine[1] to every 25 c.c. of glycerine. Very minute objects, which cannot be transferred with forceps, are embedded in the original watch glass after careful changes of paraffin. If iced water is used for cooling, the block will usually float. A total of half-an-hour in the oven is long enough for small pieces of material; the temperature should be just above the melting point of the paraffin employed. For fungi which are to be sectioned at temperatures of 15° to 20° C., paraffin melting at 52° C., or a mixture of this with paraffin melting at 60°, has proved most satisfactory. The melting point can be ascertained by putting a drop of liquid paraffin on the bulb of a warmed thermometer, and, as the thermometer cools, noting the temperature of solidification. Material embedded in paraffin can be kept indefinitely. It is convenient to embed with the material a slip of paper on which the genus and species are written in ink with a number or other symbols indicating the previous treatment.

Very minute objects, for which the numerous transfers between the fixing fluid and paraffin are difficult, may be placed, after washing in water, in 10 per cent. aqueous glycerine; the water is allowed to evaporate in a desiccator till the glycerine is undiluted. This can be tested by the addition of a drop of pure glycerine which causes rings or ripples in the fluid if water is still present. The material is next well washed in absolute alcohol, transferred to 10 per cent. cedar oil in absolute alcohol, and the mixture allowed to evaporate until pure cedar oil remains; shavings of paraffin are then added, first at room temperature, then on the top of the oven, and then inside, and the material is embedded as before, special care being taken to transfer as little cedar oil as possible to the embedding saucer, and two changes of fresh paraffin being used in succession if required.

The ribbon of microtome sections is now prepared in the usual way, sections being cut from 3μ to 20μ thick according to the objects to be studied and the stain to be used. In reasonably transparent material such cytological stains as Breinl's combination give good results at 10 or 15μ and nuclei may then be studied entire. The ribbons of sections are laid out on white paper with the shiny side down; it is well to avoid as far as possible lifting the rows of sections over each other, either by laying out the first

[1] Broadhead, Q. E., unpublished.

strip of ribbon at the bottom of the paper and starting from the left-hand end to fill the top of the slide, or, if the ribbon is started from the upper edge of the paper, to take first the last cut sections from the right-hand end of the bottom row and place them at the lower edge of the slide. In dealing with minute objects it is advisable to take especial pains to lay the sections evenly on the slide. Before use the middle two inches of the slide are lightly smeared with albumen mixture and this is covered with three or four drops of water; strips of the ribbon 1½ inches long are then arranged on the slide, any vacant space above and below is filled with lengths of blank ribbon, and the slide is gently warmed till the ribbon lies flat. The superfluous water is drawn off and the slide left in a warm place till dry. Slides may with advantage remain at this stage for two or three days, being supported with the sections downwards and carefully protected against dust. Now, for the first time, the slide is warmed sufficiently to allow the albumen to coagulate, attaching the sections, the slide is placed in the oven for a few minutes to melt the paraffin, and is afterwards immersed in a jar of xylol. The xylol may be washed off with 90 or 95 per cent. spirit, and the slide washed at once in water. In our experience the sections, once attached to the slide, can be transferred direct from alcohol to water and from water to alcohol without injury to the most delicate structures.

The sequence of the slides should be carefully preserved, either by numbering them with a glass pencil as they are covered with sections, or by subsequently matching the pattern of the sections as seen with the naked eye and numbering on the label. If at any time during these processes sections are lost, so that the sequence of sections on the slide, or from slide to slide, is interrupted, a note to that effect should be made; otherwise an object under the microscope cannot be followed with confidence from one section to another.

When there is much food stored in the structures to be studied it is sometimes desirable to treat the material with gastric juice, which digests the contents of the cytoplasm, leaving the nuclei unaffected. As albumen is destroyed by this treatment the sections must be attached to the slide without its help. The slide is very carefully cleaned, and should not be used unless it is covered by an even film of moisture when breathed upon. Three or four drops of distilled water are then spread over the middle two inches of the slide, the sections are added and the usual procedure followed, very special precautions being taken to dry the slide slowly and thoroughly before placing it in the oven.

Sections of material fixed in a fluid containing osmic acid should be bleached to remove the blackening due to this reagent. They may be placed in a solution of chlorine or of hydrogen peroxide in spirit. The time required depends on the material and can only be ascertained by microscopic examination. Ten minutes is a minimum for chlorine, twenty minutes for hydrogen peroxide; the action of the former is improved if the jar stands in sunlight. With refractory material, bleaching should be repeated in a fresh solution; it is useless to leave the material indefinitely in the old. When bleached the sections are well washed in spirit and may be left in spirit till required.

Staining. Most of the stains in ordinary use may be employed for fungi, but it must be borne in mind with fungi, as with other material, that a definite relation exists between the fixative and the stain. Iron haematoxylin, for instance, though admirable after osmic fixatives, gives poor results on specimens fixed in Merkel's fluid; Congo red is particularly satisfactory for hyphae treated with acetic alcohol. The stains described below can safely be used after such fixatives as Flemming's fluid and 2 BD; directions for preparing them will be found at the end of this chapter. All stains require careful preparation, many need time in which to ripen, most should be filtered before use and may with advantage be filtered again if they have been left standing. Since stains dissolved in clove oil give better results in the presence of small quantities of absolute alcohol, it is preferable, as well as economical, to use them again and again, the surplus stain being drained back into the bottle after use. A loss of brilliance in the colour of the stain will show when it should be discarded. In the following description of methods the term spirit is used when the presence of water is necessary or unimportant, the term absolute alcohol when lack of water is essential. Commercial absolute alcohol, though containing some minor impurities, is free from water and has proved thoroughly satisfactory for cytological work.

It is advisable to use pipette bottles for absolute alcohol, cedar oil, pure xylol, clove oil and the clove oil stains. When the preparation consists of sections attached to a slide, the latter is shaken almost dry by rapid jerks of the wrist and held in a sloping position while the reagent is applied to the upper end and allowed to run over the sections. If the process is quickly carried out, a very small quantity of reagent is sufficient for each slide. Unless the material and stain are so well known that the results can be judged with a one-sixth objective, it is advisable to mount for examination under a one-twelfth in cedar oil, which can be renewed under the cover-slip as required by standing the slide on its edge in a shallow bath, and not to transfer to the mounting medium till the stain is found to be as good as possible. A minimum quantity of cedar oil should be used, as any surplus is liable to leak out and clog the bearings of the microscope. Before a permanent mount is made, all traces of clove oil should be carefully removed, either with cedar oil or xylol, for clove oil dissolves a number of stains. If cleanly mounted, even such stains as erythrosin or Congo red retain their brilliant colour for thirty years or more. Sections are usually mounted in canada balsam dissolved in xylol, but the Sira medium, prepared by Stafford Allen and Sons, London, according to the directions of the Scientific and Industrial Research Association, is preferable for delicate, cytological work.

Whenever the stain selected requires washing out under the microscope, or microscopic examination while the slide is wet, the stage of the microscope should be covered by a piece of glass. A blank lantern slide or photographic plate serves well for this purpose. Not only is the stage kept clean, but the glass plate is readily moved, carrying the wet slide without interference from capillary attraction.

Safranin and Light Green. For general work, and especially for parasitic fungi when part of the host tissue has to be stained, the sections may be washed with water, placed in safranin for about five minutes, washed in water to remove the surplus stain, washed in spirit, washed in absolute alcohol, transferred to a drop of solution of light green in clove oil, and watched under the microscope till the safranin remains only in the nuclei and in the lignified walls of the host, the walls of the hyphae and the cellulose walls of the host being stained green. The clove oil is then washed away with cedar oil or xylol, and the sections mounted in cedar oil or canada balsam. The stain serves well to differentiate fungal hyphae in the xylem of the host.

Methylene Blue and Erythrosin. A similar procedure with methylene blue followed by solution of erythrosin in clove oil is also effective, and is preferable if much resin is present in the tissues of the host; the nuclei in this combination are stained blue, and the walls of the hyphae red.

If hand sections are stained by either of these methods, they should be left in absolute alcohol for some minutes, and care should be taken to cover the alcohol, so as to prevent the absorption of water from the breath of the operator or otherwise.

Haematoxylin. For microtome sections designed to show general cytological characters, Heidenhain's haematoxylin is satisfactory, and may be followed by light green, erythrosin or orange G in clove oil. The sections are washed in distilled water, left in 4 per cent. aqueous solution of iron alum for about twenty minutes, well washed in changes of distilled water, placed in haematoxylin for twenty-four hours, and again thoroughly rinsed in distilled water before the stain is washed out under the microscope. For washing out, a Petri dish is convenient, containing a solution of iron alum in distilled water to which five or six drops of glacial acetic acid have been added. The solution should be frequently renewed. The concentration of iron alum to be employed depends on the material, 4 or 8 per cent. may be tried as a first experiment; rapid differentiation increases the sharpness of the stain. When the stain is satisfactory the slide is placed for ten or fifteen minutes in running tap water, the faint alkalinity of which neutralises the acidity of the solution employed for washing out and changes the colour of the stain from brown to a good black. The slide is then taken through absolute alcohol to clove oil, or, if a counter stain is not required, to cedar oil or xylol.

For critical cytological work a 0·05 per cent. solution of haematoxylin is to be preferred to the stronger solution of Heidenhain. The slide is mordanted, as before, in 4 per cent. aqueous solution of iron alum for about twenty minutes, well rinsed in distilled water and left in haematoxylin for about twenty-four hours. It is again thoroughly washed in distilled water, and differentiated under the microscope in saturated aqueous solution of iron alum to which a few drops of glacial acetic acid have been added. The subsequent treatment is the same as for Heidenhain's haematoxylin. In view of the rapid action of the solution for washing out, it is often wise before immersing the slide in iron alum to examine it in distilled water and to choose a section by observation of which during differentiation the quality of the stain can be judged. In localities where tap water is alkaline it is important between mordanting and washing out to use distilled water only, no tap water at all. The dilute haematoxylin and concentrated washing out solution give a clear and very delicate stain.

The tendency with either method is to wash out insufficiently; it is necessary to bear in mind the darkening of the stain in alkaline tap water and to aim at nearly colourless cytoplasm. If the stain is unsatisfactory at the first attempt the slide should be discarded. It is useless, once the sections have been immersed in tap water, to wash away the stain and try again. An indifferent result may be obtained, but the time is better spent on cutting fresh sections.

Safranin, Gentian Violet and Orange G. Flemming's triple stain, or gentian violet and orange G alone, are also useful. The sections are placed in safranin for three to twelve hours, washed with water to remove the surplus stain, washed with spirit, examined, further washed out, if necessary with acid alcohol, washed with spirit, washed with water, placed in gentian violet for a few minutes, and again washed with water; they may then be covered for a few seconds with an aqueous solution of orange G and taken up to cedar oil, or they may be taken at once through spirit and absolute alcohol to clove oil in which orange G has been dissolved; the passage through spirit and absolute alcohol should be rapid, or much of the stain will be lost.

Safranin, Polychrome Methylene Blue and Orange Tannin. If the nuclei are reasonably large, by far the best cytological stain in our experience is the combination originated by Breinl. The slide is placed for fifteen minutes in a solution of iodine and potassium iodide in spirit, washed with water, transferred for thirty minutes or longer to equal parts of aqueous and of alcoholic safranin, washed with water, placed for ten minutes in polychrome methylene blue, and again washed with water; it is then dipped into orange tannin and in this fluid examined under the microscope till the orange has displaced the blue in the cytoplasm, washed with water, well washed with 90 or 95 per cent. spirit, quickly transferred through absolute alcohol to commercial aniline oil, and again watched under the microscope till any excess of blue is removed. In our experience pure aniline oil does not act on the stain, the required reaction being presumably due to an impurity in the commercial substance. Finally the aniline oil is replaced by cedar oil and the sections covered in readiness for a preliminary examination under high powers, before permanent mounting in balsam. In good preparations the spireme is brilliant blue-black, the cytoplasm faintly yellow, and the chromosomes of the meiotic metaphase bright red. The stain is rather difficult to use and failures in its first application are inevitable, but a successful preparation, when obtained, will be found to justify perseverance; as it is very transparent, sections as thick as 15μ can be employed.

Gentian Violet and Light Green. For general work, and also where dense storage substances make the use of haematoxylin inadvisable, sections may be stained for a few minutes in gentian violet, washed with water, taken rapidly through spirit and absolute alcohol, and examined in solution of light green in clove oil. The chromatin should be purple, the walls green, and the cytoplasm almost colourless. This and the following stain, if unsuccessful, can be washed away without injury by immersion in spirit, or in spirit to which a few drops of hydrochloric acid have been added. The same slide can be stained again and again till the desired result is obtained.

Iodine Gentian Violet. This valuable cytological stain is due to W. C. F. Newton.[1] Sections are placed in 1 per cent. aqueous solution of gentian violet which has been boiled and filtered; after four to seven minutes they are rinsed in water, placed for thirty seconds in 80 per cent. spirit containing 1 per cent. iodine and 1 per cent. potassium iodide, rinsed in water, taken quickly through spirit and absolute alcohol and differentiated in clove oil. If the stain proves difficult to wash out, the slide, after the iodine solution has been washed off with alcohol, may be placed for a few seconds in 1 per cent. aqueous solution of chromic acid[2] and taken up through alcohol to clove oil as before.

Congo Red. Congo red is a useful counter-stain for rusts after haematoxylin, since it colours the walls of the parasite but not of the host; it acts excellently after fixation with acetic alcohol, and, if the sections are well washed before transfer to balsam, will remain bright for many years,

[1] Newton, W. F. C., 1927.
[2] La Cour, 1931.

CULTURE MEDIA

1. Potato Agar.

About 250 grm. of unpeeled potatoes are carefully washed, cut into small pieces and boiled for about an hour. The extract is allowed to cool and settle; as much as possible of the clear, supernatant liquor is poured off and made up to 1000 c.c. with tap water. It is placed in a flask with 25 grm. of agar and sterilised. The quantity is sufficient for about three dozen Petri dishes.

2. Dung Agar.

About 1000 grm. of horse, cow, or rabbit dung are soaked in cold water for three days, the liquid is poured off and diluted till of the colour of straw; 2·5 grm. of agar are added for every 100 c.c. of the diluted fluid.

3. Pea Agar.

400 dried peas are boiled for an hour, the liquid is poured off, made up to 1000 c.c. and 25 grm. of agar are added.

4. Prune Agar.

25 good prunes are simmered for an hour, care being taken not to burst the skins; the liquid is poured off, and made up to 1000 c.c.; 200 grm. cane sugar and 50 grm. agar are added. If a larger amount of sugar is desired, decinormal solution of sodium carbonate may be added till the prune juice is brown instead of reddish. The agar will then set satisfactorily.

5. Barnes' Medium.

Tripotassium phosphate, K_3PO_4	0·1 grm.
Ammonium nitrate, NH_4NO_3	0·1 grm.
Potassium nitrate, KNO_3	0·1 grm.
Glucose	0·1 grm.
Agar	2·5 grm.
Distilled water	100·0 c.c.

6. Medium M, as no. 5 above, but with magnesium sulphate ($MgSO_4$) instead of potassium nitrate.

FIXATIVES

1. Flemming's fluid, strong:

1 % chromic acid in water	75 c.c.
Glacial acetic acid	5 c.c.
2 % osmic acid in water	20 c.c.

Flemming's strong fluid diluted with an equal volume of water (Flemming ½ S.) is to be preferred for delicate objects.

2. Flemming's fluid, weak:

1 % chromic acid in water	25 c.c.
1 % acetic acid in water	10 c.c.
1 % osmic acid in water	10 c.c.
Water	55 c.c.

3. 2 BD[1]:

1 % chromic acid in water	100 c.c.
1 % potassium bichromate in water	100 c.c.
2 % osmic acid in water	30 c.c.
5 % acetic acid in water	30 c.c.
Saponine	0·1 grm.

4. Merkel's fluid:

1 % chromic acid in water	25 c.c.
1 % platinic chloride in water	25 c.c.
Water	50 c.c.

5. Chromacetic mixture, strong:

1 % chromic acid in water	140 c.c.
Glacial acetic acid	1 c.c.

6. Chromacetic mixture, weak:

1 % chromic acid in water	140 c.c.
Glacial acetic acid	1 c.c.
Water	60 c.c.

7. Acetic alcohol:

Glacial acetic acid	20 or 25 c.c.
Absolute alcohol	80 or 75 c.c.

ALBUMEN

White of egg	50 c.c.
Glycerine	50 c.c.
Sodium salicylate	1 grm.

An egg is cautiously broken, the white is separated from the yolk, the sodium salicylate is dissolved in a very little water, the other ingredients are added and the whole well beaten and filtered.

GASTRIC JUICE

Benger's liquor pepticus: a few drops in 0·2 % hydrochloric acid.

STAINS

1. Safranin, Polychrome methylene blue and Orange tannin (Breinl's stain):

Safranin O, saturated aqueous solution	50 c.c.
safranin, alcohol soluble, saturated solution in absolute alcohol	50 c.c.
aniline oil	a few drops.

The mixture is improved by keeping for three months to ripen.

[1] La Cour, 1931.

Polychrome methylene blue	7 grm.
sodium carbonate	0·5 grm.
water	100 c.c.

This also is improved by keeping; ripening may be accelerated in an incubator.

Orange tannin: may be used as supplied by British Drug Houses.

Iodine	1 grm.
potassium iodide	1 grm.
80 % spirit	100 c.c.

2. Safranin, Gentian violet and Orange G (Flemming's triple stain):

Safranin, alcohol soluble, saturated alcoholic solution	50 c.c.
water	50 c.c.
aniline oil	3 drops.
Gentian violet	1 grm.
20 % spirit	100 c.c.

Orange G: saturated aqueous solution *or* saturated solution in clove oil.

3. Iron haematoxylin:

Haematoxylin	1 grm.
absolute alcohol	10 *or* 100 c.c.

This stock solution is ripened by keeping for some months or by the addition of a drop of hydrogen peroxide; 5 c.c. are then made up to 100 c.c. with water, and a drop of the mordant is added before use.

Iron alum, (mordant) $(NH_4)_2SO_4$, $Fe_2(SO_4)_3$, $24H_2O$	4 grm.
warm distilled water	100 c.c.

Iron alum (for washing out): A saturated solution is prepared by leaving excess of crystals at the bottom of a bottle containing distilled water. Care should be taken to keep the mouth of the bottle clean or to use a loose cover, as the evaporating solution fixes a stopper firmly in place.

Congo red:

Congo red	1 grm.
water	100 c.c.
ammonia	a few drops.

or Saturated solution of congo red in clove oil.

5. Safranin and light green:

Safranin: either of the methods of preparation described above; that for Breinl's stain is on the whole the better.

Light green: saturated solution in clove oil diluted with clove oil as required.

Saturated solutions are conveniently prepared by covering the bottom of a bottle with the dry stain and filling with clove oil; the saturated solution can be decanted off from time to time and filtered if necessary, and the bottle refilled till all the stain is dissolved.

6. Methylene blue and Erythrosin:

Methylene blue	1 grm.
water	100 c.c.

Erythrosin: saturated solution in clove oil diluted with clove oil as required.

7. Cotton blue:

Saturated solution of cotton blue in spirit ...	10 c.c.
pure glycerine	10 c.c.
water	80 c.c.

The material is left in the stain for about three days during which some of the water and spirit evaporate; it is then washed in dilute glycerine.

8. Erythrosin glycerine:

Saturated aqueous solution of erythrosin ...	50 c.c.
pure glycerine	100 c.c.
water	50 c.c.

The material is stained for about 20 minutes, well washed in 50% glycerine, and left in 50% glycerine for one or two days.

CLEANING SLIDES AND COVERSLIPS

Both slides and coverslips should be cleaned before use; in view of the greater delicacy of the coverslips they should be treated separately.

1. The slides or coverslips are placed in hot 5 % solution of potassium bichromate in water and a little concentrated sulphuric acid is added at intervals of 5 or 10 minutes; the mixture is kept bubbling for not less than 30 minutes. The slides or coverslips are then washed in running water for 12 hours, washed in distilled water and kept till required in 95 % spirit.

2. New coverslips may need only polishing with a silk rag; if this is inadequate they are usually sufficiently cleaned by keeping for about a week in 95 % spirit to which a few drops of hydrochloric acid have been added. They may then be washed in water or spirit and transferred to alcohol till needed, or may be washed in spirit and used at once.

3. Single coverslips are quickly and effectively cleaned by rubbing with soap and water between the thumb and finger, washing in water and washing in spirit.

BLEACHING AGENTS

1. Chlorine:

Potassium chlorate	a few crystals
concentrated hydrochloric acid	1 drop
When a yellow-green cloud of chlorine is visible, add 60 % spirit	60 to 100 c.c.

Prolonged washing in spirit is required after this reagent.

2. Hydrogen peroxide:

Hydrogen peroxide, 20 vol. solution	40 c.c.
60 or 80 % spirit	60 c.c.

LIST OF LITERATURE

The bibliography is limited to books and papers referred to in the text. The figures in square brackets after each title indicate the page or pages on which reference is made.

ALLEN, R. F. (1933). The spermatia of corn rust, Puccinia sorghi. Phytopath. XXIII, p. 923. **[307, 313.]**

—— (1934 i). A Cytological Study of Heterothallism in the Flax Rust. Journ. Agric. Research, XLIX, p. 765. **[313.]**

—— (1934 ii). A Cytological Study of Heterothallism in Puccinia sorghi. Journ. Agric. Research, XLIX, p. 1047. **[313.]**

—— (1935). A Cytological Study of Puccinia malvacearum from the Sporidium to the Teliospore. Journ. Agric. Research, LI, p. 801. **[313.]**

AMES, L. M. (1932). An hermaphrodite self-sterile but cross-fertile condition in *Pleurage anserina*. Bull. Torrey Bot. Club, LIX, p. 341. **[259.]**

—— (1934). Hermaphroditism involving self-sterility and cross-fertility in the Ascomycete *Pleurage anserina*. Mycologia, XXVI, p. 392. **[259.]**

ANDRUS, C. F. (1931). The Mechanism of Sex in Uromyces appendiculatus and U. vignae. Journ. Agric. Research, XLII, p. 559. **[313.]**

—— (1936). Cell relations in the perithecium of Ceratostomella multiannulata. Mycologia, XXVIII, p. 135. **[264.]**

ANDRUS, C. F. and HARTER, L. L. (1933). Morphology of Reproduction in Ceratostomella fimbriata. Journ. Agric. Research, XLVI, p. 1059. **[262.]**

APSTEIN, C. (1911). *Synchaetophagus balticus*, ein in *Synchaeta* lebender Pilz. Wissenschaftliche Meeresuntersuchungen, Abt. Kiel, Neue Folge, XII, p. 163. **[13, 110.]**

ARNAUD, G. (1913). La Mitose chez *Capnodium meridionale* et chez *Coleosporium Senecionis*. Bull. Soc. Myc. de France, XXIX, p. 345. **[303.]**

—— (1920). La famille des Parodiellinacées (Pyrenomycètes). C.R. Acad. Sci. CLXX, p. 202. **[187, 243.]**

ARNAUDOW, N. (1923). Ein neuer Rädertiere (Rotatoria)-fangender Pilz. (*Sommerstorffia spinosa* nov. gen., nov. sp.) Flora, CXVI, p. 109. **[13, 110.]**

—— (1925). Untersuchung über den Tiere fangenden Pilz Zoophagus insidians Som. Flora, CXVIII–CXIX, p. 1. **[13.]**

ARTHUR, J. C. (1917). Orange Rusts of *Rubus*. Bot. Gaz. LXIII, p. 501. **[320.]**

ASHBY, S. F. (1922). Oospores in Cultures of *Phytophthora Faberi*. Kew Bull. IX, p. 257. **[119.]**

ASHBY, S. F. and NOWELL, W. (1926). The Fungi of Stigmatomycosis. Ann. Bot. XL, p. 69. **[172.]**

ASHWORTH, D. (1931). *Puccinia Malvacearum* in Monosporidial Culture. Trans. Brit. Myc. Soc. XVI, p. 177. **[312.]**

—— (1935). The Receptive Hyphae of Rust Fungi. Ann. Bot. XLIX, p. 95. **[313.]**

ASHWORTH, J. H. (1923). On *Rhinosporidium seeberi* (Wernicke 1903), with special reference to its Sporulation and Affinities. Trans. Roy. Soc. Edin. LIII, p. 301. **[56, 64.]**

ASTHANA, R. P. and HAWKER, L. E. (1936). The Influence of Certain Fungi on the Sporulation of Melanospora destruens Shear and of Some Other Ascomycetes. Ann. Bot. L, p. 325. **[32.]**

ATANASOFF, D. (1919). A Novel Method of Ascospore Discharge. Mycologia, XI, p. 125. **[266.]**

ATKINS, D. (1929). On a Fungus Allied to the Saprolegniaceae found in the Pea-crab *Pinnotheres*. Journ. Marine Biol. Assoc. Plymouth, XVI, p. 203. [**110**.]

ATKINSON, G. F. (1894 i). Preliminary note on the swarm spores of *Pythium* and *Ceratiomyxa*. Bot. Gaz. XIX, p. 375. [**115**.]

—— (1894 ii). *Completoria complens* Lohde. Bot. Gaz. XIX, p. 467. [**145**.]

—— (1903). The genus *Harpochytrium* in the United States. Ann. Myc. I, p. 479. [**77**.]

—— (1905). Life-History of *Hypocrea alutacea*. Bot. Gaz. XL, p. 401. [**250**.]

—— (1906). The Development of *Agaricus campestris*. Bot. Gaz. XLII, p. 241. [**340**.]

—— (1909 i). Some fungus parasites of algae. Bot. Gaz. XLVIII, p. 321. [**71, 83**.]

—— (1909 ii). Some Problems in the Evolution of the Lower Fungi. Ann. Myc. VII, p. 441. [**55, 71, 105**.]

—— (1911). The origin and taxonomic value of the "veil" in *Dictyophora* and *Ithyphallus*. Bot. Gaz. L, p. 1. [**350**.]

—— (1914). The Development of *Agaricus arvensis* and *A. comtulus*. Am. Journ. Bot. I, p. 3. [**340**.]

—— (1915). Phylogeny and Relationships in the Ascomycetes. Ann. Miss. Bot. Gard. II, p. 315. [**165**.]

—— (1916). Origin and development of the lamellae in *Coprinus*. Bot. Gaz. LXI, p. 89. [**340**.]

AYERS, T. T. (1933). Growth of Dispira cornuta in Artificial Culture. Mycologia, XXV, p. 333. [**139**.]

—— (1935). Parasitism of Dispira cornuta. Mycologia, XXVII, p. 235. [**139**.]

AYYAR, T. S. R. (1928). Pythium aphanidermatum (Eds.) Fitz. on Opuntia dillenii Haw. Mem. Dept. Agric. India Bot. XVI, p. 191. [**114**.]

BACKUS, M. P. (1933). The development of the ascus and the occurrence of giant ascospores in Coccomyces hiemalis. Bull. Torrey Bot. Club, LX, p. 611. [**240**.]

—— (1934). Initiation of the ascocarp and associated phenomena in Coccomyces hiemalis. Contributions from Boyce Thompson Inst. of Research, VI, p. 339. [**240**.]

BAGCHEE, K. (1925). Cytology of the Ascomycetes. Pustularia bolarioides Ramsb. I. Spore Development. Ann. Bot. XXXIX, p. 127. [**216**.]

BAILEY, M. A. (1920). Puccinia malvacearum and the Mycoplasm Theory. Ann. Bot. XXXIV, p. 173. [**302**.]

BAINIER, G. (1903). Sur quelques espèces de Mucorinées nouvelles ou peu connues. Bull. Soc. Myc. France, XIX, p. 153. [**133**.]

BAIRD, E. A. (1924). The structure and behaviour of the nucleus in the life-history of Phycomyces nitens (Agardh) Kunze and Rhizopus nigricans Ehrbg. Trans. Wisconsin Acad. Sci. Arts and Letters, XXI, p. 357. [**132**.]

BALLS, W. L. (1908). Temperature and Growth. Ann. Bot. XXII, p. 558. [**31, 33**.]

BALLY, W. (1912). Cytologische Studien an Chytridineen. Jahrb. f. wiss. Bot. L, p. 95. [**65**.]

—— (1913). Die Chytridineen im Lichte der neueren Kernforschung. Myc. Centralbl. II, p. 289. [**65**.]

—— (1919). Einige Bemerkungen zu den amitotischen Kernteilungen der Chytridineen. Ber. d. deutsch. bot. Ges. XXXVII, p. 103. [**65**.]

BARCLAY, A. (1891). On the Life-History of a remarkable Uredine on *Jasminum grandiflorum*. Trans. Linn. Soc. Bot. ii, CXLI, p. 141. [**319**.]

BARGER, G. (1932). Ergot and Ergotism. Gurney and Jackson, London and Edinburgh. [**253**.]

Barker, B. T. P. (1901). A conjugating "Yeast". Proc. Roy. Soc. lxviii, p. 345. [**174**.]

—— (1903 i). The Morphology and Development of the Ascus in Monascus. Ann. Bot. xvii, p. 167. [**183, 225**.]

—— (1903 ii). The Development of the Ascocarp in Ryparobius. Report Brit. Assoc. p. 849. [**225**.]

—— (1904). Further Observations on the Ascocarp of Ryparobius. Report Brit. Assoc. p. 825. [**225**.]

Barnes, B. (1924). A New Aspect of the Dung Flora. Report Imp. Bot. Conf. p. 346. [**39**.]

—— (1925). A Contribution to the Morphology and Physiology of Acmosporium Corda and Lachnea Fries. Report Brit. Assoc. p. 358. [**39**.]

—— (1928). Variations in Eurotium herbariorum (Wigg.) Link, induced by the Action of High Temperatures. Ann. Bot. xlii, p. 783. [**42**.]

—— (1930). Variations in Botrytis cinerea Pers., Induced by the Action of High Temperatures. Ann. Bot. xliv, p. 825. [**42**.]

—— (1933). *Helotium ciliatosporum* Boudier. Trans. Brit. Myc. Soc. xviii, p. 76. [**227**.]

—— (1934 i). Spore Discharge in Basidiobolus ranarum Eidam. Ann. Bot. xlviii, p. 453. [**146**.]

—— (1934 ii). The biology of aquatic fungi. Report Brit. Assoc. p. 320. [**13**.]

—— (1935 i). On Variation in *Thamnidium elegans* Link, induced by the Action of High Temperatures. Trans. Brit. Myc. Soc. xix, p. 291. [**42**.]

—— (1935 ii). Induced Variation. Trans. Brit. Myc. Soc. xx, p. 17. [**42**.]

Barnes, B. and Melville, R. (1932). Notes on British Aquatic Fungi. Trans. Brit. Myc. Soc. xvii, p. 82. [**13, 90, 96, 97, 103, 110**.]

Barrett, J. T. (1912 i). Development and Sexuality of some Species of Olpidiopsis. Ann. Bot. xxvi, p. 209. [**53, 55, 59, 68**.]

—— (1912 ii). The Development of *Blastocladia strangulata*. Bot. Gaz. liv, p. 352. [**91**.]

Barton-Wright, E. C. and Boswell, J. G. (1931). The Biochemistry of Dry-rot in Wood. II. An Investigation of the Products of Decay of Spruce Wood rotted by *Merulius lachrymans*. Biochem. Journ. xxv, p. 494. [**14**.]

de Bary, A. (1853). Untersuchungen über die Brandpilze und die durch sie verursachten Krankheiten der Pflanzen. Müller, Berlin. [**292**.]

—— (1863 i). Recherches sur le développement de quelques champignons parasites. Ann. Sci. Nat. Bot. 4me sér. xx, p. 5. [**122**.]

—— (1863 ii). Ueber die Fruchtentwickelung der Ascomyceten. Engelmann, Leipzig. [**191, 204**.]

—— (1864). Syzygites megalocarpus. Beitr. z. Morph. u. Physiol. d. Pilze, i, p. 74. [**126**.]

—— (1870). *Eurotium*, *Erysiphe*, *Cincinnobolus* nebst Bemerkungen über die Geschlechtsorgane der Ascomyceten. Beitr. z. Morph. u. Physiol. d. Pilze, iii, p. 1. [**180**.]

—— (1876). Researches into the nature of the Potato fungus, *Phytophthora infestans*. Journ. Bot. Lond. xiv, pp. 105, 149; and Journ. Roy. Soc. Agric. ser. 2, xii, p. 239. [**115, 118**.]

—— (1881 i). Untersuchungen über die Peronosporeen und Saprolegnieen und die Grundlagen eines natürlichen Systems der Pilze. Beitr. z. Morph. u. Physiol. d. Pilze, iv, p. 85. [**57, 104, 112**.]

—— (1881 ii). Zur Kenntnis der Peronosporeen. Bot. Zeit. xxxix, pp. 521, 537, 553, 569, 585, 601, 617. [**112**.]

—— (1887). Comparative Morphology and Biology of the Fungi, Mycetozoa and Bacteria. Eng. Trans. Clarendon Press, Oxford. [**1, 10, 55, 73, 302, 347**.]

de Bary, A. (1888). Species der Saprolegnieen. Bot. Zeit. xlvi, pp. 597, 613, 629, 645. [**108**.]

de Bary, A. and Woronin, M. (1863). Beitrag zur Kenntnis der Chytridineen. Ber. über die Verhandlungen der Naturf. Ges. z. Freiburg, iii, p. 22. [**65**.]

Bauch, R. (1923). Über *Ustilago longissima* und ihre Varietät *macrospora*. Zeitschr. f. Bot. xv, p. 241. [**288, 294**.]

—— (1925). Untersuchungen über zweisporige Hymenomyceten. I. Haploide Parthenogenesis bei Camarophyllus virgineus. Zeitschr. f. Bot. xviii, p. 337. [**328**.]

—— (1927). Untersuchungen über zweisporige Hymenomyceten. II. Kern-degeneration bei einigen *Clavaria*-Arten. Arch. f. Protistenk. lviii, p. 285. [**328**.]

Bayliss, J. S. (1911). Observations on *Marasmius oreades* and *Clitocybe gigantea* as parasitic fungi causing "Fairy-Rings". Journ. Econ. Biol. vi, p. 111. [**15**.]

de Beauchamp, P. (1914). L'évolution et les affinités des Protistes du genre *Dermocystidium*. C.R. Acad Sci. clviii, p. 1359. [**56, 64**.]

Bennett, F. T. (1935). Corticium Disease of Turf. Journ. of the Board of Greenkeeping Research, iv, p. 32. [**334**.]

Bensaude, M. (1918). Recherches sur le cycle évolutif et la sexualité chez les Basidiomycètes. Bouloy, Nemours. [**159, 287, 289**.]

Berkeley, M. J. (1842). On an edible Fungus from Tierra del Fuego and an allied Chilian species. Trans. Linn. Soc. xix, p. 37. [**229**.]

—— (1843). On some Entomogenous Sphaeriae. Journ. Bot. Lond. ii, p. 205. [**250**.]

—— (1847 i). Propagation of Bunt. Trans. Hort. Soc. Lond. ii, p. 113. [**292**.]

—— (1847 ii). Fungi. Hooker's Flora Antarctica, ii, p. 453. [**229**.]

—— (1848). Decades of Fungi. Journ. Bot. Lond. ii, vii, p. 576. [**229**.]

Bernard, N. (1909). L'évolution dans la symbiose des Orchidées et leurs champignons commensaux. Ann. Sci. Nat. Bot. 9me sér. ix, p. 1. [**23, 24**.]

—— (1911 i). Sur la fonction fungicide des bulbes d'Ophrydées. Ann. Sci. Nat. Bot. 9me sér. xiv, p. 223. [**23**.]

—— (1911 ii). Les mycorhizes des Solanum. Ann. Sci. Nat. Bot. 9me sér. xiv, p. 235. [**23, 27**.]

Bertrand, G. (1902). Sur le bleuissement de certains champignons du genre Boletus. Ann. Inst. Pasteur, xvi, p. 179. [**344**.]

Bezssonof, N. (1914). Sur quelques faits relatifs à la formation du périthèce et la délimitation des ascospores chez les Erysiphaceae. C.R. Acad. Sci. clviii, p. 1123. [**192, 194**.]

—— (1918 i). Über die Bildung der Fruchtkörper des *Penicillium glaucum* in konzentrierten Zuckerlösungen. Ber. d. deutsch. bot. Ges. xxxvi, p. 225. [**40, 182**.]

—— (1918 ii). Über das Wachstum der Aspergillaceen und anderer Pilze auf stark zuckerhaltigen Nährböden. (Vorl. Mitt.) Ber. d. deutsch. bot. Ges. xxxvi, p. 646. [**40**.]

—— (1919). Über die Züchtung von Pilzen auf hochkonzentrierten rohr-zuckerhaltigen Nährböden und über die Chondriomfrage. Ber. d. deutsch. bot. Ges. xxxvii, p. 136. [**40**.]

Biffen, R. H. (1899). A Fat-Destroying Fungus. Ann. Bot. xiii, p. 363. [**17**.]

—— (1901). On the Biology of Bulgaria polymorpha Wett. Ann. Bot. xv, p. 119. [**229**.]

—— (1907). Studies in the Inheritance of Disease Resistance. I. Journ. Agric. Sci. ii, p. 109. [**29**.]

—— (1908). First Record of Two Species of Laboulbeniaceae for Britain. Trans. Brit. Myc. Soc. iii, p. 83. [**274**.]

—— (1912). Studies in the Inheritance of Disease Resistance. II. Journ. Agric. Sci. iv, p. 421. [**29**.]

BLAAUW, A. H. (1914). Licht und Wachstum, I. Zeitschr. f. Bot. VI, p. 641. [**35**.]

BLACKMAN, V. H. (1904). On the Fertilization, Alternation of Generations and general Cytology of the Uredineae. Ann. Bot. XVIII, p. 323. [**303, 310**.]

BLACKMAN, V. H. and FRASER, H. C. I. (1905). Fertilization in Sphaerotheca. Ann. Bot. XIX, p. 567. [**191**.]

—— —— (1906 i). Further Studies on the Sexuality of the Uredineae. Ann. Bot. XX, p. 35. [**303, 310**.]

—— —— (1906 ii). On the Sexuality and Development of the Ascocarp in *Humaria granulata*. Proc. Roy. Soc. LXXVII, p. 354. [**212**.]

BLACKMAN, V. H. and WELSFORD, E. J. (1912). The Development of the Perithecium of Polystigma rubrum. Ann. Bot. XXVI, p. 761. [**248**.]

—— —— (1916). Studies in the Physiology of Parasitism. II. Infection by Botrytis cinerea. Ann. Bot. XXX, p. 389. [**19**.]

BLAKESLEE, A. F. (1904). Sexual Reproduction in the Mucorineae. Proc. Am. Acad. Arts and Sci. XL, p. 205. [**126, 131, 133, 135, 136, 137**.]

—— (1906 i). Zygospore Germinations in the Mucorineae. Ann. Myc. IV, p. 1. [**132**.]

—— (1906 ii). Two Conidia-bearing Fungi. Bot. Gaz. XL, p. 161. [**136**.]

—— (1907). Heterothallism in the Bread Mould *Rhizopus nigricans*. Bot. Gaz. XLIII, p. 415. [**135**.]

—— (1913). Conjugation in the Heterogamic Genus *Zygorhynchus*. Myc. Centralbl. II, p. 241. [**135**.]

—— (1915). Sexual Reactions between Hermaphroditic and Dioecious Mucors. Woods Hole Biol. Bull. XIX, p. 87. [**133**.]

—— (1920). Sexuality in Mucors. Science, N.S. LI, pp. 375, 403. [**133**.]

BLAKESLEE, A. F., CARTLEDGE, J. L. and WELCH, D. S. (1921). Sexual Dimorphism in *Cunninghamella*. Bot. Gaz. LXXII, p. 185. [**136**.]

BLOMFIELD, J. E. and SCHWARTZ, E. J. (1910). Some Observations on the Tumours of Veronica Chamaedrys. Ann. Bot. XXIV, p. 35. [**46, 48**.]

BOAS, F. (1918). Zur Ernährungsphysiologie einiger Pilze. Ann. Myc. XVI, p. 229. [**40**.]

—— (1919 i). Bemerkungen über konidienbildende Stoffe bei Pilzen. Ber. d. deutsch. bot. Ges. XXXVII, p. 57. [**40**.]

—— (1919 ii). Selbstvergiftung bei *Aspergillus niger*. Ber. d. deutsch. bot. Ges. XXXVII, p. 63. [**40**.]

BOEDIJN, K. (1923). On the Development of Stigmatomyces. Nederlandische Mycologische Vereeniging, XIII, p. 91. [**276**.]

BOUDIER, E. (1905–10). Icones Mycologicae. Lhomme, Paris. [**337**.]

—— (1907). Histoire et Classification des Discomycètes d'Europe. Paris (Libraire des Sci. Nat.). [**200**.]

BOULANGER, E. (1904). La culture artificielle de la Truffe. Bull. Soc. Myc. de France, XX, p. 77. [**236**.]

—— (1906 i). Note sur la Truffe. Bull. Soc. Myc. de France, XXII, p. 42. [**236**.]

—— (1906 ii). Germination de la spore échinulée de la Truffe. Bull. Soc. Myc. de France, XXII, p. 138. [**236**.]

BOYLE, C. (1921). Studies in the Physiology of Parasitism. VI. Infection by Sclerotinia Libertiana. Ann. Bot. XXXV, p. 337. [**19**.]

BRACHER, R. (1924). Notes on *Rhytisma Acerinum* and *Rhytisma Pseudoplatani*. Trans. Brit. Myc. Soc. IX, p. 183. [**241**.]

BREFELD, O. (1874). Botanische Untersuchungen über Schimmelpilze. II. *Penicillium*, p. 1. Felix, Leipzig. [**182**.]

—— (1877). Botanische Untersuchungen über Schimmelpilze. III. Basidiomyceten, p. 2. Felix, Leipzig. [**2, 36, 331**.]

—— (1883). Botanische Untersuchungen über Hefenpilze. V. Die Brandpilze. Felix, Leipzig. [**292**.]

BREFELD, O. (1884). Botanische Untersuchungen über Myxomyceten und Entomophthoreen. Untersuch. Gesammt. Mycol. VI, p. 34. [**145**.]

—— (1888). Untersuchungen aus dem Gesammtgebiete der Mykologie. VII. Basidiomyceten, II. Kön. bot. Inst. in Münster i.W. [**324, 333, 337**.]

—— (1891). Die Hemiasci und die Ascomyceten. Untersuch. Gesammt. Mycol. IX. [**225**.]

BRIERLEY, W. B. (1913). The Structure and Life-History of *Leptosphaeria Lemaneae* (Cohn). Mem. and Proc. Manchester Lit. and Phil. Soc. LVII, 2, p. 1. [**13, 267**.]

—— (1915). The "Endoconidia" of Thielavia basicola Zopf. Ann. Bot. XXIX, p. 483. [**178**.]

—— (1917). Spore Germination in Onygena equina Willd. Ann. Bot. XXXI, p. 12. [**184**.]

BRODIE, H. J. (1931). The Oidia of Coprinus lagopus and their Relation with Insects. Ann. Bot. XLV, p. 315. [**289**.]

—— (1932). Oidial Mycelia and the Diploidisation Process in Coprinus lagopus. Ann. Bot. XLVI, p. 727. [**289**.]

—— (1935 i). The Heterothallism of *Paneolus subbalteatus* Berk., a Sclerotium-producing Agaric. Canadian Journ. Res. XII, p. 657. [**290**.]

—— (1935 ii). The Oidia of *Psilocybe coprophila* and the Pairing Reactions of Monosporous Mycelia. Canadian Journ. Res. XII, p. 661. [**290**.]

BROOKS, F. T. (1909). Notes on Polyporus squamosus Huds. New Phyt. VIII, p. 350. [**344**.]

—— (1910). The Development of Gnomonia erythrostoma, the Cherry-Leaf-Scorch Disease. Ann. Bot. XXIV, p. 585. [**268**.]

—— (1911–12). "Silver-leaf" Disease. Journ. Agric. Sci. IV, p. 133. [**335**.]

—— (1912–13). Silver-leaf Disease. II. Journ. Agric. Sci. V, p. 288. [**335**.]

—— (1923). Some Present-day Aspects of Mycology. Trans. Brit. Myc. Soc. IX, p. 14. [**20**.]

—— (1924). Epidemic Plant Diseases. Trans. Brit. Myc. Soc. IX, p. 229. [**20**.]

—— (1925). *Polyporus adustus* (Willd.) Fr. as a Wound Parasite of Apple Trees. Trans. Brit. Myc. Soc. X, p. 225. [**29, 344**.]

—— (1935). Some Aspects of Plant Pathology. Report Brit. Assoc. p. 169. [**20, 30**.]

BROOKS, F. T. and BAILEY, M. A. (1919). Silver-leaf Disease. III. Journ. Agric. Sci. IX, p. 189. [**335**.]

BROOKS, F. T. and BRENCHLEY, G. H. (1929). Infection Experiments on Plum Trees in Relation to *Stereum purpureum* and Silver-leaf Disease. New Phyt. XXVIII, p. 218. [**335**.]

BROOKS, F. T. and HANSFORD, C. G. (1923). Mould Growths upon Cold-store Meat. Trans. Brit. Myc. Soc. VIII, p. 113. [**39**.]

BROOKS, F. T. and MOORE, W. C. (1926). Silver-leaf Disease. V. Journ. Pomology and Hort. Sci. V, p. 61. [**335**.]

BROOKS, F. T. and SHARPLES, A. (1923). Pink Disease of Plantation Rubber. Ann. Applied Biol. II, p. 58. [**334**.]

BROOKS, F. T. and STOREY, H. H. (1923). Silver-leaf Disease. IV. Journ. Pomology and Hort. Sci. III, p. 1. [**335**.]

BROWN, A. M. (1932). Diploidisation of Haploid by Diploid Mycelia of *Puccinia Helianthi* Schw. Nature, CXXX, p. 777. [**289**.]

—— (1935). A study of coalescing haploid pustules in Puccinia Helianthi. Phytopath. XXV, p. 1085. [**314**.]

BROWN, H. B. (1913). Studies in the Development of Xylaria. Ann. Myc. XI, p. 1. [**273**.]

BROWN, H. P. (1915). A Tuber Rot accompanying *Hymenochaete rubiginosa* (Schrad.) Lév. Mycologia, VII, p. 1. [**335**.]

BROWN, W. (1915). Studies in the Physiology of Parasitism. I. The Action of Botrytis cinerea. Ann. Bot. XXIX, p. 313. [**19**.]

—— (1916). Studies in the Physiology of Parasitism. III. On the Relation between the "Infection Drop" and the underlying Host Tissue. Ann. Bot. XXX, p. 399. [**19**.]

—— (1917 i). On the Physiology of Parasitism. New Phyt. XVI, p. 109. [**19**.]

—— (1917 ii). Studies in the Physiology of Parasitism. IV. On the Distribution of Cytase in Cultures of Botrytis cinerea. Ann. Bot. XXXI, p. 489. [**19**.]

—— (1922 i). Studies in the Physiology of Parasitism. VIII. On the Exosmosis of Nutrient Substances from the Host Tissue into the Infection Drop. Ann. Bot. XXXVI, p. 101. [**19, 33**.]

—— (1922 ii). On the Germination and Growth of Fungi at various Temperatures and in various Concentrations of Oxygen and Carbon Dioxide. Ann. Bot. XXXVI, p. 285. [**19, 31, 40**.]

—— (1922 iii). Studies in the Physiology of Parasitism. IX. The Effect on the Germination of Fungal Spores of Volatile Substances arising from Plant Tissues. Ann. Bot. XXXVI, p. 285. [**19**.]

—— (1923). Experiments on the Growth of Fungi in Culture Media. Ann. Bot. XXXVII, p. 105. [**19, 31, 32, 40**.]

—— (1925). Studies in the Genus Fusarium. II. An Analysis of Factors which determine the Growth-forms of Certain Strains. Ann. Bot. XXXIX, p. 373. [**31, 40**.]

—— (1926). Studies in the Genus Fusarium. IV. On the Occurrence of Saltations. Ann. Bot. XL, p. 223. [**40, 43**.]

BROWN, W. and HORNE, A. S. (1926). Studies in the Genus Fusarium. III. An Analysis of Factors which determine Certain Microscopic Features of Fusarium Strains. Ann. Bot. XL, p. 203. [**40**.]

BROWN, W. H. (1909). Nuclear Phenomena in *Pyronema confluens*. Johns Hopkins Univ. Circ. VI, p. 42. [**204**.]

—— (1910). The Development of the Ascocarp in *Leotia*. Bot. Gaz. L, p. 443. [**158, 233**.]

—— (1911). The Development of the Ascocarp in *Lachnea scutellata*. Bot. Gaz. LII, p. 275. [**158**.]

—— (1915). The Development of *Pyronema confluens*, var. *inigneum*. Am. Journ. Bot. II, p. 289. [**206**.]

BRUNSWIK, H. (1924). Untersuchungen über die Geschlechts- und Kernverhältnisse bei der Hymenomyzetengattung Coprinus. Botan. Abhandl. herausgeg. v. Goebel, V, p. 152. [**288**.]

BUCHANAN, J. (1885). On *Cyttaria Purdiei*. Trans. New Zealand Inst. XVII, p. 317. [**229**.]

BUCHNER, P. (1930). Tier und Pflanze in Symbiose. Borntraeger, Berlin. [**14, 27**.]

BUCHOLTZ, F. (1903). Zur Morphologie und Systematik der Fungi hypogaei. Ann. Myc. I, p. 152. [**237**.]

—— (1908). Zur Entwicklung der Choiromyces-Fruchtkörper. Ann. Myc. VI, p. 539. [**236**.]

—— (1910). Zur Entwicklungsgeschichte des Balsamiaceen-Fruchtkörpers, nebst Bemerkungen zur Verwandtschaft der Tuberineen. Ann. Myc. VIII, p. 121. [**236**.]

—— (1912). Beiträge zur Kenntnis der Gattung Endogone Link. Beih. Bot. Centralbl. XXIX, 2, p. 147. [**138**.]

BUCKLEY, J. J. C. and CLAPHAM, P. A. (1929). The Invasion of Helminth Eggs by Chytridiacean Fungi. Journ. Helminthology, VII, p. 1. [**81**.]

BUHR, H. (1932). Untersuchungen über zweisporige Hymenomyceten. Arch. f. Protistenk. LXXVII, p. 125. [**328**.]

BULLER, A. H. R. (1905). The Destruction of Wooden Paving Blocks by the Fungus Lentinus lepideus. Journ. Econ. Biol. I, p. 2. [**14, 343.**]

—— (1906). *Polyporus squamosus* as a Timber-destroying Fungus. Journ. Econ. Biol. I, p. 101. [**14, 344.**]

—— (1909 i). The Destruction of Wood by Fungi. Sci. Prog. XI, p. 1. [**14.**]

—— (1909 ii). Researches on Fungi, vol. I. Longmans, Green and Co., London. [**14, 35, 36, 37, 342.**]

—— (1921). Upon the Ocellus Function of the Subsporangial Swelling of *Pilobolus*. Trans. Brit. Myc. Soc. VII, p. 61. [**34, 131.**]

—— (1922 i). The Basidial and Oidial Fruit-bodies of *Dacryomyces deliquescens*. Trans. Brit. Myc. Soc. VII, p. 226. [**289, 328, 337.**]

—— (1922 ii). Researches on Fungi, vol. II. Longmans, Green and Co., London. [**15, 285, 329, 331, 333.**]

—— (1924). Researches on Fungi, vol. III. Longmans, Green and Co., London. [**286, 306, 308, 330.**]

—— (1931). Researches on Fungi, vol. IV. Longmans, Green and Co., London. [**289, 290.**]

—— (1933). Researches on Fungi, vol. V. Longmans, Green and Co., London. [**4, 286, 300.**]

BULLER, A. H. R. and VANTERPOOL, T. C. (1925). Violent Spore-Discharge in *Tilletia Tritici*. Nature, CXVI, p. 934. [**300.**]

VON BÜREN, G. (1915). Die schweizerischen Protomycetaceen mit besonderer Berücksichtigung ihrer Entwicklungsgeschichte und Biologie. Beitr. z. Kryptogamenflora d. Schweiz, V, i, p. 1. [**20, 87, 88, 89.**]

—— (1922). Weitere Untersuchungen über die Entwicklungsgeschichte und Biologie der Protomycetaceen. Beitr. z. Kryptogamenflora d. Schweiz, V, iii, p. 1. [**87, 88, 89.**]

BURGEFF, H. (1909). Die Wurzelpilze der Orchideen, ihre Kultur und ihr Leben in der Pflanze. Fischer, Jena. [**22.**]

—— (1914). Untersuchungen über Variabilität, Sexualität und Erblichkeit bei Phycomyces nitens Kunze. Flora, CVII, p. 259. [**132.**]

—— (1915). Untersuchungen über Variabilität, Sexualität und Erblichkeit bei Phycomyces nitens Kunze. II. Flora, CVIII, p. 353. [**132.**]

—— (1920 i). Über den Parasitismus des Chaetocladium und die heterocaryotische Natur der von ihm auf Mucorineen erzeugten Gallen. Zeitschr. f. Bot. XII, p. 1. [**137.**]

—— (1920 ii). Sexualität und Parasitismus bei Mucorineen. Ber. d. deutsch. bot. Ges. XXXVIII, p. 318. [**133.**]

—— (1924). Untersuchungen über Sexualität und Parasitismus bei Mucorineen. Bot. Abhandlungen, IV, p. 1. [**133, 137.**]

—— (1925). Über Arten und Artkreuzung in der Gattung Phycomyces Kunze. Flora, CXVII–CXIX, p. 40. [**132.**]

—— (1928). Variabilität, Vererbung und Mutation bei Phycomyces Blakesleeanus Bgff. Zeitschr. Induk. Abstamm. u. Vererbungslehre, XLIX, p. 26. [**132.**]

—— (1930). Parasitismus, Wasserbewegung und Stofftransport. Ein Beitrag zur Physiologie der Mucorineenparasiten (Sicyonten) mit einem Vorwort über das Sexualitätsproblem. Zeitschr. f. Bot. XXIII, p. 589. [**133.**]

—— (1931). Organisation und Entwicklung tropischer Orchideen-Saprophyten. (Vorläufige Mitteilung.) Ber. d. deutsch. bot. Ges. XLIX, p. (46). [**23.**]

BURGEFF, H. and SEYBOLD, A. (1927). Zur Frage der biochemischen Unterscheidung der Geschlechter. Zeitschr. f. Bot. XIX, p. 497. [**128.**]

BURGER, O. F. (1919). Sexuality in *Cunninghamella*. Bot. Gaz. LXVIII, p. 134. [**136.**]

BURT, E. A. (1894). A South American Anthurus, its structure and development. Mem. Boston Soc. Nat. Hist. III, p. 497. [**325, 347.**]

BURT, E. A. (1896 i). The Development of Mutinus caninus (Huds.) Fr. Ann. Bot. X, p. 343. **[350.]**

—— (1896 ii). The Phalloidineae of the United States. I and II. Bot. Gaz. XXII, pp. 273, 379. **[347.]**

—— (1897). The Phalloidineae of the United States. III. Bot. Gaz. XXIV, p. 73. **[347.]**

—— (1901). Structure and Nature of Tremella mycetophila Peck. Bull. Torrey Bot. Club, XXVIII, p. 285. **[325.]**

—— (1915). The Thelephoraceae of North America. IV. Exobasidium. Ann. Miss. Bot. Gard. II, p. 656. **[336.]**

BUTLER, E. J. (1907). An Account of the Genus *Pythium* and some Chytridiaceae. Mem. Dept. Agric. India Bot. i, V, p. 107. **[13, 14, 53, 54, 55, 59, 64, 69, 79, 105, 112, 114, 115, 117.]**

—— (1911). On Allomyces, a new aquatic Fungus. Ann. Bot. XXV, p. 1023. **[13, 91, 93.]**

—— (1913). Pythium de Baryanum Hesse. Mem. Dept. Agric. India Bot. V, V, p. 262. **[55, 112, 115.]**

—— (1928). Morphology of the Chytridiacean Fungus, Catenaria anguillulae, in Liver-Fluke Eggs. Ann. Bot. XLII, p. 813. **[81.]**

BUTLER, E. J. and KULKARNI, G. S. (1913). Studies in Peronosporaceae. *Colocasia* Blight caused by *Phytophthora Colocasiae* Rac. Mem. Dept. Agric. India Bot. V, V, p. 233. **[115.]**

BUTLER, J. B. and BUCKLEY, J. J. C. (1927). *Catenaria anguillulae* as a parasite of the ova of *Fasciola hepatica*. Sci. Proc. Roy. Dublin Soc. XVIII, p. 497. **[13, 81.]**

BUTLER, J. B. and HUMPHRIES, A. (1932). On the Cultivation in Artificial Media of *Catenaria Anguillulae*, a Chytridiacean Parasite of the Ova of the Liver Fluke, *Fasciola Hepatica*. Sci. Proc. Roy. Dublin Soc. XX, p. 301. **[81.]**

CARRUTHERS, D. (1911). Contributions to the Cytology of Helvella crispa. Ann. Bot. XXV, p. 243. **[232.]**

CAULLERY, M. and MESNIL, F. (1905). Recherches sur les Haplosporidies. Arch. Zool. Expt. et Gén. iv, IV, p. 101. **[56.]**

CAVERS, F. (1903). On Saprophytism and Mycorhiza in the Hepaticae. New Phyt. II, p. 30. **[23.]**

—— (1904). On the Structure and Biology of Fegatella conica. Ann. Bot. XVIII, p. 95. **[23.]**

CAYLEY, D. M. (1921). Some Observations on the Life-history of Nectria galligena Bres. Ann. Bot. XXXV, p. 79. **[248.]**

—— (1923 i). Fungi associated with "Die Back" in Stone Fruit Trees. Ann. Applied Biol. X, p. 253. **[268.]**

—— (1923 ii). The Phenomenon of Mutual Aversion between Mono-Spore Mycelia of the same Fungus (*Diaporthe perniciosa* Marchal). Journ. Genetics, XIII, p. 353. **[155.]**

—— (1929). Some observations on Mycetozoa of the genus *Didymium*. Trans. Brit. Myc. Soc. XIV, p. 227. **[45.]**

CHATTON, E. (1908). Sur la reproduction et les affinités du *Blastulidium paedophthorum* C. Perez. C.R. Soc. Biol. Paris, LXIV, p. 34. **[56.]**

CHATTON, E. and BRODSKY, A. (1909). Le Parasitisme d'une Chytridinée du Genre *Sphaerita* Dang., chez *Amoeba limax* Dujard. Arch. f. Protistenk. XVII, p. 1. **[56, 64.]**

CHATTON, E. and ROUBAUD, E. (1909). Sur un *Amoebidium* du rectum des larves des *Simulies*. (*Simulium argyreatum* Meig. et *S. fasciatum* Meig.) C.R. Soc. Biol. Paris, LXVI, p. 701. **[13, 56.]**

CHAUDHURI, H. (1924). A Description of Colletotrichum biologicum n.sp., and Observations on the Occurrence of a Saltation in the Species. Ann. Bot. XXXVIII, p. 735. **[43.]**

CHOW, C. H. (1934). Contribution à l'Étude du développement des Coprins. Botaniste, XXVI, p. 89. [**289**.]

CHRISTMAN, A. H. (1905). Sexual Reproduction in the Rusts. Bot. Gaz. XXXIX, p. 267. [**310**.]

—— (1907 i). The Nature and Development of the Primary Uredospore. Trans. Wisconsin Acad. Sci. Arts and Letters, XV, p. 517. [**314**.]

—— (1907 ii). Alternation of Generations and the Morphology of the Spore-forms in Rusts. Bot. Gaz. XLIV, p. 81. [**319, 321**.]

CHRISTOPH, H. (1921). Untersuchungen über die mykotrophen Verhältnisse der "Ericales" und die Keimung der Pirolaceen. Beih. Bot. Centralbl. XXXVIII, p. 115. [**24**.]

CHUPP, C. (1917). Studies in Clubroot of Cruciferous Plants. Cornell Univ. Agric. Expt. Stat. Coll. of Agric. Bull. 387, p. 421. [**47, 48**.]

CHURCH, A. H. (1893). A Marine Fungus. Ann. Bot. VII, p. 399. [**13**.]

CLARK, J. F. (1902). On the toxic properties of some copper compounds, with special reference to Bordeaux Mixture. Bot. Gaz. XXXIII, p. 45. [**31**.]

CLAUSSEN, P. (1905). Zur Entwickelungsgeschichte der Ascomyceten. *Boudiera*. Bot. Zeit. LXIII, p. 1. [**202**.]

—— (1908). Über Eientwicklung und Befruchtung bei *Saprolegnia monoica*. Ber. d. deutsch. bot. Ges. XXVI (Festschrift), p. 144. [**108**.]

—— (1912). Zur Entwicklungsgeschichte der Ascomyceten. *Pyronema confluens*. Zeitschr. f. Bot. IV, p. 1. [**158, 159, 204**.]

—— (1921). Entwicklungsgeschichtliche Untersuchungen über den Erreger der als "Kalkbrut" bezeichneten Krankheit der Bienen. Arbeiten aus der biol. Reichsanstalt f. Land- und Forstwirtschaft, X, p. 469. [**171**.]

CLÉMENCET, M. (1932). Contribution à l'étude du développement et de l'anatomie des Ascomycètes hypogés. Les Elaphomycètacées. Botaniste, XXIV, p. 1. [**186**.]

CLINTON, G. P. (1902). *Cladochytrium Alismatis*. Bot. Gaz. XXXIII, p. 49. [**78**.]

—— (1911). Oospores of potato blight, *Phytophthora infestans*. Biennial Report, Conn. Agric. Expt. Stat. p. 753. [**118**.]

COKER, W. C. (1923). The Saprolegniaceae, with notes on other Water Molds. Univ. of North Carolina Press, p. 1. [**10, 13, 73, 90, 91, 103, 104, 105, 108**.]

—— (1927). Other water molds from the soil. Journ. Elisha Mitchell Sci. Soc. XLII, p. 207. [**14, 103**.]

COKER, W. C. and BRAXTON, H. H. (1926). New water molds from the soil. Journ. Elisha Mitchell Sci. Soc. XLII, p. 139. [**14, 103**.]

COLLEY, R. G. (1918). Parasitism, Morphology and Cytology of *Cronartium ribicola*. Journ. Agric. Research, XV, p. 619. [**323**.]

COLLINS, E. J. (1935). The Problem of Immunity to Wart Disease (Synchytrium endobioticum (Schilb.) Perc.) in the Potato. Ann. Bot. XLIX, p. 479. [**65**.]

COLLINS, M. I. (1920). Note on certain variations of the sporangia in a species of Saprolegnia. Proc. Linn. Soc. New South Wales, XLV, p. 277. [**38, 90**.]

COLSON, B. (1934). The Cytology and Morphology of Neurospora tetrasperma Dodge. Ann. Bot. XLVIII, p. 212. [**246**.]

—— (1935). The Cytology of the Mushroom Psalliota campestris Quel. Ann. Bot. XLIX, p. 1. [**326**.]

CONARD, H. S. (1915). The structure and development of Secotium agaricoides. Mycologia, VII, p. 94. [**343**.]

CONTE, A. and FAUCHERON, L. (1907). Présence de levures dans le corps adipeux de divers Coccides. C.R. Acad. Sci. CXLV, p. 1223. [**27**.]

COOK, W. R. I. (1933). A monograph of the Plasmodiophorales. Arch. f. Protistenk. LXXX, p. 179. [**46**.]

COOK, W. R. I. (1935). The Genus *Lagenidium* Schenk, with special reference to *L. Rabenhorstii* Zopf and *L. entophytum* Zopf. Arch. f. Protistenk. LXXXVI, p. 58. **[83.]**

COOK, W. R. I. and NICHOLSON, W. H. (1933). A Contribution to Our Knowledge of Woronina polycystis Cornu. Ann. Bot. XLVII, p. 851. **[69.]**

COOKE, M. C. (1881–3). Illustrations of British Fungi. Williams and Norgate, London. **[340, 342.]**

COOKSON, I. (1928). The Structure and Development of the Perithecium in Melanospora zamiae, Corda. Ann. Bot. XLII, p. 255. **[245.]**

CORNU, M. (1872). Monographie des Saprolegniées. Ann. Sci. Nat. Bot. 5me sér. XV, p. 5. **[69, 94.]**

COSTANTIN, J. and DUFOUR, L. (1920). Sur la Biologie du *Goodyera repens*. Rev. Gén. de Bot. XXXII, p. 529. **[23.]**

COTNER, F. B. (1930). Cytological Study of the zoospores of Blastocladia. Bot. Gaz. LXXXIX, p. 295. **[94.]**

COTTON, A. D. (1908). Notes on Marine Pyrenomycetes. Trans. Brit. Myc. Soc. III, p. 92. **[13.]**

—— (1911). On the Structure and Systematic Position of *Sparassis*. Trans. Brit. Myc. Soc. III, p. 333. **[336.]**

COUCH, J. N. (1924). Some observations on spore formation and discharge in Leptolegnia, Achlya and Aphanomyces. Journ. Elisha Mitchell Sci. Soc. XL, p. 27. **[105.]**

—— (1925). A New Dioecious Species of Choanephora. Journ. Elisha Mitchell Sci. Soc. XLI, p. 141. **[136.]**

—— (1926). Heterothallism in Dictyuchus, a Genus of the Water Moulds. Ann. Bot. XL, p. 849. **[108.]**

—— (1927). Some new water fungi from the soil with observations on spore formation. Journ. Elisha Mitchell Sci. Soc. XLII, p. 227. **[14, 103.]**

—— (1930–1 i). Observations on Some Species of Water Molds Connecting Achlya and Dictyuchus. Journ. Elisha Mitchell Sci. Soc. XLVI, p. 225. **[14, 103.]**

—— (1930–1 ii). Micromyces Zygogonii Dang., Parasitic on Spirogyra. Journ. Elisha Mitchell Sci. Soc. XLVI, p. 231. **[67.]**

—— (1932). The development of the sexual organs in *Leptolegnia caudata*. Amer. Journ. Bot. XIX, p. 584. **[108.]**

—— (1933). Basidia of Septobasidium (Glenospora) Curtisii. Journ. Elisha Mitchell Sci. Soc. XLIX, p. 156. **[325.]**

—— (1935 i). A new saprophytic species of Lagenidium, with notes on other forms. Mycologia, XXVII, p. 376. **[84.]**

—— (1935 ii). Septobasidium in the United States. Journ. Elisha Mitchell Sci. Soc. LI, p. 1. **[325.]**

CRAIGIE, J. H. (1927 i). Experiments on Sex in Rust Fungi. Nature, CXX, p. 116. **[314.]**

—— (1927 ii). Function of the Pycnia of the Rust Fungi. Nature, CXX, p. 765. **[308, 313.]**

—— (1928). On the occurrence of pycnia and aecia in certain rust fungi. Phytopath. XVIII, p. 1005. **[312.]**

—— (1933). Union of Pycniospores and Haploid Hyphae in *Puccinia Helianthi* Schw. Nature, CXXXI, p. 25. **[313.]**

CUNNINGHAM, D. D. (1879). On the Occurrence of Conidial Fructifications in the *Mucorini*, illustrated by *Choanephora*. Trans. Linn. Soc. ii, I, p. 409. **[136.]**

—— (1895). A new and parasitic species of Choanephora. Ann. Roy. Bot. Gard. Calcutta, VI, p. 163. **[136.]**

CUNNINGHAM, G. C. (1912). The Comparative Susceptibility of Cruciferous Plants to *Plasmodiophora Brassicae*. Phytopath. II, p. 138. **[48.]**

CUNNINGHAM, G. H. (1927). The Development of *Geaster velutinus*. Trans. Brit. Myc. Soc. XII, p. 12. **[352.]**

CURREY, F. (1873). On a new Genus in the Order Mucedines. Journ. Linn. Soc. Bot. XIII, p. 333. **[136.]**

CURTIS, K. M. (1921). The Life History and Cytology of *Synchytrium endobioticum* (Schilb.) Perc., the cause of Wart Disease in Potato. Phil. Trans. Roy. Soc. B., CCX, p. 409. **[55, 59, 65.]**

CUTTING, E. M. (1909). On the Sexuality and Development of the Ascocarp in Ascophanus carneus. Ann. Bot. XXIII, p. 399. **[217.]**

CZAPEK, F. (1899). Zur Biologie der holzbewohnenden Pilze. Ber. d. deutsch. bot. Ges. XVII, p. 166. **[14.]**

DALE, E. (1903). Observations on the Gymnoascaceae. Ann. Bot. XVII, p. 571. **[176, 177.]**

—— (1909). On the Morphology of *Aspergillus repens* de Bary. Ann. Myc. VII, p. 215. **[181.]**

—— (1912). On the Fungi of the Soil. Ann. Myc. X, p. 452. **[14.]**

DANGEARD, P. A. (1886). Recherches sur les Organismes inférieures. Ann. Sci. Nat. Bot. 7me sér. IV, p. 241. **[56, 64, 85.]**

—— (1889). Mémoire sur les Chytridinées. Botaniste, I, p. 55. **[67.]**

—— (1894 i). Recherches histologiques sur la famille des Ustilaginées. Botaniste, III, p. 240. **[292, 300.]**

—— (1894 ii). La reproduction sexuelle des Ascomycètes. Botaniste, IV, p. 30. **[158.]**

—— (1897). Second Mémoire sur la reproduction sexuelle des Ascomycètes. Botaniste, V, p. 245. **[192.]**

—— (1900). Recherches sur la structure du *Polyphagus Euglenae* Nowak., et sa reproduction sexuelle. Botaniste, VII, p. 213. **[59, 76.]**

—— (1906). La fécondation nucléaire chez les Mucorinées. C.R. Acad. Sci. CXLII, p. 645. **[135.]**

—— (1907). Recherches sur le développement du périthèce chez les Ascomycètes. Botaniste, X, p. 1. **[160, 170, 177, 182, 192, 206, 217, 223, 260.]**

—— (1911). Un nouveau genre de Chytridiacées. Bull. Soc. Myc. de France, XXVII, p. 200. **[81.]**

DAS GUPTA, S. N. (1933). Formation of Pycnidia in Cytospora ludibunda by the Intermingling of two Infertile Strains. Ann. Bot. XLVII, p. 689. **[155.]**

DASTUR, J. F. (1913). On *Phytophthora parasitica* n.sp., a new disease of the castor oil plant. Mem. Dept. Agric. India Bot. iv, v, p. 177. **[115.]**

—— (1920). Choanephora cucurbitarum (B. and Rav.) Thaxter, on Chillies (Capsicum spp.). Ann. Bot. XXXIV, p. 399. **[136.]**

—— (1921). Cytology of Tilletia Tritici (Bjerk.) Wint. Ann. Bot. XXXV, p. 399. **[300.]**

DAUPHIN, J. (1908). Contribution à l'étude des Mortierellées. Ann. Sci. nat. Bot. ix, VIII, p. 1. **[137.]**

DAVIS, B. M. (1900). The fertilisation of *Albugo candida*. Bot. Gaz. XXIX, p. 297. **[121.]**

—— (1903). Oogenesis in *Saprolegnia*. Bot. Gaz. XXXV pp. 233, 320. **[108.]**

—— (1905). Fertilisation in the Saprolegniales. Bot. Gaz. XXXIX, p. 61. **[108.]**

DAWSON, M. (1900). On the Biology of Poronia punctata (L.). Ann. Bot. XIV, p. 245. **[36, 269.]**

VON DECKENBACH, C. (1903). *Coenomyces consuens* nov. gen., nov. spec. Ein Beitrag zur Phylogenie der Pilze. Flora, XCII, p. 253. **[13.]**

DELITSCH, H. (1926). Zur Entwicklungsgeschichte der coprophilen Ascomyceten. Inaug. Dis. Leipzig. **[224.]**

DENIS, M. (1926). La Castration des Spirogyres par des Champignons parasites. Revue Algologique, III, p. 14. **[67.]**

DERX, H. G. (1926). Heterothallism in the genus *Penicillium*. Trans. Brit. Myc. Soc. XI, p. 108. **[182.]**

—— (1930). Étude sur les Sporobolomycètes. Ann. Myc. XXVIII, p. 1. **[286.]**

DEY, P. K. (1919). Studies in the Physiology of Parasitism. V. Infection by Colletotrichum Lindemuthianum. Ann. Bot. XXXIII, p. 305. **[19.]**

DICKINSON, S. (1927). Experiments on the Physiology and Genetics of the Smut Fungi.—Hyphal-Fusion. Proc. Roy. Soc. B, CI, p. 126. **[294.]**

—— (1932). The Nature of Saltation in Fusarium and Helminthosporium. Rept. Univ. Minnesota Agric. Expt. Stat. 1932, p. 1. **[43.]**

—— (1933). The Nature of Saltation in Certain Species of Helminthosporium and Fusarium. Ann. Bot. XLVII, p. 920. **[43.]**

DICKSON, H. (1932). The Effects of X-rays, Ultra-violet Light, and Heat in producing Saltants in Chaetomium cochliodes and other Fungi. Ann. Bot. XLVI, p. 389. **[42.]**

—— (1933). Saltation Induced by X-rays in Seven Species of Chaetomium. Ann. Bot. XLVII, p. 735. **[42.]**

DIEDICKE, H. (1902). Ueber den Zusammenhang zwischen Pleospora- und Helminthosporium-Arten. Centralbl. f. Bakt. Abt. ii, IX, p. 317. **[29, 266.]**

DIETEL, P. (1903). Ueber die Leguminosen lebenden Rostpilze und die Verwandtschaftsverhältnisse der Gattungen der Pucciniaceen. Ann. Myc. I, p. 3. **[321.]**

DILLON WESTON, W. A. P. (1925). A Preliminary Note on the Perithecia of *Nectria galligena*. Ann. Applied Biol. XII, p. 398. **[248.]**

DITTSCHLAG, E. (1910). Zur Kenntnis der Kernverhältnisse von *Puccinia Falcariae*. Centralbl. f. Bakt. Abt. ii, XXVIII, p. 28. **[314.]**

DIXON-STEWART, D. (1932). Species of *Mortierella* isolated from Soil. Trans. Brit. Myc. Soc. XVII, p. 208. **[137.]**

DODGE, B. O. (1912). Methods of Culture and the Morphology of the Archicarp in certain species of the Ascobolaceae. Bull. Torrey Bot. Club, XXXIX, p. 139. **[223.]**

—— (1914). The morphological relationships of the Floridae and the Ascomycetes. Bull. Torrey Bot. Club, XLI, p. 165. **[165.]**

—— (1918 i). Studies in the genus *Gymnosporangium*. I. Notes on the distribution of the mycelium, buffer cells and the germination of the aecidiospore. Brooklyn Bot. Gard. Mem. I, p. 129. **[317.]**

—— (1918 ii). Studies in the genus Gymnosporangium. III. The origin of the teleutospore. Mycologia, X, p. 182. **[317.]**

—— (1920). Life History of Ascobolus magnificus. Mycologia, XII, p. 115. **[219.]**

—— (1922). A Lachnea with a botryose conidial stage. Bull. Torrey Bot. Club, XLIX, p. 301. **[208.]**

—— (1923). The distribution of the orange rusts of Rubus. Phytopath. XIII, p. 61. **[320.]**

—— (1927). Nuclear Phenomena Associated with Heterothallism and Homothallism in the Ascomycete Neurospora. Journ. Agric. Research, XXXV, p. 289. **[245, 246.]**

—— (1928 i). Production of fertile hybrids in the Ascomycete Neurospora. Journ. Agric. Research, XXXVI, p. 1. **[245, 246.]**

—— (1928 ii). Unisexual conidia from bisexual mycelia. Mycologia, XX, p. 226. **[245.]**

—— (1932). The non-sexual and sexual functions of microconidia of Neurospora. Bull. Torrey Bot. Club, LIX, p. 347. **[246.]**

—— (1933). The perithecium and ascus of Penicillium. Mycologia, XXV, p. 90. **[182.]**

DODGE, B. O. (1935 i). A recessive factor lethal for ascus formation in Neurospora. Bull. Torrey Bot. Club, LXII, p. 117. [**246**.]

—— (1935 ii). The mechanics of sexual reproduction in Neurospora. Mycologia, XXVII, p. 418. [**246**.]

—— (1936). Spermatia and nuclear migrations in Pleurage anserina. Mycologia, XXVIII, p. 284. [**259**.]

DODGE, B. O. and GAISER, L. O. (1926). The question of nuclear fusions in the Blackberry Rust, *Caeoma nitens*. Journ. Agric. Research, XXXII, p. 1003. [**320**.]

DOMARADSKY, M. (1908). Zur Fruchtkörperentwickelung von *Aspergillus Fischeri* Wehm. Ber. d. deutsch. bot. Ges. XXVI*a*, p. 14. [**181**.]

DOWDING, E. S. (1931 i). The Sexuality of the Normal, Giant and Dwarf Spores of Pleurage anserina (Ces.), Kuntze. Ann. Bot. XLV, p. 1. [**259**.]

—— (1931 ii). The Sexuality of Ascobolus stercorarius and the Transportation of the Oidia by Mites and Flies. Ann. Bot. XLV, p. 621. [**223**.]

—— (1933). Gelasinospora, a new genus of Pyrenomycetes with pitted spores. Canadian Journ. Res. IX, p. 294. [**258**.]

DOWSON, W. J. (1925). A Die-back of Rambler Roses due to *Gnomonia rubi* Rehm. Journ. Roy. Hort. Soc. L, p. 55. [**268**.]

DRAYTON, F. L. (1932). The sexual function of the microconidia in certain Discomycetes. Mycologia, XXIV, p. 345. [**227**.]

—— (1934). The sexual mechanism of Sclerotinia Gladioli. Mycologia, XXVI, p. 46. [**227**.]

DRECHSLER, C. (1925). The Cottony Leak of Cucumbers Caused by *Pythium aphanidermatum*. Journ. Agric. Research, XXX, p. 1035. [**114**.]

—— (1929). The Beet Water Mold and Several Related Root Parasites. Journ. Agric. Research, XXXVIII, p. 309. [**110, 114**.]

—— (1930 i). Repetitional Diplanetism in the Genus Phytophthora. Journ. Agric. Research, XL, p. 557. [**55**.]

—— (1930 ii). Some new species of *Pythium*. Journ. Wash. Acad. Sci. XX, p. 398. [**114**.]

—— (1931). A crown-rot of hollyhocks caused by *Phytophthora megasperma* n.sp. Journ. Wash. Acad. Sci. XXI, p. 513. [**115**.]

—— (1932). A species of *Pythiogeton* isolated from decaying leaf-sheaths of the common cat-tail. Journ. Wash. Acad. Sci. XXII, p. 421. [**119**.]

—— (1933 i). Morphological diversity among fungi capturing and destroying nematodes. Journ. Wash. Acad. Sci. XXIII, p. 138. [**141**.]

—— (1933 ii). Morphological features of some fungi capturing and killing amoebae. Journ. Wash. Acad. Sci. XXIII, p. 200. [**141**.]

—— (1933 iii). Morphological features of some more fungi that capture and kill nematodes. Journ. Wash. Acad. Sci. XXIII, p. 267. [**141**.]

—— (1933 iv). Several more fungi that prey on nematodes. Journ. Wash. Acad. Sci. XXIII, p. 355. [**141**.]

—— (1934). Organs of capture in some fungi preying on nematodes. Mycologia, XXVI, p. 135. [**141**.]

—— (1935 i). Some conidial Phycomycetes destructive to terricolous Amoebae. Mycologia, XXVII, p. 6. [**141, 142**.]

—— (1935 ii). Some non-catenulate conidial Phycomycetes preying on terricolous Amoebae. Mycologia, XXVII, p. 176. [**141, 142**.]

—— (1935 iii). A new species of conidial Phycomycete preying on nematodes. Mycologia, XXVII, p. 206. [**141**.]

DUFF, G. H. (1922). Development in the Geoglossaceae. Bot. Gaz. LXXIV, p. 264. [**233, 234**.]

DUFRENOY, J. (1917). The Endotrophic Mycorhiza of Ericaceae. New Phyt. XVI, p. 222. [**24**.]

DUGGAR, B. M. and DAVIS, A. R. (1916). Studies in the Physiology of Fungi. I. Nitrogen Fixation. Ann. Miss. Bot. Gard. III, p. 413. [**15.**]

DUMÉE, P. and MAIRE, R. (1902). Remarques sur le *Zaghouania Phillyreae* Pat. Bull. Soc. Myc. de France, XVIII, p. 17. [**324.**]

DUNKERLY, J. S. (1914). *Dermocystidium pusula* Perez, parasitic in *Trutta fario*. Zool. Anzeig. XLIV, p. 179. [**13, 56, 64.**]

DURAND, E. J. (1908). The Geoglossaceae of North America. Ann. Myc. VI, p. 387. [**233.**]

EDGERTON, C. W. (1914). Plus and Minus Strains in *Glomerella*. Am. Journ. Bot. I, p. 244. [**155.**]

EHRENBERG, C. G. (1820). Syzygites, eine neue Schimmelgattung, nebst Beobachtungen über sichtbare Bewegungen in Schimmeln. Verhandl. d. Ges. Naturf. Freunde Berlin, I, p. 98. [**126.**]

EIDAM, E. (1880). Beitrag zur Kenntniss der Gymnoasceen. Cohn's Beitr. z. Biol. d. Pflanzen, III, p. 267. [**177.**]

—— (1883). Zur Kenntniss der Entwickelung bei den Ascomyceten. Cohn's Beitr. z. Biol. d. Pflanzen, III, p. 385. [**167.**]

—— (1886). *Basidiobolus*, eine neue Gattung der *Entomophthoraceen*. Cohn's Beitr. z. Biol. d. Pflanzen, IV, p. 181. [**146.**]

ELFVING, F. (1917). Phycomyces und die sogenannte physiologische Fernwirkung. Ofvers. af Finsk. Vet. Soc. Förhandl. LIX, Afd. A, No. 18, p. 1. [**33.**]

ELLIOTT, J. A. (1925). A cytological study of Ceratostomella fimbriata (E. & H.) Elliott. Phytopath. XV, p. 417. [**262.**]

EMMONS, C. W. (1935). The ascocarp in species of Penicillium. Mycologia, XXVII, p. 128. [**182.**]

ENGLER, A. and PRANTL, K. Die natürlichen Pflanzenfamilien, I. i. (1897) and I. i.** (1900). Engelmann, Leipzig. [**10, 334, 349, 354.**]

ERIKSSON, J. (1894). Ueber die Specialisirung des Parasitismus bei den Getreiderostpilzen. Ber. d. deutsch. bot. Ges. XII, p. 292. [**302.**]

—— (1898). A General Review of the Principal Results of Swedish Research into Grain Rust. Bot. Gaz. XXV, p. 26. [**302.**]

—— (1905). On the Vegetative Life of some Uredineae. Ann. Bot. XIX, p. 53. [**302.**]

FAIRCHILD, D. G. (1897). Ueber Kernteilung und Befruchtung bei *Basidiobolus ranarum* Eidam. Jahrb. f. wiss. Bot. XXX, p. 283. [**146.**]

FALCK, R. (1912). Die Merulius-Fäule des Bauholzes. A. Möller's Hausschwammforschungen, p. 1. Fischer, Jena. [**344.**]

FARLOW, W. G. (1876). On the American Grape-Vine Mildew. Bull. Bussey Inst. p. 415. [**123.**]

FAULL, J. H. (1905). Development of Ascus and Spore Formation in Ascomycetes. Proc. Boston Soc. Nat. Hist. XXXII, p. 77. [**164.**]

—— (1911). The Cytology of the Laboulbeniales. Ann. Bot. XXV, p. 649. [**274.**]

—— (1912). The Cytology of Laboulbenia chaetophora and L. Gyrinidarum. Ann. Bot. XXVI, p. 325. [**274, 280.**]

FAVORSKY, W. (1910). Nouvelle recherche sur le développement et la cytologie du *Plasmodiophora Brassicae* Wor. Mém. de la Soc. des Naturalistes de Kieff, XX, p. 149. [**48, 61.**]

FEDERLEY, H. (1904). Die Copulation der Conidien bei *Ustilago Tragopogi pratensis* Pers. Ofvers. af Finsk. Vet. Soc. Förhandl. LXVI, No. 2, p. 1. [**299.**]

FERDINANDSEN, C. and WINGE, O. (1920). A Phyllochorella Parasitic on Sargassum. Mycologia, XII, p. 102. [**13, 254.**]

FISCH, C. (1882). Beiträge zur Entwickelungsgeschichte einiger Ascomyceten. Bot. Zeit. XL, p. 850. [**248.**]

FISCHER VON WALDHEIM, A. (1867). Sur la structure des spores des Ustilaginées. Bull. Soc. Imp. des Nat. de Moscou, XL, p. 242. [**292**.]

FISCHER, A. (1882). Untersuchungen über die Parasiten der Saprolegnieen. Jahrb. f. wiss. Bot. XIII, p. 286. [**13**, **69**.]

FISCHER, C. C. E. (1909). On the Development of the Fructification of Armillaria mucida Schrad. Ann. Bot. XXIII, p. 503. [**340**.]

FISCHER, E. (1886). Zur Kenntniss der Pilzgattung *Cyttaria*. Bot. Zeit. XLVI, p. 812. [**229**.]

—— (1910). Beiträge zur Morphologie und Systematik der Phalloideen. Ann. Myc. VIII, p. 314. [**347**.]

—— (1923 i). Mykologische Beiträge 27. Zur vergleichenden Morphologie der Fruchtkörper von Staheliomyces, Xylophallus und Mutinus. Mitteil. Naturf. Ges. Bern, p. 39. [**347**, **350**.]

—— (1923 ii). Pilze. Ber. d. Schweiz. Bot. Ges. XXXII, p. 1. [**138**.]

—— (1936). Neue Beiträge zur Kenntnis der Verwandtschaftsverhältnisse der Gastromyceten. Ber. d. Schweiz. Bot. Ges. XLV, p. 231. [**346**.]

FITZPATRICK, H. M. (1918). Sexuality in *Rhizina undulata* Fries. Bot. Gaz. LXV, p. 201. [**231**.]

—— (1923). Generic concepts in the Pythiaceae and Blastocladiaceae. Mycologia, XV, p. 166. [**115**, **117**.]

FITZPATRICK, R. E. (1934). The life history and parasitism of *Taphrina deformans*. Sci. Agric. XIV, p. 305. [**199**.]

FOËX, M. (1912). Les conidiophores des Erysiphacées. Rev. Gén. de Bot. XXIV, p. 200. [**188**.]

FORBES, E. J. (1935). Observations on some British Water Moulds (Saprolegniales and Blastocladiales). Trans. Brit. Myc. Soc. XIX, p. 221. [**119**.]

FRANK, B. (1883). Ueber einige neue oder wenige bekannte Pflanzenkrankheiten. *Polystigma rubrum*. Ber. d. deutsch. bot. Ges. I, p. 54. [**248**.]

—— (1885). Ueber die auf Wurzelsymbiose beruhende Ernährung gewisser Bäume durch unterirdische Pilze. Ber. d. deutsch. bot. Ges. III, p. 128. [**22**, **27**.]

—— (1886). Ueber *Gnomonia erythrostoma*, die Ursache, etc. Ber. d. deutsch. bot. Ges. IV, p. 200. [**268**.]

—— (1891). Ueber die auf Verdauung von Pilzen abzielende Symbiose der mit endotrophen Mycorhizen begabten Pflanzen, so wie der Leguminosen und Erlen. Ber. d. deutsch. bot. Ges. IX, p. 244. [**22**.]

FRASER, H. C. I. (1907). On the Sexuality and Development of the Ascocarp in Lachnea stercorea Pers. Ann. Bot. XXI, p. 349. [**212**.]

—— (1908). Contributions to the Cytology of Humaria rutilans. Ann. Bot. XXII, p. 35. [**158**, **160**, **214**.]

FRASER (GWYNNE-VAUGHAN), H. C. I. (1913). The Development of the Ascocarp in Lachnea cretea. Ann. Bot. XXVII, p. 554. [**208**.]

FRASER, H. C. I. and BROOKS, W. E. St J. (1909). Further Studies on the Cytology of the Ascus. Ann. Bot. XXIII, p. 537. [**160**, **164**, **223**.]

FRASER, H. C. I. and CHAMBERS, H. S. (1907). The Morphology of *Aspergillus herbariorum*. Ann. Myc. V, p. 419. [**38**, **180**.]

FRASER, H. C. I. and WELSFORD, E. J. (1908). Further Contributions to the Cytology of the Ascomycetes. Ann. Bot. XXII, p. 465. [**160**, **164**, **216**.]

FRASER, L. (1933). An Investigation of the Sooty Moulds of New South Wales. I. Historical and Introductory Account. Proc. Linn. Soc. N.S.W. LVIII, p. 375. [**195**.]

—— (1934). An Investigation of the Sooty Moulds of New South Wales. II. An Examination of the Cultural Behaviour of Certain Sooty Mould Fungi. Proc. Linn. Soc. N.S.W. LIX, p. 123. [**195**.]

FRASER, L. (1935 i). An Investigation of the Sooty Moulds of New South Wales. III. The Life Histories and Systematic Positions of Aithaloderma and Capnodium, together with Descriptions of New Species. Proc. Linn. Soc. N.S.W. LX, p. 97. [**195.**]

—— (1935 ii). An Investigation of the Sooty Moulds of New South Wales. IV. The Species of the Eucapnodieae. Proc. Linn. Soc. N.S.W. LX, p. 159. [**195.**]

—— (1935 iii). An Investigation of the Sooty Moulds of New South Wales. V. The Species of the Chaetothyrieae. Proc. Linn. Soc. N.S.W. LX, p. 280. [**195.**]

FREEMAN, D. L. (1910). Untersuchungen über die Stromabildung der *Xylaria Hypoxylon* in künstlichen Kulturen. Ann. Myc. VIII, p. 192. [**36, 39, 273.**]

FREEMAN, E. M. and JOHNSON, E. C. (1911). The Rusts of Grain in the United States. U.S. Dept. Agric. Bureau of Plant Industry Bull. 216. [**29.**]

FREY, C. N. (1924). The Cytology and Physiology of Venturia inaequalis (Cooke) Winter. Trans. Wisconsin Acad. Sci. Arts and Letters, XXI, p. 303. [**267.**]

FRIES, R. E. (1899). *Basidiobolus myxophilus* en ny Phycomycet. Bihang till K. Svensk. Vet. Akad. Handl. XXV, III, 3, p. 1. [**146.**]

FRITSCH, F. E. (1935). The Structure and Reproduction of the Algae. Cambridge, p. 501. [**52, 77.**]

FROMME, F. D. (1912). Sexual fusions and spore development of the flax rust. Bull. Torrey Bot. Club, XXXIX, p. 113. [**307, 310.**]

—— (1914). The morphology and cytology of the aecidium cup. Bot. Gaz. LVIII, p. 1. [**310.**]

FULTON, H. R. (1906). Chemotropism of Fungi. Bot. Gaz. XLI, p. 81. [**31, 34.**]

GADD, C. H. (1924). *Phytophthora faberi* Maubl. Ann. Roy. Bot. Gard. Peradeniya, IX, p. 47. [**119.**]

GALLAUD, I. (1905). Études sur les mycorhizes endotrophes. Rev. Gén. de Bot. XVII, p. 1. [**23.**]

GÄUMANN, E. (1926). Vergleichende Morphologie der Pilze. Fischer, Jena. [**10.**]

GHATAK, P. N. (1936). On the Development of the Perithecium of Microeurotium albidum. Ann. Bot. L, p. 849. [**181.**]

GILBERT, J. H. (1875). Note on the Occurrence of "Fairy Rings." Journ. Linn. Soc. XV, p. 17. [**15, 38.**]

GODDARD, H. N. (1913). Can Fungi living in Agricultural Soil Assimilate free Nitrogen? Bot. Gaz. LVI, p. 249. [**22.**]

GODFREY, G. H. (1923). A Phytophthora Footrot of Rhubarb. Journ. Agric. Research, XXIII, p. 1. [**115.**]

GOLDSTEIN, B. (1929). A cytological study of the fungus Massospora cicadina, parasitic on the 17-year cicada, Magicicada septendecim. Amer. Journ. Bot. XVI, p. 394. [**146.**]

GRAVES, A. H. (1916). Chemotropism in *Rhizopus nigricans*. Bot. Gaz. LXII, p. 337. [**31, 33.**]

GREEN, E. (1925). Note on the Occurrence of Clamp-Connections in Hirneola auricula judae. Ann. Bot. XXXIX, p. 214. [**324.**]

—— (1927). The Life-history of Zygorhynchus Moelleri. Ann. Bot. XLI, p. 419. [**135.**]

—— (1931). Observations on certain Ascobolaceae. Trans. Brit. Myc. Soc. XV, p. 321. [**223, 224, 225.**]

GREGOR, M. J. F. (1932). A study of heterothallism in Ceratostomella pluriannulata, Hedgcock. Ann. Myc. XXX, p. 1. [**264.**]

—— (1935). A disease of Bracken and other ferns caused by *Corticium anceps* (Bres. et Syd.) Gregor. Phytopath. Zeitschr. VIII, p. 401. [**334.**]

GRIGGS, R. F. (1910). *Monochytrium*: a new Genus of the Chytridiales, its Life-History and Cytology. Ohio Nat. X, p. 44. [**61.**]

GRIGORAKI, L. D. (1925). Recherches cytologiques et taxonomiques sur les dermatophytes et quelques autres champignons parasites. Ann. Sci. Nat. Bot. 10me sér. VII, p. 165. [**177**.]

GROVE, W. B. (1913 i). The Evolution of the Higher Uredineae. New Phyt. XII, p. 89. [**321**.]

—— (1913 ii). The British Rust Fungi. Camb. Univ. Press. [**302, 310**.]

GUILLIERMOND, A. (1901). Recherches sur la sporulation des Schizosaccharomycètes. C.R. Acad. Sci. CXXXIII, p. 242. [**173, 174**.]

—— (1903). Recherches sur la germination des spores chez le *Schizosaccharomyces Ludwigii*. Bull. Soc. Myc. de France, XIX, p. 18. [**173**.]

—— (1905 i). Remarques sur le Karyokinèse des Ascomycètes. Ann. Myc. III, p. 35. [**214**.]

—— (1905 ii). Recherches sur la germination des spores et la conjugaison chez les levures. Rev. Gén. de Bot. XVII, p. 337. [**173**.]

—— (1909). Recherches cytologiques et taxonomiques sur les Endomycetacées. Rev. Gén. de Bot. XXI, p. 353. [**167, 168, 175**.]

—— (1910 i). Quelques remarques sur la copulation des levures. Ann. Myc. VIII, p. 287. [**173**.]

—— (1910 ii). Sur un curieux exemple de parthénogénèse observé dans une levure. C.R. Soc. Biol. Paris, LXVIII, p. 363. [**173, 176**.]

—— (1913). Les Progrès de la Cytologie des Champignons. Prog. Rei Bot. IV, p. 389. [**10, 168, 214**.]

—— (1920 i). *Zygosaccharomyces Pastori*, nouvelle espèce de levures à copulation hétérogamique. Bull. Soc. Myc. de France, XXXVI, p. 203. [**173**.]

—— (1920 ii). The Yeasts. Am. trans. by F. W. Tanner. John Wiley and Sons, New York. [**18, 173**.]

—— (1922). Observation cytologique sur un *Leptomitus* et en particulier sur le mode de formation et la germination des zoospores. C.R. Acad. Sci. CLXXV, p. 377. [**103**.]

—— (1927 i). Cytologie et sexualité du *Spermophthora Gossypii*. C.R. Acad. Sci. CLXXXIV, p. 1189. [**172**.]

—— (1927 ii). Sur la cytologie des *Nematospora*. C.R. Acad. Sci. CLXXXV, p. 1510. [**173**.]

—— (1928 i). Quelques faits nouveaux relatifs au développement du *Spermophthora Gossypii*. C.R. Acad. Sci. CLXXXVI, p. 161. [**172**.]

—— (1928 ii). Études cytologiques et taxonomiques sur les Levures du genre *Sporobolomyces*. Bull. Soc. myc. de France, XLIII, p. 245. [**286**.]

—— (1928 iii). Recherches sur quelques Ascomycètes inférieurs isolés de la stigmatomycose des graines de cotonnier. Essai sur la phylogénie des Ascomycètes. Rev. Gén. de Bot. XL, pp. 328, 397, 474, 555, 606, 690. [**173**.]

—— (1931). Recherches sur l'homothallisme chez les levures. Rev. Gén. de Bot. XLIII, p. 49. [**173**.]

GUILLIERMOND, A. and PÉJU (1920). Une nouvelle espèce de levures de genre *Debaryomyces, D. Klockerii* n.sp. Bull. Soc. Myc. de France, XXXVI, p. 164. [**173**.]

GWYNNE-VAUGHAN, H. C. I.[1] (1922). Fungi. Ascomycetes, Ustilaginales, Uredinales. University Press, Cambridge. [**166**.]

—— (1926). Determination of Sex in Plants. Report Brit. Assoc. p. 406. [**42**.]

—— (1928). Sex and Nutrition in the Fungi. Report Brit. Assoc. p. 222. [**42**.]

—— (1935). Problems of Sex in the Higher Fungi. Sci. Prog. XXIX, p. 593. [**42**.]

[1] For earlier references see Fraser, H. C. I.

GWYNNE-VAUGHAN, H. C. I. (1937). Contributions to the Study of Lachnea melaloma. Ann. Bot. New series, I, p. 99. [**209.**]

GWYNNE-VAUGHAN, H. C. I. and BROADHEAD, Q. E. (1936). Contributions to the Study of Ceratostomella fimbriata. Ann. Bot. L, p. 747. [**262.**]

GWYNNE-VAUGHAN, H. C. I. and WILLIAMSON, H. S. (1927). Germination in Lachnea cretea. Ann. Bot. XLI, p. 489. [**208.**]

—— —— (1930). Contributions to the Study of Humaria granulata, Quél. Ann. Bot. XLIV, p. 127. [**160, 162, 212.**]

—— —— (1931). Contributions to the Study of Pyronema confluens. Ann. Bot. XLV, p. 355. [**4, 160, 204.**]

—— —— (1932). The Cytology and Development of Ascobolus magnificus. Ann. Bot. XLVI, p. 653. [**4, 160, 162, 219.**]

—— —— (1933 i). The Asci of Lachnea scutellata. Ann. Bot. XLVII, p. 375. [**160, 210.**]

—— —— (1933 ii). Notes on the Ascobolaceae. Trans. Brit. Myc. Soc. XVIII, p. 127. [**150, 224, 225.**]

—— —— (1934). The Cytology and Development of Ascophanus Aurora. Ann. Bot. XLVIII, p. 261. [**160, 162, 217.**]

HAENICKE, A. (1916). Vererbungsphysiologische Untersuchungen an Arten von Penicillium und Aspergillus. Zeitschr. f. Bot. VIII, p. 225. [**42.**]

HAKE, W. L. (1923). British Laboulbeniaceae. A catalogue of the British Specimens in the Thaxter Collection at the British Museum. Trans. Brit. Myc. Soc. IX, p. 78. [**274.**]

HALL, A. D. (1920). The Soil. John Murray, London. 3rd ed. [**14.**]

HANNA, W. F. (1925). The Problem of Sex in Coprinus lagopus. Ann. Bot. XXXIX, p. 431. [**288, 289.**]

—— (1929). Nuclear Association in the Aecidium of *Puccinia graminis*. Nature, CXXIV, p. 267. [**313.**]

HANSEN, E. C. (1901). Grundlinien zur Systematik der Saccharomyceten. Centralbl. f. Bakt. Abt. ii, XII, p. 529. [**173.**]

HARPER, R. A. (1895). Die Entwickelung des Peritheciums bei *Sphaerotheca Castagnei*. Ber. d. deutsch. bot. Ges. XIII, p. 475. [**159, 191.**]

—— (1896). Über das Verhalten der Kerne bei der Fruchtentwickelung einiger Ascomyceten. Jahrb. f. wiss. Bot. XXIX, p. 655. [**192, 223.**]

—— (1897). Kernteilung und freie Zellbildung im Ascus. Jahrb. f. wiss. Bot. XXX, p. 249. [**191.**]

—— (1899 i). Nuclear Phenomena in certain stages in the development of the Smuts. Trans. Wisconsin Acad. Sci. Arts and Letters, XII, p. 475. [**299.**]

—— (1899 ii). Cell Division in Sporangia and Asci. Ann. Bot. XIII, p. 467. [**53, 67, 130.**]

—— (1900). Sexual Reproduction in Pyronema confluens and the Morphology of the Ascocarp. Ann. Bot. XIV, p. 321. [**204.**]

—— (1902). Binucleate Cells in Certain Hymenomycetes. Bot. Gaz. XXXIII, p. 1. [**285, 327, 334.**]

—— (1905). Sexual Reproduction and the Organisation of the Nucleus in Certain Mildews. Publ. Carn. Inst. Washington. [**164, 192.**]

HARTIG, R. (1880). Der Eichenwurzeltodter *Rosellina quercina*. Untersuch. aus dem forst-botanischen Inst. zu München, III, p. 1. [**261.**]

HARVEY, J. V. (1925). A study of the water molds and Pythiums occurring in the soils of Chapel Hill. Journ. Elisha Mitchell Sci. Soc. XLI, p. 151. [**14, 103, 108.**]

—— (1927 i). A Survey of the Watermolds Occurring in the Soils of Wisconsin as Studied during the Summer of 1926. Trans. Wisconsin Acad. Sci. Arts and Letters, XXIII, p. 551. [**14, 103.**]

—— (1927 ii). Brevilegnia diclina n.sp. Journ. Elisha Mitchell Sci. Soc. XLII, p. 243. [**14, 103.**]

HASSELBRING, H. (1907). Gravity as a Form-Stimulus in Fungi. Bot. Gaz. XLIII, p. 251. [**37**.]

HATCH, W. R. (1933–4). Sexuality of Allomyces arbuscula Butler. Journ. Elisha Mitchell Sci. Soc. XLIX, p. 163. [**89, 91, 93**.]

—— (1935). Gametogenesis in Allomyces arbuscula. Ann. Bot. XLIX, p. 623. [**89, 91, 93**.]

HAYES, H. K., STAKMAN, E. C. and AAMODT, O. S. (1925). Inheritance in wheat of resistance to black stem rust. Phytopath. XV, p. 371. [**29**.]

HEIN, I. (1927). Studies on morphogenesis and development of the ascocarp of Sphaerotheca castagnei. Bull. Torrey Bot. Club, LIV, p. 383. [**191**.]

HENDERSON M. W. (1919). A comparative study of the structure and saprophytism of the Pyrolaceae and Monotropaceae with reference to their derivation from the Ericaceae. Contrib. Bot. Lab. Univ. of Pennsylvania, V, p. 42. [**25**.]

HENRARD, P. (1934). Polarité, hérédité et variation chez diverses espèces d'Aspergillus. Cellule, XLIII, p. 351. [**155**.]

HIGGINS, B. B. (1914). Life-History of a new species of *Sphaerella*. Myc. Centralbl. IV, p. 187. [**266**.]

—— (1920). Morphology and Life-History of some of the Ascomycetes with special reference to the presence and function of Spermatia. Am. Journ. Bot. VII, p. 435. [**248, 266**.]

HILEY, W. E. (1919). The Fungal Diseases of the Common Larch. Clarendon Press, Oxford. [**227**.]

HIRMER, M. (1920). Zur Kenntnis der Vielkernigkeit der Autobasidiomyzeten. I. Zeitschr. f. Bot. XII, p. 657. [**331**.]

HOFFMANN, A. W. H. (1912). Zur Entwicklungsgeschichte von *Endophyllum Sempervivi*. Centralbl. f. Bakt. Abt. ii, XXXII, p. 137. [**319**.]

HOFFMANN, H. (1856). Die Pollinarien und Spermatien von Agaricus. Bot. Zeit. XIV, p. 153. [**2**.]

HOLDEN, R. J. and HARPER, R. A. (1903). Nuclear Divisions and Nuclear Fusion in Coleosporium Sonchi-arvensis Lév. Trans. Wisconsin Acad. Sci. Arts and Letters, XIV, pt i, p. 63. [**303**.]

HORNE, A. S. (1930). Nuclear Division in the Plasmodiophorales. Ann. Bot. XLIV, p. 199. [**47**.]

HORNE, A. S. and DAS GUPTA, S. N. (1929). Studies in the Genera Cytosporina, Phomopsis and Diaporthe. I. On the Occurrence of an "Ever-saltating" Strain in Diaporthe. Ann. Bot. XLIII, p. 417. [**43**.]

HOWARTH, W. O. and CHIPPINDALE, H. G. (1930–1). Notes on the Biology of *Coryne sarcoides* Jacq. and *C. urnalis* Nyl. Manchester Memoirs, LXXV, p. 47. [**229**.]

HUBER-PESTALOZZI, G. (1931). Infektion einer *Mougeotia*-Population durch *Micromyces Zygogonii* Dangeard an einem alpinen Standort. Hedwigia, LXXI, p. 88. [**67**.]

HUTTON, J. (1790). On certain natural appearances of the ground on the Hill of Arthur's Seat. Trans. Roy. Soc. Edin. II, p. 1. [**15**.]

IKENO, S. (1903 i). Die Sporenbildung von *Taphrina*-Arten. Flora, XCII, p. 1. [**197**.]

—— (1903 ii). Über die Sporenbildung und systematische Stellung von *Monascus purpureus* Went. Ber. d. deutsch. bot. Ges. XXI, p. 259. [**183**.]

INGOLD, C. T. (1932). The Sporangiophore of *Pilobolus*. New Phyt. XXXI, p. 58. [**131**.]

—— (1933). Spore Discharge in the Ascomycetes. I. Pyrenomycetes. New Phyt. XXXII, p. 175. [**257, 260, 261, 264**.]

—— (1934). The Spore Discharge Mechanism in *Basidiobolus ranarum*. New Phyt. XXXIII, p. 274. [**147**.]

JACKSON, H. S. (1931). Present Evolutionary Tendencies and the Origin of Life Cycles in the Uredinales. Mem. Torrey Bot. Club, XVIII, p. 1. [**321**.]

JAHN, E. (1911). Myxomycetenstudien, 8. Ber. d. deutsch. bot. Ges. XXIX, p. 231. [**45**, **46**.]

—— (1919). Lebensdauer und Alterserscheinungen eines Plasmodiums (Myxomycetenstudien, 10). Ber. d. deutsch. bot. Ges. XXXVII, p. (18). [**46**.]

JAHN, T. L. (1933). On certain parasites of Phacus and Euglena: *Sphaerita phaci* sp. nov. Arch. f. Protistenk. LXXIX, p. 349. [**64**.]

JOCHEMS, S. C. J. (1927). The occurrence of Blakeslea trispora Thaxter in the Dutch East Indies. Phytopath. XVII, p. 181. [**136**.]

JOHNSON, M. E. M. (1920). On the Biology of *Panus stypticus*. Trans. Brit. Myc. Soc. VI, p. 348. [**343**.]

JOHNSON, T. and NEWTON, M. (1933). Hybridization between *Puccinia Graminis Tritici* and *Puccinia Graminis Avenae*. Proc. World's Grain Exhib. and Conf., Canada, II, p. 219. [**322**.]

JOHNSON, T., NEWTON, M. and BROWN, A. M. (1934). Further Studies on the inheritance of spore colour and pathogenicity in crosses between physiologic forms of *Puccinia Graminis Tritici*. Sci. Agric. XIV, p. 360. [**322**.]

JOLIVETTE, H. D. M. (1914). Studies in the Reactions of *Pilobolus* to Light Stimuli. Bot. Gaz. LVII, p. 89. [**34**.]

JONES, F. R. and DRECHSLER, C. (1920). Crownwart of alfalfa caused by *Urophlyctis alfalfae*. Journ. Agric. Research, XX, p. 295. [**79**.]

—— —— (1925). Root Rot of Peas in the United States caused by *Aphanomyces euteiches* (n.sp.). Journ. Agric. Research, XXX, p. 293. [**110**.]

JONES, S. G. (1925). Life-history and Cytology of Rhytisma acerinum (Pers.) Fries. Ann. Bot. XXXIX, p. 41. [**241**.]

—— (1935). The Structure of Lophodermium pinastri (Schrad.) Chev. Ann. Bot. XLIX, p. 699. [**242**.]

JUEL, H. O. (1901). *Pyrrhosorus*, eine neue marine Pilzgattung. Bihang till K. Svensk. Vet. Akad. Handl. iii, XXVI, p. 14. [**13**.]

—— (1901–2). *Taphridium*, eine neue Gattung der Protomycetaceen. Bihang till K. Svensk. Vet. Akad. Handl. iii, XXVII, p. 1. [**197**.]

—— (1902). Über Zellinhalt, Befruchtung und Sporenbildung bei *Dipodascus*. Flora, XCI, p. 47. [**170**.]

—— (1921). Cytologische Pilzstudien. II. Zur Kenntnis einiger Hemiasceen. Nova Acta Regiae Scientiarum Upsaliensis, iv, v, p. 3. [**88**, **167**, **168**, **170**, **197**.]

KANOUSE, B. B. (1925). Physiology and Morphology of *Pythiomorpha gonapodioides*. Bot. Gaz. LXXIX, p. 196. [**38**, **119**.]

—— (1927). A Monographic Study of Special Groups of the Water Molds. I. Blastocladiaceae. II. Leptomitaceae and Pythiomorphaceae. Amer. Journ. Bot. XIV, p. 287. [**94**, **98**, **119**.]

KARLING, J. S. (1930). Studies in the Chytridiales. IV. A further study of *Diplophlyctis intestina* (Schenk) Schroeter. Amer. Journ. Bot. XVII, p. 770. [**58**.]

—— (1931). Further studies in the Chytridiales. VI. The occurrence and life history of a new species of *Cladochytrium* in cells of *Eriocaulon septangulale*. Amer. Journ. Bot. XVIII, p. 526. [**78**.]

—— (1932). Studies in the Chytridiales. VII. The organization of the chytrid thallus. Amer. Journ. Bot. XIX, p. 41. [**58**.]

—— (1934). A saprophytic species of Catenaria isolated from roots of Panicum variegatum. Mycologia, XXVI, p. 528. [**81**.]

—— (1935). A further study of Cladochytrium replicatum with special reference to its distribution, host range and culture on artificial media. Amer. Journ. Bot. XXII, p. 439. [**78**.]

KASANOWSKY, V. (1911). *Aphanomyces laevis* de Bary. I. Entwicklung der Sexualorgane und Befruchtung. Ber. d. deutsch. bot. Ges. XXIX, p. 210. [**108.**]

KAUFFMAN, C. H. (1906). *Cortinarius* as a Mycorhiza producing fungus. Bot. Gaz. XLII, p. 208. [**28.**]

—— (1908). A Contribution to the Physiology of the Saprolegniaceae. Ann. Bot. XXII, p. 361. [**38, 90.**]

KEENE, M. L. (1914). Cytological Studies of the Zygospores of Sporodinia grandis. Ann. Bot. XXVIII, p. 455. [**135.**]

KEILIN, D. (1921). On a new Saccharomycete *Monosporella unicuspidata* gen. n. nom., n. sp., parasitic in the body cavity of a dipterous larva (*Dasyhelia obscura* Winwertz). Parasitology, XIII, p. 83. [**173.**]

KEMPTON, F. E. (1919). Origin and Development of the Pycnidium. Bot. Gaz. LXVIII, p. 233. [**1.**]

KEVORKIAN, A. G. (1935). Studies in the Leptomitaceae. II. Cytology of Apodachlya brachynema and Sapromyces Reinschii. Mycologia, XXVII, p. 274. [**98, 101.**]

KIDSTON, R. and LANG, W. H. (1921). On Old Red Sandstone Plants showing Structure from the Rhynie Chert Bed, Aberdeenshire. Trans. Roy. Soc. Edin. LII, p. 855. [**1.**]

KIHLMANN, O. (1885). Zur Entwickelungsgeschichte der Ascomyceten. *Melanospora parasitica*. Acta Soc. Fennicae, XIV, p. 313. [**245.**]

KILLIAN, C. (1917). Über die Sexualität von Venturia inaequalis (Cooke) Ad. Zeitschr. f. Bot. IX, p. 353. [**267.**]

—— (1918). Morphologie, Biologie und Entwicklungsgeschichte von Cryptomyces Pteridis (Rebent.) Rehm. Zeitschr. f. Bot. X, p. 49. [**241.**]

—— (1919). Sur la sexualité de l'Ergot du Seigle, le *Claviceps purpurea* (Tulasne). Bull. Soc. Myc. de France, XXXV, p. 182. [**253.**]

—— (1922). Le développement du *Stigmatea Robertiani* Fries. Rev. Gén. de Bot. XXXIV, p. 577. [**266.**]

—— (1924). Le développement du *Graphiola Phoenicis* Poit. et ses affinités. Rev. Gén. de Bot. XXXVI, pp. 385, 451. [**292.**]

—— (1931). Biologie et développement du "Placosphaeria onobrychidis". Ann. Sci. Nat. Bot. XIII, p. 403. [**254.**]

KING, C. A (1903). Observations on the Cytology of *Araiospora pulchra* Thaxter. Proc. Boston Soc. Nat. Hist. XXXI, p. 211. [**100.**]

KLEBAHN, H. (1918). Haupt- und Nebenfruchtformen der Askomyzeten. Erster Teil. Eigene Untersuchungen. Gebr. Borntraeger, Leipzig. [**149.**]

KLEBS, G. (1896). Die Bedingungen der Fortpflanzung bei einigen Algen und Pilzen. Fischer, Jena. [**38.**]

—— (1899). Zur Physiologie der Fortpflanzung einiger Pilze. II. *Saprolegnia mixta*. Jahrb. f. wiss. Bot. XXXIII, p. 513. [**38, 90.**]

—— (1900). Zur Physiologie der Fortpflanzung einiger Pilze. III. Allgemeine Betrachtungen. Jahrb. f. wiss. Bot. XXXV, p. 153. [**38.**]

KLØCKER, A. (1903). Om Slaegten *Penicilliums* Plads i Systemet og Beskrivelse af en ny ascusdannende Art. Meddelelser fra Carlsberg Laboratoriet, VI, p. 84. [**182.**]

—— (1909 i). *Endomyces Javanensis* n.sp. C.R. des trav. du lab. de Carlsberg, VII, p. 267. [**167.**]

—— (1909 ii). Deux nouveaux Genres de la Famille des Saccharomycètes. C.R. des trav. du lab. de Carlsberg, VII, p. 273. [**173.**]

KLØCKER, A. and SCHIÖNNING, H. (1895). Experimentelle Untersuchungen über die vermeintliche Umbildung des Aspergillus oryzae in einem Saccharomyceten. Centralbl. f. Bakt. Abt. ii, I, p. 777. [**18.**]

—— —— (1896). Experimentelle Untersuchungen über die vermeintliche Umbildung verschiedener Schimmelpilze in Saccharomyceten. Centralbl. f. Bakt. Abt. ii, II, p. 185. [**18.**]

KLUYVER, A. J. and VAN NIEL, C. B. (1924). Spiegelbilder erzeugende Hefenarten und die neue Hefengattung *Sporobolomyces*. Centralbl. f. Bakt. Abt. ii, LXIII, p. 1. [**286**.]

KNIEP, H. (1915). Beiträge zur Kenntnis der Hymenomyceten. III. Zeitschr. f. Bot. VII, p. 369. [**334**.]

—— (1916). Beiträge zur Kenntnis der Hymenomyceten. IV. Zeitschr. f. Bot. VIII, p. 353. [**159**.]

—— (1917). Beiträge zur Kenntnis der Hymenomyceten. V. Zeitschr. f. Bot. IX, p. 81. [**334**.]

—— (1919). Untersuchungen über den Antherenbrand (Ustilago violacea Pers.). Ein Beitrag zum Sexualitätsproblem. Zeitschr. f. Bot. XI, p. 257. [**294**.]

—— (1920). Über morphologische und physiologische Geschlechtsdifferenzierung. Verhandl. d. Physikal.-Med. Gesellsch. zu Wurzburg, XLVI, p. 1. [**287, 288**.]

—— (1922). Über Geschlechtsbestimmung und Reductionsteilung. Verhandl. d. Physikal.-Med. Gesellsch. zu Wurzburg, XLVII, p. 1. [**288**.]

—— (1923). Über erbliche Änderungen von Geschlechtsfaktoren bei Pilzen. Zeitschr. f. induktive Abstammungs- und Vererbungslehre, XXXI, p. 170. [**288**.]

—— (1926). Über Artkreuzungen bei Brandpilzen. Zeitschrift f. Pilzkunde, X, p. 217. [**288, 294**.]

—— (1928). Die Sexualität der niederen Pflanzen. Fischer, Jena. [**288**.]

—— (1929–30). *Allomyces javanicus* n.sp., ein anisogamer Phycomycet mit Planogameten. Ber. d. deutsch. bot. Ges. XLVII, p. 199. [**89, 91, 92**.]

—— (1930). Über den Generationswechsel von Allomyces. Zeitschr. f. Bot. XXII, p. 433. [**89, 92**.]

KNOLL, F. (1912). Untersuchungen über den Bau und die Function der Cystiden und verwandter Organe. Jahrb. f. wiss. Bot. L, p. 453. [**330**.]

KNOWLES, E. K. (1887). The Curl of Peach Leaves: a Study of the Abnormal Structure produced by *Exoascus deformans*. Bot. Gaz. XII, p. 216. [**197**.]

KNUDSON, L. (1922). Non-symbiotic Germination of Orchid Seeds. Bot. Gaz. LXXIII, p. 1. [**24**.]

KOMINAMI, K. (1914). *Zygorhynchus japonicus*—une nouvelle Mucorinée hétérogame, isolée du sol du Japon. Myc. Centralbl. V, p. 1. [**135**.]

KRAFCZYK, H. (1931). Die Zygosporenbildung bei *Pilobolus cristallinus*. Ber. d. deutsch. bot. Ges. XLIX, p. 141. [**133**.]

KRÜGER, F. (1910). Beitrag zur Kenntnis der Kernverhältnisse von *Albugo candida* und *Peronospora Ficariae*. Centralbl. f. Bakt. Abt. ii, XXVII, p. 186. [**121, 122**.]

KULKARNI, G. S. (1913). Observations on the downy mildew (Sclerospora graminicola (Sacc.) Schroet.) of Bajri and Jowar. Mem. Dept. Agric. India Bot. V, V, p. 268. [**125**.]

KUNKEL, L. O. (1914). Nuclear behaviour in the promycelia of *Caeoma nitens* Burrill and *Puccinia Peckiana* Howe. Am. Journ. Bot. I, p. 37. [**319**.]

—— (1915). A contribution to the life history of *Spongospora subterranea*. Journ. Agric. Research, IV, p. 265. [**47, 48**.]

—— (1918). Tissue Invasion by *Plasmodiophora Brassicae*. Journ. Agric. Research, XIV, p. 543. [**48**.]

KURSANOV L. (1914). Über die Peridienentwicklung im Aecidium. Ber. d. deutsch. bot. Ges. XXXII, p. 317. [**309**.]

—— (1917). Recherches morphologiques et cytologiques sur les Urédinées. Bull. Soc. Imp. des Nat. de Moscou, XXXI, p. 1. [**320**.]

KUSANO, S. (1911). *Gastrodia elata* and its Symbiotic Association with *Armillaria mellea*. Journ. Coll. Agric. Tokyo, IV, p. 1. [**23, 25**.]

KUSANO, S. (1912). On the Life-History and Cytology of a new *Olpidium* with special reference to the Copulation of motile Isogametes. Journ. Coll. Agric. Tokyo, IV, p. 141. [**59, 60**.]

—— (1930). The Life-History and Physiology of *Synchytrium fulgens* Schroet., with Special Reference to its Sexuality. Japanese Journ. Bot. v, p. 35. [**67**.]

KUYPER, H. P. (1905). Die Perithecienentwicklung von *Monascus purpureus* Went. und *Monascus Barkeri* Dangeard, sowie die systematische Stellung dieser Pilze. Ann. Myc. III, p. 32. [**183**.]

LA COUR, L. (1931). Improvements in Everyday Technique in Plant Cytology. Journ. R. Micr. Soc. LI, p. 119. [**377, 379**.]

LAFFERTY, H. A. and PETHYBRIDGE, G. H. (1922–4). On a Phytophthora Parasitic on Apples which has both Amphigynous and Paragynous Antheridia: and on Allied Species which show the same Phenomenon. Sci. Proc. Roy. Dublin Soc. N.S. XVII, p. 29. [**118**.]

DE LAGERHEIM, A. (1892). *Dipodascus albidus*, eine neue geschlechtliche Hemiascee. Jahrb. f. wiss. Bot. XXIV, p. 549. [**170**.]

—— (1900). Mycologische Studien. II. Untersuchungen über die Monoblepharideen. Bihang till K. Svensk. Vet. Akad. Handl. iii, XXV, p. 1. [**94**.]

LAIBACH, F. (1927). Zytologische Untersuchungen über die Monoblepharideen. Jahrb. f. wiss. Bot. LXVI, p. 596. [**96**.]

LAKON, G. (1926). Über die systematische Stellung der Pilzgattung *Basidiobolus* Eidam. Jahrb. f. wiss. Bot. LXV, p. 388. [**146**.]

LAMB, I. M. (1934 i). On the Morphology and Cytology of *Puccinia Prostii*, Moug., a Micro-form with Pycnidia. Trans. Roy. Soc. Ed. LVIII, p. 143. [**319**.]

—— (1934 ii). Entwicklungsgeschichtliche Untersuchung einer morphologisch abweichenden Puccinia-Art (*Pucc. Sonchi* Rob.). Hedwigia, LXXIV, p. 181. [**319**.]

—— (1935). The Initiation of the Dikaryophase in Puccinia phragmites (Schum.) Körn. Ann. Bot. XLIX, p. 403. [**313**.]

LANDER, C. A. (1933). The morphology of the developing fruiting body of *Lycoperdon gemmatum*. Amer. Journ. Bot. XX, p. 204. [**351**.]

LANG, W. H. (1902). On the Prothalli of Ophioglossum pendulum and Helminthostachys zeylanica. Ann. Bot. XVI, pp. 26, 36. [**26**.]

LAWES, J. B., GILBERT, J. H. and WARRINGTON, R. (1883). Contribution to the Chemistry of "Fairy Rings". Journ. Chem. Soc. Lond. XLIII, p. 208. [**15**.]

LECHMERE, E. A. (1910). An Investigation of a Species of Saprolegnia. New Phyt. IX, p. 305. [**38, 90**.]

—— (1911). Further Investigation of the Methods of Reproduction in the Saprolegniaceae. New Phyt. X, p. 167. [**38, 90**.]

LEDEBOER, M. J. S. (1934). Physiologische onderzoekingen over Ceratostomella ulmi (Schwarz) Buisman. Proefschrift, Baarn. [**264**.]

LEGER, L. (1908). Mycétozoaires endoparasites des Insectes. I. *Sporomyxa Scauri* n.g., n.sp. Arch. f. Protistenk. XII, p. 109. [**56**.]

—— (1914). Sur un nouveau Protiste du genre *Dermocystidium*, parasite de la Truite. C.R. Acad. Sci. CLVIII, p. 807. [**56**.]

—— (1927). Sur la nature et l'évolution des 'spherules' décrites chez les Ichthyophones, Phycomycètes parasites de la truite. C.R. Acad. Sci. CLXXXIV, p. 1268. [**147**.]

LEGER, L. and DUBOSCQ, O. (1909). Sur les *Chytridiopsis* et leur évolution. Arch. Zool. Expt. et Gén. V, I, N. et R. p. ix. [**56**.]

LEGER, L. and HESSE, E. (1909). Sur un nouveau Entophyte parasite d'un Coléoptère. C.R. Acad. Sci. CXLIX, p. 303. [**56**.]

—— —— (1923). Sur un Champignon du type *Ichthyophonus*, parasite de l'intestin de la Truite. C.R. Acad. Sci. CLXXVI, p. 420. [**147**.]

LEHFELDT, W. (1923). Über die Entstehung des Paarkernmyzels bei heterothallischen Basidiomyceten. Hedwigia, LXIV, p. 30. [**337**.]

LEITGEB, H. (1869). Neue Saprolegnieen. Jahrb. f. wiss. Bot. VII, p. 357. [**55**, **105**.]

—— (1882). *Completoria complens* Lohde, ein in Farnprothallien schmarotzender Pilz. Sitzungsber. d. K. Akademie d. Wissenschaften Wien, LXXXIV, i, p. 288. [**145**.]

LENDNER, A. (1908). Les Mucorinées de la Suisse. Matériaux pour la Flore Cryptogamique Suisse, III. Wyss, Berne, p. 1. [**125**, **133**.]

—— (1910). Observations sur les Zygospores des Mucorinées. Bull. Soc. Bot. Genève, ii, II, p. 56. [**133**.]

—— (1923). Une Mucorinée nouvelle du genre Absidia. Bull. Soc. Bot. Genève, ii, XV, p. 147. [**133**.]

LEONIAN, L. H. (1925). Physiological studies on the genus Phytophthora. Am. Journ. Bot. XII, p. 444. [**40**, **115**.]

—— (1926). The morphology and the pathogenicity of some Phytophthora mutations. Phytopath. XVI, p. 723. [**115**.]

—— (1927). The effect of different hosts upon the sporangia of some Phytophthoras. Phytopath. XVII, p. 483. [**115**.]

—— (1930). Differential growth of Phytophthoras under the action of malachite green. Amer. Journ. Bot. XVII, p. 671. [**115**.]

—— (1934). Identification of Phytophthora Species. Agr. Expt. Stat. West Virginia University, Bull. 262, p. 1. [**115**.]

LEVINE, M. (1913). Studies in the Cytology of the Hymenomycetes, especially Boleti. Bull. Torrey Bot. Club, XL, p. 137. [**331**.]

LEVISOHN, I. (1927). Beitrag zur Entwicklungsgeschichte und Biologie von *Basidiobolus ranarum* Eidam. Jahrb. f. wiss. Bot. LXVI, p. 513. [**147**.]

LEWIS, C. E. (1906). The Development of the Spores in *Amanita bisporigera*. Bot. Gaz. XLI, p. 348. [**328**.]

LEWIS, I. M. (1911). The Development of the Spores in *Pleurage zygospora*. Bot. Gaz. LI, p. 369. [**257**.]

LEWTON-BRAIN, L. (1901). Cordyceps ophioglossoides. Ann. Bot. XV, p. 522. [**252**.]

LIKHITÉ, V. (1925). Développement et biologie de quelques Ascomycètes. Thèses présentées à la Faculté des Sciences de Strasbourg, p. 1. [**268**.]

LINDEGREN, C. C. (1932 i). The genetics of Neurospora. I. The inheritance of response to heat-treatment. Bull. Torrey Bot. Club, LIX, p. 85. [**246**.]

—— (1932 ii). The genetics of Neurospora. II. Segregation of sex factors in the asci of N. crassa, N. sitophila and N. tetrasperma. Bull. Torrey Bot. Club, LIX, p. 119. [**246**.]

—— (1933). The genetics of Neurospora. III. Pure bred stocks and crossings over in N. crassa. Bull. Torrey Bot. Club, LX, p. 133. [**246**.]

—— (1934). The genetics of Neurospora. VI. Bisexual and akaryotic ascospores from N. crassa. Genetica, XVI, p. 315. [**246**.]

—— (1935). The Genetics of *Neurospora*. VII. Developmental competition between different Genotypes within the Ascus. Zeitschr. f. inductive Abstammungs- und Vererbungslehre, LXVIII, p. 331. [**246**.]

—— (1936). A Six-Point Map of the Sex-Chromosome of *Neurospora crassa*. Journ. Genetics, XXXII, p. 243. [**246**.]

LINDFORS, T. (1924). Studien über den Entwicklungsverlauf bei einigen Rostpilzen aus zytologischen und anatomischen Gesichtspunkten. Svensk. Bot. Tidskr. XVIII, p. 1. [**307**, **317**.]

LINE, J. (1922). The Parasitism of *Nectria cinnabarina*. Trans. Brit. Myc. Soc. VIII, p. 22. [**20**, **247**.]

LISTER, A. and LISTER, G. (1925). A Monograph of the Mycetozoa, a descriptive catalogue of the species in the herbarium of the British Museum. 3rd ed. Trustees of the British Museum, London. [**45**.]

LISTER, G. (1933). Field notes on Mycetozoa. Trans. Brit. Myc. Soc. XVIII, p. 18. [**45**.]

LOEWENTHAL, W. (1904–5). Weitere Untersuchungen an Chytridiaceen. Arch. f. Protistenk. V, p. 221. [**59, 75**.]

LOHWAG, H. (1926). Sporobolomyces—kein Basidiomyzet. Ann. Myc. XXIV, p. 194. [**286**.]

LUDI, R. (1901). Beiträge zur Kenntnis der Chytridiaceen. (Fortsetzung.) Hedwigia, XLI, Beibl. p. (1). [**65**.]

LUND, A. (1934). Studies on Danish Freshwater Phycomycetes and Notes on their Occurrence particularly relative to the Hydrogen Ion Concentration of the Water. Mém. Acad. Roy. Sci. et Lett. Danemark, ix, VI, p. 1. [**12**.]

LUPO, P. (1922). Stroma and formation of perithecia in *Hypoxylon*. Bot. Gaz. LXXIII, p. 486. [**274**.]

LUTMAN, B. F. (1911). Some contributions to the life-history and cytology of the smuts. Trans. Wisconsin Acad. Sci. Arts and Letters, XVI, p. 1191. [**292**.]

—— (1913). Studies in Club Root. I. The relation of *Plasmodiophora Brassicae* to its host and the structure and growth of its plasmodium. Vermont Agric. Expt. Stat. Bull. 175, p. 1. [**48**.]

McBETH, I. G. and SCALES, F. M. (1913). Destruction of Cellulose by Bacteria and Fungi. U.S. Dept. of Agric. Bureau of Plant Industry, Bull. 266. [**15**.]

McCORMICK, F. A. (1911). Homothallic Conjugation in *Rhizopus*. Bot. Gaz. LI, p. 229. [**135**.]

McCREA, A. (1931). The reactions of Claviceps purpurea to variations of environment. Amer. Journ. Bot. XVIII, p. 50. [**252**.]

McCUBBIN, W. A. (1910). Development of the Helvellinaceae. I. *Helvella elastica*. Bot. Gaz. XLIX, p. 105. [**158, 232**.]

McDOUGALL, W. B. (1914). On the mycorrhizas of forest trees. Am. Journ. Bot. I, p. 51. [**27**.]

—— (1919). Development of *Stropharia epimyces*. Bot. Gaz. LXVII, p. 258. [**340**.]

McLENNAN, E. (1920). The endophytic fungus of *Lolium*. I. Proc. Roy. Soc. Victoria, XXXII (N.S.), p. 252. [**26**.]

—— (1926). The Endophytic Fungus of Lolium. II. The mycorrhiza on the Roots of Lolium temulentum L., with a Discussion on the Physiological Relationships of the Organism concerned. Ann. Bot. XL, p. 43. [**22, 26**.]

—— (1932). Some "Crinoline" Fungi—Papuan Species of *Dictyophora* Desvaux. Victorian Naturalist, XLIX, p. 3. [**351**.]

McRAE, W. (1918). *Phytophthora Meadii* n.sp., on *Hevea brasiliensis*. Mem. Dept. Agric. India Bot. IX, p. 219. [**115**.]

MAGNUS, P. (1905). Ueber die Gattung zu der *Rhizidium dicksonii* Wright gehört. Hedwigia, XLIV, p. 347. [**64**.]

MAGNUS, W. (1900). Studien an der endotrophen Mycorrhiza von *Neottia Nidus-Avis*. Jahrb. f. wiss. Bot. XXXV, p. 205. [**23**.]

MAGROU, J. (1921). Symbiose et Tubérisation. Ann. Sci. Nat. Bot. 10me sér. IV, p. 181. [**27**.]

—— (1925). La symbiose chez les Hépatiques—Le *Pellia epiphylla* et son Champignon commensal. Ann. Sci. Nat. Bot. 10me sér. VII, p. 725. [**23**.]

MAIRE, R. (1902). Recherches cytologiques et taxonomiques sur les Basidiomycètes. Bull. Soc. Myc. de France, XVIII, p. 1 (supplément). [**325**.]

—— (1905). Recherches cytologiques sur quelques Ascomycètes. Ann. Myc. III, p. 123. [**160, 232**.]

MAIRE, R. (1911). La Biologie des Urédinales. Prog. Rei Bot. IV, p. 109. **[302.]**

—— (1912). Contribution à l'Étude des Laboulbéniales de l'Afrique du Nord. Bull. Soc. Hist. Nat. de l'Afrique du Nord, IV, p. 3. **[274.]**

—— (1916). Deuxième contribution à l'Étude des Laboulbéniales de l'Afrique du Nord. Bull. Soc. Hist. Nat. de l'Afrique du Nord, VII, p. 6. **[274.]**

MAIRE, R. and CHEMIN, E. (1922). Un nouveau Pyrénomycète marin. C.R. Acad. Sci. CLXXV, p. 319. **[13.]**

MAIRE, R. and TISON, A. (1909). La cytologie des Plasmodiophoracées et la classe des Phytomyxinae. Ann. Myc. VII, p. 226. **[46, 48.]**

—— —— (1911). Nouvelles recherches sur les Plasmodiophoracées. Ann. Myc. IX, p. 226. **[46, 48.]**

MAJMONE, B. (1914). Parasitismus und Vermehrungsformen von *Empusa elegans* n.sp. Centralbl. f. Bakt., Abt. ii, XL, p. 98. **[143.]**

MANGENOT, G. (1919). Sur la formation des asques chez *Endomyces lindneri*. C.R. Soc. Biol. Paris, LXXXII, pp. 230, 477. **[169.]**

—— (1922). À propos de quelques formes peu connues de l'Endomycetaceae. Bull. Soc. Myc. de France, XXXVIII, p. 42. **[167.]**

MARCHAL, E. (1902–3). De la spécialisation du parasitisme chez l'*Erysiphe Graminis*. C.R. Acad. Sci. CXXXV, p. 210 and CXXXVI, p. 1280. **[29.]**

MARCHAND, H. (1913). La conjugaison des spores chez les levures. Rev. Gén. de Bot. XXV, p. 207. **[173.]**

MARRYAT, D. C. E. (1907). Notes on the Infection and Histology of two Wheats immune to the Attacks of *Puccinia glumarum*, yellow Rust. Journ. Agric. Sci. II, p. 129. **[29.]**

MARSH, R. W. (1926). Additional Records of *Ctenomyces serratus* Eidam. Trans. Brit. Myc. Soc. X, p. 314. **[177.]**

MARTENS, P. (1932). L'origine du crochet et de l'anse d'anastomose chez les champignons supérieurs. Zeitschr. f. Pflanzenernährung, Düngung und Bodenkunde, XXXI, p. 208. **[159.]**

MARTENS, P. and VANDENDRIES, R. (1932). Le cycle conidien haploïde et diploïde chez Pholiota aurivella. Cellule, XLI, p. 337. **[289.]**

MARTIN, E. M. (1924). Cytological studies of *Taphrina coryli* Nishida on *Corylus americana*. Trans. Wisconsin Acad. Sci. Arts and Letters, XXI, p. 345. **[197.]**

—— (1925). Cultural and morphological studies of some species of *Taphrina*. Phytopath. XV, p. 67. **[197.]**

MARTIN, G. W. (1925). Morphology of *Conidiobolus villosus*. Bot. Gaz. LXXX, p. 311. **[145.]**

MASON, E. (1928). Note on the Presence of Mycorrhiza in the Roots of Salt Marsh Plants. New Phyt. XXVI, p. 193. **[22.]**

MASSEE, G. (1895). A revision of the Genus *Cordyceps*. Ann. Bot. II, p. 207. **[250.]**

—— (1897). A Monograph of the Geoglossaceae. Ann. Bot. XI, p. 225. **[233.]**

—— (1909). The Structure and Affinities of British Tuberaceae. Ann. Bot. XXIII, p. 243. **[236.]**

MASSEE, I. (1914). Observations on the Life-History of *Ustilago Vaillantii* Tul. Journ. Econ. Biol. IX, p. 7. **[297.]**

MATRUCHOT, L. (1903). Une Mucorinée purement conidienne, *Cunninghamella africana*. Étude éthologique et morphologique. Ann. Myc. I, p. 45. **[136, 137.]**

MATSUURA, H. and GONDO, A. (1935). A Karyological Study on *Peziza subumbrina* Boud., with Special Reference to a Heteromorphic Pair of Chromosomes. Journ. Faculty of Sci., Hokkaido Imp. Univ. Ser. V, III, p. 205. **[216.]**

MAUPAS, E. (1915). Sur un Champignon parasite des *Rhabditis*. Bull. Soc. Hist. Nat. de l'Afrique du Nord, VII, p. 34. **[86.]**

MAYO, J. K. (1925). The Enzymes of *Stereum purpureum*. New Phyt. XXIV, p. 162. [**335**.]

MAYR, H. (1882). Ueber die Parasitismus von *Nectria cinnabarina*. Untersuchungen aus dem forst-botanischen Inst. zu München, III, p. 1. [**247**.]

MELHUS, I. E. (1914). A species of *Rhizophidium* parasitic on the oospores of various Peronosporaceae. Phytopath. IV, p. 55. [**71**.]

MELIN, E. (1921). On the Mycorrhiza of *Pinus sylvestris* L. and *Picea Abies* Karst. A preliminary note. Journ. of Ecology, IX, p. 254. [**27**.]

MEYEN, F. J. F. (1841). Pflanzenpathologie. Berlin, p. 143. [**307**.]

MILBRAITH, D. G. (1923). Downy Mildew on Lettuce in California. Journ. Agric. Research, XXIII, p. 989. [**123**.]

VON MINDEN, M. (1916). Beiträge zur Biologie und Systematik einheimischer submerser Phycomyceten. Mykologische Unters. u. Berichte von Dr R. Falck, II, p. 146. [**13, 82, 94, 97, 98, 119**.]

MIRANDE, R. (1920). *Zoophagus insidians* Sommerstoff, capteur de Rotifères vivants. Bull. Soc. Myc. de France, XXXVI, p. 47. [**13**.]

MITTMANN, G. (1932). Kulturversuche mit Einsporstämmen und zytologische Untersuchungen in der Gattung Ceratostomella. Jahrb. f. wiss. Bot. LXVII, p. 185. [**262, 264**.]

MIX, A. J. (1935). The life history of Taphrina deformans. Phytopath. XXV, p. 41. [**199**.]

MIYAKE, K. (1901). The fertilisation of Pythium de Baryanum. Ann. Bot. XV, p. 653. [**113**.]

MIYOSHI, M. (1894). Ueber Chemotropismus der Pilze. Bot. Zeit. LII, p. 1. [**33**.]

MÖLLER, A. (1895). Brasilische Pilzblumen. Pt 7 of Bot. Mitteilungen aus den Tropen, ed. by A. F. Schimper. Fischer, Jena. [**347**.]

MOLLIARD, M. (1909). Le cycle de développement du *Crucibulum vulgare* Tul. et de quelques Champignons supérieurs obtenus en cultures pures. Bull. Soc. Bot. de France, LVI, p. 91. [**354**.]

—— (1910). De l'action du *Marasmius oreades* Fr., sur la végétation. Bull. Soc. Bot. de France, LVII, p. 62. [**15**.]

—— (1918). Sur la vie saprophytique d'un *Entomophthora* (*E. Henrici* n.sp.). C.R. Acad. Sci. Paris, CLXVII, p. 958. [**143**.]

—— (1920). L'influence d'une dose réduite de potassium sur les caractères physiologiques de *Sterigmatocystis nigra*. C.R. Acad. Sci. CLXVII, p. 302. [**40**.]

MORAL, H. (1913). Ueber das Auftreten von *Dermocystidium pusula* (Perez), einem einzelligen Parasiten der Haut des Molches bei *Triton cristatus*. Arch. f. Mikrosk. Anat. LXXXI, p. 381. [**13**.]

MOREAU, F. (1911). Les phénomènes intimes de la reproduction sexuelle chez quelques Mucorinées. Bull. Soc. Bot. de France, LVIII, p. 618. [**135**.]

—— (1912). Sur la reproduction sexuée de *Zygorhynchus Moelleri*. C.R. Soc. Biol. Paris, LXXIII, p. 14. [**135**.]

—— (1913 i). Recherches sur la reproduction des Mucorinées et de quelques autres Thallophytes. Botaniste, XIII, p. 1. [**125**.]

—— (1913 ii). Une nouvelle Mucorinée du sol, *Zygorhynchus Bernardi* nov. sp. Bull. Soc. Bot. de France, LX, p. 256. [**135**.]

—— (1914). Sur le développement du périthèce chez une Hypocréale le *Peckiella laterita* (Fries) Maire. R. Bull. Soc. Bot. de France, LXI, p. 160. [**245**.]

—— (1920). À propos du nouveau genre *Kunkelia* Arthur. Bull. Soc. Myc. de France, XXXVI, p. 101. [**319, 320**.]

—— (1925). Recherches sur quelques lichens des genres *Parmelia*, *Physcia* et *Anaptychia*. Rev. Gén. de Bot. XXXVII, p. 385. [**159**.]

MOREAU, F. and MOREAU, MME F. (1922). Le mycélium à boucles chez les Ascomycètes. C.R. Acad. Sci. CLXXIV, p. 1072. [**159**.]

—— —— (1930). Le développement du périthèce chez quelques Ascomycètes. Rev. Gén. de Bot. XLII, p. 65. [**206**.]

—— —— (1931). Existe-t-il une double réduction chromatique chez les Ascomycètes? Rev. Gén. de Bot. XLIII, p. 465. [**206**.]

MOREAU, MME F. (1914). Les phénomènes de la sexualité chez les Urédinées. Botaniste, XIII, p. 145. [**303**.]

MOSS, E. H. (1923). Developmental Studies in the Genus Collybia. Trans. Roy. Canadian Inst. XIV, pt 2, p. 321. [**340**.]

—— (1926). The Uredo Stage of the Pucciniastreae. Ann. Bot. XL, p. 813. [**315**.]

MOUNCE, I. (1921). Homothallism and the production of Fruit Bodies by Monosporous Mycelia in the Genus Coprinus. Trans. Brit. Myc. Soc. VII, p. 198. [**287**.]

—— (1922). Homothallism and Heterothallism in the Genus Coprinus. Trans. Brit. Myc. Soc. VII, p. 256. [**287**.]

—— (1923). The Production of Fruit Bodies of *Coprinus comatus* in Laboratory Cultures. Trans. Brit. Myc. Soc. VIII, p. 221. [**287**.]

MÜCKE, M. (1908). Zur Kenntnis der Eientwicklung und Befruchtung von *Achlya polyandra* de Bary. Ber. d. deutsch. bot. Ges. XXVI *a*, p. 367. [**108**.]

MÜLLER, F. (1911). Untersuchungen über die chemotaktische Reizbarkeit der Zoosporen von Chytridiaceen und Saprolegniaceen. Jahrb. f. wiss. Bot. XLIX, p. 421. [**30, 33, 34**.]

MURPHY, P. A. (1918). The Morphology and Cytology of the Sexual Organs of Phytophthora erythroseptica Pethyb. Ann. Bot. XXXII, p. 115. [**115**.]

NADSON, G. A. and KONOKOTIN, A. G. (1910). *Guilliermondia*, un nouveau genre de la famille des Saccharomycetaceae à copulation hétérogamique. Bull. du Jard. Imp. de St Pétersbourg, XI, p. 117. [**173**.]

NADSON, G. A. and PHILLIPOV, G. S. (1925 i). Une nouvelle Mucorinée *Mucor Guilliermondii* nov. sp. et ses formes-levures. Rev. Gén. de Bot. XXXVII, p. 450. [**18, 125**.]

—— —— (1925 ii). Influence des rayons X sur la sexualité et la formation des mutantes chez les champignons inférieurs (Mucorinées). C.R. Soc. Biol. Paris, CXIII, p. 473. [**42**.]

—— —— (1928). De la formation de nouvelles races stables chez les champignons inférieurs sous l'influence des rayons X. C.R. Acad. Sci. CLXXXVI, p. 1566. [**42**.]

NAMYSLOWSKI, B. (1920). État actuel des recherches sur les phénomènes de la sexualité des Mucorinées. Rev. Gén. de Bot. XXXII, p. 193. [**125**.]

NAOUMOFF, M. N. (1924). Les bases morphologiques de la systématique dans la famille des Mucoracées. Bull. Soc. Myc. de France, XL, p. 86. [**125**.]

NAWASCHIN, S. (1899). Beobachtungen über den feineren Bau und Umwandlungen von *Plasmodiophora Brassicae* Wor. im Laufe ihres intrazellularen Lebens. Flora, LXXXVI, p. 404. [**48**.]

NĚMEC, B. (1911 i). Zur Kenntnis der niederen Pilze. I. Eine neue Chytridiazee. Bull. Int. de l'Acad. des Sci. de Bohème, XVI, p. 67. [**56**.]

—— (1911 ii). Zur Kenntnis der niederen Pilze. II. Die Haustorien von *Uromyces Betae* Pers. Bull. Int. de l'Acad. des Sci. de Bohème, XVI, p. 118. [**303**.]

—— (1911 iii). Über eine Chytridiazee der Zuckerrübe. Ber. d. deutsch. bot. Ges. XXIX, p. 48. [**56**.]

—— (1912). Zur Kenntnis der niederen Pilze. IV. *Olpidium Brassicae* Wor., und zwei *Entophlyctis*-Arten. Bull. Int. de l'Acad. des Sci. de Bohème, XVII, p. 16. [**61**.]

NĚMEC, B. (1913). Zur Kenntnis der niederen Pilze. V. Über die Gattung *Anisomyxa Plantaginis* n.g., n.sp. Bull. Int. de l'Acad. des Sci. de Bohème, XVIII, p. 18. [**56**.]

—— (1922). O pohlavnosti u *Olpidium Brassicae*. Zolast. otisk ze Sborniku Klubu, Praze, p. 1. [**61**.]

NERESHEIMER, E. and CLODI, C. (1914). *Ichthyophonus Hoferi* Plehn et Mulsow, der Erreger der Taumelkrankheit der Salmoniden. Arch. f. Protistenk. XXXIV, p. 217. [**148**.]

NEWTON, D. E. (1926 i). The Bisexuality of Individual Strains of Coprinus Rostrupianus. Ann. Bot. XL, p. 105. [**288, 289, 331**.]

—— (1926 ii). The Distribution of Spores of Diverse Sex on the Hymenium of Coprinus lagopus. Ann. Bot. XL, p. 891. [**289**.]

NEWTON, M. and BROWN, A. M. (1934). Studies on the nature of disease resistance in cereals. I. The reactions to rust of mature and immature tissues. Canadian Journ. Res. XI, p. 564. [**322**.]

NEWTON, M. and JOHNSON, T. (1936). Stripe rust, *Puccinia glumarum*, in Canada. Canadian Journ. Res. XIV, p. 89. [**322**.]

NEWTON, W. C. F. (1927). Chromosome Studies in Tulipa and some related Genera. Journ. Linn. Soc. XLVII, p. 339. [**377**.]

NICHOLS, M. A. (1896). The morphology and development of certain pyrenomycetous fungi. Bot. Gaz. XXII, p. 301. [**1, 245, 264**.]

NIENBERG, W. (1914). Zur Entwickelungsgeschichte von *Polystigma rubra*. Zeitschr. f. Bot. VI, p. 369. [**248**.]

NOBLE, M. (1937). The Morphology and Cytology of Typhula Trifolii Rostr. Ann. Bot. New series, I, p. 67. [**2, 337**.]

NOWAKOWSKI, L. (1876). Beitrag zur Kenntnis der Chytridiaceen. II. *Polyphagus Euglenae*, eine Chytridiacee mit geschlechtlicher Fortpflanzung. Cohn's Beitr. z. Biol. d. Pflanzen, II, p. 73. [**59, 76**.]

OBEL, P. (1910). Researches on the conditions of the forming of oogonia in Achlya. Ann. Myc. VIII, p. 421. [**38, 90**.]

OJERHOLM, E. (1934). Multiciliate zoospores in Physoderma Zeae-maydis. Bull. Torrey Bot. Club, XLI, p. 13. [**79**.]

OLIVE, E. W. (1905). The Morphology of *Monascus purpureus*. Bot. Gaz. XXXIX, p. 56. [**183**.]

—— (1907). Cell and Nuclear Division in *Basidiobolus*. Ann. Myc. V, p. 404. [**146**.]

—— (1908). Sexual Cell-Fusions and Vegetative Nuclear Divisions in the Rusts. Ann. Bot. XXII, p. 331. [**310**.]

—— (1911). Origin of Heteroecism in Rusts. Phytopath. I, p. 139. [**321**.]

OLIVE, E. W. and WHETZEL, H. H. (1917). Endophyllum-like rusts of Porto Rico. Am. Journ. Bot. IV, p. 44. [**320**.]

OLIVER, F. W. (1890). On Sarcodes sanguinea Torr. Ann. Bot. IV, p. 303. [**25**.]

OLTMANNS, F. (1887). Ueber die Entwickelung der Perithecien in der Gattung *Chaetomium*. Bot. Zeit. XLV, p. 193. [**257**.]

—— (1922). Morphologie und Biologie der Algen. II. Phaeophyceae-Rhodophyceae. Jena, p. 70. [**52**.]

ORBAN, G. (1919). Untersuchungen über die Sexualität von *Phycomyces nitens*. Beih. z. bot. Centralbl. XXXVI, p. 1. [**132**.]

ORTON, C. R. (1924). Studies in the Morphology of the Ascomycetes. I. Mycologia, XVI, p. 49. [**254**.]

—— (1927). A working hypothesis on the origin of rusts, with special reference to the phenomenon of heteroecism. Bot. Gaz. LXXXIV, p. 113. [**321**.]

OSBORN, T. G. B. (1911). Spongospora subterranea (Wallr.) Johnson. Ann. Bot. XXV, p. 327. [**48**.]

OVERTON, J. B. (1906). The Morphology of the Ascocarp and Spore-formation in the many spored Ascus of *Thecotheus Pelletieri*. Bot. Gaz. LXII, p. 450. [**225**.]

PAGE, W. M. (1925). Contributions to the Study of the Lower Pyrenomycetes. Report Brit. Assoc. p. 359. [**257, 261**.]

—— (1933). A Contribution to the Life-History of *Sordaria fimicola* (Four-spored Form) with Special Reference to the Abnormal Spores. Trans. Brit. Myc. Soc. XVII, p. 296. [**258**.]

—— (1936). Note on Abnormal Spores in *Podospora minuta*. Trans. Brit. Myc. Soc. XX, p. 186. [**259**.]

PARR, R. (1918). The Response of Pilobolus to Light. Ann. Bot. XXXII, p. 177. [**34, 131**.]

PASCHER, A. (1918). Über die Myxomyceten. Ber. d. deutsch. bot. Ges. XXXVI, p. 359. [**45**.]

PAULSON, R. (1921). Sporulation of Gonidia in the Thallus of *Evernia Prunastri*. Trans. Brit. Myc. Soc. VII, p. 41. [**22**.]

—— (1924). Tree Mycorrhiza. Trans. Brit. Myc. Soc. IX, p. 213. [**27, 28**.]

PETCH, T. (1908). The Phalloideae of Ceylon. Ann. Roy. Bot. Gard. Peradeniya, IV, p. 162. [**347**.]

—— (1921). Studies in Entomogenous Fungi. Trans. Brit. Myc. Soc. VII, pp. 89, 133. [**248**.]

PETERSEN, H. E. (1903). Note sur les Phycomycètes observés dans les téguments vides des nymphes de Phryganées. Journ. de Bot. XVII, p. 214. [**13, 59, 74**.]

—— (1905). Contributions à la connaissance des Phycomycètes marins (Chytridineae Fischer). Övers. Danske Videnskabernes Selskabs Forh. p. 439. [**13, 63, 64**.]

—— (1910). An Account of Danish Freshwater Phycomycetes, with biological and systematic remarks. Ann. Myc. VIII, p. 494. [**13, 59, 74, 85, 97, 119**.]

PETHYBRIDGE, G. H. (1913). On the Rotting of Potato Tubers by a new Species of *Phytophthora* having a Method of Sexual Reproduction hitherto undescribed. Sci. Proc. Roy. Dublin Soc. N.S. XIII, p. 529. [**115, 116**.]

—— (1922). Some recent work on the potato blight. Rept. Int. Potato Conference, 1921, Roy. Hort. Soc. p. 112. [**115**.]

PETHYBRIDGE, G. H. and LAFFERTY, H. A. (1919). A disease of Tomato and other Plants caused by a new species of *Phytophthora*. Sci. Proc. Roy. Dublin Soc. N.S. XV, p. 487. [**118**.]

PETHYBRIDGE, G. H. and MURPHY, P. A. (1913). On Pure Cultures of *Phytophthora infestans* de Bary, and the Development of Oospores. Sci. Proc. Roy. Dublin Soc. N.S. XIII, p. 566. [**118**.]

PEYRONEL, B. (1921). Nouveaux cas de rapports mycorhiziques entre Phanérogames et Basidiomycètes. Bull. Soc. Myc. de France, XXXVII, p. 143. [**27**.]

PILLAY, T. P. (1923). Zur Entwicklungsgeschichte von Sphaerobolus stellatus Tode. Inaug. Diss. Jahrb. d. Phil. Fakultät II der Universität, Bern, p. 197. [**358**.]

DU PLESSIS, S. J. (1933). The Life-history and Morphology of Olpidiopsis Ricciae, nov. sp., Infecting Riccia Species in South Africa. Ann. Bot. XLVII, p. 755. [**67**.]

PLOWRIGHT, C. B. (1889). A Monograph of the British Uredineae and Ustilagineae. Kegan Paul, Trench and Co., London. [**292, 302**.]

POIRAULT, G. and RACIBORSKI, M. (1895). Sur les noyaux des Urédinées. Journ. de Bot. IX, pp. 318, 381. [**303**.]

POLE EVANS, I. B. (1911). South African Cereal Rusts, with Observations on the Problem of Breeding Rust-resisting Wheats. Journ. Agric. Sci. IV, p. 241. [**29**.]

PRATT, C. (1924). The Staling of Fungal Cultures. Ann. Bot. XXXVIII, p. 599. [**31, 32**.]

PREVOST, B. (1807). Mémoire sur la cause immédiate de la Carie. Fontanel, Montauban. [**292**.]

PRINGSHEIM, N. (1858). Beiträge zur Morphologie und Systematik der Algen. II. Die Saprolegnieen. Jahrb. f. wiss. Bot. I, p. 284. [**103, 108**.]

—— (1873). Weitere Nachträge zur Morphologie und Systematik der Saprolegnieen. Jahrb. f. wiss. Bot. IX, p. 191. [**52**.]

QUINTANILHA, A. (1933). Le problème de la sexualité chez les champignons. Recherches sur le genre "Coprinus". Impressa da Universidade, Coimbra. [**288, 289**.]

—— (1935). Cytologie et génétique de la sexualité chez les Hymenomycètes. Boletim da Soc. Broteriana, X, p. 5. [**289**.]

RABENHORST, L. (1881–1920). Kryptogamen-Flora. Kummer, Leipzig. [**10**.]

RACIBORSKI, M. (1896). Ueber den Einfluss äusserer Bedingungen auf die Wachsthumsweise des Basidiobolus ranarum. Flora, LXXXII, p. 107. [**146**.]

RAMLOW, G. (1906). Zur Entwicklungsgeschichte von *Thelebolus stercoreus*. Bot. Zeit. LXIV, p. 85. [**225**.]

—— (1914). Beiträge zur Entwicklungsgeschichte der Ascoboleen. Myc. Centralbl. V, p. 177. [**217**.]

RAMSBOTTOM, J. (1912). Some notes on the history of the Classification of the Uredinales with full list of British Uredinales. Trans. Brit. Myc. Soc. IV, p. 77. [**302**.]

—— (1913). Some Notes on the History of the Classification of the Discomycetes. Trans. Brit. Myc. Soc. IV, p. 382. [**200**.]

—— (1914). The Generic Name *Protascus*. Trans. Brit. Myc. Soc. V, p. 143. [**167**.]

—— (1922). Orchid Mycorrhiza. Trans. Brit. Myc. Soc. VIII, p. 28. [**23, 24**.]

—— (1923). A Handbook of the Larger Fungi. Trustees of the British Museum, London. [**10**.]

—— (1936). The Uses of Fungi. Report Brit. Assoc. p. 189. [**18**.]

RATHAY, E. (1883). Untersuchungen über die Spermogonien der Rostpilze. Denkschriften der kaiserlichen Akad. d. Wissenschaft. Wien, XLVI, Pt 2, p. 1. [**308**.]

RAWITSCHER, F. (1912). Beiträge zur Kenntnis der Ustilagineen. Zeit. f. Bot. IV, p. 673. [**296, 297**.]

—— (1914). Zur Sexualität der Brandpilze *Tilletia tritici*. Ber. d. deutsch. bot. Ges. XXXII, p. 310. [**300**.]

—— (1922). Beiträge zur Kenntnis der Ustilagineen. II. Zeitschr. f. Bot. XIV, p. 273. [**301**.]

RAYMOND, J. R. (1934). Contribution à la connaissance cytologique des Ascomycètes. Botaniste, XXVI, p. 371. [**192, 206, 212**.]

RAYNER, M. C. (1915). Obligate Symbiosis in Calluna vulgaris. Ann. Bot. XXIX, p. 97. [**22, 24, 25**.]

—— (1916). Recent Developments in the Study of Endotrophic Mycorrhiza. New Phyt. XV, p. 161. [**24**.]

—— (1922). Mycorrhiza in the Ericaceae. Trans. Brit. Myc. Soc. VIII, p. 61. [**24**.]

REA, C. (1922). British Basidiomycetae. University Press, Cambridge. [**10**.]

REHSTEINER, H. (1892). Beiträge zur Entwicklungsgeschichte der Fruchtkörper einiger Gastromyceten. Bot. Zeit. L, p. 760. [**347**.]

REINSCH, P. F. (1878). Beobachtungen über einige neue Saprolegnieae, über die Parasiten in Desmidien-Zellen, und über die Stachelkugeln in Achlya-Schläuchen. Jahrb. f. wiss. Bot. II, p. 205. [**100**.]

REISSEK, S. (1847). Ueber Endophyten der Pflanzenzelle, eine gesetzmasse den Samenfäden oder beweglichen Spiralfasern analoge Erscheinung. Naturwiss. Abhandl. v. W. Haidinger, I, p. 31. [**22**.]

REXHAUSEN, L. (1920). Über die Bedeutung der ektotropen Mykorrhiza für die höhern Pflanzen. Cohn's Beit. z. Biol. der Pflanzen, XIV, p. 19. [**28**.]

RIDDLE, L. W. (1906). On the Cytology of the Entomophthoraceae. Am. Acad. Arts and Sci. XLIII, p. 177. [**143**.]

RIDLER, W. F. F. (1922). The Fungus present in *Pellia epiphylla* (L.) Corda. Ann. Bot. XXXVI, p. 193. [**23**.]

—— (1923). The Fungus present in *Lunularia cruciata* (L.) Dum. Trans. Brit. Myc. Soc. IX, p. 82. [**23**.]

ROBERTSON, M. (1909). Notes on an *Ichthyosporidium* causing a fatal disease in Sea Trout. Proc. Zool. Soc. London, p. 399. [**148**.]

ROBINSON, W. (1914). Some Experiments on the Effect of External Stimuli on the Sporidia of Puccinia malvacearum (Mont). Ann. Bot. XXVIII, p. 331. [**34, 36**.]

—— (1926 i). On some features of Growth and Reproduction in *Sporodinia grandis* Link. Trans. Brit. Myc. Soc. X, p. 307. [**39**.]

—— (1926 ii). The conditions of growth and development in Pyronema confluens Tul. Ann. Bot. XL, p. 1. [**38, 39**.]

ROLLAND, L. (1910). Atlas des Champignons de France, Suisse et Belgique. Klincksieck, Paris. [**351**.]

ROSE, L. (1910). Beiträge zur Kenntniss der Organismen in Eichenschleimfluss. Inaug. Diss. Univ. of Berlin. [**176**.]

ROSENBAUM, J. (1917). Studies of the genus *Phytophthora*. Journ. Agric. Research, VIII, p. 233. [**54, 115**.]

ROSENBERG, O. (1903). Über die Befruchtung von *Plasmopara alpina* (Johans). Bihang till K. Svensk. Vet. Akad. Handl. XXVIII, p. 1. [**123**.]

—— (1933). Askokarp-utvecklingen hos Humaria aggregata (B. & Br.) Sacc. K. Svenska Vetenskapsakademiens Årsbok för År 1933, p. 345. [**214**.]

ROSTOWZEW, S. J. (1903). Beiträge zur Kenntnis der Peronosporeen. Flora, XCII, p. 405. [**122, 123**.]

ROTHERT, W. (1903). Die Sporenentwicklung bei *Aphanomyces*. Flora, XCII, p. 293. [**53**.]

ROUPPERT, C. (1909). Revision du genre Sphaerosoma. Bull. Acad. Sci. de Cracovie, p. 75. [**232**.]

ROZE, E. and CORNU, M. (1869). Sur deux nouveaux types génériques pour les familles des Saprolegniées et les Péronosporées. Ann. Sci. Nat. Bot. 5me sér. XI, p. 72. [**123**.]

RUHLAND, W. (1904). Studien über die Befruchtung der *Albugo Lepigoni* und einiger Peronosporeen. Jahrb. f. wiss. Bot. XXXIX, p. 135. [**123**.]

RYTZ, W. (1907). Beiträge zur Kenntnis der Gattung *Synchytrium*. Centralbl. f. Bakt. ii, XVIII, pp. 634; 799. [**67**.]

—— (1917). Beiträge zur Kenntnis der Gattung *Synchytrium*. I. Fortsetzung. Die cytologischen Verhältnisse bei *Synchytrium taraxaci*. Beih. z. bot. Centralbl. XXXIV, ii, p. 343. [**65**.]

SACHS, J. (1855). Morphologie des *Crucibulum vulgare* Tulasne. Bot. Zeit. XIII, p. 833. [**355**.]

SALMON, E. S. (1900). A Monograph of the Erysiphaceae. Mem. Torrey Bot. Club, IX, p. 1. [**187**.]

—— (1903 i). On Specialization of Parasitism in the *Erysiphaceae*. Beih. z. bot. Centralbl. XIV, p. 261. [**187**.]

—— (1903 ii). Infection Powers of Ascospores in the Erysiphaceae. Journ. Bot. Lond. XLI, pp. 159, 204. [**187**.]

—— (1904 i). On Erysiphe Graminis D.C. and its adaptive Parasitism within the Genus Bromus. Ann. Myc. II, p. 1. [**29, 187**.]

—— (1904 ii). Cultural Experiments with Biologic Forms of the Erysiphaceae. Phil. Trans. Roy. Soc. B, CXCVII, p. 107. [**29, 187**.]

SALMON, E. S. (1905). On Endophytic Adaptation shown by *Erysiphe graminis* D.C. under cultural conditions. Phil. Trans. Roy. Soc. B, CXCVIII, p. 87. [**29, 187**.]

—— (1906). On Oidiopsis taurica (Lév.), an endophytic member of the Erysiphaceae. Ann. Bot. XX, p. 187. [**187**.]

—— (1907). Notes on the Hop Mildew. Journ. Agric. Sci. II, p. 327. [**187**.]

—— (1913 i). Spraying Experiments against the American Gooseberry Mildew. Journ. S. E. Agric. Coll. XXII, p. 404. [**190**.]

—— (1913 ii). Observations on the Life-History of the American Gooseberry Mildew. Journ. S. E. Agric. Coll. XXII, p. 433. [**190**.]

—— (1913 iii). Observations on the Perithecial Stage of the American Gooseberry Mildew. Journ. S. E. Agric. Coll. XXII, p. 440. [**190**.]

SAMPSON, K. (1925). Some Infection Experiments with Loose and Covered Smuts of Oats which Indicate the Existence in them of Biological Species. Ann. Applied Biol. XII, p. 314. [**29**.]

SANDS, M. C. (1907). Nuclear Structure and Spore Formation in *Microsphaera Alni*. Trans. Wisconsin Acad. Sci. Arts and Letters, XV, p. 733. [**187**.]

SAPPIN-TROUFFY, P. (1896). Recherches histologiques sur la famille des Urédinées. Botaniste, V, p. 59. [**302**.]

SARTORIS, G. B. (1927). A Cytological Study of *Ceratostomella adiposum* (Butl.) Comb. Nov., the Black Rot Fungus of Sugar Cane. Journ. Agric. Research, XXXV, p. 577. [**264**.]

SARTORY, A. (1918). Sporulation par symbiose chez les champignons inférieurs. C.R. Acad. Sci. CLXVII, p. 302. [**40**.]

SATAVA, J. (1918 i). Die Geschlechtsformen der Hefen. Sonderabdruck aus der Oesterreichischen Brauer- u. Hopfen-Zeitung, XXXI, No. 14. [**173, 174**.]

—— (1918 ii). O redukovanych formach kvasinek. Selbstverlag, Prague. [**173, 174**.]

—— (1934). Les formes sexuelles et asexuelles des levures et leur pouvoir fermentatif. IIIme Congrès internat. technique et chimique des industries agricoles, Paris. [**174**.]

SATINA, S. and BLAKESLEE, A. F. (1925). Studies on biochemical differences between (+) and (−) sexes in Mucors. I. Tellurium salts as indicators of the reduction reaction. Proc. Nat. Acad. Sci. XI, p. 528. [**125, 128**.]

—— —— (1926 i). Studies on biochemical differences between (+) and (−) sexes in Mucors. II. A Preliminary report on the Manoilov reaction and other tests. Proc. Nat. Acad. Sci. XII, p. 191. [**125, 128**.]

—— —— (1926 ii). Biochemical differences between sexes in green plants. Proc. Nat. Acad. Sci. XII, p. 197. [**125, 128**.]

—— —— (1926 iii). The Mucor parasite Parasitella in relation to sex. Proc. Nat. Acad. Sci. XII, p. 202. [**125, 128**.]

—— —— (1927). Further studies on biochemical differences between sexes in plants. Proc. Nat. Acad. Sci. XIII, p. 115. [**125, 128**.]

—— —— (1928 i). Studies on the biochemical differences between sexes in Mucors. 4. Enzymes which act upon carbohydrates and their derivatives. Proc. Nat. Acad. Sci. XIV, p. 229. [**125, 128**.]

—— —— (1928 ii). Studies on biochemical differences between sexes in Mucors. 5. Quantitative determinations of sugars in (+) and (−) races. Proc. Nat. Acad. Sci. XIV, p. 308. [**125, 128**.]

—— —— (1929). Criteria of male and female in bread moulds (Mucors). Proc. Nat. Acad. Sci. XV, p. 735. [**125, 128**.]

—— —— (1930). Imperfect sexual reactions in homothallic and heterothallic Mucors. Bot. Gaz. XC, p. 299. [**125, 128**.]

SAWYER, W. H. (1917). Development of some species of *Pholiota*. Bot. Gaz. LXIV, p. 206. [**340**.]

SAWYER, W. H. (1929). Observations on some entomogenous members of the Entomophthoraceae in artificial culture. Amer. Journ. Bot. XVI, p. 87. [**143.**]

—— (1933). The Development of Entomophthora sphaerosperma upon Rhopobota vacciniana. Ann. Bot. XLVII, p. 799. [**143.**]

SCALES, F. M. (1915). Some filamentous fungi tested for cellulose destroying power. Bot. Gaz. LX, p. 149. [**15.**]

SCHERFFEL, A. (1925 i). Endophytische Phycomyceten-Parasiten der Bacillariaceen und einige neue Monadinen. Ein Beitrag zur Phylogenie der Oomyceten (Schröter). Arch. f. Protistenk. LII, p. 1. [**44, 56, 64.**]

—— (1925 ii). Zur Sexualität der Chytridineen. (Der "Beiträge zur Kenntnis der Chytridineen", Teil I.) Arch. f. Protistenk. LIII, p. 1. [**59, 72, 75.**]

—— (1926 i). Einiges über neue oder ungenügend bekannte Chytridineen. (Der "Beiträge zur Kenntnis der Chytridineen", Teil II.) Arch. f. Protistenk. LIV, p. 167. [**56, 64, 72, 77.**]

—— (1926 ii). Beiträge zur Kenntnis der Chytridineen. Teil III. Arch. f. Protistenk. LIV, p. 510. [**52, 56.**]

SCHIEMANN, E. (1912). Mutationen bei *Aspergillus niger* van Tieghem. Zeitschr. Induk. Abstam. u. Vererbungslehre, VIII, p. 1. [**42.**]

SCHIKORRA, W. (1909). Über die Entwickelungsgeschichte von *Monascus*. Zeitschr. f. Bot. I, p. 379. [**183.**]

SCHIÖNNING, H. (1895). Nouvelle et singulière formation de l'ascus dans une levure. Meddelelser fra Carlsberg Laboratoriet, IV, p. 30. [**174.**]

VON SCHRENK, H. (1902). Decay of Timber and the Means of Preventing it. U.S. Dept. of Agric. Bureau of Plant Industry, Bull. 14, New York. [**13.**]

SCHUSSNIG, B. (1929). Beiträge zur Entwicklungsgeschichte der Protophyten. IV. Zur Entwicklungsgeschichte der Pseudosporeen. Arch. f. Protistenk. LXVIII, p. 555. [**43, 44, 56, 64.**]

SCHWARTZ, E. J. (1914). The Plasmodiophoraceae and their relationship to the Mycetozoa and the Chytridiales. Ann. Bot. XXVIII, p. 227. [**46, 48.**]

SCHWARZE, C. A. (1922). The Method of Cleavage in the Sporangia of Certain Fungi. Mycologia, XIV, p. 143. [**53, 105, 130.**]

SCHWEIZER, G. (1923). Ein Beitrag zur Entwicklungsgeschichte und Biologie von Ascobolus citrinus nov. spec. Zeitschr. f. Bot. XV, p. 529. [**224.**]

—— (1931). Zur Entwicklungsgeschichte von Ascobolus strobilinus nov. spec. Planta, XII, p. 588. [**219.**]

SCHWEIZER, J. (1918). Die Spezialisation von *Bremia lactucae* Regel. Verhandl. Schweiz. Naturf. Ges. XCIX, p. 224. [**123.**]

SEAVER, F. J. (1909). Notes on North American Hypocreales. Mycologia, I, p. 41. [**243.**]

SERBINOW, J. L. (1907). Beiträge zur Kenntnis der Phycomyceten. Organisation und Entwicklungsgeschichte einiger Chytridineen-Pilze (Chytridineae Schröter). Script. Hort. Bot. Univ. Imp. Petrop. XXIV, p. 5. [**59, 67, 72.**]

SHANTZ, H. L. and PIEMEISEL, R. L. (1917). Fungus Fairy Rings in Eastern Colorado and their effects on vegetation. Journ. Agric. Research, XI, p. 191. [**15.**]

SHEAR, C. L. (1925). The Life History of the Texas Root Rot Fungus, Ozonium omnivorum, Shear. Journ. Agric. Research, XXX, p. 475. [**359.**]

SHEAR, C. L. and DODGE, B. O. (1927). Life-histories and heterothallism of the red bread mould fungi of the *Monilia sitophila* group. Journ. Agric. Research, XXXIV, p. 1019. [**245.**]

SHERBAKOFF, C. D. (1917). Buckeye rot of Tomato Fruit. Phytop. VII, p. 119. [**115.**]

SKUPIENSKI, F. X. (1918). Sur la sexualité chez une espèce de Myxomycète Acrasiée, *Dictyostelium mucoroides*. C.R. Acad. Sci. CLXVII, p. 960. [**45.**]

—— (1920). Recherches sur le cycle évolutif de certains Myxomycètes. Flinikowski, Paris. [**45.**]

SMITH, J. H. (1923). On the Apical Growth of Fungal Hyphae. Ann. Bot. XXXVII, p. 341. [**1**.]

SMITH, W. G. (1908). Synopsis of the British Basidiomycetes. Trustees of the British Museum, London. [**10, 352, 353, 354, 356, 357**.]

ZU SOLMS-LAUBACH, H. (1887). *Ustilago Treubii* Solms. Ann. du jard. bot. de Buitenzorg, VI, p. 79. [**292**.]

SOMMERSTORFF, H. (1911). Ein Tiere fangender Pilz (*Zoophagus insidians* nov. gen., nov. spec.). Österr. Bot. Zeitschr. LXI, p. 361. [**13**.]

SOWERBY, J. (1797). Coloured Figures of English Fungi or Mushrooms. Davis, London. [**335**.]

SPARROW, F. K. (1929). A note on the occurrence of two Rotifer-capturing Phycomycetes. Mycologia, XXI, p. 90. [**13, 110**.]

—— (1931). Two New Species of Pythium Parasitic in Green Algae. Ann. Bot. XLV, p. 257. [**13, 55, 90**.]

—— (1932). Observations on the aquatic fungi of Cold Spring Harbor. Mycologia, XXIV, p. 268. [**13, 69**.]

—— (1933 i). The Monoblepharidales. Ann. Bot. XLVII, p. 517. [**13, 94, 96, 97**.]

—— (1933 ii). Observations on operculate chytridiaceous fungi collected in the vicinity of Ithaca, N.Y. Amer. Journ. Bot. XX, p. 63. [**13, 72, 80**.]

—— (1933 iii). Inoperculate Chytridiaceous Organisms Collected in the Vicinity of Ithaca, N.Y., with Notes on Other Aquatic Fungi. Mycologia, XXV, p. 513. [**13, 59, 72**.]

—— (1934). Observations on Marine Phycomycetes collected in Denmark. Dansk. Bot. Arkiv, VIII, no. 6, p. 1. [**13, 63, 64, 115**.]

—— (1935). Recent Contributions to our Knowledge of the Aquatic Phycomycetes. Biol. Reviews, X, p. 152. [**13, 58**.]

—— (1936 i). A contribution to our knowledge of the aquatic Phycomycetes of Great Britain. Journ. Linn. Soc. L, p. 417. [**13, 50, 72, 73, 119**.]

—— (1936 ii). Biological Observations on the Marine Fungi of Woods Hole Waters. Biol. Bull. LXX, p. 236. [**13**.]

SPEARE, A. T. (1921). Massospora cicadina Peck. Mycologia, XIII, p. 72. [**146**.]

SPEGAZZINI, C. (1925). Un nuevo genero de las Helvellaceas. Mycologia, XVII, p. 210. [**232**.]

STAHL, E. (1900). Der Sinn der Mycorrhizenbildung. Jahrb. f. wiss. Bot. XXXIV, p. 539. [**25, 26**.]

STAKMAN, E. C., PIEMEISEL, F. J. and LEVINE, M. N. (1918). Plasticity of Biologic Forms of *Puccinia graminis*. Journ. Agric. Research, XV, p. 361. [**29**.]

STEVENS, F. L. (1899). The compound oosphere of *Albugo Bliti*. Bot. Gaz. XXVIII, pp. 149, 225. [**121**.]

—— (1901). Gametogenesis and fertilisation in *Albugo*. Bot. Gaz. XXXII, pp. 77, 157, 238. [**121**.]

—— (1902). Studies in the fertilisation of Phycomycetes. Bot. Gaz. XXXIV, p. 420. [**123, 125**.]

—— (1904). Oogenesis and fertilisation in *Albugo Ipomoeae-panduranae*. Bot. Gaz. XXXVIII, p. 300. [**121**.]

—— (1907). Some Remarkable Nuclear Structures in *Synchytrium*. Ann. Myc. V, p. 480. [**65**.]

—— (1922). The *Helminthosporium* Foot-rot of Wheat, with Observations on the Morphology of *Helminthosporium* and on the Occurrence of Saltation in the Genus. Illinois Nat. Hist. Survey Bull. XIV, p. 77. [**43**.]

—— (1928). Effects of ultra-violet radiation on various fungi. Bot. Gaz. LXXXVI, p. 210. [**36**.]

—— (1930). The response to ultra-violet irradiation shown by various races of *Glomerella cingulata*. Amer. Journ. Bot. XVII, p. 870. [**36**.]

STEVENS, F. L. and HALL, J. C. (1909). Variations of Fungi due to Environment. Bot. Gaz. XLVIII, p. 1. [**31**.]

STEVENS, F. L. and STEVENS, A. C. (1903). Mitosis of the Primary Nucleus in *Synchytrium decipiens*. Bot. Gaz. XXXV, p. 405. [**65, 67**.]

STOPPEL, R. (1907). *Eremascus fertilis* nov. spec. Flora, XCVII, p. 332. [**167**.]

STRASBURGER, E. (1878). Wirkung des Lichtes und der Wärme auf Schwärmsporen. Jenaische Zeitschr. f. Naturw. XII, p. 551. [**34**.]

—— (1880). Zellbildung und Zelltheilung. Jena. [**4**.]

STREETER, S. G. (1909). The Influence of Gravity on the Direction of Growth of *Amanita*. Bot. Gaz. XLVIII, p. 414. [**35, 37**.]

SUTHERLAND, G. K. (1914). New Marine Pyrenomycetes. Trans. Brit. Myc. Soc. v, p. 147. [**13**.]

—— (1915 i). New Marine Fungi on *Pelvetia*. New Phyt. XIV, p. 33. [**13**.]

—— (1915 ii). Additional Notes on Marine Pyrenomycetes. New Phyt. XIV, p. 183. [**13**.]

SWANTON, E. W. (1909). Fungi and How to Know Them. Methuen, London. [**10**.]

SWINGLE, D. B. (1934). Fertilization in *Ascodesmis nigricans* Van Tiegh. Amer. Journ. Bot. XXI, p. 519. [**202**.]

TABOR, R. J. and BUNTING, R. H. (1923). On a Disease of Cocoa and Coffee Fruits caused by a Fungus hitherto undescribed. Ann. Bot. XXXVII, p. 153. [**119**.]

TANDY, G. (1927). The Cytology of Pyronema domesticum (Sow.) Sacc. Ann. Bot. XLI, p. 321. [**160, 206**.]

TERNETZ, C. (1907). Über die Assimilation des atmosphärischen Stickstoffes durch Pilze. Jahrb. f. wiss. Bot. XLIV, p. 353. [**22, 24**.]

THAXTER, R. (1888). The Entomophthoraceae of the United States. Mem. Boston Soc. Nat. Hist. IV, p. 133. [**143**.]

—— (1895 i). New or peculiar aquatic fungi. 1. *Monoblepharis*. Bot. Gaz. XX, p. 433. [**13, 94**.]

—— (1895 ii). New or peculiar aquatic fungi. 2. *Gonapodya* Fischer and *Myrioblepharis* nov. gen. Bot. Gaz. XX, p. 477. [**13, 97**.]

—— (1895 iii). New or peculiar American Zygomycetes. 1. *Dispira*. Bot. Gaz. XX, p. 513. [**139**.]

—— (1896 i). Contribution towards a Monograph of the Laboulbeniaceae, Pt I. Mem. Am. Acad. Arts and Sci. XII, p. 195. [**274, 279**.]

—— (1896 ii). New or peculiar aquatic fungi. 3. *Blastocladia*. Bot. Gaz. XXI, p. 45. [**13, 94**.]

—— (1896 iii). New or peculiar aquatic fungi. 4. *Rhipidium*, *Sapromyces* and *Araiospora* nov. gen. Bot. Gaz. XXI, p. 317. [**13, 98, 100**.]

—— (1897). New or peculiar Zygomycetes. 2. *Syncephalastrum* and *Syncephalis*. Bot. Gaz. XXIV, p. 1. [**140, 141**.]

—— (1903). Mycological Notes, 1. A New England Choanephora. Rhodora, v, p. 97. [**136**.]

—— (1908). Contribution towards a Monograph of the Laboulbeniaceae, Pt II. Mem. Am. Acad. Arts and Sci. XIII, p. 219. [**274**.]

—— (1914 i). On certain peculiar fungus parasites of living insects. Bot. Gaz. LVIII, p. 235. [**276**.]

—— (1914 ii). New or peculiar Zygomycetes. 3. *Blakeslea*, *Dissophora*, *Haplosporangium*, Nova Genera. Bot. Gaz. LVIII, p. 353. [**136, 137, 138**.]

—— (1922). A Revision of the Endogoneae. Proc. Am. Acad. Arts and Sci. Boston, LVII, p. 291. [**138, 139**.]

—— (1924). Contribution towards a monograph of the Laboulbeniaceae. III. Mem. Am. Acad. Arts and Sci. XIV, p. 313. [**274**.]

—— (1926). Contribution towards a monograph of the Laboulbeniaceae. IV. Mem. Am. Acad. Arts and Sci. XV, p. 431. [**274**.]

THEISSEN, F. (1913). Über einige Mikrothyriaceen. Mycologia, XI, p. 493. [**197.**]
—— (1914). Über Membranstructuren bei die Microthyriaceen. Myc. Centralbl. III, p. 273. [**197.**]
THOM, C. (1910). Cultural Studies of Species of Penicillium. U.S. Dept. Agric. Bureau Animal Industry, Bull. 118. [**182.**]
THOM, C. and CHURCH, M. B. (1926). The Aspergilli. Baillière, Tindall and Cox, London. [**178.**]
THURSTON, H. W., Jr (1923). Intermingling gametophytic and sporophytic mycelium in *Gymnosporangium bermudianum*. Bot. Gaz. LXXV, p. 225. [**310.**]
TIEGS, E. (1919). Beiträge zur Oekologie der Wasserpilze. Ber. d. deutsch. bot. Ges. XXXVII, p. 496. [**13.**]
TOBLER, G. (1913). Die Synchytrien. Studien zu einer Monographie der Gattung. Arch. f. Protistenk. XXVIII, p. 141. [**20, 65, 67.**]
TORREY, G. S. (1921). Les conidies de *Cunninghamella echinulata* Thaxter. Bull. Soc. Myc. de France, XXXVII, p. 93. [**136.**]
TRANZSCHEL, W. (1904). Ueber die Möglichkeit die Biologiewertwechseln der Rostpilze auf Grund morphologischer Merkmale vorauszusehen. Arb. Kais. Petersburg Naturf. Ges. XXXV, p. 1. [**302, 321.**]
TROW, A. H. (1899). Observations on the Biology and Cytology of a New Variety of Achlya americana. Ann. Bot. XIII, p. 131. [**108.**]
—— (1901). Observations on the Biology and Cytology of Pythium ultimum. Ann. Bot. XV, p. 269. [**89, 115.**]
—— (1905). Fertilisation in the Saprolegniales. Bot. Gaz. XXXIX, p. 300. [**108.**]
TRUSCOTT, J. H. L. (1933). Observations on Lagena radicicola. Mycologia, XXV, p. 263. [**86.**]
TULASNE, L. R. (1853). Observations sur l'organisation des Trémellinées. Ann. Sci. Nat. Bot. 3me sér. XIX, p. 193. [**326.**]
TULASNE, L. R. and TULASNE, C. (1847). Mémoire sur les Ustilaginées comparées aux Urédinées. Ann. Sci. Nat. Bot. 3me sér. VII, p. 12. [**292.**]
—— —— (1851). Fungi hypogaei. Histoire et Monographie des Champignons hypogés. Klincksieck, Paris. [**184, 185.**]
—— —— (1854). Second Mémoire sur les Urédinées et Ustilaginées. Ann. Sci. Nat. Bot. 4me sér. II, p. 77. [**292, 302.**]
—— —— (1861–5). Selecta Fungorum Carpologia. Imperiali Typograph., Paris. [**10, 250, 252, 269, 270.**]
—— —— (1866). Note sur les phénomènes de copulation que présentent quelques Champignons. Ann. Sci. Nat. Bot. 5me sér. VI, p. 211. [**204.**]
VALKANOV, A. (1932). Nachtrag zu meiner Arbeit über rotatorienbefallende Pilze. Arch. f. Protistenk. LXXVIII, p. 485. [**110.**]
VALLORY, M. J. (1911). Sur la formation du périthèce dans le *Chaetomium Kunzeanum* Zopf. var. *chlorinum* Mich. C.R. Acad. Sci. CLIII, p. 1012. [**257.**]
VANDENDRIES, R. (1922). Recherches sur la sexualité des Basidiomycètes. C.R. Soc. Biol. Paris, LXXXVI, p. 513. [**287.**]
—— (1923 i). Nouvelles Recherches sur la Sexualité des Basidiomycètes. Bull. Soc. Roy. Bot. de Belg. LVI, p. 73. [**287.**]
—— (1923 ii). Recherches sur le déterminisme sexuel des Basidiomycètes. Mém. de l'Acad. roy. de Belg. p. 64. [**287.**]
—— (1925 i). Contribution nouvelle à l'étude de la sexualité des Basidiomycètes. Cellule, XXXV, p. 129. [**287, 289.**]
—— (1925 ii). Les Mutations sexuelles des Basidiomycètes. I. Identification de Coprinus radians Desm. Bull. Soc. Roy. Bot. de Belg. LVIII, p. 1. [**287, 289.**]
—— (1925 iii). Recherches expérimentales prouvant la fixité du sexe dans *Coprinus radians* Desm. Bull. Soc. Myc. de France, XLI, p. 358. [**287.**]

VANDENDRIES, R. (1927). Les mutations sexuelles. L'hétéro-homothallisme et la stérilité entre races géographiques de "Coprinus micaceus". Mém. Acad. Roy. de Belgique, IX, p. 3. **[289.]**

—— (1932). La tétrapolarité sexuelle de Pleurotus columbinus. Cellule, XLI, p. 267. **[288.]**

VANDENDRIES, R. and BRODIE, H. J. (1933). Nouvelles investigations dans le Domaine de la Sexualité des Basidiomycètes et étude experimentale des barrages sexuels. Cellule, XLII, p. 165. **[288, 289.]**

VANTERPOOL, T. C. and LEDINGHAM, G. A. (1930). Studies on browning root rot of cereals. I. The association of Lagena radicicola, n.gen., n.sp. with root injury of wheat. Canadian Journ. Research, II, p. 171. **[86.]**

VARITCHAK, B. (1931). Contribution à l'étude du développement des Ascomycètes. Botaniste, XXIII, p. 1. **[264.]**

—— (1933). Deuxième contribution à l'étude du développement des Ascomycètes. L'évolution nucléaire dans le sac sporifère de *Pericystis apis* Maassen et sa signification pour la phylogénie des Ascomycètes. Botaniste, XXV, p. 343. **[171.]**

VAVILOV, N. I. (1914). Immunity to Fungous Diseases as a Physiological Test in Genetics and Systematics, exemplified in Cereals. Journ. Genetics, IV, p. 49. **[29.]**

VITTADINI, C. (1831). Monographia Tuberacearum. Rusconi, Mediolani. **[236, 347.]**

VOKES, M. M. (1931). Nuclear division and development of sterigmata in *Coprinus atramentarius*. Bot. Gaz. XCI, p. 194. **[343.]**

VUILLEMIN, P. (1902). Les Cephalidées, section physiologique de la famille des Mucorinées. Bull. Soc. Sci. Nancy, iii, III, p. 21. **[139.]**

—— (1922). Une nouvelle espèce de Syncephalastrum; affinités de ce genre. C.R. Acad. Sci. CLXXIV, p. 986. **[140.]**

WADHAM, S. M. (1925). Observations on Clover Rot (*Sclerotinia trifoliorum* Eriks.). New Phyt. XXIV, p. 50. **[227.]**

WAGER, H. (1896). On the structure and reproduction of Cystopus candidus Lév. Ann. Bot. X, p. 295. **[121.]**

—— (1900). On the fertilisation of Peronospora parasitica. Ann. Bot. XIV, p. 263. **[123.]**

—— (1913). The Life-History and Cytology of Polyphagus Euglenae. Ann. Bot. XXVII, p. 173. **[59, 76.]**

—— (1914 i). Movements of Aquatic Micro-Organisms in Response to External Forces. The Naturalist, No. 689, p. 171. **[34.]**

—— (1914 ii). The Development of the Basidium in Tremella and Dacryomyces. The Naturalist, London, p. 364. **[325, 337.]**

WAGER, H. and PENISTON, A. (1910). Cytological Observations on the Yeast Plant. Ann. Bot. XXIV, p. 45. **[173.]**

WAHRLICH, W. (1893). Zur Anatomie der Zelle bei Pilzen und Fadenalgen. Script. Hort. Bot. Univ. Imp. Petropol. IV, p. 101. **[4.]**

WAKSMAN, S. A. (1918). The Importance of Mold Action in the Soil. Soil Sci. VI, p. 137. **[15.]**

—— (1931). Principles of Soil Microbiology. London. **[13, 14, 15.]**

WALKER, L. B. (1919). Development of *Pluteus admirabilis* and *Tubaria furfuracea*. Bot. Gaz. LXVIII, p. 1. **[340.]**

—— (1920). Development in *Cyathus fascicularis*, *C. striatus* and *Crucibulum vulgare*. Bot. Gaz. LII, p. 1. **[354, 356.]**

WALKER, L. B. and ANDERSEN, E. N. (1925). The Relation of Glycogen to Spore Ejection. Mycologia, XVII, p. 154. **[150, 357.]**

WALTER, H. (1921). Wachstumsschwankungen und hydrotropische Krümmungen bei Phycomyces nitens. Zeitschr. f. Bot. XIII, p. 673. **[33, 132.]**

WANG, D. T. (1934). Contribution à l'étude des Ustilaginées. Botaniste, XXVI, p. 539. [**292.**]

WARD, H. MARSHALL (1888 i). The Structure and Life History of *Entyloma Ranunculi* (Bonorden). Phil. Trans. Roy. Soc. B, CLXXVIII, p. 173. [**292.**]

—— (1888 ii). Some recent Publications bearing on the Sources of Nitrogen in Plants. Ann. Bot. I, p. 325. [**22.**]

—— (1892). The Ginger-Beer Plant and the Organisms composing it. Phil. Trans. Roy. Soc. B, CLXXXIII, p. 125. [**18.**]

—— (1897). On the Biology of *Stereum hirsutum*. Phil. Trans. Roy. Soc. B, CLXXXIX, p. 123. [**335.**]

—— (1898). Penicillium as a Wood-destroying Fungus. Ann. Bot. XII, p. 565. [**13.**]

—— (1899). *Onygena equina* Willd. A Horn-destroying Fungus. Phil. Trans. Roy. Soc. B, CXCI, p. 269. [**184.**]

—— (1902). On the Relations between Host and Parasite in the Bromes and their Brown Rust, Puccinia dispersa (Erikss.). Ann. Bot. XVI, p. 233. [**29.**]

—— (1903 i). Experiments on the Effect of Mineral Starvation on the Parasitism of the Uredine Fungus, Puccinia dispersa, on species of Bromus. Proc. Roy. Soc. LXXI, p. 138. [**29.**]

—— (1903 ii). Further Observations on the Brown Rust of the Bromes, *Puccinia dispersa* (Erikss.) and its adaptive Parasitism. Ann. Myc. I, p. 132. [**29.**]

—— (1905). Recent Researches on the Parasitism of Fungi. Ann. Bot. XIX, p. 1. [**29.**]

WARE, A. M. (1926). *Pseudoperonospora Humuli* and its Mycelial Invasion of the Host Plant. Trans. Brit. Myc. Soc. XI, p. 91. [**122.**]

WEBB, P. C. R. (1935). The Cytology and Life-history of Sorosphaera Veronicae. Ann. Bot. XLIX, p. 41. [**46, 47.**]

WEHMER, C. (1913–14). Versuche über die Bedingungen der Holzansteckung und -Zersetzung durch *Merulius* (Hausschwammstudien IV.). 1. Infektionsversuche im Laboratorium und Keller. Mycol. Centralbl. III, p. 321. [**14.**]

WEIMER, J. L. and HARTER, L. L. (1923). Temperature Relations of Eleven Species of *Rhizopus*. Journ. Agric. Research, XXIV, p. 1. [**39.**]

WEISS, F. E. (1904). A Mycorhiza from the Lower Coal Measures. Ann. Bot. XVIII, p. 255. [**22.**]

—— (1933). On the Germination and the Seedlings of Gentiana. Journ. Roy. Hort. Soc. LVIII, p. 296. [**25.**]

WELCH, D. S., HERRICK, G. W. and CURTIS, R. W. (1934). The Dutch Elm Disease. Cornell Extension Bull. 290, p. 3. [**264.**]

WELSFORD, E. J. (1907). Fertilization in *Ascobolus furfuraceus*. New Phyt. VI, p. 156. [**223.**]

—— (1915). Nuclear Migrations in Phragmidium violaceum. Ann. Bot. XXIX, p. 293. [**310.**]

—— (1921). Division of the nuclei in Synchytrium endobioticum Perc. Ann. Bot. XXXV, p. 298. [**66.**]

WERTH, E. and LUDWIGS, K. (1912). Zur Sporenbildung bei Rost und Brandpilzen. Ber. d. deutsch. bot. Ges. XXX, p. 522. [**299, 319.**]

WEST, C. (1917). On Stigeosporium Marattiacearum and the Mycorrhiza of the Marattiaceae. Ann. Bot. XXXI, p. 77. [**26, 120.**]

WESTON, W. H. (1918). The Development of Thraustotheca, a Peculiar Water-Mould. Ann. Bot. XXXII, p. 155. [**105.**]

—— (1919). Repeated zoospore emergence in *Dictyuchus*. Bot. Gaz. LXVIII, p. 287. [**55, 105.**]

—— (1923). Production and Dispersal of Conidia in the Philippine Sclerosporas of Maize. Journ. Agric. Research, XXIII, p. 239. [**125.**]

WESTON, W. H. (1924). Nocturnal Production of conidia by *Sclerospora graminicola*. Journ. Agric. Research, XXVII, p. 771. **[125.]**

VON WETTSTEIN, F. (1921). Das Vorkommen von Chitin und seine Verwertung als systematisch-phylogenetisches Merkmal im Pflanzenreich. Sitzber. Akad. Wiss. Wien, Math.-Nat. Kl. i, CXXX, p. 3. **[4.]**

WHITE, J. H. (1919). On the Biology of *Fomes applanatus*. Trans. Roy. Canadian Inst. XII, p. 133. **[344.]**

WIEBEN, M. (1927). Die Infektion, die Myzelüberwinterung und die Kopulation bei Exoasceen. Forsch. auf d. Gebiete der Pflanzenkrankheiten u. d. Immunität im Pflanzenreich, III, p. 139. **[199.]**

WILCOX, M. S. (1928). The sexuality and arrangement of the spores in the ascus of *Neurospora sitophila*. Mycologia, XX, p. 3. **[246.]**

WILSON, M. (1915). The Life-History and Cytology of *Tuburcinia primulicola* Rostrup. Report Brit. Assoc. p. 730. **[301.]**

WILSON, M. and CADMAN, E. J. (1928). The Life-history and Cytology of *Reticularia Lycoperdon* (Bull.). Trans. Roy. Soc. Edin. LV, p. 555. **[45.]**

WINGARD, S. A. (1922). Yeast spot of Lima beans. Phytopath. XII, p. 525. **[173.]**

—— (1925). Studies on the pathogenicity, morphology and cytology of Nematospora Phaseoli. Bull. Torrey Bot. Club, LII, p. 249. **[173.]**

WINGE, O. (1911). Encore le *Sphaerotheca Castagnei*. Bull. Soc. Myc. de France, XXVII, p. 211. **[192, 194.]**

—— (1913). Cytological Studies in the Plasmodiophoraceae. Arkiv f. Botanik, XII, p. 1. **[46, 48.]**

—— (1935). On Haplophase and Diplophase in some Saccharomycetes. C.R. Trav. du Lab. Carlsberg, série Physiol. XXI, p. 77. **[174.]**

WOLF, F. A. (1912). Spore Formation in *Podospora anserina* (Rabh.) Wint. Ann. Myc. X, p. 60. **[260.]**

VAN DER WOLK, P. C. (1913). *Protascus colorans*, a new genus and a new species of the Protascineae Group, the source of the "Yellow-Grains" in Rice. Myc. Centralbl. III, p. 153. **[167.]**

WORMALD, H. (1919). The "Brown Rot" Disease of Fruit Trees, with Special Reference to two Biologic Forms of Monilia cinerea Bon. I. Ann. Bot. XXXIII, p. 361. **[29, 227.]**

—— (1921). On the Occurrence in Britain of the Ascigerous Stage of a "Brown-Rot" Fungus. Ann. Bot. XXXV, p. 125. **[227.]**

—— (1922). Further Studies on the "Brown Rot" Fungi. Ann. Bot. XXXVI, p. 305. **[227.]**

WORONIN, M. (1878). *Plasmodiophora Brassicae*, Urheber der Kohlpflanzen-Hernie. Jahrb. f. wiss. Bot. XI, p. 548. **[46, 48, 61.]**

—— (1886). *Sphaeria Lemaneae*, *Sordaria fimiseda*, *Sordaria coprophila* und *Arthrobotrys oligospora*. Beitr. Morph. und Phys. der Pilze, III, p. 325. **[257, 267.]**

—— (1904). Beitrag zur Kenntnis der Monoblepharideen. Mém. de l'Acad. Imp. de St Pétersbourg, viii, XVI, p. 1. **[94.]**

YOUNG, E. M. (1931). The Morphology and Cytology of Monascus ruber. Amer. Journ. Bot. XVIII, p. 499. **[183.]**

ZICKLER, H. (1934). Genetische Untersuchungen an einem heterothallischen Ascomyzeten (Bombardia lunata nov. spec.). Planta, XXII, p. 1. **[261.]**

ZOPF, W. (1881). Zur Entwickelungsgeschichte der Ascomyceten. Chaetomium. Nova Acta Acad. C. Leop.-Carol. G. Nat. Cur. XLII, p. 199. **[257.]**

—— (1884). Zur Kenntnis der Phycomyceten. I. Zur Morphologie und Biologie der Ancylisteen und Chytridiaceen, zugleich ein Beitrag zur Phytopathologie, Nova Acta Acad. C. Leop. Carol. G. Nat. Cur. XLVII, p. 143. **[63, 64, 68, 71, 73, 74, 83, 84.]**

ZOPF, W. (1888). Zur Kenntnis der Infectionskrankheiten niederer Thiere und Pflanzen. Nova Acta Acad. C. Leop.-Carol. G. Nat. Cur. LII, p. 317. [101.]

ZUKAL, H. (1887). Über Kultur der Askenfrucht von *Penicillium crustaceum*. K. K. zool.-bot. Gesells. in Wien, XXX, VII, p. 66. [182.]

ZYCHA, H. (1935). Mucorineae. Kryptogamenfl. d. Mark Brandenburg, VI*a*, p. 1. [125.]

INDEX

A glossary has not been prepared for this volume, but the page on which the definition of a technical term will be found is shown in the index in clarendon type, and the same method is used to indicate the principal reference to a family or species. Names of authors will be found in the list of literature on pp. 383 to 424.